FOLIATIONS
GEOMETRY AND DYNAMICS

Proceedings of the Euroworkshop on

FOLIATIONS
GEOMETRY AND DYNAMICS

29 May – 9 June 2000 Warsaw, Poland

Editors

Paweł Walczak
Uniwersytet Łódzki, Łódź, Poland

Lawrence Conlon
Washington University, St. Louis, Missouri, USA

Rémi Langevin
Université de Bourgogne, Dijon, France

Takashi Tsuboi
University of Tokyo, Japan

World Scientific
New Jersey • London • Singapore • Hong Kong

Published by

World Scientific Publishing Co. Pte. Ltd.

P O Box 128, Farrer Road, Singapore 912805

USA office: Suite 1B, 1060 Main Street, River Edge, NJ 07661

UK office: 57 Shelton Street, Covent Garden, London WC2H 9HE

British Library Cataloguing-in-Publication Data
A catalogue record for this book is available from the British Library.

FOLIATIONS: GEOMETRY AND DYNAMICS
Proceedings of the Euroworkshop

ISBN 981-02-4796-6

Printed in Singapore by World Scientific Printers (S) Pte Ltd

PREFACE

This volume contains surveys and research papers contributed by the participants of the Euroworkshop *Foliations: Geometry and Dynamics* held in Warsaw, at the Stefan Banach International Mathematical Centre and the Mathematical Institute of the Polish Academy of Sciences, between May 29 and June 9, 2000. The Workshop was sponsored by the European Commission grant (HPCFCT-2000-00051), the Japan Society for Promotion in Science (Grant-in-Aid for Scientific Research 1204003), the University of Łódź and the Banach Centre. Participants numbered in approximately 70, more than 50 of them from abroad (Brazil, Canada, France, Germany, Italy, Japan, Romania, Russia, Spain, Switzerland, Ukraine, USA and Uzbekhistan).

The organizing committee consisted of four people who became editors of this volume. They were supported by the staff of the Banach Centre and a group of young mathematicians from the University of Łódź.

The papers contained in this volume are closely related to the lectures given at the Workshop, which was designed to cover various aspects of the theory of foliations, focusing on topology, geometry and dynamics of such objects. Some of the surveys presented here contain lecture notes of excellent mini-courses given by Yoshihiko Mitsumatsu and Hitoshi Moriyoshi.

All the papers contained in this volume were refereed by experts. The research papers contain original results. The surveys bring light to the current state of some aspects of the foliation theory. We hope that both, the Workshop and this volume of proceedings, should make a significant contribution to the progress of our filed of science. We express our gratitude to the participants, the contributors of this volume, the sponsors and all the colleagues who helped us while organizing the Workshop and preparing the volume for publication. In particular, we would like to mention Marek Badura, Maciej Czarnecki and Mariusz Frydrych (who organized the www-page of the conference as well as the participants data base) and Zofia Walczak (who worked a lot with the TeX-files of the articles contained here).

The Editors

CONTENTS

SURVEY ARTICLES

T. Asuke: Some results on secondary characteristic classes of transversely holomorphic foliations 3

H. Colman: LS-categories for foliated manifolds 17

S. Hurder: Dynamics and the Godbillon-Vey class: a history and survey 29

R. Langevin: Similarity and conformal geometry of foliations ... 61

Y. Mitsumatsu: Foliations and contact structures on 3-manifolds 75

H. Moriyoshi: Operator algebras and the index theorem on foliated manifolds 127

RESEARCH PAPERS

J. Álvarez López and Y. Kordyukov: Distributional Betti numbers of transitive foliations of codimension one........ 159

S. Aranson and E. Zhuzhoma: Circle at infinity influences on the smoothness of surface flows........ 185

A. Biś and P. Walczak: Entropies of hyperbolic groups and some foliated spaces........ 197

M. Brittenham: Tautly foliated 3-manifolds with no R-covered foliations........ 213

J. Cantwell and L. Conlon: Endests of exceptional leaves; a theorem of G. Duminy........ 225

M. Frydrych and J. Kalina: Some remarks on partially holomorphic foliations 263

A. Haefliger: Foliations and compactly generated pseudogroups .. 275

J. Heitsch: Traces and invariants for non-compact manifolds 297

M. Hilsum: Hilbert modules of foliated manifolds with boundary 315

V. Kaimanovich: Non-Euclidean affine laminations 333

R. Langevin and P. Walczak: Transverse Lusternik-Schnirelmann category and non-proper leaves 351

V. Medvedev and E. Zhuzhoma: Structurally stable diffeomorphisms have no codimension one Plykin attractors on 3-manifolds 355

T. Mizutani: On exact Poisson manifolds of dimension 3 371

Y. Nakae: Foliation cones corresponding to some pretzel links 387

T. Noda and T. Tsuboi: Regular projectively Anosov flows without compact leaves 403

T. Tsuboi: On the perfectness of groups of diffeomorphisms of the interval tangent to the identity at the endpoints 421

R. Wolak: Basic distribution for singular Riemannian foliations .. 441

List of participants 449

Program 451

SURVEY ARTICLES

Proceedings of
FOLIATIONS: GEOMETRY AND DYNAMICS
held in Warsaw, May 29–June 9, 2000
ed. by Paweł WALCZAK *et al.*
World Scientific, Singapore, 2002
pp. 3–16

SOME RESULTS ON SECONDARY CHARACTERISTIC CLASSES OF TRANSVERSELY HOLOMORPHIC FOLIATIONS

TARO ASUKE

Department of Mathematics, Faculty of Sciences, Hiroshima University, 1-3-1 Kagamiyama, Higashi-Hiroshima, Hiroshima 739-8526, Japan, e-mail: asuke@math.sci.hiroshima-u.ac.jp

Transversely holomorphic foliations with non-trivial Godbillon-Vey classes are discussed. We introduce some examples of such foliations and recall a formula which relates the Bott and the Godbillon-Vey classes. Finally, we introduce the following our recent result, namely, associated with such foliations, there exists a natural S^1-bundle which has interesting properties, for example, the Bott class is always well-defined for the lifted foliation and the Godbillon-Vey class of the original foliation is obtained by integrating the 'absolute value' of the Bott class along the fibre.

1 Introduction

Secondary characteristic classes are one of the main tools to study foliations. Among them, the Godbillon-Vey class is the most important one and studied by many people.

It is known that this class is non-trivial and it can moreover vary continuously if we deform foliations. On the other hand, if we consider foliations with some transverse structures, this invariant is often rigid, or even trivial [11, 8, 20]. In this article we first review the following results by following our previous papers [1, 2, 4]:

1) There are transversely holomorphic foliations whose Godbillon-Vey classes are non-trivial in any codimensional cases.
2) The Godbillon-Vey class is rigid under smooth deformations in any codimensional cases.

For transversely holomorphic foliations, we can consider not only the real secondary classes but the complex secondary classes. We recall a formula which relates these classes, in particular, it turns out that the Godbillon-Vey class is written in terms of the Bott class and the first Chern class of the complex normal bundle (Theorem 3.16). The rigidity of the Godbillon-Vey class is shown by using this formula. We will mention the residuality of the elements of WU_q. An example of Heitsch [19] shows that the imaginary part of the Bott class is not residual in general.

Finally, we introduce the following our resent result. By using the above formula, we can define a natural S^1-bundle with the following properties (see Theorem 5.2 for the precise and detailed statements):

1) The imaginary part of the Bott class of the lifted foliation remains non-trivial while the Godbillon-Vey class of the lifted foliation is trivial.

2) The Godbillon-Vey class of the original foliation is obtained by integrating the 'absolute value' of the Bott class with respect to a natural trivialization of the normal bundle of the lifted foliation.

This article is organized as follows: first we introduce some examples of transversely holomorphic foliations. Then in the third section we recall the definitions of the Godbillon-Vey class and related notions. The formula which relates the real and the complex secondary classes is also introduced there. The forth section concerns the rigidity and the residuality. Finally, in the last section we define a natural S^1-bundle over manifolds equipped with transversely holomorphic foliations and introduce the above result.

2 Examples

We introduce some examples of transversely holomorphic foliations. The number q will always be the complex codimension of the foliations.

We begin with a trivial one.

Example 2.1 Consider the foliation of $\boldsymbol{R}^p \times \boldsymbol{C}^q$ whose leaves are the subspaces $\boldsymbol{R}^p \times \{z\}$, where $z \in \boldsymbol{C}^q$. This foliation is restricted to a foliation of $(\boldsymbol{R}^p \times \boldsymbol{C}^q) \setminus \{(0,0)\}$, which is invariant by the mapping ρ defined by $\rho(x,z) = (2x, 2z)$. By taking the quotient we obtain a transversely holomorphic foliation of $S^1 \times S^{p+2q-1}$.

The following example is a classical one due to Bott [7, 9].

Example 2.2 Consider $\boldsymbol{C}^{q+1} \setminus \{0\}$ and the holomorphic vector field X_λ on it defined by the formula

$$X_\lambda = \lambda_0 z_0 \frac{\partial}{\partial z_0} + \lambda_1 z_1 \frac{\partial}{\partial z_1} + \cdots + \lambda_q z_q \frac{\partial}{\partial z_q},$$

where $(z_0, z_1, \cdots, z_q)$ is the standard coordinate of $\boldsymbol{C}^{q+1}$. We suppose that none of λ_i is equal to zero, then by integrating this vector field, we obtain a transversely holomorphic (and holomorphic) foliation $\widetilde{\mathcal{H}}_\lambda$ of $\boldsymbol{C}^{q+1} \setminus \{0\}$.

We suppose moreover that none of the ratios λ_i/λ_j is a negative real number, then the foliation $\widetilde{\mathcal{H}}_\lambda$ is transversal to the unit sphere S^{2q+1}. Hence we obtain a transversely holomorphic foliation $\mathcal{H}_\lambda$ of S^{2q+1} by restricting $\widetilde{\mathcal{H}}_\lambda$.

Particularly when all λ_i's are the same, then the foliation $\mathcal{H}_\lambda$ is nothing but the Hopf fibration. In this case, we can consider a kind of suspension. This is a complex version of Heitsch's [18] and already appeared in [24] in the case where $q = 1$. We set $\lambda_i = 1$ for all i and denote by X the vector field $X_{(1,1,\cdots,1)}$.

Example 2.3 We construct a foliation $\widetilde{\mathcal{G}}$ of $\mathrm{SL}(q+1, \boldsymbol{C}) \times (\boldsymbol{C}^{q+1} \setminus \{0\})$ by suspending the Hopf fibration, namely, since the vector field X is preserved under the natural action of $\mathrm{SL}(q+1, \boldsymbol{C})$ on $\boldsymbol{C}^{q+1}$, this vector field together with the orbits of the right action of $\mathrm{SL}(q+1, \boldsymbol{C})$ on $\mathrm{SL}(q+1, \boldsymbol{C}) \times (\boldsymbol{C}^{q+1} \setminus \{0\})$ given by $(g_0, v_0) \cdot g = (g_0 g, g^{-1} v_0)$ defines a foliation, which we call $\widetilde{\mathcal{G}}$. Since $\widetilde{\mathcal{G}}$ is transversal to $\mathrm{SL}(q+1, \boldsymbol{C}) \times S^{2q+1}$, it induces a transversely holomorphic foliation $\widehat{\mathcal{G}}$ of $\mathrm{SL}(q+1, \boldsymbol{C}) \times S^{2q+1}$.

This foliation is invariant under the left action of $\mathrm{SL}(q+1, \boldsymbol{C})$ defined by $g \cdot (g_0, v_0) = (g g_0, v_0)$. Hence for any cocompact lattice Γ of $\mathrm{SL}(q+1, \boldsymbol{C})$, $\widehat{\mathcal{G}}$ induces a foliation $\widehat{\mathcal{G}}(\Gamma)$ of $\Gamma \backslash \mathrm{SL}(q+1, \boldsymbol{C}) \times S^{2q+1}$. We include formally the case where Γ is trivial and consider $\widehat{\mathcal{G}}(\{e\}) = \widehat{\mathcal{G}}$. The foliation $\widehat{\mathcal{G}}(\Gamma)$ is invariant under the right action of $\mathrm{SL}(q+1, \boldsymbol{C})$ defined as above, so it induces a foliation $\mathcal{G}(\Gamma)$ of $\Gamma \backslash \mathrm{SL}(q+1, \boldsymbol{C}) \underset{\mathrm{SU}(q+1)}{\times} S^{2q+1}$.

Notice that there is an S^1-action on $\Gamma \backslash \mathrm{SL}(q+1, \boldsymbol{C}) \times S^{2q+1}$ given by $[(g_0, v_0)] \cdot z = [(g_0, v_0 z)]$, where we consider $S^1 = \{z \in \boldsymbol{C} \mid |z| = 1\}$. This action preserves the leaves of $\widehat{\mathcal{G}}(\Gamma)$, therefore the foliation $\widehat{\mathcal{G}}(\Gamma)$ projects down to a foliation $\widehat{\mathcal{F}}(\Gamma)$ of $\Gamma \backslash \mathrm{SL}(q+1, \boldsymbol{C}) \times \boldsymbol{C}P^q$. Similarly, the foliation $\mathcal{G}(\Gamma)$ induces a foliation $\mathcal{F}(\Gamma)$ of $\Gamma \backslash \mathrm{SL}(q+1, \boldsymbol{C}) \underset{\mathrm{SU}(q+1)}{\times} \boldsymbol{C}P^q$.

The following example is a kind of locally homogeneous foliations.

Example 2.4 Let G be a complex Lie group and H be its complex subgroup. Suppose that K is a closed subgroup of G contained in H: $G \supset H \supset K$. Then the natural foliation of G by the left cosets of H, namely, the foliation of G whose leaves are $\{gH\}_{g \in G}$, induces a foliation $\mathcal{F}_K$ of G/K. Suppose moreover that G/K admits a cocompact lattice, namely, there exists a discrete subgroup Γ of G such that $\Gamma \backslash G/K$ is a Hausdorff manifold.

Then $\Gamma\backslash G/K$ is naturally equipped with a foliation $\mathcal{F}_K(\Gamma)$ induced by $\mathcal{F}_K$, because the foliation $\mathcal{F}_K$ is invariant under the natural left action of G. The foliations $\mathcal{F}_K$ are of complex codimension $\dim_{\boldsymbol{C}} G - \dim_{\boldsymbol{C}} H$.

If G is semisimple and K is its compact subgroup, then it is well-known that there always exist cocompact lattices Γ of G/K. As an example, we can consider a complex version of Roussarie's example [10].

Example 2.5 We consider the case where $G = \mathrm{SL}(q+1, \boldsymbol{C})$ and

$$H = \left\{ \begin{pmatrix} a & * \\ 0 & C \end{pmatrix} \; ; \; C \in \mathrm{GL}(q, \boldsymbol{C}) \right\}.$$

Then the above construction works for any compact subgroups K contained in H and we obtain foliations $\mathcal{F}_K$, which are complex codimension q. In particular, $\mathcal{F}_{T^1 \times \mathrm{SU}(q)}(\Gamma)$ and $\mathcal{F}_{\mathrm{SU}(q)}(\Gamma)$ are isomorphic to $\mathcal{F}(\Gamma)$ and $\mathcal{G}(\Gamma)$ in Example 2.3, respectively (cf. [24]).

There is another important example of this kind.

Example 2.6 We consider the case where $G = \mathrm{SO}(q+2, \boldsymbol{C})$. We denote by E_{ij} the $(q+2) \times (q+2)$-matrix whose (i,j)-entry is equal to one and the others are zero, where we count rows and columns from zero. We consider the subgroup H as the Lie group which corresponds to the Lie algebra

$$\mathfrak{h} = \left\langle X_{01}, X_{0k} + \sqrt{-1} X_{1k}, X_{ij} \; ; \; 2 \leq k \leq q+1, \, 2 \leq i < j \leq q+1 \right\rangle_{\boldsymbol{C}},$$

where $X_{ij} = E_{ij} - E_{ji}$. As in the above example, if K is a compact subgroup contained in H, the subgroup H defines a transversely holomorphic foliation $\mathcal{F}_K$ of complex codimension q.

As we will see in the next section, the Godbillon-Vey class of some of the above examples is non-trivial.

3 Definitions

In this section, we recall the some basic notions, especially the definitions of secondary classes of transversely holomorphic foliations. Some details of basic notions on transversely holomorphic foliations are also found in [12, 15]. We briefly mention also the real secondary classes but we refer the readers to [10] for general real case.

Definition 3.1 A foliation $\mathcal{F}$ of a manifold M is a transversely holomorphic foliation of complex codimension q if we can find a local coordinate system of the form $\{(U_i \times V_i, (x_i, z_i), f_{ji})\}$ such that

1) The foliation $\mathcal{F}$ restricted to each $U_i \times V_i$ is given by $\{U_i \times \{z\}\}_{z \in V_i}$.

2) Each V_i is an open subset of $\boldsymbol{C}^q$.

3) If we write $f_{ji}(x_i, z_i) = (\varphi_{ji}(x_i, z_i), \gamma_{ji}(z_i))$, then γ_{ji} are biholomorphic local diffeomorphisms.

After [12] we call such a chart adapted.

It is well-known that there are connections of special kind associated with foliations, namely, the Bott (or basic) connections. The definition of the Bott connections can be found in [10] in the real foliation and in [7] holomorphic foliations. We will here recall the definition of the Bott connections for transversely holomorphic foliations. For this purpose we introduce several vector bundles over M.

Definition 3.2 We denote by $Q_{\boldsymbol{R}}(\mathcal{F})$ the normal bundle of $\mathcal{F}$ as a real foliation: $Q_{\boldsymbol{R}}(\mathcal{F}) = TM/T\mathcal{F}$. Since the foliation is transversely holomorphic, we have a natural decomposition of the complexification $Q(\mathcal{F})$ of $Q_{\boldsymbol{R}}(\mathcal{F})$ into the holomorphic part and anti-holomorphic part as usual: $Q(\mathcal{F}) = Q^{(1,0)}(\mathcal{F}) \oplus Q^{(0,1)}(\mathcal{F})$. We define a bundle $\widetilde{Q}^{(0,1)}(\mathcal{F})$ (resp. $\widetilde{Q}^{(1,0)}(\mathcal{F})$) as a subbundle of $TM \otimes \boldsymbol{C}$ locally spanned by $\frac{\partial}{\partial x}$ and $\frac{\partial}{\partial \bar{z}}$ (resp. $\frac{\partial}{\partial x}$ and $\frac{\partial}{\partial z}$), where we consider an adapted chart (we omitted subindices).

Note that the bundles $\widetilde{Q}^{(1,0)}(\mathcal{F})$ and $\widetilde{Q}^{(0,1)}(\mathcal{F})$ are involutive, namely, they define virtually 'foliations'. Note also that

$$Q^{(1,0)}(\mathcal{F}) = TM \otimes \boldsymbol{C}/\widetilde{Q}^{(0,1)}(\mathcal{F})$$

and that

$$Q^{(0,1)}(\mathcal{F}) = TM \otimes \boldsymbol{C}/\widetilde{Q}^{(1,0)}(\mathcal{F}).$$

In the following we discuss $Q^{(1,0)}(\mathcal{F})$, but the arguments apply also to $Q^{(0,1)}(\mathcal{F})$ by taking complex conjugate.

With the above preparations we are able to proceed as in the real case to define the Bott connections on $Q^{(1,0)}(\mathcal{F})$. Let π denote the natural projection from $TM \otimes \boldsymbol{C}$ to $Q^{(1,0)}(\mathcal{F})$.

Definition 3.3 A Bott connection ∇ on $Q^{(1,0)}(\mathcal{F})$ is a connection which satisfies the equation $\nabla_X Y = \pi[X, \widetilde{Y}]$ for any section X of $\widetilde{Q}^{(0,1)}(\mathcal{F})$ and Y of $Q^{(1,0)}(\mathcal{F})$, where $\widetilde{Y}$ is any lift of Y to $TM \otimes \boldsymbol{C}$.

One can easily show that the Bott connections on $Q^{(1,0)}(\mathcal{F})$ are transversely of type $(1, 0)$, namely, locally we can write $\nabla = d + \zeta$ where ζ involves only dz, and hence the Bott (basic) connections for holomorphic foliations are also the Bott connections in our sense. Conversely, connections transversely of type $(1, 0)$ are Bott connections.

We remark that the natural projection $TM \otimes \boldsymbol{C}$ to $Q(\mathcal{F})$ induces a natural projection $\widetilde{Q}^{(1,0)}(\mathcal{F})$ to $Q^{(1,0)}(\mathcal{F})$.

Let $\boldsymbol{C}[v_1, \cdots, v_q]$ be the polynomial algebra in v_i, where $1 \leq i \leq q$. We set the degree of v_i as $2i$. Let I_q denote the ideal of this algebra generated by the monomials of degree greater than $2q$, and define the algebra $\boldsymbol{C}_q[v_1, \cdots, v_q]$ as the quotient of $\boldsymbol{C}[v_1, \cdots, v_q]$ by I_q. We define $\boldsymbol{C}_q[\overline{v}_1, \cdots, \overline{v}_q]$ in the obvious way. We define also $\boldsymbol{R}_{2q}[c_1, \cdots, c_{2q}]$ as the quotient of $\boldsymbol{R}[c_1, \cdots, c_{2q}]$ by the ideal $I_{2q}^{\boldsymbol{R}}$ generated by the monomials of degree greater than $4q$, where the degree of c_i is equal to $2i$.

The most important theorem in the theory of secondary classes of foliation is the following theorem of Bott.

Theorem 3.4 (Bott vanishing theorem [10])

- *Given a codimension-2q foliation, we can choose Bott connections so that the Chern-Weil homomorphism defined on $\boldsymbol{R}[c_1, \cdots, c_{2q}]$ annihilates $I_{2q}^{\boldsymbol{R}}$.*
- *Given a complex codimension-q transversely holomorphic foliation, we can choose complex Bott connections so that the Chern-Weil homomorphism defined on $\boldsymbol{C}[v_1, \cdots, v_q]$ annihilates I_q.*

The Bott vanishing theorem leads us to the following definition.

Definition 3.5 We define differential graded algebras WU_q and WO_{2q} by setting

$$\mathrm{WU}_q = \boldsymbol{C}_q[v_1, \cdots, v_q] \otimes \boldsymbol{C}_q[\overline{v}_1, \cdots, \overline{v}_q] \otimes \bigwedge[\widetilde{u}_1, \cdots, \widetilde{u}_q],$$
$$\mathrm{WO}_{2q} = \boldsymbol{R}_{2q}[c_1, \cdots, c_{2q}] \otimes \bigwedge[h_1, h_3, \cdots, h_{2q-1}].$$

We equip these algebras with an exterior differential determined by $d\widetilde{u}_i = v_i - \overline{v}_i$, $dv_i = d\overline{v}_i = 0$, $dh_i = c_i$ and $dc_i = 0$, respectively. We define the degree of the elements $\widetilde{u}_i$ and h_i as $2i-1$, respectively. Finally, for an index set $J = (j_1, \cdots, j_q)$ which consists of non-negative integers we set $v_J = v_1^{j_1} \cdots v_q^{j_q}$ and $|J| = j_1 + 2j_2 + \cdots + qj_q$. We define $\overline{v}_J$ in the same way.

Remark 3.6 As we mentioned, associated with a transversely holomorphic foliation are there two virtual 'foliations' given by $\widetilde{Q}^{(1,0)}(\mathcal{F})$ and $\widetilde{Q}^{(0,1)}(\mathcal{F})$. The definition of WU_q reflects this fact.

The Bott vanishing theorem implies that if M is a smooth manifold equipped with a transversely holomorphic foliation $\mathcal{F}$, then we have a homomorphism $\chi_{\boldsymbol{C}}$ from $H^*(\mathrm{WU}_q)$ to $H^*_{\mathrm{DR}}(M; \boldsymbol{C})$. A homomorphism $\chi_{\boldsymbol{R}}$ from $H^*(\mathrm{WO}_{2q})$ to $H^*_{\mathrm{DR}}(M; \boldsymbol{R})$ is also defined by using the Bott vanishing theorem. It is well-known that $\chi_{\boldsymbol{R}}$ and $\chi_{\boldsymbol{C}}$ depend only on the foliation. We denote the image of an element ω of $H^*(\mathrm{WU}_q)$ or $H^*(\mathrm{WO}_{2q})$ by $\omega(\mathcal{F})$.

Definition 3.7 An element of $H^*(\mathrm{WU}_q)$ is called a complex secondary class if it involves one of the $\widetilde{u}_i$'s. Similarly, an element of $H^*(\mathrm{WO}_{2q})$ is called a real secondary class if it involves one of the h_i's. Their images by χ_C or χ_R are also called secondary classes.

A set of basis of $H^*(\mathrm{WO}_{2q})$ is well-known [14]. The author does not know general description of sets of bases of $H^*(\mathrm{WU}_q)$, but we have an algorithm to find them [2].

We introduce two important classes.

Definition 3.8 We define respectively the class ξ_q of $H^*(\mathrm{WU}_q)$ and the class GV_{2q} of $H^*(\mathrm{WO}_{2q})$ by the following formulae:

$$\begin{aligned}\xi_q &= \sqrt{-1}\widetilde{u}_1(v_1^q + v_1^{q-1}\overline{v}_1 + \cdots + \overline{v}_1^q) \\ \mathrm{GV}_{2q} &= h_1 c_1^{2q}\end{aligned}$$

The class GV_{2q} is called the Godbillon-Vey class, and the class ξ_q is the imaginary part of the Bott class. We denote by Chern_1 the first Chern class $(v_1 + \overline{v}_1)/2$ of the complex normal bundle of the foliation.

If Chern_1 is trivial, then we can define the Bott class as follows. First we choose a Bott connection ∇_b and a trivialization σ of the line bundle $\bigwedge^q Q^{(1,0)}(\mathcal{F})$ so that $du_1(\nabla_b, \sigma) = v_1(\nabla_b)$, where $v_1(\nabla_b)$ denotes the first Chern form calculated by using ∇_b and $u_1(\nabla_b, \sigma)$ is the differential form obtained by comparing on $\bigwedge^q Q^{(1,0)}(\mathcal{F})$ the connection ∇_b and the flat connection with respect to σ. The Bott vanishing theorem then implies that the differential form $u_1(\nabla_b, \sigma)v_J(\nabla_b)$ is closed if $|J| = q$. In general the class defined by this differential form depends on the homotopy class of σ, but if $v_J = v_1^q$, then the resulting class is independent of σ.

Definition 3.9 We denote by $u_1 v_J(\mathcal{F}, \sigma)$ the class defined by

$$u_1(\nabla_b, \sigma)v_J(\nabla_b).$$

Particularly when $v_J = v_1^q$, then we set $\mathrm{Bott}_q(\mathcal{F}) = u_1(\nabla_b, \sigma)v_1(\nabla_b)^q$ and call it the Bott class.

Remark 3.10 The class ξ_q is in fact a real cohomology class, while the Bott class is a complex cohomology class. Indeed, it is known that $\xi_q(\mathcal{F}) = -2\,\mathrm{Im}\,\mathrm{Bott}_q(\mathcal{F})$. We refer the readers to [9, 10, 1] for the details. We also remark that it is the Bott class which was firstly found, not the Godbillon-Vey class.

The definitions which we made have certainly a sense, namely, we have the following. The proof appeared in [4] and we omit it.

Theorem 3.11 ([4])

1) *The Godbillon-Vey classes of the foliations $\mathcal{F}_K(\Gamma)$ in Examples 2.5 and 2.6 (we assume in the latter case that q is odd) are non-trivial if K*

contains a maximal torus $T^{\max}$. In the former case, $T^{\max} = T^q$ and in the latter case, $T^{\max} = T^{\frac{q+1}{2}}$.

2) *If K is contained in $\mathrm{SU}(q)$ (resp. $\mathrm{SO}(q)$), the Godbillon-Vey class of the foliation in Example 2.5 (resp. 2.6) is trivial.*

3) *The Γ-structures defined by Examples 2.5 and 2.6 are not the same if the complex codimension is greater than one. These foliations are different even as real foliations.*

Remark 3.12 We can find similar examples with non-vanishing Godbillon-Vey class by using $\mathrm{Sp}(n, \boldsymbol{C})$, $n > 1$ and G_2 [4]. The example obtained by using G_2, which is of complex codimension 5, is also different from the ones in Examples 2.5 and 2.6. On the other hand, the example obtained by using $\mathrm{Sp}(n, \boldsymbol{C})$ is the pull-back of the foliation of $\mathrm{SL}(2n, \boldsymbol{C})$ in Example 2.5 (hence of complex codimension equal to $2n-1$). It seems that the semisimplicity of the groups is essential.

On the other hand, there is a classical result asserting that the class ξ_q varies continuously:

Theorem 3.13 **([9])** *We have the following formulae, namely,*

$$\begin{aligned} \mathrm{Bott}_q(\mathcal{H}_\lambda, \sigma_0) &= \frac{(\lambda_0 + \cdots + \lambda_q)^{q+1}}{\lambda_0 \lambda_1 \cdots \lambda_q} \mathrm{vol}_{S^{2q+1}}, \\ \xi(\mathcal{H}_\lambda) &= -2\mathrm{Im} \frac{(\lambda_0 + \cdots + \lambda_q)^{q+1}}{\lambda_0 \lambda_1 \cdots \lambda_q} \mathrm{vol}_{S^{2q+1}}, \end{aligned}$$

where $\mathcal{H}_\lambda$ is the foliation of S^{2q+1} given in Example 2.2 and σ_0 is the natural trivialization of the bundle $\bigwedge^q Q^{(1,0)}(\mathcal{H}_\lambda)$ (see [7] for details).

Given a transversely holomorphic foliation, it is natural to compare the complex and the real secondary classes. This can be done by considering the natural mapping $B\lambda$ from $B\Gamma_q^{\boldsymbol{C}}$ to $B\Gamma_{2q}$ which corresponds to forgetting the transverse holomorphic structure, where $B\Gamma_q^{\boldsymbol{C}}$ denotes the classifying space of transversely holomorphic foliations of complex codimension q and $B\Gamma_{2q}$ denotes the classifying space of foliations of codimension $2q$. The mapping $B\lambda$ induces a mapping λ from $H^*(\mathrm{WO}_{2q})$ to $H^*(\mathrm{WU}_q)$ which is characterized by the equation $\chi_{\boldsymbol{C}} \circ \lambda = \chi_{\boldsymbol{R}}$. A mapping, say λ_0, from WO_{2q} to WU_q which induces λ is given as follows [1]:

Definition 3.14 We define a mapping λ_0 from WO_{2q} to WU_q by the formulae

$$\lambda_0(c_k) = (\sqrt{-1})^k \sum_{j=0}^{k} (-1)^j v_{k-j}\overline{v}_j,$$

$$\lambda_0(h_{2k+1}) = \frac{(-1)^k}{2}\sqrt{-1}\sum_{j=0}^{2k+1}(-1)^j \widetilde{u}_{2k-j+1}(v_j + \overline{v}_j).$$

Remark 3.15 We regard v_i's as the Chern classes and c_i's as the Pontryagin classes, respectively. If we regard them as the Chern and the Pontryagin characters, the above formulae are slightly modified but essentially remain the same.

Note that the kernel of the mapping λ can be viewed as the obstructions to foliations being transversely holomorphic. For example in the case of complex codimension one, the formula $\lambda(h_1c_1^2) = 2\lambda(h_1c_2)$ holds [1]. This shows that if a real foliation of real codimension 2 do not satisfy the above equation, this foliation cannot be transversely holomorphic. In [2] we completely determine the kernel, image, cokernel and coimage of the mapping λ in the case where $q \leq 3$.

The Bott class and the Godbillon-Vey class are related by the following theorem.

Theorem 3.16 ([1]) *The formula*

$$\lambda(\mathrm{GV}_{2q}) = \frac{(2q)!}{q!q!}\xi_q(\mathrm{Chern}_1)^q$$

holds.

Proof. It follows from the fact that $\lambda_0(h_1) = \sqrt{-1}\widetilde{u}_1$ and $\lambda_0(c_1) = \sqrt{-1}(v_1 - \overline{v}_1)$. □

This theorem shows that if we consider transversely holomorphic foliations with non-vanishing Godbillon-Vey class, there is a strict restriction to the cohomology of ambient manifolds, namely, it must be similar to the cohomology of $S^{2q+1} \times \boldsymbol{C}P^q$. For example, the Godbillon-Vey class of any transversely holomorphic foliations of $S^1 \times S^4$ is trivial (such an example is obtained from Example 2.1 by setting $p = 3$ and $q = 1$). On the other hand, it can be shown that there is a family of real foliations of $S^1 \times S^4$ which is of real codimension two and whose Godbillon-Vey class varies continuously [1].

We remark also that $H^*(\mathfrak{sl}(q+1, \boldsymbol{C})_{\boldsymbol{R}}, \mathfrak{t}^1 \oplus \mathfrak{su}(q, \boldsymbol{C}); \boldsymbol{C}) \cong H^*(S^3 \times \cdots \times S^{2q+1} \times \boldsymbol{C}P^q; \boldsymbol{C})$. This is used in [4] to show the non-triviality of the Godbillon-Vey class of $\mathcal{F}_{T^1 \times \mathrm{SU}(q)}$ of Example 2.5.

4 Rigidity and residuality

The formula in Theorem 3.16 together with Theorem 3.13 seems to suggest that the Godbillon-Vey class can vary continuously even in the category of transversely holomorphic foliations. But this is not the case, namely, the Godbillon-Vey class is rigid in the category of transversely holomorphic foliations.

First we introduce the following mapping.

Definition 4.1 We denote by ρ the homomorphism from WU_{q+1} to WU_q defined by the following formulae:

$$\rho(\widetilde{u}_i) = \begin{cases} \widetilde{u}_i & \text{if } i \neq q+1 \\ 0 & \text{if } i = q+1 \end{cases}$$

$$\rho(v_i) = \begin{cases} v_i & \text{if } i \neq q+1 \\ 0 & \text{if } i = q+1 \end{cases}, \quad \rho(\overline{v}_i) = \begin{cases} \overline{v}_i & \text{if } i \neq q+1 \\ 0 & \text{if } i = q+1 \end{cases}$$

We denote by ρ_* the induced homomorphism from $H^*(\mathrm{WU}_{q+1})$ to $H^*(\mathrm{WU}_q)$.

Some arguments essentially parallel to that of the real case by Heitsch [16] show the following. Combining with the formula in Theorem 3.16, we obtain the rigidity of the Godbillon-Vey class. We refer the readers to [4, 1, 16] for the details.

Theorem 4.2 ([4], cf. [16]) *The complex secondary classes which belong to the image of ρ_* are rigid under smooth deformations. In particular, the Godbillon-Vey class of transversely holomorphic foliation is rigid under smooth deformations.*

Compared with the real case, the space $H^*(\mathrm{WU}_q)$ and the cokernel of ρ_* seems rather complicated [2]. For example, we have the following.

Proposition 4.3 *When $q = 1$, $\operatorname{coker}\rho_*$ is generated by $\widetilde{u}_1(v_1 + \overline{v}_1)$. When $q = 2$, $\operatorname{coker}\rho_*$ is generated by the elements of $H^*(\mathrm{WU}_2)$ of degree 5, 10 and 12, namely, the following elements:*

5	$\widetilde{u}_1(v_1^2 + v_1\overline{v}_1 + \overline{v}_1^2), \widetilde{u}_1(v_2 + \overline{v}_2) + \widetilde{u}_2(v_1 + \overline{v}_1)$
10	$\widetilde{u}_1\widetilde{u}_2 v_1\overline{v}_1(v_1 + \overline{v}_1)$
12	$\widetilde{u}_1\widetilde{u}_2 v_1^2\overline{v}_1^2, \widetilde{u}_1\widetilde{u}_2 v_1^2\overline{v}_2, \widetilde{u}_1\widetilde{u}_2 v_2\overline{v}_1^2, \widetilde{u}_1\widetilde{u}_2 v_2\overline{v}_2$

Here the number in the left column stands for the degree of the elements in the same row.

Remark 4.4 The Baum-Bott example (Example 2.2) shows that the elements of the lowest degree can vary (cf. Theorem 3.13). We do not know how about the other classes. Such classes are product classes except for $\widetilde{u}_1\widetilde{u}_2(v_1^2\overline{v}_2 - v_2\overline{v}_1^2)$, namely, they are written as a product of secondary classes. As was pointed out in [27], it is difficult to study the product classes.

For transversely holomorphic foliations, the notion of Γ-vector fields is as follows:

Definition 4.5 A vector field X on M is a Γ-vector field for the foliation $\mathcal{F}$ if $[Y, X]$ is a section of $\widetilde{Q}^{(0,1)}$ for any section Y of $\widetilde{Q}^{(0,1)}$.

In terms of adapted charts, X is a Γ-vector field if and only if X is projectable to V_i and the projection is holomorphic (cf. Heitsch[17], Ghys–Gomez-Mont–Saludes [13]).

As in the real case, we can show the following. We omit the proof because it is identical to the real case, which was done by Heitsch [18].

Theorem 4.6 *Let α be an element of* WU_q *which is in the kernel of ρ, then α is residual in the sense of Heitsch [18].*

If α does not belong to the kernel, the situation become subtle. For example, the imaginary part of the Bott class ξ_q is no longer residual in general. Indeed, an example in [19] shows that if the first Chern class of the normal bundle of the foliation is trivial, then the class ξ_q is residual but the residue does depend on the choice of the trivialization. This means that we cannot define the residue of the class ξ_q in the sense of Heitsch depending only on the foliation.

5 Another relation between the Bott class and the Godbillon-Vey class

In this section we introduce a result which relates the Bott class and the Godbillon-Vey class in another way than the formula given in Theorem 3.16. Let M be a manifold equipped with a transversely holomorphic foliation $\mathcal{F}$ with non-trivial Godbillon-Vey class. As we remarked after Theorem 3.16, such a manifold is of some specific cohomological type. On the other hand, among the examples which we introduced, any example of transversely holomorphic foliation whose Godbillon-Vey class is non-trivial can be obtained by taking a quotient of a transversely holomorphic foliation by an S^1-action. Moreover, we can choose the latter foliation so that its Godbillon-Vey class is trivial.

Recently in [5] we showed that this is true in general. As a preparation, we introduce the following bundle.

Definition 5.1 We denote by $M^*(\mathcal{F})$ the principal S^1-bundle associated with $\bigwedge^q Q^{(1,0)}(\mathcal{F})$. We denote by $\mathcal{F}^*$ the foliation of $M^*(\mathcal{F})$ obtained by pulling-pack $\mathcal{F}$.

If there existed a nowhere vanishing section of $\bigwedge^q Q^{(1,0)}(\mathcal{F})$, it would give a global transverse complex volume form Ω such that $\ker \Omega = \widetilde{Q}^{(0,1)}(\mathcal{F})$. But as we assumed that the Godbillon-Vey class of $\mathcal{F}$ is non-trivial, this is not the case. Nevertheless, it is true that the line bundle $\bigwedge^q Q^{(1,0)}(\mathcal{F}^*)$ is trivial. Moreover, we choose a natural trivialization σ of $\bigwedge^q Q^{(1,0)}(\mathcal{F}^*)$ closely related to the class $\xi_q(\mathcal{F})$. By using this trivialization, we can show the following. The proof will appear in [5].

Theorem 5.2 ([5]) *Let $(M, \mathcal{F})$ be a transversely holomorphic foliation of complex codimension q. Assume that $\mathrm{GV}_{2q}(\mathcal{F})$ is non-trivial and let $\widehat{M}$ be an S^1-bundle over M which corresponds to $c\,\mathrm{Chern}_1(\mathcal{F})$, where $c \neq 0$ and $c\,\mathrm{Chern}_1(\mathcal{F})$ is assumed to be an integral class (for example, we can set $c = 1$). We denote by π the natural projection, then $\widehat{M}$ has the following properties, namely,*

1. *The first Chern class of the complex normal bundle of the lifted foliation $\widehat{\mathcal{F}}$ of $\widehat{M}$ is trivial. Hence the Godbillon-Vey class of $\widehat{\mathcal{F}}$ is trivial.*

2. *The imaginary part of the Bott class of $\widehat{\mathcal{F}}$ is non-trivial and there exists an element ω of $H^{2q+2}(\widehat{M})$ such that $\pi_!(\omega) = \mathrm{GV}_{2q}(\mathcal{F})$.*

3. *Suppose that $(\widehat{M}, \widehat{\mathcal{F}}) = (M^*(\mathcal{F}), \mathcal{F}^*)$, then we have the formula*

$$\pi_!(\sqrt{-1}\, u_1 \bar{u}_1 v_J \bar{v}_K(\mathcal{F}^*, \sigma)) = \sqrt{-1}\, \tilde{u}_1 v_J \bar{v}_K(\mathcal{F})$$

if $|J| = |K| = q$, where $\pi_!$ denotes the integration along the fibre. Consequently, $\pi_!(\sqrt{-1}\,\mathrm{Bott}_q(\mathcal{F}^*)\,\overline{\mathrm{Bott}_q(\mathcal{F}^*)}) = \mathrm{GV}_{2q}(\mathcal{F})$.

Example 5.3 Consider the bundle $M^*(\mathcal{F})$. Since it is associated with the bundle $\bigwedge^q Q^{(1,0)}(\mathcal{F})$ whose first Chern class is $q\,\mathrm{Chern}_1(\mathcal{F})$, it has the properties as above.

Example 5.4 The foliation $\widehat{\mathcal{G}}(\Gamma)$ in Example 2.3 is obtained from the foliation $\widehat{\mathcal{F}}(\Gamma)$ in the above way. As we explained in [4], the first Chern class of the complex normal bundle of $\widehat{\mathcal{F}}(\Gamma)$ is essentially the first Chern class of the tangent bundle of $\boldsymbol{CP}^q$, which is a certain integer multiple of the Euler class of the Hopf fibration. Hence we can choose the 'fiberwise' Hopf fibration as the S^1-bundle appeared in the statement.

Acknowledgment

The author is partially supported by the JSPS Postdoctoral Fellowships for Research Abroad (1999–2000)

This article is based on the author's talk at the conference 'Foliations: Geometry and Dynamics' held between May 29 and June 9, 2000, and written during his stay at the Unité de Mathématiques Pures et Appliquées, École Normale Supérieure de Lyon. The author would like to express his gratitude to the members to UMPA, especially to Professor Étienne Ghys.

References

1. T. Asuke, *On the real secondary classes of transversely holomorphic foliations*, Ann. Inst. Fourier, **50**, No. 3 (2000), 995–1017.
2. T. Asuke, *On the real secondary classes of transversely holomorphic foliations* II, preprint.
3. T. Asuke, *A Remark on the Bott class*, to appear in Ann. Fac. Sci. Toulouse.
4. T. Asuke, *The Godbillon-Vey class of transversely holomorphic foliations*, preprint.
5. T. Asuke, *Localization and residue of the Bott class*, preprint.
6. D. Baker, *On a class of foliations and the evaluation of their characteristic classes*, Comment. Math. Helvetici, **53** (1978), 334–363.
7. P. Baum and R. Bott, *Singularities of Holomorphic Foliations*, J. Diff. Geom., **7** (1972), 279–342.
8. C. Benson and B. Ellis, *Characteristic classes of transversely homogeneous foliations*, Trans. Amer. Math. Soc., **289** (1985), 849–859.
9. R. Bott, *On the Lefschetz Formula and Exotic Characteristic Classes*, Symposia Math., **10** (1972), 95–105.
10. R. Bott, S. Gilter and I.M. James, Lectures on Algebraic and Differential Topology, Lecture Notes in Mathematics, No. 279, Springer-Verlag, 1972.
11. R. Brooks and W. Goldman, *The Godbillon-Vey Invariant of a Transversely Homogeneous Foliation*, Trans. Amer. Math. Soc., **286** (1984), 651–664.
12. T. Duchamp and M. Kalka, *Deformation Theory for Holomorphic foliations*, J. Diff. Geom., **14** (1979), 317–337.
13. E. Ghys, X. Gomez-Mont and J. Saludes, *Fatou and Julia Components of Transversely Holomorphic Foliations*, preprint.
14. C. Godbillon, *Cohomologies d'algèbres de Lie de champs de vecteurs formels*, Séminaire Bourbaki, 1972/73, n° 421, Lecture Notes in Math.,

383, Springer Verlag, 1974, 69–87.
15. X. Gomez-Mont, *Transversal holomorphic structures*, J. Diff. Geom., **15** (1980), 161–185.
16. J.L. Heitsch, *Deformations of secondary characteristic classes*, Topology, **12** (1973), 381–388.
17. J.L. Heitsch, *A cohomology for foliated manifolds*, Comment. Math. Helv. **50** (1975), 197–218.
18. J.L. Heitsch, *Independent variation of secondary classes*, Annals of Math., **108** (1978), 421–460.
19. J.L. Heitsch, *A Residue Formula for Holomorphic Foliations*, Michigan Math. J., **27** (1980), 181–194.
20. J.L. Heitsch, *Secondary Invariants of Transversely Homogeneous Foliations*, Michigan Math. J., **33** (1986), 47–54.
21. S. Hurder *Secondary classes, Weil measures and the geometry of foliations*, J. Diff. Geom., **20** (1984), 291-309.
22. S. Hurder *Independent Rigid Secondary Classes for Holomorphic Foliations*, Invent. Math., **66** (1982), 313–323.
23. F.W. Kamber and P. Tondeur, Foliated Bundles and Characteristic Classes, Lecture Notes in Math., 493, Springer-Verlag, 1975.
24. S. Morita, *Discontinuous invariants of foliations*, in Foliations (Tokyo 1993), Adv. Stud. Pure Math., 5, North-Holland, 1985, 169–193.
25. H.V. Pittie, *Characteristic classes of foliations*, Research Notes in Mathematics; 10, Pitman Publishing, 1976.
26. H.V. Pittie, *The secondary characteristic classes of parabolic foliations*, Comment. Math. Helvetici, **54** (1979), 601–614.
27. O.H. Rasmussen, *Exotic Characteristic Classes for Holomorphic Foliations*, Invent. Math., **46** (1978), 153–171.

Received October 24, 2000.

Proceedings of
FOLIATIONS: GEOMETRY AND DYNAMICS
held in Warsaw, May 29–June 9, 2000
ed. by Paweł WALCZAK *et al.*
World Scientific, Singapore, 2002
pp. 17–28

LS-CATEGORIES FOR FOLIATED MANIFOLDS

HELLEN COLMAN
Department of Mathematics (m/c 249), University of Illinois at Chicago
851 S. Morgan St., Chicago, IL 60607-7045 USA,
e-mail: hcolman@math.uic.edu

The purpose of this paper is to introduce some foliated generalizations of category, in the sense of Lusternik and Schnirelmann. We generalize the classical LS-theory to foliations and prove a generalization of the Lusternik and Schnirelmann's main result for compact-Hausdorff foliations.

1 Introduction

The LS-category was introduced in the course of research into the calculus of variations in 1930 [17]. The motivation was that it gives a lower bound for the number of critical points of any smooth function on a compact manifold. For a survey in LS-category see [14, 15].

There are many extensions of the original concept adapted to various contexts such as the *fibrewise category*, introduced by I.M. James and J. Morris [16] and the *equivariant category*, by E. Fadell [9].

A good category for foliations should be an invariant of homotopy type, for some type of homotopy compatible with the foliated structure.

We propose on the one hand the *tangential category*, $\mathrm{cat}_t(M, \mathcal{F})$, which coincides with the category of the leaves when the foliation is a product and on the other hand the *transverse category*, $\mathrm{cat}_{\pitchfork}(M, \mathcal{F})$, that coincides with the category of the leaf space if the foliation is a fibration.

In general, we can compare these foliated categories with the ordinary category of the ambient manifold, the leaf space and the leaves.

Cohomological lower bounds for the tangential and transverse category can be obtained in terms of leafwise and basic cohomology respectively.

Finally, we give a generalization of Lusternik-Schnirelmann's main result. Precisely, if $(M, \mathcal{F})$ is a compact-Hausdorff foliation and f is a basic function on M then the number of critical leaves is at least $\mathrm{cat}_{\pitchfork}(M, \mathcal{F})$.

2 LS-category

Given a topological space X, a subset U of X is categorical if U is contractible in X. The category of X, $\mathrm{cat}X$, is the least number of categorical open sets required to cover X. If no such covering exists, the category is said to be infinite.

Category is an invariant of homotopy type.

If X is connected and $\mathrm{cat}X$ is finite then $\mathrm{cat}X \leq 1 + \dim X$, where dim denotes covering dimension. To obtain lower bounds for category we consider a cohomology theory with any coefficient ring. We have $\mathrm{cat}X \geq \mathrm{nil}\tilde{H}^*(X)$.

The Lusternik-Schnirelmann's main theorem asserts that if M is a compact manifold and f is a smooth function on M then the number of critical points of f is at least $\mathrm{cat}X$.

3 Fibrewise category

The following outline summarizes basic facts of the fibrewise context. A *fibrewise space* is a topological space X together with a projection p_X over a given base space B. A *fibrewise constant map* between fibrewise spaces is a map $c\colon X \to Y$ such that there exists a section $s\colon B \to Y$ of p_Y with $c = sp_X$. A *fibrewise nullhomotopy* is a fibre-preserving homotopy into a fibrewise constant map.

A subset U of a fibrewise space X is said to be fibrewise categorical if the inclusion $U \subset X$ is fibrewise nullhomotopic. The *fibrewise category* of X, $\mathrm{cat}_B X$, is the least number of fibrewise categorical open sets required to cover X. If no such covering exists, the fibrewise category is said to be infinite.

Fibrewise category is an invariant of fibrewise homotopy type.

To obtain cohomological lower bounds for fibrewise category we take the quotient of the reduced cohomology $\tilde{H}^*(X)$ by the ideal generated by the subring $p^*\tilde{H}^*(B)$. We have that

$$\mathrm{cat}_B X \geq \mathrm{nil}\frac{\tilde{H}^*(X)}{\langle p^*\tilde{H}^*(B)\rangle}.$$

The fibrewise category of a product coincides with the category of the fibre. Consider the space $F \times B$ for any space F. If $U \subset F$ is a categorical subset of F then $U \times B$ is fibrewise categorical. For any fibrewise categorical subset $V \subset F \times B$ the fibrewise nullhomotopy restricted to F gives a contraction of V in F. So $\mathrm{cat}_B(F \times B) = \mathrm{cat}F$.

If the projection has no global sections, fibrewise category is not finite. In order to avoid such requirement we give a generalization that only requires local sections [2]. We describe a subset U of a fibrewise space X to be local fibrewise categorical if there exists a local section $s\colon p(U) \to X$ such that the inclusion $U \subset X$ is fibrewise homotopic to the map $c = sp_U$, where $p_U : U \to p(U)$ is the restriction of p. So, the *local fibrewise category* of X, $\mathrm{cat}_p X$, defined in the same way, only requires the existence of local sections.

We assert that if $p\colon X \to B$ is a locally trivial bundle with fibre F then $\mathrm{cat}F \leq \mathrm{cat}_p X \leq \mathrm{cat}F\mathrm{cat}B$. Therefore for all non-trivial principal bundles the local fibrewise category is finite whereas the fibrewise category is infinite.

But the existence of local sections implies the fibres are closed. So, for all maps $p\colon X \to B$ in which $p^{-1}(b), b \in B$ is not closed we have infinite fibrewise and local fibrewise category. This fact motivated the search for a better notion of category suitable to any foliation.

4 Equivariant category

There are several different developments of an equivariant version of the classical theory [9, 18, 6]. They all coincide in the initial step of definitions and some basic facts.

Let G be a compact Lie group acting on a topological space X. A homotopy between G-spaces is equivariant if $H_t(gx) = gH_t(X)$, $x \in X, g \in G$. An invariant subset $U \subset X$ is said to be G-categorical if there exists an equivariant homotopy $H : U \times I \to X$ of the inclusion such that $H_1 : U \to X$ has image in a single orbit. The *equivariant category* of X is the least number of G-categorical open sets required to cover X.

Equivariant category is an invariant of equivariant homotopy type.

If G acts freely on M, then the equivariant category coincides with the ordinary category of the orbit space. When the action is not necessarily free it is just bounded below by the category of the orbit space, $\mathrm{cat}X/G \leq \mathrm{cat}_G X$.

To obtain cohomological lower bounds, take the Borel cohomology $H_G(X) = H(EG \times_G X)$, the latter being the Čech cohomology of the orbit space of the product $EG \times X$ with respect to the diagonal action. Then

Proposition 4.1 *If* $\tilde{H}^*_G(orb(x)) = 0$, $\forall x \in X$ *so* $\mathrm{cat}_G X \geq \mathrm{nil}\tilde{H}^*_G(X)$.

5 Foliated categories

We propose two generalizations of category adapted to the field of foliations that provide two new invariants of the (integrable, foliated) homotopy type.

We will assume manifolds and foliations are smooth.

Let $(M, \mathcal{F})$ and $(M', \mathcal{F}')$ be foliated manifolds. A smooth map $f\colon (M, \mathcal{F}) \to (M', \mathcal{F}')$ is said to be foliated if it sends leaves into leaves. A homotopy $H\colon M \times \mathbf{R} \to M'$ is said to be *integrable* if H is a foliated map considering $M \times \mathbf{R}$ foliated by leaves $L \times \mathbf{R}$, $L \in \mathcal{F}$. The notation $\simeq_{\mathcal{F}\times\mathbf{R}}$ will denote integrable homotopy. And the homotopy is *foliated* if it is a foliated map considering $M \times \mathbf{R}$ foliated by leaves L, $L \in \mathcal{F}$. We use $\simeq_{\mathcal{F}}$ to denote foliated homotopy. An integrable homotopy induces an homeomorphism between leaf spaces whereas a foliated homotopy induces just a homotopy, in the ordinary sense.

5.1 Tangential category

An open subset U of M is tangentially categorical if the inclusion map $(U, \mathcal{F}_U) \hookrightarrow (M, \mathcal{F})$ is integrably homotopic to a map constant on each leaf of $\mathcal{F}_U$. Here U is regarded as a foliated manifold with the foliation $\mathcal{F}_U$ induced by $\mathcal{F}$ on U.

Definition 5.1 The tangential category $\mathrm{cat}_t(M, \mathcal{F})$ of a foliated manifold $(M, \mathcal{F})$ is the least number of tangentially categorical open sets required to cover M. If no such covering exists, let $\mathrm{cat}_t(M, \mathcal{F}) = \infty$.

When $\mathcal{F}$ is a foliation by a single leaf, an open subset is tangentially categorical if and only if it is categorical, so $\mathrm{cat}_t(M, \mathcal{F}) = \mathrm{cat}M$. For a foliation by points, we have $\mathrm{cat}_t(M, \mathcal{F}) = 1$.

A distinguished open set of a foliated chart is always categorical, so $\mathrm{cat}_t(M, \mathcal{F})$ is finite in compact manifolds.

Tangential category is an invariant of integrable homotopy type.

We have that $\mathrm{cat}_t(M, \mathcal{F}) \geq \mathrm{cat}L$ for any leaf $L \in \mathcal{F}$ since the integrable homotopy on a tangentially categorical subset U restricted to a leaf of $\mathcal{F}_U$ gives a contraction on the corresponding leaf of $\mathcal{F}$. We will say that a foliated manifold is *tangentially contractible* if $\mathrm{cat}_t(M, \mathcal{F}) = 1$. In this case every leaf is contractible and closed. Thus, if M is a compact manifold, $(M, \mathcal{F})$ is tangentially contractible iff $\mathcal{F}$ is by points. The contractibility of every leaf is not sufficient to contract tangentially the manifold, as shown by the example of the linear foliation on the torus with irrational slope.

If the projection $\pi\colon M \to M/\mathcal{F}$ onto the space of leaves has a global section, we can compare the tangential category with the fibrewise category of M as a fibrewise space over $M/\mathcal{F}$. It is clear that

Proposition 5.2 $\mathrm{cat}_t(M,\mathcal{F}) \leq \mathrm{cat}_{M/\mathcal{F}}M$.

Finally we can obtain cohomological lower bounds considering the tangential cohomology [12].

An r-form $\omega \in \Omega^r(M)$ is a tangential form if it can be written locally as

$$\omega = \sum f(x,y)dx^{i_1} \wedge \cdots \wedge dx^{i_r}$$

where $(x_1, \ldots, x_p, y)$ is a distinguished open set of a foliated chart.

The space of tangential forms $\Omega_{\mathcal{F}}(M)$ has a differential $d_{\mathcal{F}}$ defined as one does classically for an r-form with y playing the role of a parameter. We call its cohomology $H_{\mathcal{F}}(M)$ the *tangential cohomology* of $(M,\mathcal{F})$. If the foliation is by points, the tangential cohomology is 0 in positive degrees.

Proposition 5.3 *Let U be a tangentially categorical open subset of M. The map $i^*\colon \tilde{H}_{\mathcal{F}}(M) \to \tilde{H}_{\mathcal{F}_U}(U)$ induced by the inclusion $U \hookrightarrow M$ is null.*

We can define a relative tangential cohomology considering the complex

$$\Omega^r_{\mathcal{F}}(M,U) = \Omega^r_{\mathcal{F}}(M) \oplus \Omega^{r-1}_{\mathcal{F}}(U)$$

with differential $d_{\mathcal{F}}(\omega,\theta) = (d_{\mathcal{F}}\omega, \omega|_U - d_{\mathcal{F}}\theta)$. We have the exact sequence of the pair (M,U)

$$\cdots \to H^{r-1}_{\mathcal{F}_U}(U) \to H^r_{\mathcal{F}}(M,U) \to H^r_{\mathcal{F}}(M) \to H^r_{\mathcal{F}_U}(U) \longrightarrow \cdots$$

We define a product [5] in relative tangential cohomology and prove:

Theorem 5.4 $\mathrm{cat}_t(M,\mathcal{F}) \geq \mathrm{nil}\tilde{H}^*_{\mathcal{F}}(M)$.

The tangential cohomology of a manifold is in general hard to calculate. W.Singhof and E.Vogt [22] gave a cohomological bound for the tangential category using ordinary cohomology instead of foliated cohomology which allows explicit calculation of the tangential category in many cases.

The same authors prove also a generalization of the classical estimate of category by the dimension of the space.

Proposition 5.5 ([22]) $\mathrm{cat}_t(M,\mathcal{F}) \leq \dim \mathcal{F} + 1$.

This estimate together with the lower bound by the category of the leaves, give the exact value of the tangential category for a large number of foliations, namely all foliations containing a leaf L such that $\mathrm{cat}L = \dim L + 1$. Classes of manifolds verifying this condition have been studied (see J.Oprea and J.Walsh [19]) and include symplectic manifolds (M^{2n},ω) with $\omega|_{\pi_2 M} = 0$, aspherical and hyper-aspherical manifolds.

In particular,

1. For any flow $\mathcal{F}$ in a compact manifold M, $\mathrm{cat}_t(M,\mathcal{F}) = 2$.

2. For any 2-dimensional Seifert fibration $\mathcal{F}$, $\text{cat}_t(M,\mathcal{F}) = 3$.
3. Any 2-dimensional foliation in the 3-sphere has tangential category 3.

5.2 Transverse category

An open subset U of M is transversely categorical if the inclusion map $(U,\mathcal{F}_U) \hookrightarrow (M,\mathcal{F})$ factors through a leaf up to foliated homotopy.

Definition 5.6 The transverse category $\text{cat}_{\pitchfork}(M,\mathcal{F})$ of a foliated manifold $(M,\mathcal{F})$ is the least number of transversely categorical open sets required to cover M. If no such covering exists, let $\text{cat}_{\pitchfork}(M,\mathcal{F}) = \infty$.

When $\mathcal{F}$ is a foliation by points, $\text{cat}_{\pitchfork}(M,\mathcal{F}) = \text{cat}M$. If $\mathcal{F}$ is a foliation by a single leaf, $\text{cat}_{\pitchfork}(M,\mathcal{F}) = 1$.

Transverse category is an invariant of foliated homotopy type which is finite on compact manifolds.

We will say that a foliated manifold is *transversely contractible* if $\text{cat}_{\pitchfork}(M,\mathcal{F}) = 1$. This implies the contractibility of the leaf space (in the ordinary sense). The contractibility of the leaf space is not sufficient to contract the manifold transversely, as shown again by the linear foliation on the torus. If the slope is irrational, the leaf space is contractible since the quotient topology is the coarse topology. The transverse category is 3 since the torus admits a covering by 3 distinguished open sets and the transverse category is bounded below by the category of the torus (because the leaves are contractible).

Even when the leaf space is a better space, its contractibility does not imply the transverse contractibility of the foliated manifold. Consider the Seifert fibration on the Klein bottle $K = S^1 \times_{Z_2} S^1$, $\mathbf{Z_2}$ acting by reflection in the second factor. The leaves of $\mathcal{F}$ are the images in K of $S^1 \times \{y\}$, $y \in S^1$. The leaf space is $K/\mathcal{F} = [-1,1]$. So $(K,\mathcal{F})$ is not transversely contractible whereas the leaf space is contractible.

If the foliation is defined by an action of a compact Lie group, i.e. if the action is locally free, we can compare the transverse category with the equivariant category of M as a G-space. We claim

Proposition 5.7 $\text{cat}_{\pitchfork}(M,\mathcal{F}) \leq \text{cat}_G M$.

Finally we can obtain cohomological lower bounds considering the basic cohomology [21].

An r-form $\omega \in \Omega^r(M)$ is a basic form if it can be written locally as

$$\omega = \sum g(y) dy^{j_1} \wedge \cdots \wedge dy^{j_r}$$

where $(x, y_1, \ldots, y_q)$ is a distinguished open set of a foliated chart.

We denote $\Omega_b(M)$ the complex of basic forms with the restriction of the differential on $\Omega(M)$. Its cohomology $H_b(M)$ is the *basic cohomology* of $(M,\mathcal{F})$. If the foliation is by one leaf, the basic cohomology is 0 in positive degrees.

Proposition 5.8 *Let U be a transversely categorical open subset of M. The map $i^*\colon \tilde{H}_b(M) \to \tilde{H}_b(U)$ induced by the inclusion in basic cohomology is null.*

We define a relative basic cohomology considering the complex [1]:

$$\Omega_b^r(M,U) = \Omega_b^r(M) \oplus \Omega_b^{r-1}(U)$$

with differential $d(\omega,\theta) = (d\omega, \omega|_U - d\theta)$. In this way we have a long exact sequence in basic cohomology associated to the pair (M,U).

Theorem 5.9 $\mathrm{cat}_{\pitchfork}(M,\mathcal{F}) \geq \mathrm{nil}\tilde{k}^* H_b^*(M)$ *where k^* is the induced by the inclusion map $k\colon \Omega_b(M) \hookrightarrow \Omega(M)$.*

We generalize the classical result of Varadarajan [24] about the categories of the spaces involved in a fibration.

Theorem 5.10 *Let $L_0 \in \mathcal{F}$ be a leaf such that $\mathrm{cat}L_0 \geq \mathrm{cat}L$. Then* $\mathrm{cat}M \leq \mathrm{cat}L_0\, \mathrm{cat}_{\pitchfork}(M,\mathcal{F})$.

Corollary 5.11 *For any foliation $\mathcal{F}$ in M,*

$$\mathrm{cat}M \leq \mathrm{cat}_t(M,\mathcal{F})\, \mathrm{cat}_{\pitchfork}(M,\mathcal{F}).$$

5.3 Saturated transverse category

We can define a transverse category in a stronger form requiring M to be covered by saturated open categorical sets.

Definition 5.12 The saturated transverse category $\mathrm{cat}_{\pitchfork}^s(M,\mathcal{F})$ of a foliated manifold $(M,\mathcal{F})$ is the least number of saturated transversely categorical open sets required to cover M. If no such covering exists, let $\mathrm{cat}_{\pitchfork}^s(M,\mathcal{F}) = \infty$.

Obviously the transverse category cannot exceed the saturated transverse category. The linear foliation on T^2 with irrational slope shows that transverse category can be finite even when saturated transverse category is infinite.

This invariant coincides with the category of the leaf space for a foliation defined by a fibration. In general the saturated transverse category is a subtle invariant.

5.3.1 Compact-Hausdorff foliations

Saturated transverse category provides a lower bound for the number of critical leaves of any basic function on a compact manifold endowed with a compact-Hausdorff foliation.

A compact-Hausdorff foliation is a foliation with all leaves compact and their leaf space is Hausdorff. The structure of compact-Hausdorff foliations has been studied by D.B.A. Epstein [8], R. Edwards, K. Millet and D. Sullivan [7] and others. The basic structure theorem for compact-Hausdorff foliations states that for any leaf L there exists a finite subgroup G of the orthogonal group $O(n)$, $n = \operatorname{codim}\mathcal{F}$; a homomorphism $h\colon \pi_1(L) \to G$ and a leaf preserving diffeomorphism of $\tilde{L} \times_G D$ onto a neighbourhood of L, where D is an open ball in $\mathbf{R}^n$ and $\tilde{L}$ is the covering space associated to the kernel of h. We call this neighbourhood a *standard* neighbourhood of L. This local model implies that $M/\mathcal{F}$ is a Satake manifold [23].

The standard neighbourhoods of the local model are transversely categorical since the radial contraction map $R_t\colon D \to D$ defined by multiplying the points of D by t commutes with the action of G and so induces a foliated homotopy.

Proposition 5.13 *Let $\mathcal{F}$ be a compact-Hausdorff foliation of a compact manifold M. Then* $\operatorname{cat}^s_{\pitchfork}(M, \mathcal{F}) < \infty$.

One of the open problems is to understand for what other classes of foliations must $\operatorname{cat}^s_{\pitchfork}(M, \mathcal{F})$ be finite.

The transverse saturated category can be finite even though $\mathcal{F}$ has non-compact leaves, and even when there is an exceptional minimal set for $\mathcal{F}$. Examples of such foliations with $\operatorname{cat}^s_{\pitchfork}(M, \mathcal{F})$ finite are given in [13], where the following is shown:

Theorem 5.14 *If M is a compact manifold and* $\operatorname{cat}^s_{\pitchfork}(M, \mathcal{F}) < \infty$ *then $\mathcal{F}$ has a compact leaf.*

Let $f\colon M \to \mathbf{R}$ denote a basic function, that is, constant along the leaves. Consider the critical set K of f. It is saturated. The leaves $L \subset K$ are called critical leaves of f. For any $c \in \mathbf{R}$ we write $M_c = f^{-1}(-\infty, c]$ and $K_c = K \cap f^{-1}(c)$. Our aim is to determine the number of critical leaves of f.

Proposition 5.15 *Let $\mathcal{F}$ be a compact-Hausdorff foliation of a compact manifold M and $f\colon M \to \mathbf{R}$ a basic function. Then*

1. *If c is a regular value of f then for some $\epsilon > 0$ there is a foliated homotopy H_t of M with $H_1(M_{c+\epsilon}) = M_{c-\epsilon}$.*

2. *If c is an isolated critical value of f there are arbitrarily small neighbourhoods U of K_c and arbitrarily small $\epsilon > 0$ such that there exists a foliated homotopy H_t of M with $H_1(M_{c+\epsilon} - U) \subset M_{c-\epsilon}$.*

The proof is based on an extension of the standard methods of integrating the (foliated) gradient field ∇f (see [4] for details of the proof of the Proposition 5.15).

Theorem 5.16 *Let $f\colon M \to \mathbf{R}$ be a basic function where M is a compact manifold with a compact-Hausdorff foliation $\mathcal{F}$. Then, f has at least $\mathrm{cat}^s_{\pitchfork}(M, \mathcal{F})$ critical leaves.*

Write $\mathrm{cat}^s_{\pitchfork}(A, M)$ for the minimal number of saturated categorical open sets *of* M required to cover A, where A is any saturated subset of M.

For every positive integer m, we define

$$c_m(f) = \inf\{c \in \mathbf{R} |\ \mathrm{cat}^s_{\pitchfork}(M_c, M) \geq m\}.$$

We prove the theorem by proving the following

Lemma 5.17 *Let $f\colon M \to \mathbf{R}$ be a basic function where M is a compact manifold with a compact-Hausdorff foliation $\mathcal{F}$. Then*

1. *If $-\infty < c_m(f) < \infty$ then $c_m(f)$ is a critical value of f.*
2. *If $-\infty < c_m(f) = c_n(f) = c < \infty$ then $\mathrm{cat}^s_{\pitchfork}(K_c, M) \geq n - m + 1$.*

The proof of the lemma depends on a few properties that the relative category $\mathrm{cat}^s_{\pitchfork}(A, M)$ verifies, namely the following:
Let A, B be saturated subsets of M

1. Monotonicity: If $A \subset B$ then $\mathrm{cat}^s_{\pitchfork}(A, M) \leq \mathrm{cat}^s_{\pitchfork}(B, M)$.
2. Subadditivity: $\mathrm{cat}^s_{\pitchfork}(A \cup B, M) \leq \mathrm{cat}^s_{\pitchfork}(A, M) + \mathrm{cat}^s_{\pitchfork}(B, M)$.
3. Deformation invariance: If A is transversely deformable into B then $\mathrm{cat}^s_{\pitchfork}(A, M) \leq \mathrm{cat}^s_{\pitchfork}(B, M)$.

The remaining part of the proof is the same as in the classical case with the suitable obvious modifications.

5.3.2 Codimension one foliations

We can also generalize the Lusternik-Schnirelmann theorem for any foliation of codimension one.

If a basic function on a compact manifold endowed with a codimension 1 foliation $\mathcal{F}$ has a finite number of critical leaves, so $\mathcal{F}$ is a compact-Hausdorff foliation [4]. The result is based in the properties of minimal sets on codimension one foliations [11], part B.

So, either all leaves are compact in which case $\mathcal{F}$ is a compact-Hausdorff foliation [10, 20] or else f has infinitely many critical leaves. In any case we have

Theorem 5.18 *Any basic function on a compact manifold endowed with a codimension one foliation has at least* $\mathrm{cat}^s_{\pitchfork}(M,\mathcal{F})$ *critical leaves.*

5.3.3 *Computations of the saturated transverse category*

In order to understand the information contained in the saturated transverse category of a foliation, we study further properties of this new invariant in particular classes of foliations. This study was begun by S. Hurder and H. Colman [3] who proved the following results.

For any foliation, we have $\mathrm{cat} M/\mathcal{F} \leq \mathrm{cat}^s_{\pitchfork}(M,\mathcal{F})$. If the foliation is a fibration the saturated transverse category coincides with the category of the leaf space. For a compact-Hausdorff foliation the saturated transverse category can be arbitrarily large for a fixed leaf space $M/\mathcal{F}$.

Our main result provides upper and lower bounds on the saturated transverse category in terms of the codimension of the foliation and the categories of certain associated spaces.

Proposition 5.19 *Let M be a compact manifold endowed with a codimension q compact-Hausdorff foliation $\mathcal{F}$. Let $E \subset M$ be the set of exceptional leaves (leaves with non-trivial holonomy). Then*

$$\max\{\mathrm{cat} M/\mathcal{F}, \mathrm{cat}^s_{\pitchfork}(E,\mathcal{F})\} \leq \mathrm{cat}^s_{\pitchfork}(M,\mathcal{F}) \leq \mathrm{cat}^s_{\pitchfork}(E,\mathcal{F}) + q.$$

Examples show that both the lower and upper bound are realized.

Let $\{E^i\}_{i=0,\ldots,k}$ be the Epstein filtration for a compact-Hausdorff foliation, where $E^0 = M$ and E^{i+1} is the exceptional set of $\mathcal{F}|_{E^i}$. We conjecture that

$$\mathrm{cat}^s_{\pitchfork}(M,\mathcal{F}) = \max\{\mathrm{cat} B_i | \, i = 1,\ldots,k\}$$

where $B^i = \pi(E^i)$, the image of the i-term of the Epstein filtration in the leaf space.

Compact-Hausdorff foliations are a special case of *compact* foliations where all leaves are compact but the holonomy groups are allowed to be infinite. The saturated transverse category of these foliations is generally unknown.

Acknowledgments

I would like to thank for its hospitality and its support, the Stefan Banach International Mathematical Centre in Warsaw, during the Conference on

Foliations, May 29-June 9, 2000 and the University of Sheffield, EU RTN1-1999-00176, Modern Homotopy Theory.

References

1. H. Colman, *Categoría LS en foliaciones,* Publ. Depto. de Geometría y Topología, **93**, Santiago de Compostela, 1998.
2. H. Colman, *Local fibrewise category,* Comptes Rendus Acad. Sci. Paris, Ser. I **324**, no 5 (1997), 549–552.
3. H. Colman and S. Hurder, *LS-category of compact-Hausdorff foliations,* preprint.
4. H. Colman and E. Macias-Virgós, *The transverse Lusternik-Schnirelmann category of a foliated manifold,* Topology, **40** (2) (2001), 419–430.
5. H. Colman and E. Macias-Virgós, *Tangential category of a foliated manifold,* submitted.
6. M. Clapp and D. Puppe, *Invariants of the Lusternik–Schnirelmann type and the topology of critical sets,* Trans. Amer. Math. Soc., **298** (1986), 603–620.
7. R. Edwards, K. Millet and D. Sullivan, *Foliations with all leaves compact,* Topology, **16** (1977), 13–32.
8. D.B.A. Epstein, *Foliations with all leaves compact,* Ann. Inst. Fourier, **26** (1976), 265–282.
9. E. Fadell, *The equivariant Lusternik-Schnirelmann method for invariant functionals and relative cohomological index theories,* in Méthodes Topologiques en Analyse Non-Lineaire, ed. A. Granas, Montreal, 1985.
10. A. Haefliger, *Variétés feuilletés,* Ann. Scuola Norm. Pisa, (3) **16** (1962), 367–397.
11. G. Hector and U. Hirsch, Introduction to the Geometry of foliations, A and B, Vieweg, 1981–83.
12. J.L. Heitsch, *A cohomology for foliated manifolds,* Comment. Math. Helvetici, **50** (1975), 197–218.
13. S. Hurder, *Category and compact leaves,* preprint, June 2000 (available at http://www.math.uic.edu/~hurder/).
14. I.M. James, *On category, in the sense of Lusternik-Schnirelmann,* Topology, **17** (1978), 331–348.
15. I.M. James, Lusternik-Schnirelmann category, Chapter **27** Handbook of Algebraic Topology, Elsevier Science, Amsterdam, 1995, 1293–1310.
16. I.M. James and J. Morris, *Fibrewise category,* Proceedings of the Royal Society of Edinburgh, **119A** (1991), 177–190.

17. L. Lusternik and L. Schnirelmann, Méthodes topologiques dans les Problèmes Variationnels, Hermann, Paris, 1934.
18. W. Marzantowicz, *A G-Lusternik-Schnirelmann category of space with an action of a compact Lie group,* Topology, **28** (1989), 403–412.
19. J. Oprea and J. Walsh, *Quotient maps, group actions and Lusternik-Schnirelmann category,* to appear in Topology and its Aplications.
20. G. Reeb, *Sur certaines propiétés topologiques des variétés feuilletés,* Act. Sci. et Ind., **1183** (1952), 91–154.
21. B. Reinhart, *Foliated manifolds with bundle-like metrics,* Ann. of Math., **69** (1959), 119–132.
22. W. Singhof and E. Vogt, *Tangentiale Kategorie,* preprint in German.
23. I. Satake, *On a generalization of the notion of manifold,* Proc. Nat. Acad. Sci. U.S.A., **42** (1956), 359–363.
24. K. Varadarajan, *On fibrations and category,* Math. Z., **88** (1965), 267–273.

Received October 31, 2000.

Proceedings of
FOLIATIONS: GEOMETRY AND DYNAMICS
held in Warsaw, May 29–June 9, 2000
ed. by Paweł WALCZAK *et al.*
World Scientific, Singapore, 2002
pp. 29–60

DYNAMICS AND THE GODBILLON-VEY CLASS. A HISTORY AND SURVEY

STEVEN HURDER
Department of Mathematics (m/c 249), University of Illinois at Chicago
851 S. Morgan St., Chicago, IL 60607-7045 USA,
e-mail: hurder@uic.edu, Web: http://www.math.uic.edu/~hurder/

We survey thirty years of study of the relations between dynamics and the Godbillon-Vey invariant of codimension one foliations. We include a section on open problems.

1 A simple definition

In 1971, C. Godbillon and J. Vey introduced the invariant of foliations named after them. Previously, the study of foliations was considered either as an area of topology, viewing a foliation is a generalized fibration structure on a manifold (cf. [76]), or as an area within differential equations, whose key results concerned recurrence and limit sets (cf. [97]). With the advent of the Godbillon-Vey class and its generalization to the other secondary classes, the field obtained a unique focus, characterized by the interplay of geometry, topology, dynamical systems, and eventually ergodic theory. The study of the Godbillon-Vey invariant for codimension one foliations, in particular, illustrates the breadth of the field of foliations, and it is still a developing subject. We discuss here the path of ideas and results from "a simple definition", to our current understanding of how the Godbillon-Vey class depends on the geometry and dynamics of a foliation, concluding with a selection of open problems. This work updates and expands on parts of the Seminaire Bourbaki "Sur l'invariant de Godbillon-Vey" by Ghys [31], and the problem session from the 1992 foliations meeting in Rio de Janeiro [72]. The bibliography attempts to be comprehensive and up-to-date.

Let M be a C^∞–manifold with a codimension–one foliation $\mathcal{F}$ which is transversally C^2, and with C^∞ leaves. We assume that the normal bundle to $\mathcal{F}$ is oriented, so there is a 1-form ω on M whose kernel defines the leaf tangent bundle $T\mathcal{F}$. The Frobenius Theorem implies there is a 1–form η on M such that $d\omega = \eta \wedge \omega$. Godbillon and Vey [41] observed that 3-form $\eta \wedge d\eta$ is closed, and its cohomology class $GV(\mathcal{F}) = [\eta \wedge d\eta] \in H^3(M)$ is an invariant under diffeomorphism and foliated concordance [77].

The simplicity of the definition of the Godbillon-Vey invariant $GV(\mathcal{F}) = [\eta \wedge d\eta]$ gives little indication of its geometric or dynamical meaning. The 1-form η entering in the definition is sometimes called the *modular form* for $\mathcal{F}$. It has long been known that the restriction of η to leaves of $\mathcal{F}$ are closed forms, and for a C^1-foliation there is a well-defined leafwise cohomology class $[\eta] \in H^1(M, \mathcal{F})$, called the *Reeb class*, and the integral of η along leafwise curves measures the transverse holonomy expansion [95] (see also [43, 96] and § 2, [31]) . The idea can even be traced back to the work of Poincaré [93]. So $[\eta]$ is a cohomology invariant of $\mathcal{F}$ which measures the transverse expansion.

The study of dynamics of flows and foliations have both been profoundly influenced by the special examples derived from Anosov flows on compact manifolds [1, 91, 92], especially the Anosov flows which are the geodesic flow of a compact manifold with constant negative curvature. For example, let M denote the unit tangent bundle to a metric of constant curvature on a closed orientable surface Σ_g of genus $g > 1$, and $\mathcal{F}$ the weak-stable foliation of the geodesic flow on M. The Roussarie calculation, included in [41], evaluated $GV(\mathcal{F})$ on the fundamental class, $\langle GV(\mathcal{F}), [M]\rangle = \pm 4\pi(1-g)$, giving the first example of a codimension one foliation with non-vanishing $GV(\mathcal{F})$. Thurston's celebrated construction [106] of a family of smooth foliations $\{\mathcal{F}_t \mid t > 0\}$ on the 3-sphere, for which $\langle GV(\mathcal{F}_t), [M]\rangle = t$, is constructed from the weak stable foliation starting with a punctured surface. The leaves of the weak stable foliation of an Anosov flow have exponential growth and are transversally expansive. Moussu and Pelletier [86] and Sullivan (in [102]) asked whether this is always true - if $GV(\mathcal{F}) \neq 0$ must $\mathcal{F}$ have leaves of exponential growth? This conjecture was supported by all of the examples. Moreover, in Thurston's examples the Godbillon-Vey class assumed a continuous range of values, suggesting that a geometric interpretation of $GV(\mathcal{F})$ should involve dynamical information such as "entropy".

Thurston's note [106] also contained a geometric interpretation of the Godbillon-Vey invariant for 3-manifolds as measuring "helical wobble". Reinhart and Wood [98] gave a formula expressing the the Godbillon-Vey class of a foliation of a Riemannian 3-manifold in terms of the curvatures of the leaves and normal bundle, which can be interpreted as the helical wobble. (See [31] for a nice discussion on this point.) As pointed out by

Langevin, the Reinhart-Wood formula is suggestive of an integral geometry interpretation of the Godbillon-Vey invariant and hints at possible dynamical connections [3, 73]. Sullivan [105] gave another interpretation of the Godbillon-Vey invariant as the "second derivative" of the intersection of certain foliation currents. Sullivan's calculation is also an interpretation of helical wobble, but more in the spirit of dynamical systems.

2 Structure theory

During the period 1976–1982, there were a succession of works by various authors which proved the Moussu-Pelletier and Sullivan conjecture for increasingly general classes of foliations. The approach to the conjecture was to prove a "vanishing theorem", where the leaves of $\mathcal{F}$ are assumed to have a dynamical property which implies subexponential growth, and then show this forces $GV(\mathcal{F}) = 0$.

The first result was by Herman [53] in 1976, who showed that $GV(\mathcal{F}) = 0$ if $\mathcal{F}$ is the suspension of an action of $\mathbb{Z}^2$ on the circle. This was extended soon after by Wallet [121], so that one knew that $GV(\mathcal{F}) = 0$ for a codimension one C^2-foliation $\mathcal{F}$ transverse to a circle bundle over a compact Riemann surface, if no leaf of $\mathcal{F}$ has holonomy. Wallet's result was in turn extended by Morita and Tsuboi [84] in 1978, who showed that $GV(\mathcal{F}) = 0$ if $\mathcal{F}$ is a foliation without holonomy on any compact manifold. This result introduced C^2-dynamics and Sacksteder's Theorem into proofs of the "vanishing theorem" for the first time, for a corollary of the Sacksteder theorem is that a foliation without holonomy is defined by a transverse invariant measure. The vanishing of $GV(\mathcal{F})$ is a consequence of constructing a sequence of smooth transverse 1-forms approximating this measure.

A foliation $\mathcal{F}$ is *almost without holonomy* if the only leaves with holonomy are compact. The simplest example of such is the Reeb foliation, though in general these foliations can have very complicated topology, as they are depth one with the non-compact leaves spiralling in on the compact leaves. Their study has a long history [18, 44, 48, 83, 85, 86, 88], with the basic structure theorem for C^2-foliations proving that the holonomy of the compact leaves must be abelian, that their complement is a union of local minimal sets, and there exists a transverse vector field defined in an open neighbourhood of the compact leaves which commutes with the holonomy. Mizutani, Morita and Tsuboi [82] and Cantwell and Conlon [8] proved in 1980 that $GV(\mathcal{F}) = 0$ for foliations almost without holonomy. Both proofs combined the structure theory for foliations almost without holonomy with the techniques developed for foliations without holonomy.

Throughout the 1970's, the geometry of codimension one foliations with increasing degrees of complexity was actively studied. This research culminated in three distinct approaches to a general structure theory: Dippolito proved the "octopus decomposition" for C^0-foliations; Nishimori [89] introduced the "SRH" (Staircase, Room, Hall) decomposition; Cantwell and Conlon [9] and Hector [49] proved a Poincaré-Bendixson theorem for the asymptotic behaviour of leaves. All of these can be seen as developing the ideas of the almost without holonomy classification theory. The Poincaré-Bendixson theory also provided an approximate correspondence between the growth rates of leaves and their level of complexity [7, 10, 11, 47, 110, 111, 112, 113, 115]. During 1980-81, a succession of authors extended the vanishing theorem for foliations almost without holonomy, to increasingly more general classes [8, 28, 89, 114]. None of these results directly related the growth of leaves to the vanishing of $GV(\mathcal{F})$, and all relied on the structure theory to estimate the form $\eta \wedge d\eta$.

Underlying these succession of works was a new idea, that the Godbillon-Vey invariant could be "localized" to an open saturated subset $U \subset M$, where the special "dynamics and geometry" of $\mathcal{F}|U$ could be used to show $GV(\mathcal{F}|U) = 0$. This idea became more explicit in the works of Cantwell-Conlon [8], Nishimori [89], Tsuchiya [114], and Duminy-Sergiescu [28].

3 Duminy's Theorem

In a brilliant work growing out of the paper [28], G. Duminy [26] introduced the *Godbillon measure* $g_{\mathcal{F}}$ on the Σ-algebra $\mathcal{B}_O(\mathcal{F})$ generated by the open saturated subsets of a foliation $\mathcal{F}$ of codimension one. Duminy's note was also highly original on two other points: The Godbillon measure is one half of the Godbillon-Vey invariant, constructed from the leaf cohomology class $[\eta] \in H^1(M, \mathcal{F})$. The second half, the "Vey class" $[d\eta] \in H^2(M/\mathcal{F})$, was considered as a fixed invariant of $\mathcal{F}$ on which the Godbillon measure could be evaluated to give $GV(\mathcal{F}|U)$ for $U \in \mathcal{B}_O(\mathcal{F})$. The other innovation, in a companion manuscript [27], was an estimation of the Godbillon measure using what are essentially techniques of ergodic theory. Duminy's work lifted the veil on the study of the relation between $GV(\mathcal{F})$ and dynamics.

Theorem 3.1 (Duminy) *If $\mathcal{F}$ is a codimension one, C^2-foliation of a compact manifold M with non-trivial Godbillon measure $g_{\mathcal{F}}$, then $\mathcal{F}$ must have a hyperbolic resilient leaf, and hence there is an open subset of M consisting of leaves with exponential growth.*

While the original manuscripts of Duminy were widely circulated, they were unfortunately never published. An account of Duminy's method was later published by Cantwell and Conlon [12], who extended the theorem to non-compact manifolds. This extension is not so simple as passing from cohomology on a compact manifold to that on a non-compact manifold, as the key estimates Duminy used required the Poincaré-Bendixson structure theory of C^2-foliations on compact manifolds [9, 49]. Cantwell and Conlon extended in the Poincaré-Bendixson theory to open manifolds in [12].

Duminy's method of proof was that, if there is no resilient leaf for $\mathcal{F}$, then by Sacksteder's Theorem [100] there are no exceptional minimal sets. Hence, by the Poincaré-Bendixson theory, all leaves of $\mathcal{F}$ either lie at finite level, or lie in "arbitrarily thin" subsets $U \in \mathcal{B}_O(\mathcal{F})$. The finite level case is analyzed analogously to the almost without holonomy case, while the analysis of the thin sets used the new techniques of [27]. Duminy's new techniques raised new questions:

1. Exactly what are the properties of the "Godbillon measure"?
2. Could the technique of the Godbillon measure also yield an approach to relating the secondary classes for higher codimension foliations to their dynamics?
3. What is the meaning of the calculation in [27] which miraculously showed just what needed to be shown?
4. Does the "Vey class" have a geometric or dynamical meaning comparable to that proved for the Godbillon measure?

Only the first two of these questions have been partially answered to date.

After Duminy's manuscripts first appeared in March 1982, these questions were discussed among Larry Conlon, André Haefliger, James Heitsch, Paul Schweitzer, and the author, who were visiting Princeton in Spring, 1982. Three of the five participants in this mini-program at the Institute for Advanced Study subsequently published papers on Duminy's work!

The first development following on Duminy's work was by Heitsch and Hurder [52]. They extended Duminy's ideas in two directions, one formal and the other more fundamental. First, they showed that for a codimension $q \geq 1$, C^1-foliation, each cohomology class $y_I \in H^*(\mathfrak{gl}(\mathrm{q}), \mathrm{SO}(\mathrm{q}))$ yields an associated *Weil measure*, denoted by $\chi_{\mathcal{F}}(y_I)$, for which the Godbillon measure $g_{\mathcal{F}} = \chi_{\mathcal{F}}(y_1)$ is the simplest one. The classes y_I are those which appear in the E^2-term of the spectral sequence calculation of the cohomology of the truncated Weil algebra $WO(q)$, hence their name. The fact that the Weil measures are defined for C^1-foliations was just an observation based on the definitions. The definition of the "Vey class", and more generally the classes $[c_J]$ corresponding to the Chern forms $c_J(\mathbf{Q})$ of the normal bundle to $\mathcal{F}$, still require a C^2-foliation for their definition.

The more fundamental point of the work [52] was that the measures were defined on the Σ-algebra $\mathcal{B}(\mathcal{F})$ of all measurable, saturated subsets of M, and that the leafwise forms used to define the measures $\chi_{\mathcal{F}}(y_I)$ need only be transversally measurable. It may happen that the only *open* saturated subsets for a foliation are either empty, or M itself, so the extension of the Godbillon measure to the algebra of measurable sets $\mathcal{B}(\mathcal{F})$ widened the application of the ideas. On the other hand, the ability to calculate these new invariants using measurable data had a more important impact. Dynamical hypotheses on a foliation, combined with asymptotic techniques, often yields leafwise smooth, but only transversally measurable data, so that this extension allowed techniques of ergodic theory to be applied in the evaluation of the Godbillon and Weil measures.

For example, the Godbillon measure $g_{\mathcal{F}}$ can be calculated using a transversally measurable, transverse volume form for $\mathcal{F}$. A foliation with all leaves compact is easily shown to have a transversally measurable, closed transverse volume form, hence its Godbillon-Vey classes must vanish. This argument applies in any codimension. Cantwell and Conlon used the modification of Duminy's approach to give a simpler proof of Duminy's Theorem [12, 21]. Hurder [55] showed that for a foliation of arbitrary codimension, if all leaves are compact, then all of the Weil measures must vanish since such a foliation admits a transversally measurable, holonomy invariant transverse Riemannian metric. This was the first "vanishing theorem" for the other secondary classes of foliations, for codimension greater than one.

Duminy's papers [26, 27] were a demarcation in the study of the dynamics of foliations. The fundamental conjecture posed in 1974 was solved, while the reformulation of his ideas transformed the study of the relation between dynamics of foliations and the Godbillon-Vey and other secondary classes, into questions of ergodic theory. The study of foliation dynamics was afterwards motivated by results and techniques of the ergodic theory for smooth maps [103, 71] and group actions [125]. Connections to the theory of ergodic equivalence relations [29] and cocycles [101] became fundamental.

4 Ergodic theory

One of the "often discovered"' facts in ergodic theory is that a diffeomorphism $f\colon M \to M$ of a compact Riemannian manifold M cannot expand volume on an invariant set of positive measure. This basic fact was first formulated by Schmidt [101] in cocycle language as saying that the additive Radon-Nikodym cocycle $\nu\colon \mathbb{Z} \times M \to \mathbb{R}$ defined by $\nu(n,x) = \log\{|(f^n)'(x)|\}$ has subexponential growth for almost every $x \in M$. The key to the proof is that the group $\mathbb{Z}$ has subexponential growth, while if $\limsup_{n\to\infty} \nu(n,x)/n \geq a > 0$ on a set $E \subset M$ of positive Lebesgue measure, then the volume of the iterates $f^n(E)$ grows exponentially, and there is "no room" for all the volume in the space $\mathbb{Z} \times M$.

On the other hand, if Γ is a group with exponential word growth acting smoothly on M, then the additive Radon-Nikodym cocycle $\nu\colon \Gamma \times M \to \mathbb{R}$ can have exponential growth. This is exactly what happens, for example, with the action of a surface group on the circle at infinity in the original examples calculated by Roussarie. This balance between the growth of the group and of the Radon-Nikodym cocycle underlies the next advance in the study of the Godbillon-Vey invariants.

We need a digression into the growth rates of leaves. For a finitely generated group, the limit $gr(\Gamma) = \lim_{n\to\infty} \log\{\#\Gamma_n\}/n$ always exists, where $\Gamma_n = \{\gamma \mid \|\gamma\| \leq n\}$, as Γ is a homogeneous metric space for the word metric. Given a leaf $L \subset M$ of $\mathcal{F}$ for Riemannian manifold M, and $x \in L$, we say L has *subexponential growth* if

$$\limsup_{R\to\infty} \log\{Vol_L(x,R)\}/R = 0,$$

where $Vol_L(x,R)$ denotes the volume in the leaf metric of a ball centred at $x \in L$ with leaf radius R. If

$$\liminf_{R\to\infty} \log\{Vol_L(x,R)\}/R = 0,$$

we say that the leaf L has *non-exponential growth*, while L has *exponential growth* if

$$\liminf_{R\to\infty} \log\{Vol_L(x,R)\}/R > 0.$$

A foliation can have leaves which have nonexponential growth but not subexponential growth [47].

In 1984, Hurder [58] proved that for the pseudogroup of a codimension $q \geq 1$ foliation of a compact manifold, if almost all orbits have subexponential growth, then the Radon-Nikodym cocycle has subexponential growth. This gave the first direct relation between the growth of leaves of a foliation and the asymptotic growth rate of the modular class $[\eta]$. However, subexponential growth of the Radon-Nikodym cocycle does not imply the modular form has any local growth estimates. A cocycle with subexponential growth may still oscillate wildly on small scale, and only when averaged for large "time" does it behave in a subexponential manner.

One consequence of the Heitsch-Hurder reformulation of Duminy's methods was that it is possible to renormalize the transverse volume form using a leafwise smooth, transversally measurable change of scale function. If such a change of scale function can be chosen so that the new transverse volume form has uniform arbitrarily slow growth, then the modular form η is uniformly estimated arbitrarily close to 0, hence the Godbillon measure must vanish. This is just a measurable version of the original idea from Herman [53]! Fortunately, smooth ergodic theory had already solved the renormalization problem. A fundamental technique for the study of non-uniform hyperbolic dynamical systems is the Lyapunov tempering procedure used in Pesin theory [90]. Hurder and Katok [66] extended this tempering procedure from actions of $\mathbb{Z}$ to the context of metric equivalence relations. When applied to a foliation with almost every leaf of subexponential growth, this yields for each $\epsilon > 0$ a transverse volume form which is uniformly ϵ-invariant. Thus, the Godbillon measure must vanish. This was the key idea behind the proof in [58] of the generalization of the Moussu-Pelletier and Sullivan Conjecture to all codimensions:

Theorem 4.1 *Let $\mathcal{F}$ be a C^1 foliation of codimension $q \geq 1$ such that almost every leaf has subexponential growth rate. Then the Godbillon measure $g_{\mathcal{F}} = 0$. If $\mathcal{F}$ is C^2 then the Godbillon-Vey class $GV(\mathcal{F}) = 0$.*

The theorem implies that if $GV(\mathcal{F}) \neq 0$ then there exists a set of leaves with positive measure that do not have subexponential growth. It is not known if the set of leaves with non-exponential, but not subexponential growth can have positive Lebesgue measure. Alternately, one can ask if the method of proof of the theorem can be extended to leaves with non-exponential growth.

This discussion is focussed on the development of codimension one theory, but we detour for a moment to describe some related issues that are best observed in higher codimensions. The structure theory for codimension one foliations has no counterpart in higher codimensions. Even for a single diffeomorphism of a compact manifold M of dimension greater than one, it is impossible to give a good classification. The best that has been achieved is to restrict to classes of dynamics, and make a structure theory for diffeomorphisms within these restricted classes. Classes of examples include pseudo-Anosov maps of surfaces, Anosov maps of manifolds, affine actions, projective actions, actions with all orbits finite, distal and equicontinuous actions, and so forth. Similar classes of dynamical systems have been defined and studied for group actions on manifolds with some measure of success, and to some extent also for foliations.

An alternate approach to a classification theory for arbitrary diffeomorphisms, or groups of diffeomorphisms, of a compact manifold M is to attempt to classify the derivative cocycle of the action. Associated to a group action $\varphi\colon \Gamma \times M \to M$ is the derivative cocycle $D\varphi\colon \Gamma \times M \to \mathrm{GL}(\mathbb{R}^q)$, where we have chosen some bounded measurable trivialization of the tangent bundle TM. For a codimension q foliation $\mathcal{F}$, there is an analogous cocycle $D\varphi\colon \mathcal{G}_{\mathcal{F}} \to \mathrm{GL}(\mathbb{R}^q)$ where $\mathcal{G}_{\mathcal{F}}$ is the holonomy groupoid of $\mathcal{F}$. Rather than attempt the seemingly impossible step to classify the group actions or foliation dynamics, one studies their associated cocycles up to an appropriate notion of measurable coboundary. This yields a cohomology theory which is the direct generalization of the idea of the modular class $[\eta] \in H^1(M, \mathcal{F})$. The classification of cocycles up to measurable equivalence is a well-studied problem from smooth ergodic theory of group actions, and there are many techniques, and several celebrated theorems.

One invariant associated to a measurable cocycle is the Lyapunov band spectrum which consists of a band of asymptotic (or generalized) eigenvalues for the range of $D\varphi$. For example, these invariants have a prominent role in the study of asymptotic behaviour of the solutions of differential equations, especially in the Sacker-Sell theory. Another more delicate invariant consists of the algebraic hull [127] of $D\varphi$, which is the smallest algebraic subgroup $G \subset \mathrm{GL}(\mathbb{R}^q)$ so that $D\varphi$ is cohomologous to a cocycle with values in G.

The existence of an invariant ergodic measure allows deeper analysis of $D\varphi$. For example, if a smooth action admits an ergodic invariant measure and the group Γ is a higher rank lattice, then Zimmer's Cocycle Superrigidity Theorem [125, 79] reduces the classification of $D\varphi$ to a problem in representation theory. There remains the problem of translating information about the cocycle $D\varphi$ into dynamical information for $\mathcal{F}$, which frequently also requires the hypothesis (or construction) of an invariant measure for the action or foliation. For example, see the applications by Zimmer [123, 124, 126] of cocycle superrigidity to the smooth classification of group actions.

In codimension one, classification theory naturally evolved from an understanding of a certain collection of examples or "models". In higher codimensions, there is a far greater collection of models. The modular form η is replaced by the transverse derivative cocycle $D\varphi$, whose behaviour is far more complicated. In spite of these additional considerations, there are several results relating vanishing of secondary classes to foliation dynamics and the classification of the derivative cocycle in higher codimensions. Hurder [56] showed in 1984 that for the linear holonomy of a leaf in a C^2-foliation $\mathcal{F}$, if the algebraic hull is not amenable, then $\mathcal{F}$ has leaves of exponential growth. Hurder and Katok [66] showed in 1987 that if a foliation $\mathcal{F}$ defines an amenable equivalence relation, and hence the algebraic hull of the cocycle $D\varphi\colon \mathcal{G}_{\mathcal{F}} \to \mathrm{GL}(\mathbb{R}^q)$ is amenable, then many of its secondary classes of $\mathcal{F}$ vanish. The University of Chicago thesis of Stuck [104] extended these vanishing results. Analysis of the cocycle $D\varphi$, combined with possible additional hypotheses on the transverse geometry of the foliation $\mathcal{F}$, are expected to yield a better understanding of the dynamics of foliations in higher codimensions. For example, the geometric entropy of a foliation can be combined with information derived from $D\varphi$ to produce new results in all codimensions, as we discuss next.

5 Geometric entropy

One of the most fundamental invariants of the dynamics of a diffeomorphism of a compact manifold is its topological entropy. When positive, it implies the orbits of f exhibit an exponential amount of "chaos". When zero, the map f is somehow not typical, and has unusual regularity. For the study of group actions and foliation dynamics, it is natural to look for a corresponding entropy invariant of the system.

There are been several definitions of topological entropy for group actions, motivated by the need to have a definition for lattice dynamics where the group is typically $\mathbb{Z}^n$. These definitions admit extensions to actions of amenable groups, but for non-amenable groups, these definitions all encounter difficulties of one sort or another. The main problem with adapting these definitions for the study of foliations, however, is that they all vanish for C^1-actions whose leaves have growth rate greater than linear, so would vanish for all but the simplest or most pathological foliations!

The introduction by Ghys, Langevin and Walczak [37] of the geometric entropy $h(\mathcal{F})$ for a C^1–foliation $\mathcal{F}$ marked another transformation in the study of foliation dynamics. Their definition is completely natural: $h(\mathcal{F})$ measures the exponential rate of growth for (ϵ, n)-separated sets in the analogue of the Bowen metrics for the holonomy pseudogroup $\mathcal{G}_{\mathcal{F}}$ of $\mathcal{F}$. (Chapter 13 of [6] gives an excellent introduction and discussion of foliation entropy.) Thus, $h(\mathcal{F})$ is a measure of the complexity of the transverse dynamics of $\mathcal{F}$. The precise value of $h(\mathcal{F})$ depends upon a variety of choices, but the property $h(\mathcal{F}) = 0$ or $h(\mathcal{F}) > 0$ is well-defined. A codimension one foliation with a transverse invariant measure has entropy zero, while the entropy of the Roussarie example is positive. The authors established two key relations between $h(\mathcal{F})$ and dynamics:

Theorem 5.1 (Theorem 6.1, [37]) *If $\mathcal{F}$ is a C^2–foliation of codimension one, then $h(\mathcal{F}) > 0$ if and only if $\mathcal{F}$ has a resilient leaf.*

This theorem, combined with Duminy's Theorem, implies that if

$GV(\mathcal{F}) \neq 0$ then $h(\mathcal{F}) > 0$.

Theorem 5.2 (Theorem 5.1, [37]) *If $\mathcal{F}$ is a C^1-foliation of codimension $q \geq 1$ with $h(\mathcal{F}) = 0$, then $\mathcal{F}$ has a non-trivial holonomy invariant transverse measure.*

A number of problems and questions are suggested by these two theorems (see § 7, [37]):

1. Can one show the implication "if $GV(\mathcal{F}) \neq 0$ then $h(\mathcal{F}) > 0$" directly?
2. Show that if some secondary class is non-zero for $q > 1$, then $h(\mathcal{F}) > 0$.
3. Can one define an analog of metric entropy for a foliation, and use it to estimate $h(\mathcal{F})$?
4. What is the relation between $\mathcal{F}$-harmonic measures and $h(\mathcal{F})$?

These problems have been the continued focus of research since the appearance of the paper [37], and influenced research in many areas of the study of codimension one foliations. We will discuss the progress and updated formulations for these questions.

The problem of defining $h(\mathcal{F})$ using harmonic measures remains open. The closest result to obtaining such a relation, and this is only at the level of intuition, is the paper by Ghys [33] which proves key results about random walks on leaves and recurrence along the ends of leaves. This suggests that $\mathcal{F}$-harmonic measures should almost surely be influenced by the separation of leaves as they tend to infinity, as the entropy measures a type of expansion of the leaves, which should influence the properties of the convergence of the heat flow to harmonic measures. As both the geometric entropy and the properties of harmonic measures are natural invariants of a foliation, any relation between them would have great appeal and most likely be fundamental.

Work on the first three problems will be discussed in relation to the paper [59] by Hurder, which outlined program for the study of $h(\mathcal{F})$ using the foliation geodesic flow. A foliation with smooth leaves has a leafwise geodesic flow, defined on the unit tangent bundle $V = T_1\mathcal{F}$ to the leaves. The foliation $\mathcal{F}$ on M defines a foliation $\hat{\mathcal{F}}$ on V whose leaves cover those of $\mathcal{F}$. The foliation geodesic flow preserves $\hat{\mathcal{F}}$. The Riemannian geometry of this flow has been studied by Walczak in a series of papers [116, 117, 118, 119]. It also has a close relationship to the entropy of foliations.

Ghys, Langevin and Walczak actually defined two entropies for a foliation in [37], the pseudogroup entropy $h(\mathcal{F})$ introduced above, and another invariant called the *geometric entropy* of $\mathcal{F}$, defined by requiring points in an (ϵ, n)-separated set to be separated by the holonomy along geodesic segments of length at most n. Equivalently, the geometric entropy is that for the pseudogroup $\mathcal{G}_{\mathcal{F}}$ with the norm on maps defined via the shortest leafwise distance of a leafwise path with the same holonomy. Thus, as proved in [37], these entropies are either both zero, or non-zero.

The paper [59] proposed to reformulate the geometric entropy of $\mathcal{F}$ in terms of the entropy relative to the invariant foliation $\hat{\mathcal{F}}$ associated to the foliation geodesic flow, and then use the restriction of the normal derivative cocycle $D\varphi\colon \mathcal{G}_{\mathcal{F}} \to \mathrm{GL}(\mathbb{R}^q)$ to this flow to estimate the entropy. This would allow introduction of many techniques of smooth dynamical systems for the study of $h(\mathcal{F})$, and give a uniform approach which applied in all codimensions along the lines used in [66].

The papers [58, 66] used ergodic theory methods to prove vanishing theorems for secondary classes in terms of the range of the derivative cocycle $D\varphi$ on the metric equivalence relation defined by the pseudogroup $\mathcal{G}_{\mathcal{F}}$. A natural question following these works was to find geometric conditions which gave information about the Lyapunov band spectrum of the transverse derivative cocycle $D\varphi$, and use this information to obtain further relations between dynamics and the secondary classes. For example, the paper [57] showed that the characteristic classes must vanish for distal group actions with some additional hypotheses. A key step in [57] was to introduce invariant measures, not for the full group action, but associated to particular elements of the group along which there was hyperbolic expansion, and consider the Pesin stable manifold for these measures. The corresponding approach for foliations is to consider transversally hyperbolic invariant measures for the geodesic flow, and their stable manifolds. So, the techniques of [57, 58, 66] can be applied to the study of the relation between $h(\mathcal{F})$ and the dynamics of $\mathcal{F}$.

Two difficulties in applying these ideas to the study of foliation geometric entropy are immediately encountered: First, the entropy of the geodesic flow, relative to the invariant foliation, clearly bounds the geometric entropy from below. For foliations whose leaves have subexponential growth, the relative geodesic flow entropy and the geometric entropy are equivalent. Langevin and Walczak [74] called the relative geodesic flow entropy the *transverse entropy*, and proved that in codimension one, the geometric entropy and the transverse entropy are always equivalent. But it is not known whether these two entropies always agree for codimension greater than one.

The other difficulty is the definition and applications of a measure entropy for foliations, based on a relative measure entropy for the geodesic flow analogous to a construction well known in the ergodic theory of maps [78]. The (formidable) technical work required has never been written up, and it remains an open problem to justify this part of the program.

Langevin and Walczak [75] also studied the relations between the geometric entropy and the exponents of the transverse derivative cocycle $D\varphi$. This paper introduces the "pressure" for the dynamics of a pseudogroup. Biś and Walczak [2] showed that the geometric entropy can be calculated using pseudo-orbits.

During 1999-2000, the program of [59] for the study of $h(\mathcal{F})$ was carried out for codimension one foliations, avoiding the construction of relative measure entropy [62]. The role of the relative measure entropy was to obtain lower bound estimates on the geometric entropy in terms of the Lyapunov spectrum of invariant measures for the foliation geodesic flow. The alternative approach of [63, 64, 65] is based on the construction of special dynamical subsystems, called "ping-pong games". The existence of a ping-pong game is equivalent to the existence of a resilient leaf, and directly implies the geometric entropy of the foliation is positive. The key point is that asymptotic estimates for the restriction of $D\varphi$ along the foliation geodesic flow can be combined with stable manifold theory along transversally hyperbolic invariant measures for the geodesic flow, to produce ping-pong games, and hence prove the foliation geometric entropy must be positive.

The dynamic given by a ping-pong game represents an intermediate invariant of a foliation, and its existence can be proved by techniques similar to proving the existence of measures with positive relative entropy. A good theory of relative measure entropy should assign positive values to the dynamical subsystem determined by a ping-pong game. In the absence of such a theory, the study of the ping-pong games of $\mathcal{F}$ provides an effective method to estimate $h(\mathcal{F})$. Here are the main results from the papers [63, 64, 65]:

Theorem 5.3 (Theorem 1.1, [63]) *Suppose $\mathcal{F}$ is a codimension one, transversally C^1-foliation with $h(\mathcal{F}) > 0$, then $\mathcal{F}$ has a ping-pong game, and hence a resilient leaf.*

The proof of this uses "elementary" techniques of ergodic theory and flow dynamics, so avoids the use of C^2-structure theory.

Theorem 5.4 (Theorem 1.1, [65]) *Suppose $\mathcal{F}$ is a codimension one, transversally C^1-foliation with non-trivial Godbillon measure $g_{\mathcal{F}}$, then $\mathcal{F}$ has a ping-pong game and hence $h(\mathcal{F}) > 0$.*

The proof of this uses a new cocycle tempering method, modifying that of [58], and the basic methods for constructing ping-pong games developed in [63, 64].

For higher codimensions, this approach also yields the following result, where it is necessary to impose a Hölder condition on the transverse holonomy:

Theorem 5.5 (Theorem 1.3, [65]) *If $\mathcal{F}$ is a codimension q, transversally C^{1+a}-foliation with $h(\mathcal{F}) > 0$, then $\mathcal{F}$ is not distal. In particular, $\mathcal{F}$ cannot be a foliation with all leaves compact.*

The proofs of these results are heavily "one dimensional", but it seems likely that it will be possible to extend the techniques of proof to higher codimensions, to obtain relations between $h(\mathcal{F})$ and the secondary classes. (See also the discussion for Problem 9.5 at the end of this paper.)

6 Exceptional minimal sets

A leaf of a codimension one foliation is *semi-proper* if it approaches itself from one side. If the closure $K = \overline{L}$ of a semi-proper leaf L is minimal, then it is transversally a nowhere dense Cantor set, and we say K is an *exceptional minimal set.* The semi-proper leaves are precisely the leaves through the endpoints in the gaps of the transverse Cantor set. Semi-proper leaves and exceptional minimal sets have been understood as an essential part of the study of codimension one dynamics since the work of Denjoy, and the generalizations to foliations by Sacksteder [96, 100].

A variety of constructions of exceptional minimal sets have been given (e.g., see [99, 46, 68, 80]). The Poincaré-Bendixson classification theory for C^2-foliations [9, 49] is greatly simplified when there are no exceptional minimal sets. The proof of Duminy's theorem assumes there are no resilient leaves, and hence no exceptional minimal set, and proceeds from there.

Surprisingly, the exceptional minimal sets have defied an easy classification.

One of the open problems is to show that an exceptional minimal set for a C^2-foliation must have Lebesgue measure zero. Inaba [68] and Matsumoto [80] in 1986 gave general constructions of minimal sets for which the Lebesgue measure is zero. As remarked in [80], if an exceptional minimal set K has measure zero, then the Godbillon-Vey measure vanishes on K, so that analysis of the Godbillon-Vey invariant can ignore the contributions from exceptional minimal sets. Though unlikely, it would be an amazing result to show there exists an exceptional (local) minimal set with non-zero Godbillon-Vey measure.

Another open problem is to show that the leaves in an exceptional minimal set of a C^2-foliation must have a Cantor set of ends. A deep, unpublished work of Duminy (written up by Cantwell and Conlon in the manuscript [16]) shows that the semi-proper leaves always have a Cantor set of ends.

One can also ask when an exceptional minimal set contains only a finite number of semi-proper leaves. There exists C^1-foliations with exceptional minimal sets having a countably infinite number of semi-proper leaves, but it is conjectured that for C^2-foliations this is impossible.

In spite of the extensive study of exceptional minimal sets in codimension one dynamics, all of these questions show a key piece of the puzzle is still missing. To quote from [13], "Our very incomplete understanding of the exceptional type constitutes a major gap in the theory."

In 1988, Cantwell and Conlon [13] introduced a class of exceptional minimal sets, those of *Markov type*. A Markov minimal set K admits a finite set of expanding holonomy generators which define the foliation dynamics on it, and thus the dynamics on K are given by a quotient of a subshift of finite type. Cantwell and Conlon showed that for an exceptional minimal set of Markov type, there are only a finite number of semi-proper leaves, and K has measure zero [13]. They later showed that every leaf in a Markov minimal set K has a Cantor set of ends [15].

Inaba and Matsumoto [69] showed for transversally projective foliations, an exceptional minimal set is always Markov. This paper also gave a technical refinement of the definition of Markov property, which broadens the definition. Walczak [120] showed in fact that a Markov minimal set has Hausdorff dimension less than the dimension of M.

7 Extensions of Godbillon-Vey

The definition of the Godbillon-Vey invariant of a C^2-foliation of codimension one clearly depends upon the second derivatives of the 1-form defining $\mathcal{F}$. It is conjectured that if there is a homeomorphism $h\colon M \to M'$ mapping the leaves of a C^2 foliation $\mathcal{F}$ on M to the leaves of a C^2-foliation $\mathcal{F}'$ on M', then $h^*GV(\mathcal{F}') = GV(\mathcal{F})$. This problem has been posed in each of the foliation surveys and problems sessions [102, 31, 72] since 1978, and still remains unresolved.

The existence of a topological conjugacy h between $\mathcal{F}$ and $\mathcal{F}'$ means that the two foliations have the same topological dynamics. If $GV(\mathcal{F})$ is determined by the topological dynamics of $\mathcal{F}$, then the conjecture should be true. On the other hand, conjugation by a homeomorphism allows changing the exponents of hyperbolicity for the dynamics. For example, a homeomorphism can change the "shape" of an exceptional minimal set, and similarly distort the transverse dynamics of a foliation. Thus, it would be very surprising if the conjecture is true. However, there are no counter-examples in

C^2, and no suggestive examples to guide intuition on this question.

One approach to the problem would be to show that non-triviality of the Godbillon-Vey classes implies there is sufficient hyperbolicity in the dynamics of the foliations (at least in the support of their Godbillon-Vey measures) to prove that the homeomorphism h is actually C^1 or possibly even C^2, along the lines of [39] mentioned below.

A weaker conjecture is whether the conclusion holds when the conjugacy h has some additional regularity. For example, Raby [94] showed that $GV(\mathcal{F})$ is an invariant under C^1-diffeomorphism. Hurder and Katok [67] showed (independently, and using essentially the same methods as Raby) that if the conjugacy h and its inverse are absolutely continuous, then the conclusion is true. Natsume [87] showed that the Godbillon-Vey map in analytical K-theory of foliation C^*-algebras is also C^1-invariant.

Ghys and Tsuboi [39] used Duminy's Theorem to show that for a foliation with $GV(\mathcal{F}) \neq 0$, a C^1-conjugacy must be C^2 on the support of the Godbillon-Vey measure, thus the C^1-conjecture is only of modest interest. It is unknown if the assumption that h and its inverse are Hölder C^a-continuous, for some $a > 0$, suffices to show $h^*GV(\mathcal{F}') = GV(\mathcal{F})$.

There are counter-examples to topological invariance of extensions of the Godbillon-Vey class to foliations which are not C^2. Ghys defined in [30] a "Godbillon-Vey" type invariant for piecewise C^2-foliations in codimension one, and then showed via surgery on Anosov flows on 3-manifolds that there are homeomorphic, piecewise C^2-foliations with distinct "Godbillon-Vey" invariants. In another direction, Hurder and Katok defined in [67] a "Godbillon-Vey" type invariant for the weak-stable foliations of volume preserving Anosov flows on 3-manifolds, and showed that for the geodesic flow of a metric of variable negative curvature on a compact Riemann surface, the "Godbillon-Vey" invariant is a function of the "Mitsumatsu defect" [81], hence varies continuously and non-trivially as a function of the metric (Corollary 3.12, [67]) . The weak-stable foliations of all of these metrics are topologically conjugate, so this gives a continuous family of counter-examples. Both approaches to extending the Godbillon-Vey invariant are combined in work of Tsuboi [108, 109].

The topological invariance conjecture highlights once again how little is fundamentally known about the Godbillon-Vey class and its geometric or dynamical meaning. Perhaps, if it were possible to give a geometric or dynamic interpretation for the "Vey class", then one could determine whether the ingredients are preserved under homeomorphism, or possibly under a Hölder homeomorphism, or just under diffeomorphism.

8 Tricks and treats

While compiling this survey, and from the author's own research in the subject, several techniques and methods frequently are seen to be both essential and in some way unique to the study of the Godbillon-Vey classes and the dynamics of codimension one foliations. In this section, we compile a short sampling and brief description of selected techniques which have led to a deeper understanding of the themes of this survey. Of course, a thorough reading of the introductory texts on foliations, such as Godbillon's book [40] or the recent text by Candel and Conlon [6], reveals a far greater variety of ideas and techniques than discussed below. Still, it seems useful (and novel) to offer a list of "Tricks and Treats" for the subject, if only as an advertisement for the variety of methods which play a role in this field.

Technique 8.1 *Naive distortion lemma*

This is the most well-known method of 1-dimensional dynamics, except perhaps the well-ordering of the line. The hypothesis "$\mathcal{F}$ is C^2" often means simply that the elements of the holonomy pseudogroup satisfy this estimate, which was used in the celebrated theorems of Denjoy [24] and Sacksteder [100]. Briefly, recall that given a chain of local diffeomorphisms $g = h_{i_n} \circ \cdots h_{i_1}$ and two points u_0 and v_0 in the domain of g, then

$$|\log\{g'(u_0)\} - \log\{g'(v_0)\}| \leq \theta \sum_{p=0}^{n-1} |u_p - v_p|,$$

where θ is a constant depending on the C^2 norms of the generating elements $\{h_1, \ldots, h_N\}$ and $u_p = h_{i_p} \circ \cdots h_{i_1}(u_0)$ and $v_p = h_{i_p} \circ \cdots h_{i_1}(v_0)$. See section 8.1.A, [6]. This is usually applied in a context where the right-hand-sum is estimated by geometric considerations, as occurs when u_0 and v_0 are the endpoints of a gap for an exceptional minimal set. It is called "naive" by Sullivan, because in the modern theory of 1-dimensional dynamics there is also a "sophisticated" distortion lemma, also known as the "Schwartzian distortion lemma". This latter technique is a fundamental tool for renormalization theory of maps in 1-dimensional dynamics, but has not been used in the study of foliations, though it might prove useful for studying exceptional minimal sets.

Technique 8.2 *Octopus decomposition*

This result for C^0-foliations, which has no counterpart in classic dynamics, was introduced by Dippolito [25]. It states that an open saturated subset $U \subset M$ of a foliated compact manifold can be decomposed into a union of manifold with corners, $U = \mathcal{N} \cup A_1 \cup \cdots \cup A_n$ where $\mathcal{N}$ is the body, or nucleus, of U, and the A_i are the arms. The body $\mathcal{N}$ is a compact, connected manifold with boundary and corners, and $\mathcal{F}|\mathcal{N}$ foliates $\mathcal{N}$. Each arm A_i is a closed, non-compact submanifold with boundary and corners, and $\mathcal{F}|A_i$ is a foliated interval bundle. More technical conditions are required (see pages 130-131, [6]) but this suffices to give the idea. The power of this result lies in that the structure of foliated interval bundles is well understood, so this isolates difficulties with the study of $\mathcal{F}|U$ to the compact foliated body $\mathcal{F}|\mathcal{N}$. One imagines the arms of the octopus snaking through the various exceptional minimal sets of $\mathcal{F}$, or possibly squeezing between the proper leaves. The conjecture about whether an exceptional minimal set K for a C^2 foliation can have infinitely many semi-proper leaves is just asking how many octopi can share the set K!

Technique 8.3 *Poincaré-Bendixson theory*

The theory of levels is one of the most sophisticated tools for the study of C^2-foliations. It was developed by Cantwell and Conlon [9] and Hector [49], and defines inductively a decomposition (or hierarchy) of a foliation according to the asymptotics of the leaves. Starting with level 0, the compact minimal sets, the leaves at the next level are the local minimal sets for the complement of the previous level. The resulting structure and complications can be formidable. Fortunately, there is an introduction to the theory [14], and Chapter 8 of [6] provides a detailed and patient discussion of all aspects of the theory. One its greatest successes is the structure theory of real analytic foliations [11], where the hierarchy is finite and the structure of each stage is very well understood. When this theory is applied to codimension one foliations of 3-manifolds, one can ask about the "placement" of the leaves, or local minimal sets at various levels, within the manifold. For example, if M is Haken it is possible to formulate precise questions about the leaf placements at increasing levels and the fundamental group of M. Unfortunately, it is not possible to answer these questions yet – the ongoing work of Cantwell and Conlon [17, 18, 19, 20] have solved the placement problem for depth one, while depth two awaits.

Technique 8.4 *Micro-expansion, sheaves and quivers*

A C^1-foliation $\mathcal{F}$ with positive geometric entropy has micro-expansion, a phrase coined to suggest the explosion in the orbits of exponentially close points that has to occur when $h(\mathcal{F}) > 0$. A complete transversal $\mathcal{T}$ for $\mathcal{F}$ has finite length, so by definition of geometric entropy, for $\epsilon > 0$ small and $i \to \infty$, there exists a sequence $n_i \to \infty$ and collections of (ϵ, n_i)-separated points $\{x^i_1, \ldots, x^i_{p_i}\} \subset \mathcal{T}$ where the sequence $\{p_i\}$ has exponential growth $\exp(n_i h(\mathcal{F}))/p_i \to 1$. By a pigeon-hole principle, there must be subcollections of exponentially many points exponentially close which are

(ϵ, n_i)-separated. We can assume the set transversal $\mathcal{T}$ is a subset of the line so well-ordered, and the sets are index with $x_k^i < x_\ell^i$ for $k < \ell$. Then for each x_k^i there is an element of holonomy γ_k^i of length at most n_i which ϵ-separates x_k^i and x_{k+1}^i. One pictures this as a *sheaf* of arrows of length n_i with bases concentrated on smaller and smaller regions, the arrows representing elements of holonomy $\{\gamma_k^i\}$, and the tips pairwise separating so they end up at least ϵ apart. Moreover, the transverse derivatives along these arrows are non-uniformly hyperbolic by the mean value theorem, so these could be called *hyperbolic quivers*, which sounds even more formidable.

On the other hand, the tips of the arrows must also land somewhere on $\mathcal{T}$, and since there are exponentially many arrows, another application of the pigeon-hole principle yields exponential families of arrows (elements of holonomy) of length n_i whose bases and tips are exponentially close. These are *quivers*. Thus, there is quite a lot of dynamical information contained in the statement "$h(\mathcal{F}) > 0$". As intuitive and imprecise as the image of quivers may seem, it can be formalized in a variety of contexts to help understand the geometric entropy. For example, it is a key idea behind the proofs involving entropy in the papers [63, 64, 65]. The main question about quivers is how to quantify them. It seems likely that any scheme will be analogous to the construction of relative measure entropy for the geodesic flow of $\mathcal{F}$. There should also be a connection between quivers and the pseudo-orbit estimates of [2].

Almost the opposite concept was used in the proof of Theorem 5.1, [37] which proved that if $h(\mathcal{F}) = 0$ then there exists a holonomy invariant transverse measure. Here, the intuition is closer to a "straw mat", as the base of the arrows are uniformly spaced and the tips cluster at a subexponential rate.

Technique 8.5 *Ping-pong games and closing*

A ping-pong game for a C^0-foliation is a basic concept of topological dynamics, which is equivalent to the idea of resilient orbits, but somehow more natural to construct. The basic idea is that there should be given an interval $I_0 \subset \mathcal{T}$ in the transversal for $\mathcal{F}$, and elements of holonomy $h_1\colon I_0 \to I_1 \subset I_0$ and $h_2\colon I_0 \to I_2 \subset I_0$ so that both h_1 and h_2 are contractions. We also require that the closures of I_1 and I_2 are disjoint, and both contained in the interior of I_0. It is an exercise that the orbits of the fixed-points in a ping-pong game generate resilient leaves, and that a foliation with a ping-pong game must have $h(\mathcal{F}) > 0$. This latter fact was used in [37] to show that a foliation with a resilient leaf has $h(\mathcal{F}) > 0$. For a C^1-foliation, a hyperbolic ping-pong game is one where the maps are hyperbolic contractions on the set I_0, and thus h_1 and h_2 must each have a unique hyperbolic fixed-point. Thus, a hyperbolic ping-pong game is like having part of the hypothesis of a Markov minimal set, except that orbits of the hyperbolic fixed-points are not assumed to lie in an exceptional minimal set. A C^1-foliation with a quiver must have a ping-pong game, and hence positive entropy – a fact used in the proof of Theorem 1.1 of [65]. The existence of a quiver also implies there are attracting fixed-points, which is a type of closing lemma that says close to every quiver is a hyperbolic fixed-point. This idea is developed in detail in section 4 of [64].

Technique 8.6 *Tempering procedures*

Tempering procedures are specialty tools, which often appear to be more technical than useful, but when needed are indispensable and very powerful. Tempering is a process which converts asymptotic estimates for a cocycle into local estimates by making a change of scale (coboundary) for the initial data. For example, if $\varphi\colon \Gamma \times \mathbb{S}^1 \to \mathbb{S}^1$ is a C^1-action on the circle, then the derivative $d\varphi\colon \Gamma \times \mathbb{S}^1 \to \mathbb{R}$ is a real-valued cocycle over the action φ. We define the exponent of $d\phi$ at x by

$$\lambda(x) = \limsup_{n \to \infty} \frac{\log\left(\max\{\varphi(\gamma)'(x) \mid \|\gamma\| \leq n\}\right)}{n}.$$

If $K \subset M$ is a closed saturated subset such that $\lambda(x) \leq a$ for almost all $x \in K$, then for $\mathrm{gr}(\mathcal{F})$ the word growth rate of Γ, choose $\epsilon > a + \mathrm{gr}(\Gamma)$ and define

$$f_\epsilon(x) = \sum_{\gamma \in \Gamma} \exp\{-\epsilon \cdot \|\gamma\|\} \cdot d\phi(\gamma, x).$$

It is then an exercise that the new cocycle $\tilde{\phi}(\gamma, x) = f_\epsilon(\varphi(\gamma)(x)) \cdot \phi(\gamma, x) \cdot f_\epsilon(x)^{-1}$ has uniform local variation bounded by ϵ.

Tempering for linear cocycles over group actions and foliations was introduced by Hurder and Katok in [67], and used there and in [58] to regularize the transverse derivative cocycle $D\varphi$ for foliations. A discussion of tempering can also be found in [31]. For example, when the leaves of $\mathcal{F}$ have subexponential volume growth, the transverse Radon-Nikodym cocycle satisfies $\lambda(x) = 0$ for almost all x, so tempering constructs a transverse measure defining $\mathcal{F}$ with arbitrarily small local variation. The papers [64, 65] introduce another tempering procedure for subexponential growth cocycles over arbitrary growth foliations, which is applied in [65] to obtain new vanishing theorems for the Godbillon-Vey classes. Tempering procedures are essentially the only available method for converting transversally measurable information for a foliation (typically obtained from ergodic theory considerations) into differential geometric conclusions.

9 Open questions

"Problem sessions" were held at both the 1976 and 1992 symposia on foliations at Rio de Janeiro, with the discussion and proposed problems compiled by Paul Schweitzer [102] for the 1976 meeting, and Remi Langevin [72] for the 1992 meeting. Some of the problems remain unchanged, while a comparison between these two reports fourteen years apart illustrates some of the advances in the field, and changing emphasis in research. The survey of the Godbillon-Vey invariant by Ghys [31] also includes a number of problems with discussions about them. Here, we compile a list of questions concerning the topics of the present survey. It is not meant to be comprehensive when compared to the more general problem lists above, but does attempt to include all of the frequently mentioned problems regarding the Godbillon-Vey classes and foliation dynamics.

Problems on Godbillon-Vey invariants

Problem 9.1 *Give a* geometric *interpretation of the Godbillon-Vey invariant*

The Moussu-Pelletier and Sullivan Conjecture is a one-sided look at $GV(\mathcal{F})$, as it only relates to dynamical properties of $\mathcal{F}$ which can force the Godbillon measure to vanish. The other side is the "Vey class" which depends upon curvature properties of the leaves and normal bundle. The Reinhart-Wood formula [98] gave a *pointwise* geometric interpretation of $GV(\mathcal{F})$ for 3-manifolds. What is needed is a more global geometric property of $\mathcal{F}$ which is measured by $GV(\mathcal{F})$. The helical wobble description by Thurston [106] is a first attempt at such a result, and the Reinhart-Wood formula suitably interprets this idea locally. Langevin has suggested that possibly the Godbillon-Vey invariant can be interpreted in the context of integral geometry and conformal invariants [3, 73] as a measure in some suitable sense. The goal for any such an interpretation, is that it should provide *sufficient* conditions for $GV(\mathcal{F}) \neq 0$.

Problem 9.2 *Topological invariance of the Godbillon-Vey invariant*

Given a homeomorphism $h\colon M \to M'$ mapping the leaves of a C^2 foliation $\mathcal{F}$ on M to the leaves of a C^2-foliation $\mathcal{F}'$ on M', show $h^*GV(\mathcal{F}') = GV(\mathcal{F})$. As discussed in section 7, if h is C^1, then Raby [94] proved $h^*GV(\mathcal{F}') = GV(\mathcal{F})$, and when h and its inverse are absolutely continuous, then Hurder and Katok [67] showed this. An intermediate test case might be to assume h and its inverse are a Hölder C^a-continuous for some $a > 0$, and then prove $h^*GV(\mathcal{F}') = GV(\mathcal{F})$, using for example arguments from regularity theory of hyperbolic systems and an approach similar to

Ghys and Tsuboi [39]. Alternately, a direct proof may be possible, perhaps based on a solution to Problem 9.1.

Problem 9.3 *The Godbillon-Vey invariant and harmonic measures*

The vanishing theorems are based on relating the Godbillon measure to the existence of "almost invariant" smooth transverse measures for $\mathcal{F}$. A foliation always admits a harmonic measure, but the structure of that measure depends upon whether $\mathcal{F}$ admits transverse invariant measures, or not. Is it possible to establish relations between the values of the Godbillon-Vey invariant and the structure of harmonic measures for $\mathcal{F}$? There are other similarities in the properties of both of these invariants of $\mathcal{F}$ which suggests that such a relationship is plausible.

Problem 9.4 *What is meaning of thickness?*

The concept of "thickness" introduced by Duminy [26, 27, 12] was given in terms of the structure theory of C^2-foliations, yet its application is to show the foliation admits almost invariant transverse volume forms on an open saturated subset, which is a purely dynamical consideration. Does the thickness have an interpretation as a dynamical property of the foliation geodesic flow, or some other ergodic property of $\mathcal{F}$?

Problem 9.5 *Suppose that $\mathcal{F}$ has codimension $q > 1$ and there is some non-zero secondary class (or possibly Weil measure). Does this imply $h(\mathcal{F}) > 0$?*

Hurder [56] showed that for a C^2-foliation of codimension $q > 1$, if there is a leaf L whose linear holonomy map $D\varphi\colon \pi_1(L, x) \to \mathrm{GL}(\mathbb{R}^q)$ has non-amenable image, then $\mathcal{F}$ has leaves of exponential growth. The proof actually constructs a modified ping-pong game for $\mathcal{F}$, using the C^2-hypothesis to show that the orbits of the holonomy pseudogroup shadow the orbits of the linear holonomy group which has an actual ping-pong game by Tits [107]. Thus, it seems probable that this proof also shows $h(\mathcal{F}) > 0$ with these hypotheses. Since the Weil measures vanish for a foliation whose transverse derivative cocycle $D\varphi\colon \mathcal{G}_{\mathcal{F}} \to \mathrm{GL}(\mathbb{R}^q)$ has amenable algebraic hull [66], it should be possible to combine the methods of [56, 66, 63] to solve this problem.

Problems on minimal sets

Problem 9.6 *Let K be an exceptional minimal set for a codimension one C^2-foliation $\mathcal{F}$. Show*

1. *The Lebesgue measure of K is zero;*
2. *K has only a finite number of semi-proper leaves;*
3. *Every leaf of K has a Cantor set of ends.*

4. *Every semiproper leaf of K has germinal holonomy infinite cyclic, generated by a contraction.*
5. *K is Markov.*

The first four questions were posed at least 20 years ago. Note that K has only a finite number of semiproper ends if and only if its complement in M has only a finite number of connected open components. Duminy's Theorem [16] shows that the semiproper leaves of K must have a Cantor set of ends. A number of authors have shown the measure of K is zero for special cases [68, 80, 69]. Cantwell and Conlon showed that if K is Markov, then the first four properties follow [13, 15].

Problem 9.7 *For a codimension one C^2-foliation $\mathcal{F}$, give an example of an exceptional minimal set K with non-trivial Godbillon-Vey measure.*

This is most likely impossible, as it contradicts (9.6.1) above. If a counter-example to (9.6.1) can be constructed, then it will automatically have non-zero Godbillon measure, as an exceptional minimal set must be hyperbolic for a C^2-foliation, so it would then be plausible to ask that whether the Godbillon-Vey class localized to K is non-zero.

Problem 9.8 *For a codimension one, C^1- or C^2-foliation $\mathcal{F}$, give a structure theorem for the exceptional minimal sets of $\mathcal{F}$.*

This is asking first for an understanding of how many semiproper leaves there are, and then for some sort of generalized Markov structure on K. In other words, it is asking a lot!

Problem 9.9 *Let Γ be a* closed *subgroup of* $\mathrm{Homeo}_+(\mathbb{S}^1)$ *acting transitively. Is Γ conjugate to one of the subgroups* $\mathrm{SO}(\mathbb{R}^2)$, $\mathrm{PSL}_k(\mathbb{R}^2)$, *or* $\mathrm{Homeo}_{k,+}(\mathbb{S}^1)$ *of* $\mathrm{Homeo}_+(\mathbb{S}^1)$*?*

This question was posed by Ghys as Problem 4.4, [35]. The hypothesis the group is closed is essential, so unless the group is finite it must be non-discrete. The problem is included as an understanding of this question would surely help with understanding the minimal actions of countable groups on $\mathbb{S}^1$. (Note that the subscript k in the question indicates the k-fold covering group.)

Problems on geometric entropy

Problem 9.10 *Give a definition of the measure entropy, or some other entropy-type invariant, of a C^1-foliation $\mathcal{F}$, which can be used to establish positive lower bounds for the geometric entropy.*

This problem was asked in the original paper of Ghys, Langevin and Walczak [37]. Their earlier paper [36] gave a possible definition, but the connection to the geometric entropy is unclear. The paper by Hurder [59] proposes a definition of the measure entropy in terms of invariant measures for the associated geodesic flow. If well-defined, these measure entropies will estimate the entropy of the geodesic flow relative to the invariant foliation almost by definition. A special case is to show there exists a good definition of measure entropies for codimension 1 foliations. Another approach might be to define measure entropy for a foliation in terms of its harmonic measures.

Problems on ergodic theory

Problem 9.11 *Show the set of leaves with non-exponential growth, and not subexponential growth, has Lebesgue measure zero.*

Hector's construction in [47] of examples with leaves of this special type appear to produce a set (of such leaves) with measure zero. This growth condition, that $\limsup \neq \liminf$, implies a high degree of non-uniformity for the asymptotics of the leaf. If there exists a set of positive measure consisting of such leaves, then recurrence within the set should imply a uniformity of the growth, contradicting the hypothesis.

Problem 9.12 *Can a codimension one foliation have higher rank?*

The celebrated theorems of Burger and Monod [5] and Ghys [34] show that a higher rank group does not admit an effective C^1-action on the circle. One can view these results as about the holonomy groups of a codimension one foliation transverse to a circle bundle. Can these theorems be generalized to codimension one foliations which are not transverse to a circle bundle? Part of the problem is to give a suitable definition of higher rank for a foliation (cf. Zimmer [123].)

Problem 9.13 *Does restricted orbit equivalence preserve geometric entropy?*

Given foliated compact manifolds $(M, \mathcal{F})$ and $(M', \mathcal{F}')$, a *restricted orbit equivalence* between $\mathcal{F}$ and $\mathcal{F}'$ is a measurable isomorphism $h\colon M \to M'$ which maps the leaves of $\mathcal{F}$ to the leaves of $\mathcal{F}'$, and the restriction of h to leaves is a coarse isometry for the leaf metrics. Note that h and its inverse are assumed to preserve the Lebesgue measure class, but need not preserve the Riemannian measure. Such a map preserves the Mackey range of the Radon-Nikodym cocycle [122]. Restricted orbit equivalence also preserves the entropy positive condition, for ergodic $\mathbb{Z}^n$ actions. Does a corresponding result hold for geometric entropy: if $h(\mathcal{F}) > 0$, must $h(\mathcal{F}') > 0$ also?

Problem 9.14 *How is the flow of weights for $\mathcal{F}$ related to the dynamics of $\mathcal{F}$?*

Connes has show that the Godbillon-Vey class, or more precisely the Bott-Thurston 2-cocycle defined by it, can be calculated from the flow of weights for the von neumann algebra $\mathfrak{M}(M, \mathcal{F})$ (see [22], Chapter III.6, [23].) This gives another proof of the theorem of Hurder and Katok [66] that if $GV(\mathcal{F}) \neq 0$ then $\mathfrak{M}(M, \mathcal{F})$ has a factor of type III. The flow of weights is determined by the flow on the Mackey range of the modular cocycle, so that a type III factor corresponds to a ergodic component of the Mackey range with no invariant measure. However, almost nothing else is known about how the flow of weights is related to the topological dynamics of $\mathcal{F}$. In particular, Alberto Candel has asked whether the existence of a resilient leaf can be proven using properties of the flow of weights.

Acknowledgments

Partially supported by NSF Grant DMS-9704768.

References

1. D.V. Anosov, *Geodesic Flows on Closed Riemannian Manifolds with Negative Curvature*, Proc. Steklov Inst. Math., **90** (1967), Published by Amer. Math. Soc., Providence 1969.
2. A. Biś and P. Walczak, *Pseudo-orbits, pseudo-leaves and geometric entropy of foliations*, Ergodic Th. and Dynam. Sys., **18** (1998), 1335–1348.
3. F. Brito, R. Langevin and H. Rosenberg, *Intégrales de courbure sur une variété feuilletée*, J. Diff. Geom., **16** (1981), 19–50.
4. R. Bott, *On some formulas for the characteristic classes of group actions*, in Differential Topology, Foliations and Gelfand-Fuks cohomology, Rio de Janeiro (1976), Lect. Notes in Math., **652**, 25–61, Springer-Verlag, New York and Berlin, 1978.
5. M. Burger and N. Munod, *Bounded cohomology of lattices in higher rank Lie groups*, J. Eur. Math. Soc., (JEMS) **1** (1999), 199–235.
6. A. Candel and L. Conlon, Foliations 1, Amer. Math. Soc., Providence, RI, 2000.
7. J. Cantwell and L. Conlon, *Growth of leaves*, Comment. Math. Helv., **53** (1978), 93–111.
8. J. Cantwell and L. Conlon, *The vanishing of GV for foliations almost without holonomy*, preprint, 1981.
9. J. Cantwell and L. Conlon, *Poincaré-Bendixson theory for leaves of codimension one*, Trans. Amer. Math. Soc., **265** (1981), 181–209.

10. J. Cantwell and L. Conlon, *Non-exponential leaves at finite level*, Trans. Amer. Math. Soc., **269** (1982), 637–661.
11. J. Cantwell and L. Conlon, *Analytic foliations and the theory of levels*, Math. Ann., **265** (1983), 253–261.
12. J. Cantwell and L. Conlon, *The dynamics of open, foliated manifolds and a vanishing theorem for the Godbillon-Vey class*, Advances in Math., **53** (1984), 1–27.
13. J. Cantwell and L. Conlon, *Foliations and subshifts*, Tôhoku Math. J., **40** (1988), 165–187.
14. J. Cantwell and L. Conlon, *The theory of levels*, in Index Theory of Elliptic Operators, Foliations, and Operator Algebras, eds K. Millet et al, Contemp. Math., **70**, 1–10, Amer. Math. Soc., Providence 1988.
15. J. Cantwell and L. Conlon, *Leaves of Markov local minimal sets in foliations of codimension one*, Publications Matematiques, **33** (1989), 461–484, published by the Universitat Autòmata de Barcelona.
16. J. Cantwell and L. Conlon, *Endsets of exceptional leaves; a theorem of G. Duminy*, Notes for informal circulation, (1989).
17. J. Cantwell and L. Conlon, *Depth of knots*, Topology and Appl., **42** (1991), 277–289.
18. J. Cantwell and L. Conlon, *Isotopy of depth one foliations*, in Proceedings of the International Symposium/Workshop on Geometric Study of Foliations, 153–173, World Scientific, Singapore, 1994.
19. J. Cantwell and L. Conlon, *Isotopies of foliated 3-manifolds without holonomy*, Advances in Math., **144** (1999), 13–49.
20. J. Cantwell and L. Conlon, *Foliation Cones*, in Proceedings of the Kirbyfest, 35–86, Geometry and Topology Monographs, 1999.
21. L. Conlon, Foliations and Exotic Classes, Notes from Short Course given at Jarandilla de la Vera (Caceres), September 1985, Washington University Notes, 1986.
22. A. Connes, *Cyclic cohomology and the transverse fundamental class of a foliation*, in Geometric methods in operator algebras (Kyoto, 1983), 52–144, Pitman Res. Notes in Math., **123**, 1986.
23. A. Connes, Noncommutative Geometry, Academic Press, San Diego, CA, 1994.
24. A. Denjoy, *Sur les courbes définies par les équations différentielles à la surface du tore*, J. Math. Pures Appl., **11** (1932), 333–375.
25. P. Dippolito, *Codimension one foliations of closed manifolds*, Ann. of Math., **107** (1978), 403–457.
26. G. Duminy, *L'invariant de Godbillon–Vey d'un feuilletage se localise dans les feuilles ressort*, unpublished, Université de Lille, I, 1982.

27. G. Duminy, *Sur les cycles feuilletés de codimension un*, unpublished, Université de Lille, I, 1982.
28. G. Duminy and V. Sergiescu, *Sur la nullité de l'invariant de Godbillon-Vey*, C.R. Acad. Sci. Paris, **292** (1981), 821–824.
29. J. Feldman and C.C. Moore, *Ergodic equivalence relations, cohomology, and von Neumann algebras I*, Trans. Amer. Math. Soc., **234** (1977), 289–324.
30. É. Ghys, *Sur l'invariance topologique de la classe de Godbillon-Vey*, Annales Inst. Fourier, Grenoble, **37** (4) (1987), 59–76.
31. É. Ghys, *Sur l'invariant de Godbillon-Vey*, in Seminaire Bourbaki, March 1989. Asterisque, **177-178**, 155–181, Société Mathématique de France, 1989.
32. É. Ghys, *Rigidité différentiable des groupes fuchsiens*, Publ. Math. Inst. Hautes Etudes Sci., **78** (1993), 163–185.
33. É. Ghys, *Topologie des feuilles génériques*, Ann. of Math., **141** (1995), 387–422.
34. É. Ghys, *Actions de réseaux sue le cercle*, Invent. Math, **137** (1999), 199–231.
35. É. Ghys, *Groups acting on the circle*, preprint, IMCA, Lima, June 1999.
36. É. Ghys, R. Langevin, and P. Walczak, *Entropie mesurée et partitions de l'unité*, C. R. Acad. Sci., Paris, **303** (1986), 251–254.
37. É. Ghys, R. Langevin, and P. Walczak, *Entropie géométrique des feuilletages*, Acta Math., **168** (1988), 105–142.
38. E. Ghys and V. Sergiescu, *Sur un groupe remarquable de difféomorphismes du cercle*, Comment. Math. Helv., **62** (1987), 185–239.
39. E. Ghys and T. Tsuboi, *Différentiabilité des conjugaisons entre systèms dynamiques de dimension 1*, Ann. Inst. Fourier, Grenoble, **38** (1) (1988), 215–244.
40. C. Godbillon, Feuilletages: Etudes géométriques I, II. Publ. IRMA Strasbourg (1985-86); Progress in Math. **98**, Birkhäuser, 1991.
41. C. Godbillon and J. Vey, *Un invariant des feuilletages de codimension 1*, C.R. Acad. Sci. Paris, **273** (1971), 92–95.
42. A. Haefliger, *Structures feulletées et cohomologie à valeur dans un faisceau de groupoïdes*, Comment. Math. Helv., **32** (1958), 248–329.
43. A. Haefliger, *Variétés feulletées*, Ann. Scuola Norm. Sup. Pisa, **16** (1962), 367–397.
44. G. Hector, *Sur les feuilletages presque sans holonomie*, C.R. Acad. Sci. Paris, **274** (1972), 1703–1706.
45. G. Hector, *Sur un théorème de structure des feuilletages de codimen-*

sion un, in G. Reeb, ed. Feuilletages: Résultats anciens et nouveaux (Painlevé, Hector et Martinet) Montréal 1972, 48–54, Presses Univ. Montréal, 1974.

46. G. Hector, *Quleques exemples de feuilletages espèces rares*, Ann. Inst. Fourier, Grenoble, **26** (1976), 239–264.
47. G. Hector, *Leaves whose growth is neither exponential nor polynomial*, Topology, **16** (1977), 451–459.
48. G. Hector, *Croissance des feuilletages presque sans holonomie*, in Foliations and Gelfand-Fuks Cohomology, Rio de Janeiro (1976), Lect. Notes Math., **652**, 141–182, Springer-Verlag, New York and Berlin, 1978.
49. G. Hector, *Architecture of C^2-foliations*, Astérisque, **107-108**, 243–258, Société Mathématique de France 1983.
50. J. Heitsch, *A cohomology for foliated manifolds*, Comment. Math. Helv., **50** (1975), 197–218.
51. J. Heitsch, *Independent variation of secondary classes*, Annals of Math., **108** (1978), 421–460.
52. J. Heitsch and S. Hurder, *Secondary classes, Weil measures and the geometry of foliations*, J. Differential Geom., **20** (1984), 291–309.
53. M.R. Herman, *The Godbillon-vey invariant of foliations by planes of* $\mathbb{T}^3$, in Geometry and Topology, Rio de Janeiro (1976), Lect. Notes in Math., **597**, 294–307, Springer-Verlag, New York and Berlin, 1977.
54. M. R. Herman, *Sur la conjugaison différentiable des difféomorphisms du cercle a des rotations*, Publ. Math. Inst. Hautes Etudes Sci., **49** (1979), 5–234.
55. S. Hurder, *Vanishing of secondary classes for compact foliations*, J. London Math. Soc., **28** (1983), 175–181.
56. S. Hurder, *Foliation dynamics and leaf invariants*, Comment. Math. Helv., **60** (1985), 319–335.
57. S. Hurder, *The Jacobian Cocycle of a distal group action*, Math. Sci. Res. Inst., Preprint MSRI 12811-85, 1985.
58. S. Hurder, *The Godbillon measure of amenable foliations*, J. Diff. Geom., **23** (1986), 347–365.
59. S. Hurder, *Ergodic theory of foliations and a theorem of Sacksteder*, in Dynamical Systems: Proceedings, University of Maryland 1986-87, Lect. Notes in Math. **1342**, 291–328, Springer-Verlag, New York and Berlin, 1988.
60. S. Hurder, *Exceptional minimal sets for C^{1+a}-group actions on the circle*, Ergodic Th. and Dynam. Sys., **11** (1991), 455–467.
61. S. Hurder, *Coarse geometry of foliations*, in Geometric Study of Fo-

liations, Tokyo 1993 (eds. Mizutani et al.), 35–96, World Scientific, 1994.
62. S. Hurder, *Dynamics of C^1-foliations*, preprint, May 2000, available at http://www.math.uic.edu/~hurder/publications.
63. S. Hurder, *Entropy and dynamics of C^1 foliations*, preprint, August 2000, available at http://www.math.uic.edu/~hurder/publications.
64. S. Hurder, *Dynamics of expansive group actions on the circle*, preprint, August 2000, available at http://www.math.uic.edu/~hurder/publications.
65. S. Hurder and R. Langevin, *Dynamics and the Godbillon-Vey class of C^1 foliations*, preprint, September 2000, available at http://www.math.uic.edu/~hurder/publications.
66. S. Hurder and A. Katok, *Ergodic Theory and Weil measures for foliations*, Annals of Math., **126** (1987), 221–275.
67. S. Hurder and A. Katok, *Differentiability, rigidity and Godbillon-Vey classes for Anosov flows*, Publ. Math. Inst. Hautes Etudes Sci., **72** (1990), 5–61.
68. T. Inaba, *Examples of exceptional minimal sets*, preprint, Chiba University, 1986.
69. T. Inaba and S. Matsumoto, *Resilient leaves in transversely projective foliations*, J. Fac. Sci. Univ. Tokyo, **37** (1990), 89–101.
70. T. Inaba and P. Walczak, *Transverse Hausdorff dimension of codimension-one foliations*, Fundamenta Math., **149** (1996), 239–244.
71. A. Katok and B. Hasselblatt, Introduction to the Modern Theory of Dynamical Systems, Cambridge Univeristy Press, 1995.
72. R. Langevin, *A list of questions about foliations*, in Differential Topology, Foliations and Group Actions. Rio de Janeiro 1992, Contemp. Math. **161**, 59–80, Amer. Math. Soc., Providence 1991.
73. R. Langevin and G. Levitt, *Courbure totale des feuilletages des surfaces*, Comment. Math. Helv., **57** (1982), 175–195.
74. R. Langevin and P. Walczak, *Entropy, transverse entropy and partitions of unity*, Ergodic Th. and Dynam. Sys., **14** (1994), 551–563.
75. R. Langevin and P. Walczak, *Some invariants measuring dynamics of codimension-one foliations*, in Geometric Study of Foliations, Tokyo 1993 (eds. Mizutani et al.), 345–358, World Scientific, Singapore, 1994.
76. H.B. Lawson Jr., *Foliations*, Bull. Amer. Math. Soc., **8** (1974), 369–418.
77. H.B. Lawson Jr., *The quantitative theory of foliations*, NSF Regional Conf. Board Math. Sci., **27**, (1975).
78. F. Ledrappier and P. Walters, *A relativised variational principle for*

continuous transformations, J. London Math. Soc., **16** (1977), 568–576.

79. G.A. Margulis, Discrete Subgroups of Lie Groups, Springer-Verlag, New York and Berlin, 1991.
80. S. Matsumoto, *Measure of exceptional minimal sets of codimension one foliations*, in A Fête of Topology, 81–94. Academic Press, Boston, 1988.
81. Y. Mitsumatsu, *A relation between the topological invariance of the Godbillon-Vey invariant and the differentiability of Anosov foliations*, in Foliations, 159–167, Advanced Studies in Pure Math. **5**, North-Holland, Amsterdam, 1985.
82. T. Mizutani, S. Morita and T. Tsuboi, *The Godbillon-Vey classes of codimension one foliations which are almost without holonomy*, Annals of Math., **113** (1981), 515–527.
83. T. Mizutani, S. Morita and T. Tsuboi, *On the cobordism classes of codimension one foliation which are almost without holonomy*, Topology, **22** (1983), 325–343.
84. S. Morita and T. Tsuboi, *The Godbillon-Vey class of codimension one foliations without holonomy*, Topology, **19** (1980), 43–49.
85. R. Moussu, *Feuilletages presque sans holonomie*, C.R. Acad. Sci. Paris, **272** (1971), 114–117.
86. R. Moussu and F. Pelletier, *Sur le Theorémè de Poincaré–Bendixson*, Ann. Inst. Fourier (Grenoble), **14** (1974), 131–148.
87. T. Natsume, *The C^1-invariance of the Godbillon-Vey map in analytical K-theory*, Canad. J. Math., **39** (1987), 1210–1222.
88. T. Nishimori, *Compact leaves with abelian holonomy*, Tôhoku Math. J., **27** (1975), 259–272.
89. T. Nishimori, *SRH-decompositions of codimension one foliations and the Godbillon-Vey classes*, Tôhoku Math. J., **32** (1980), 9–34.
90. Ya.B. Pesin, *Characteristic Lyapunov exponents and smooth ergodic theory*, Russian Math. Surveys, **32**(4) (1977), 55–114.
91. J. Plante, *Anosov flows*, Amer. J. Math., **94** (1972), 729–755.
92. J. Plante and W. Thurston, *Anosov flows and the fundamental group*, Topology, **11** (1972), 147–150.
93. H. Poincaré, *Mémoires sur les courbes définies par une équation différentielle*, J. Math. Pure et Appl., (Série 3), **7** (1881), 375–422.
94. G. Raby, *Invariance de classes de Godbillon-Vey par C^1-diffeomorphismes*, Ann. Inst. Fourier, Grenoble, **38** (1) (1988), 205–213.
95. G. Reeb, *Sur certaines propritś topologiques des variétés feuilletées*, Actualité Sci. Indust. 1183, Hermann, Paris (1952).

96. G. Reeb, *Sur les structures feuilletées de codimension un et sur un théorème de M. A. Denjoy*, Ann. Inst. Fourier, Grenoble, **11** (1961), 185–200.
97. G. Reeb, Feuilletages: Résultats anciens et nouveaux, (Painlevé, Hector et Martinet) Montréal 1972, 48–54. Presses Univ. Montréal, 1974.
98. B. Reinhart and J. Wood, *A metric formula for the Godbillon-Vey invariant for foliations*, Proc. Amer. Math. Soc., **38** (1973), 427–430.
99. H. Rosenberg and R. Roussarie, *Les feuilles exceptionelles ne sont pas exceptionelles*, Comment. Math. Helv., **45** (1970), 517–523.
100. R. Sacksteder, *Foliations and pseudogroups*, Amer. J. Math., **87** (1965), 79–102.
101. K. Schmidt, Cocycles of ergodic transformation groups, MacMillan Company of India, Bombay 1977.
102. P. Schweitzer (editor), *Some problems in foliation theory and related areas*, in Differential Topology, Foliations and Gelfand-Fuks cohomology, Rio de Janeiro (1976), Lect. Notes in Math. **652**, 240–252, Springer–Verlag, New York and Berlin, 1978.
103. S. Smale, *Differentiable Dynamical Systems*, Bulletin Amer. Math. Soc., **73** (1967), 747-817.
104. G. Stuck, *On the characteristic classes of actions of lattices in higher rank Lie groups*, Trans. Amer. Math. Soc., **324** (1991), 181–200.
105. D. Sullivan, *Cycles for the dynamical study of foliated manifolds and complex manifolds*, Invent. Math., **36** (1976), 225–255.
106. W. Thurston, *Non-cobordant foliations on S^3*, Bull. Amer. Math. Soc., **78** (1972), 511–514.
107. J. Tits, *Free subgroups in linear groups*, J. of Alg., **20** (1972), 250-270.
108. T. Tsuboi, *On the Hurder-Katok extension of the Godbillon-Vey invariant*, J. Fac. Sci., Univ. of Tokyo, **37** (1990), 255–263.
109. T. Tsuboi, *Area functionals and Godbillon-Vey cocycles*, Ann. Inst. Fourier, Grenoble, **42** (1992), 421–447.
110. N. Tsuchiya, *Lower semi-continuity of growth of leaves*, J. Fac. Sci., Univ. of Tokyo, **26** (1979), 473–500.
111. N. Tsuchiya, *Growth and depth of leaves*, J. Fac. Sci., Univ. of Tokyo, **26** (1979), 465–471.
112. N. Tsuchiya, *Leaves of finite depth*, Japan J. Math., **6** (1980), 343–364.
113. N. Tsuchiya, *Leaves with nonexact polynomial growth*, Tôhoku Math. J., **32** (1980), 71–77.
114. N. Tsuchiya, *The Nishimori decompositions of codimension one foliations and the Godbillon-Vey classes*, Tôhoku Math. J., **34** (1982), 343–365.

115. N. Tsuchiya, *On decompositions and approximations of foliated manifolds*, in Foliations, 135–158, Advanced Studies in Pure Math., **5**, North-Holland, Amsterdam, 1985.
116. P. Walczak, *Dynamics of the geodesic flow of a foliation*, Ergodic Th. and Dynam. Sys., **8** (1988), 637–650.
117. P. Walczak, *On the geodesic flow of a foliation of a compact manifold of negative constant curvature*, Suppl. Rend. Circ. Mat. Palermo, **21** (1989), 349–354.
118. P. Walczak, *Jacobi operator for leaf geodesics*, Coll. Math., **45** (1993), 213–226.
119. P. Walczak, *Existence of smooth invariant measures for geodesic flows of foliations of Riemannian manifolds*, Proc. Amer. Math. Soc., **120** (1994), 903–906.
120. P. Walczak, *Hausdorff dimension of Markov invariant sets*, J. Math. Soc. of Japan, **48** (1996), 125–133.
121. G. Wallet, *Nullité de l'invariant de Godbillon-Vey d'un tore*, C.R. Acad. Sci. Paris, **283** (1976), 821–823.
122. R. Zimmer, *Orbit equivalence and rigidity of ergodic actions of Lie groups*, Ergodic Th. and Dynam. Sys., **1** (1981), 237–253.
123. R. Zimmer, *Ergodic theory, semisimple Lie groups and foliations by manifolds of negative curvature*, Publ. Math. Inst. Hautes Etudes Sci., **55** (1982), 37–62.
124. R. Zimmer, *Volume preserving actions of lattices in semisimple groups on compact manifolds*, Publ. Math. Inst. Hautes Etudes Sci., **59** (1984), 5–33.
125. R. Zimmer, Ergodic Theory and Semisimple Groups, Birkhäuser, Boston, Basel, Stuttgart, 1984.
126. R. Zimmer, *Actions of semisimple groups and discrete groups*, in Proc. Int. Congress Math., Berkeley (1986), 1247–1258, 1986.
127. R. Zimmer, *On the algebraic hull of an automorphism group of a principle bundle*, Comment. Math. Helv., **65** (1990), 375–387.

Received October 31, 2000.

Proceedings of
FOLIATIONS: GEOMETRY AND DYNAMICS
held in Warsaw, May 29–June 9, 2000
ed. by Paweł WALCZAK *et al.*
World Scientific, Singapore, 2002
pp. 61–73

SIMILARITY AND CONFORMAL GEOMETRY OF FOLIATIONS

RÉMI LANGEVIN
Laboratoire de Topologie, Departement de Mathematiques,
Université de Bourgogne, B.P. 47 870, 21078 Dijon, France,
e-mail: langevin@u-bourgogne.fr

The study of curvature integrals associated to a foliation of flat tori or constant curvature spheres started with D.Asimov's article [1]. It was continued in [2]. In the Section 2 we will look for a curvature integral invariant by homotheties. In the Section 5 we replace the affine grassmannian of the first section by the set of spheres $\mathcal{S}$. The author thanks T. Tsuboi for suggesting the study of foliations of some similarity surfaces.

1 Euclidean integral geometry for foliations

Let $W \subset \mathbb{R}^n$ be an open subset, and let $\mathcal{F}$ be a codimension 1 orientable foliation of W. As $\mathcal{F}$ is orientable, a unit normal $N(m)$ is defined at each point $m \in W$.

Definition 1 (Symmetric functions of the curvature σ_i^+ associated to an oriented codimension one foliation). As, through every point m of the foliated space there is a leaf L_m of $\mathcal{F}$, the symmetric functions of the curvatures of the leaf L_m at the point m are defined by:

$$det[Id + t(d\gamma)](m) = \sum t^i \cdot \sigma_i^+.$$

where γ is the Gauss map associated to the leaf L_m.

An integral geometric interpretation and proof of formulas involving these functions can be found at the end of [2] or in [8].

Let us recall here the euclidean integral geometric foliated exchange theorem. We need first a few geometric observations.

Let H be an affine hyperplane of $\mathbb{R}^{n+1}$. The trace $\mathcal{F}|_H$ of $\mathcal{F}$ on H is generically a foliation of $(W \cap H)$ with only isolated singularities. In fact, generically, these singularities are hyperbolic.

When the ambient space is $\mathbb{R}^3$ the singularities are of one of the two following types: center or saddle. We attribute signs to these singular points

$$\epsilon(saddle) = -1 \quad and \; \epsilon(center) = +1$$

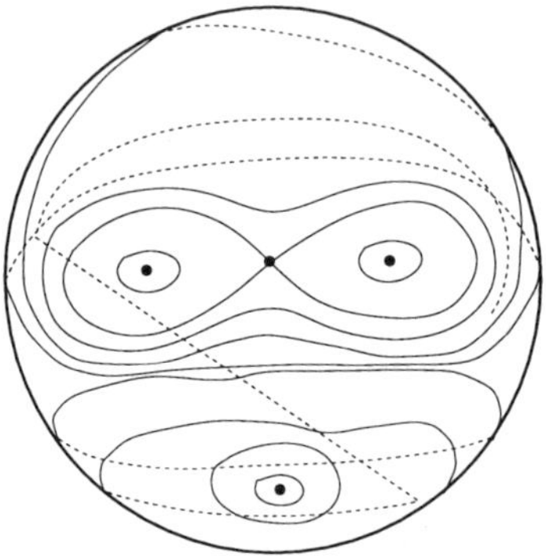

Figure 1. Center and saddle.

When the foliation is of codimension one and transversely oriented, the normals N to the leaves define a vector field with an isolated singularity at m. The sign $\epsilon(m)$ is

$$\epsilon(m) = (-1)^{index_N(m)}.$$

Definition 2 The number $|\mu|(\mathcal{F}, H)$ is the number of singular points of $\mathcal{F}|_H$. When $|\mu|(\mathcal{F}, H)$ is finite, and the singularities are all hyperbolic, the number $\mu^+(\mathcal{F}, H)$ is:

$$\mu^+(\mathcal{F}, H) = \sum_{m \in \Sigma(\mathcal{F}|_H)} \epsilon(m)$$

Remark. A singular point m of $\mathcal{F}|_H$ is a point where the leaf L_m is tangent to H. We can also locally project L_m on the normal at m to H (and to L_m). We get a function which is in general a Morse function, for which the Morse index of m satisfies:

$$(-1)^{Morse\, index\, of\, m} = \epsilon(m).$$

The sign $\epsilon(m)$ is, when the dimension of the leaves of $\mathcal{F}$ is even, the sign of the Gauss curvature of L_m at m.

We will call the integral $\int_W |K|$ when the ambient space is $\mathbb{R}^3$ (or $\int_W |k|$ when W is of dimension 2) the *total curvature of $\mathcal{F}$*.

Theorem 3 (Foliated exchange theorem)

$$\int_W |K| = \int_{\mathcal{A}(3,2)} |\mu|(\mathcal{F}, H).$$

Moreover, if one of the previous integrals are finite:

$$\int_W K = \int_{\mathcal{A}(3,2)} \mu^+(\mathcal{F}, H).$$

To prove this theorem, we need to define the polar curves of the foliation Let us recall the definition here, as we will need to make an analogous construction comparing a foliation and a family of tangent circles later.

1.1 Polar curves

The critical points of the orthogonal projection of a leaf L of $\mathcal{F}$ on a line Δ are in general isolated on L.

Definition 4 The closure of the union of these critical points:

$$\Gamma(\mathcal{F}, \Delta) = \overline{\bigcup_L crit(p_\Delta|_L)}$$

is generically almost everywhere a smooth curve (it may have singular points) [13].

Proposition 5 ([13]) *Generically the polar curve $\Gamma(\mathcal{F}, \Delta)$ is transverse to $\Delta^\perp$*

Remark. When $\Gamma(\mathcal{F}, T_m\mathcal{F}^\perp)$ is tangent to $T_m\mathcal{F}$, the Gauss curvature of the leaf L_m is zero, as, in that case, the differential of the Gauss map of the leaf L_m restricted to $T_m\Gamma(\mathcal{F}, T_m\mathcal{F}^\perp)$ is zero.

Let us first give some applications of the foliated exchange theorem in dimension 2. We note $|k|(m)$ the absolute value of the curvature of the leaf L_m of $\mathcal{F}$ through m. Let us give two easy examples; many similar others are developped in [11].

Theorem 6 *Let $D \in \mathbb{R}^2$ be the unit disc and $\mathcal{F}$ be an orientable foliation with orientable singularities, tangent to ∂D. Then*

$$\int_D |k| \geq 2\pi$$

the minimal value is achieved by the foliation by concentric circles, or any foliation of the disc with locally convex leaves.

Proof. Any affine line intersecting the disc should have at least one contact with the foliation. The measure of this set of lines is the length of the boundary circle ∂D (see [12]). □

Using Cauchy-Crofton's formula in the same way we also get

Proposition 7 *Let $A = \frac{1}{2} \leq r \leq 1$ be the annulus limited by the circle of center 0 and radius 1 and the circle of center 0 and radius $\frac{1}{2}$. The total curvature of a foliation $\mathcal{F}$ tangent to the boundary satisfies*

$$\int_A |k| \geq \pi.$$

2 Dimension-one foliations of homothety surfaces

Some results concerning the integral $\int_{\mathbf{T}^2} k^2$ of the square of the curvature of the leaves of a foliation of the flat torus $\mathbf{T}^2$ are given in [9], motivated by energy formulas for liquid crystals.

Let $\tilde{\mathcal{F}}$ be a foliation of a domain $\mathbb{R}^2$. As the 2-form $k^2 dv$ is scale invariant (invariant by similarities of the euclidean plane $\mathbb{R}^2$, if the foliation is invariant by a similarity, or in particular an homothety $\mathcal{H}$, the integral on a fundamental domain W: $\int_W k^2 dv$ is well-defined on the quotient torus: $\mathbb{R}^2/\mathcal{H}$.

Let us consider two examples on the quotient $\mathbf{T} = \mathbb{R}^2/\mathcal{H}_{\frac{1}{2}}$, where $\mathcal{H}_{\frac{1}{2}}$ is the homothety of center the origin and ratio $1/2$.

The first foliation $\mathcal{F}_1$ of $\mathbf{T}$, is the quotient of the foliation of $\mathbb{R}^2$ by horizontal lines. Notice it has two Reeb components.

As the curvature is identically zero, $\int_{\mathbf{R}} k^2 = 0$.

The second foliation $\mathcal{F}_2$ of $\mathbf{T}$, is the foliation quotient of the foliation $\tilde{\mathcal{F}}_2$ of the plane by concentric circles. The integral of k^2 is in that case:

$$I_{circles} = \int_{1/2}^{1} \int_0^{2\pi} \frac{1}{r^2} r d\theta dr = 2\pi ln(2).$$

We can also write this integral as $\int_{\mathbf{T}} 1 \cdot dw$ where $dw = \frac{1}{r} d\theta dr$ is a 2-form invariant by all the homotheties of center 0.

Let us now consider a foliation $\mathcal{F}$ of $\mathbf{T}$ admitting a closed curve which is the image in $\mathbf{T}$ of the unit circle . Using the Schwarz inequality we get

$$\int_{\mathbf{T}} r^2 \cdot k^2 dw \cdot \int_{\mathbf{T}} dw \geq [\int_{\mathbf{T}} |k| \cdot r dw]^2.$$

Integrating on the fundamental domain $A = \frac{1}{2} \geq r \geq 1$, we get a lower bound

$$\int_{\mathbf{T}} 1 \geq \int_{\mathbf{T}} |k| \cdot rdw \geq \int_{A} |k| \cdot \frac{1}{8} \geq \frac{1}{8}| \int_{\mathbf{T}} k|.$$

Using the computation made in the first section of a greatest lower bound of the total curvature of a foliation of the annulus A tangent to the boundary $\int_A |k| \geq \pi$, we get finally the lower bound

$$\int_{\mathbf{T}} k^2 dv \geq \frac{\pi}{ln(2) \cdot 128}.$$

Question.

The lower below we found has no reason to be optimal. Is $I_{circles}$ the greatest lower bound?

The same type of question can be asked about the energy of two orthogonal foliations $\mathcal{F}_1$ and $\mathcal{F}_2$, as the energy $\int_W (k_1^2 + k_2^2) dv$, where k_1 is the curvature of the leaves of $\mathcal{F}_1$, and k_2 the curvature of the leaves of $\mathcal{F}_2$, is invariant by a similarity. This is the energy $|\nabla\theta|^2$ of the section of the unit tangent bundle (here a trivial bundle) considered by Eells and Sampson [4] and [5]. This energy is also the energy of the section of $T_1 W = W \times S^1$ defined by any of the two foliation [5].

Some results for singular foliations of $\mathbb{R}^2 \setminus \{0\}$ obtained by A.S. Fawaz suggest questions for the corresponding foliation of the quotient $\mathbf{T}$ [6].

Question.

Do the images of harmonic (pairs of) foliations in $\mathbb{R}^2 \setminus 0$ in the quotient $\mathbf{T}$ minimize the energy in their isotopy classes?

Examples of such foliations are

— Foliation such that the tangent direction to the leaves makes a constant angle α with the position z. The leaves are logarithmic spirals.

— Level foliation of $\mathcal{R}e(z^(n+1))n \neq -1$ (the index is (-n)).

— Harmonic Reeb components defined by the line field associated to the multivalued vector field $X(z) = i \cdot z^{(1+i\frac{k\pi}{ln(2)})}$. The choice of the imaginary part of the exponent garantees that the foliation is well defined on the quotient $\mathbf{T}$.

3 The set of spheres

Let us first sum up some properties of the set of codimension-one spheres of S^n.

Let L be the Lorentz quadratic form

$$L(x_1, x_2, ..., x_n) = (x_1)^2 + (x_2)^2 + ... - (x_{(n+2)})^2.$$

We denote also by L the bilinear form associated to the quadratic form L. Vectors v such that $L(v) > 0$ are called space-like, vectors v such that $L(v) < 0$ are called time-like. The light cone is the isotropic cone of L. The set of points at infinity of the light cone is the union of two n-dimensional spheres S^n. The set of lines of the light cone is therefore also a sphere S^n, that we denote S_∞ and "see" in the positive half-space $x_5 \geq 0$. A natural way to see the sphere is to intersect the light cone with the affine space $x_5 = 1$. This endows S^n with a (non-canonical) riemannian metric. Intersections with other space-like affine spaces avoiding the origin give other riemannian metrics conformally equivalent to the previous one.

The space $\mathcal{S}$ of oriented codimension one spheres of S^n correspond bijectively to the quadric Λ of equation $L = 1$. In fact, let σ be a point of Λ, the hyperplane orthogonal to σ should contain a time-like vector (an L-orthogonal basis of R^{n+2} contains exactly one time like vector, and cannot contain any isotropic vector). This subspace meets then the sphere S_∞ in a (n-1)-dimensional sphere Σ. Note that a path "going to infinity" on Λ is a family of spheres whose radius goes to zero. Taking on S^n the standard metric, or any other conformally equivalent metric, a closed family of spheres of bounded radius is compact in $\mathcal{S}$.

Note that the stereographic projection of S^n on $\mathbb{R}^n$ maps the set $\mathcal{S}$ of hyperspheres in S^n to the set of hyperspheres and hyperplanes in $\mathbb{R}^n$.

The space of spheres, $\mathcal{S}$ admits a measure m invariant by the isometry group G of linear maps of $\mathbb{R}^{n+2}$ leaving L invariant. One can see it as the measure associated to the (n+1)-form on Λ: inner product of the euclidean volume form of $\mathbb{R}^{n+1}$ by the position vector of a point on Λ.

Using the stereographic projection, we can compute this measure in terms of the centers $(x_1, x_2, ..., x_n$ and radii r of spheres in $\mathbb{R}^n$

$$m = \frac{1}{r^{n+1}} |dx_1 \wedge dx_2 \wedge \cdots \wedge dx_n \wedge dr|.$$

4 Codimension-one foliation of $\mathbb{R}^3$, S^3 or $\mathbb{H}^3$

The result we mention now is contained in the paper [10].

The Theorem is formulated for foliations in $\mathbb{R}^3$, but it is clear that it can be transposed to any 3-space of constant curvature using the stereograpic projection of the sphere or of the quadric $L = 1$ on the affine plane $x_5 = 1$.

Let $\mathcal{F}$ be a codimension 1 foliation of $W \subset \mathbb{R}^3$. For any 2-dimensional generalized sphere (that is, a sphere or a plane) Σ in $\mathbb{R}^3$ denote by $N^-(\Sigma)$

the number of negative contacts with the foliation $\mathcal{F}$, i.e. the number of points of a saddle tangency of Σ and $\mathcal{F}$ (Figure 2). It is clear that the number $N^-(\Sigma)$ is conformally well-defined.

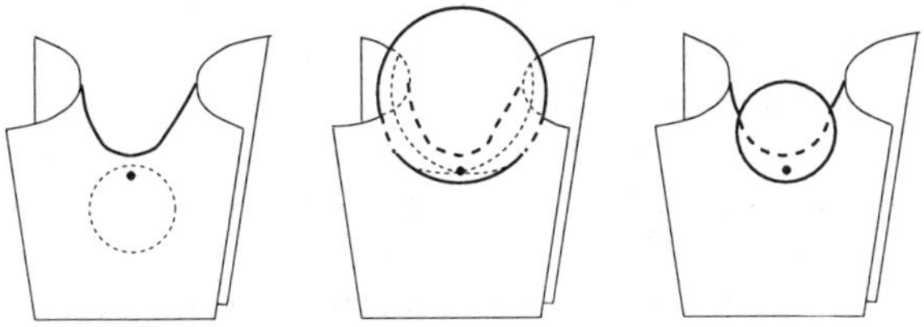

Figure 2. Saddle type contact between $\mathcal{F}$ and a sphere.

Remark. The sphere Σ has a saddle tangency with a leaf of the foliation $\mathcal{F}$ if its curvature k is in between the principal curvatures k_1 and k_2 of the leaf at the point of contact. It has a center tangency with the leaf if its curvature is not in the closed interval $[k_1, k_2]$.

Then the following theorem holds

Theorem 8 *Let $\mathcal{F}$ be a smooth foliation in a domain $W \subset \mathbb{R}^3$. Then*

$$\frac{1}{6}\int_W |k_1 - k_2|^3 dV = \int_S N^-(\Sigma)\, d\mu(\Sigma).$$

Since the right hand side is conformally well-defined, one obtains the following

Corollary 9 *Let $\mathcal{F}$ be a smooth foliation of $W \subset \mathbb{R}^3$. Then the 3-form*

$$|k_1 - k_2|^3 dV,$$

where k_i are the principal curvatures of leaves and dV is the volume element, is a conformal invariant.

Recall first the *coarea formula* [7].

Theorem 10 (Coarea formula) *Let M^l and N^q $(l \geq q)$ be Riemannian manifolds. Let dx and dy be the measure element of M and N associated to the metrics,let $\Phi : M \to N$ be a smooth map and let $f : M \to \mathbb{R}$ be a measurable function. Then*

$$\int_M f(x)\, \|Jac(\Phi)\|\, dx = \int_N \Big(\int_{\Phi^{-1}(y)} f(x)\, dx\Big) dy,$$

where the inner integral in the right hand side is computed with respect to the $(l-q)$-dimensional Hausdorff measure induced on $\Phi^{-1}(y)$ from M and

$\|Jac(\Phi)\|$ is the Radon derivative of the measure on N with respect to the image under $d\Phi$ of the q-dimensional measure on M.

Proof. Apply the coarea formula (in the case $dim(M) = dim(N)$) to the map P from the bundle of generalized spheres tangent to $\mathcal{F}$ to the space $\mathcal{S}$ of generalized spheres. More precisely, consider the normal line bundle $N\mathcal{F}$ and construct a map sending a point (x, t) to the sphere of radius $|t|$ centered at the point $x + tn$, where $t \in \mathbb{R}$ and n is a unit normal to $\mathcal{F}$ at x. Define P to be this map with the domain restricted to those values of (x, t), for which the the sphere $P(x, t)$ has a *saddle* contact with the foliation.

Taking into account that the density of the measure μ on $\mathcal{S}$ at the sphere of radius r centered at the point (x_1, x_2, x_3) is equal to $r^{-4}\, dx_1\, dx_2\, dx_3\, dr$ one can easily obtain that

$$\|JacP\| = |(1 - k_1 t)(1 - k_2 t)|\, t^{-4}.$$

The values of $t \in \overline{\mathbb{R}}$ with a saddle contact form an arc between $1/k_1$ and $1/k_2$ which does not contain a sphere of zero radius. Integrating $\|JacP\|$ along the corresponding set of the fiber one obtains that

$$\int_{\mathcal{S}} N^-(\Sigma)\, d\mu(\Sigma) = \int_U \Big(\int_{(1-k_1 t)(1-k_2 t)<0} \frac{|(1 - k_1 t)(1 - k_2 t)|dt}{t^4} \Big) dV$$

which gives the desired formula. □

5 Bilocal statements

The set $\mathcal{S}$ of circles is now 3-dimensional. Unlike the case of codimension one foliation of $\mathbb{R}^n$, $n \geq 3$, it is not possible to consider a locally defined full mesure set of spheres using only local information of a leaf for a dimension-one foliation of $W \subset \mathbb{R}^2$ (at a point the only distinguished circle is the osculating circle; these provide only a 2-dimensional family of circles). Still, the circles having more than two contact points with the foliation form a bounded family. When the foliation is oriented, we can count each circle C with a multiplicity $p(C)$, where $p(C) + 1$ is the number of points where the oriented normal to the foliation points out of the circle. Let us call $m(\mathcal{F}) = \int_{C \in \mathcal{S}} (p(C))^2$.

Taking the measure of the set of non trivial circles, or better the integral of the multiplicity, we get conformal invariants of the foliation. Let us show that we can express the second one by a bilocal integral, in fact an integral on a subset of the set of pairs of points of W.

Consider the set of circles tangent at a point x to the foliation $\mathcal{F}$. The set of other tangency points of one of these circles with the foliation is

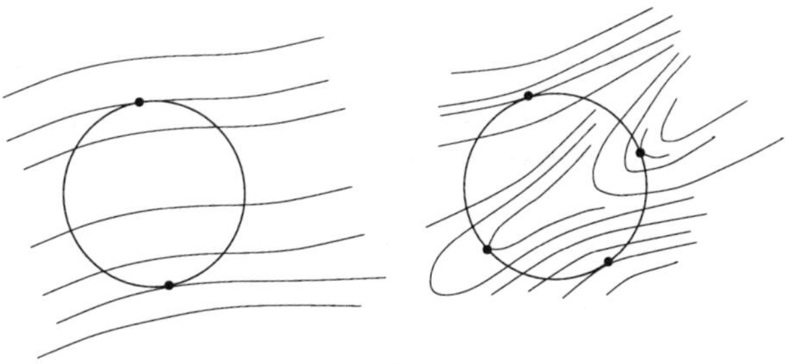

Figure 3. Trivial and non trivial circle for a foliation.

generically a curve that we will call $\Gamma_{\mathcal{F},x}$. Sending the point x to infinity, the famililiy of tangent circles is transformed into a family of parallel lines $\mathcal{L}$. The image of these tangency points is then the polar curve defined by $\mathcal{L}$ and the image of the foliation $\mathcal{F}$. The set of pairs of points (x, y) belonging to a circle tangent at x and y to the foliation is then in general a 3-dimensional subvariety $\mathcal{C}_1 = \bigcup_x \Gamma_{\mathcal{F},x}$ of $W \times W$. Call $\mathcal{C}$ the subset of $\mathcal{C}_1$ formed by pairs (x, y) such that the orientation defined by the tangent to the circle and the normal at x or y to the foliation coincide, (in other terms, turning the circle into a non-trivial one see Figure 3). Forgetting the two points we get a map G from $\mathcal{C}$ to the set $\mathcal{S}$ of oriented circles (which is 3-dimensional as $\mathcal{C}$).

The jacobian of this map depends on

— x and y,

— the tangents to the foliation and to the curve $\Gamma_{\mathcal{F},x}$ at y,

— the tangents to the foliation and to the curve $\Gamma_{\mathcal{F},y}$ at x

— the curvatures of the foliation at the two points x and y.

To fix the notations, let us call $\gamma_{x,y}$ the circle tangent at x to the leaf L_x of $\mathcal{F}$ through x, and y to the leaf L_y of $\mathcal{F}$ through y. Denote by $C_{x,y}$ the circle orthogonal to $\gamma_{x,y}$ containing x and y, θ_x the angle at x of L_x and $C_{x,y}$ and θ_y the angle at y of L_y and $C_{x,y}$.

The coarea formula gives an equality between the weighted measure of non-trivial circles and an integral on $\mathcal{C}_1$. To compute the jacobian of G, let us consider a frame tangent to $\mathcal{C} \subset W \times W$

$$v_1 = (unit\, vector\, tangent\, to\, \Gamma_y, 0)$$

$$v_2 = (0, unit\, vector\, tangent\, to\, \Gamma_x)$$

$$v_3 = vector\, tangent\, to\, E\, with\, unit\, projection\, on\, the\, first\, factor$$

where E is the curve in $L_x \times L_y \subset W \times W$ formed by pairs (z, z^*) when the two leaves are considered as the two folds of an envelope of circles.

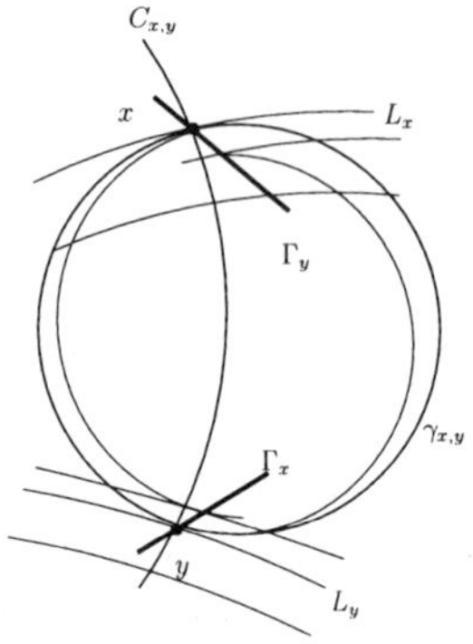

Figure 4. Frame adapted to the measure of non-trivial circles

When the pair (x, y) moves on $x \times \Gamma_x$ the corresponding circles stay tangent at x to L_x by definition of Γ_x. We can follow these circles by their second point of intersection with the circle $C_{x,y}$ (the first being x).

When the pair (x, y) moves on $\Gamma_y \times y$ the corresponding circles stay tangent at y to L_y by definition of Γ_y. We can also follow these circles by their second point of intersection with the circle $C_{x,y}$ orthogonal to $\gamma_{x,y}$ containing x and y (the first being y).

Notice that these two family of circles are light rays in the quadric Λ orthogonal to the curve E for the Lorentz form.

The area corresponding to $dG(v_1), dG(v_2)$ is the area on the two dimensional quadric Λ_2 of pairs of points on the circle $C_{x,y}$. The pencils of tangent circles at x to L_x and at y to L_y provide a projection to $C_{x,y}$. The area we are looking for is $|cos(\theta_x) \times cos(\theta_y)| \times \frac{1}{|x-y|^2}$. We can check this result using the stereographic projection of $C_{x,y}$ on the line through x and y from the pole intersection of the mediatrix of the segment $[x, y]$ with $C_{x,y}$. The measure of the pairs of points (z_1, z_2) on the line in $[t, t+dt] \times [s, s+ds]$ is $\frac{1}{|t-s|^2}|dt \wedge ds|$ and it is the same as the measure of the counter images on $C_{x,y}$. Notice that the length of the corresponding infinitesimal intervals are the same on the circle and the line.

To mesure $dG(v_3)$, which is a space-like vector, as the family of circle has an envelope, we just need to measure how they turn. The rotation speed of the circles $C_{x,y}$ is the difference of the curvature of L_x and the curvature of $C_{x,y}$.

The jacobian we are looking for is then, as the $|cos(\theta_x)|$ terms cancel out

$$|(k_x - f_{x,y})cos(\theta_y)|,$$

where $f_{x,y}$ is the curvature of the circle $C_{x,y}$.

Let us check now some examples.

Let us consider first a limit Reeb component $\mathcal{F}_1$ on a square flat torus $\mathbb{R}^2/\mathbb{Z}^2$ formed by half circles tangent to the closed leaf $\mathbb{R} \times \{O\}$ (it fails to be a foliation along the closed leaf $\mathbb{R} \times \{O\}$).

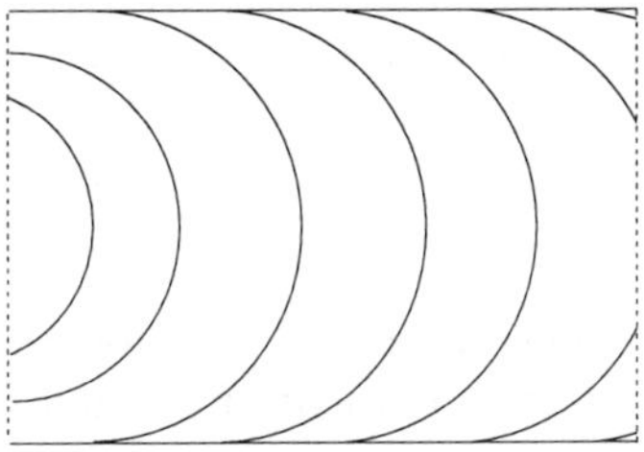

Figure 5. Limit Reeb component made of half circles.

A circle has a non trivial contact with the foliation $\tilde{\mathcal{F}}$ of $\mathbb{R}^2$ if and only if its diameter is larger than 1. The translations in $\mathbb{Z}^2$ act on the set of circles of $\mathbb{R}^2$, preserving the measure m. The measure of non trivial circles is then, using coordinates $(x, y) \in [0, 1] \times [0, 1]$ and radius r of a circle

$$m_0 = \int_{1/2}^{\infty} \frac{1}{r^3} dr \int_{\mathbb{R}^2/\mathbb{Z}^2} dx \wedge dy = 2.$$

For any foliation $\mathcal{F}$ of the torus with a Reeb component, the circles of radius larger than 1 are non-trivial, therefore

$$m(\mathcal{F}) \geq \int_{1}^{\infty} \frac{1}{r^3} dr \int_{\mathbb{R}^2/\mathbb{Z}^2} dx \wedge dy = \frac{1}{2}.$$

Question.

Is m_0 the greatest lower bound of the measure $m(\mathcal{F}_1)$ on foliations of $\mathbb{R}^2/\mathbb{Z}^2$ admitting at least one Reeb component?

Consider now two foliations of the torus $\mathbb{R}^2/\mathcal{H}_{\frac{1}{2}}$ where $\mathcal{H}_{\frac{1}{2}}$ is the homothethy of ratio $1/2$.

The first is the quotient of the foliation of $\mathbb{R}^2$ by horizontal lines. Notice it has two Reeb components. All circles of the plane are in trivial position with respect to the foliation $\tilde{\mathcal{F}}_2$; therefore $m(\mathcal{F}) = 0$.

The second foliation of the homothethy annulus, $\mathcal{F}_3$ is the foliation quotient of the foliation $\tilde{\mathcal{F}}_3$ of the plane by concentric circles. The powers of $\mathcal{H}_{\frac{1}{2}}$ act on the set of circles $\mathcal{S}$. The intersection of a fundamental domain with the set of non trivial circles for $\tilde{\mathcal{F}}$ is the zone contained between the positive cone of great circles tangent to the unit circle and the great circles tangent to the circle centered in the origin and radius 1/2. The measure of non-trivial circles (each having exactly one pair of nontrivial tangency points with the foliation) is

$$m_A = \int_{1/2}^{\infty} \int_{1}^{inf[(r-1),1]} \int_{0}^{2\pi} (1+x) \cdot \frac{1}{r^3} d\theta dx dr.$$

Question.

Is the measure m_A the minimum of $m(\mathcal{F})$ for foliations of $\mathbb{R}^2/\mathcal{H}_{\frac{1}{2}}$ admitting a closed leaf isotopic to the quotient of the unit circle?

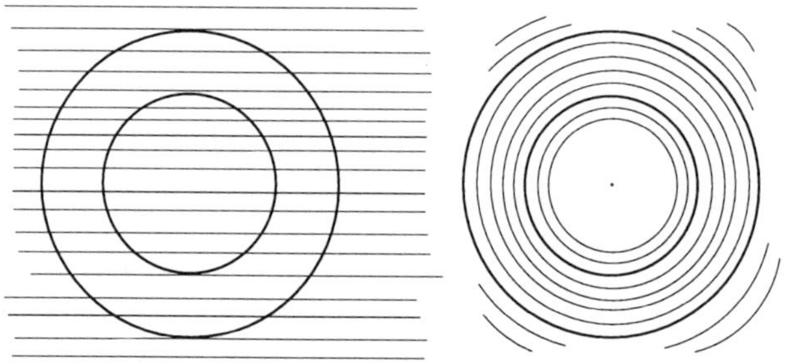

Figure 6. Two foliations of the homothety quotient torus.

The author expects some better results when the conformal integral geometry of anuli in $\mathbb{R}^2$ and tori in S^3 will be better understood. Some partial results about anuli are obtained by Y. Nikolayevsky and the author (work in progress).

References

1. D. Asimov, *Average Gaussian curvature of leaves of foliations*, Bulletin of the American Mathematical Society, **84**-1 (1978).

2. F. Brito, R. Langevin and H. Rosenberg, *Intégrales de courbure sur des variétés feuilletées*, Journal of Differential Geometry **16** (1981), 19–50.
3. R. Bryant, *A duality theorem for Willmore surfaces*, Journal of Differential Geometry **20** (1984), 23–53.
4. J. Eells and J.H. Sampson, *Harmonic mappings of riemannian manifolds*, American Journal of Mathematics, **86** (1964), 109–133.
5. J. Eells and J.H. Sampson, *Variational theory in fiber bundles*, Proc. US-Japan Seminar Diff. Geom., Kyoto (1965), 61–69.
6. A.S. Fawaz, *Energie et feuilletages*, Thèse de troisième cycle, Dijon, France, 1986.
7. H. Federer, Geometric Measure Theory, Springer Verlag, 1969.
8. R. Langevin, *Feuilletages tendus*, Bulletin de la Ssociété Mathématique de France, **107** (1979).
9. R. Langevin, *Feuilletages, énergies et cristaux liquides*, Astérisque, **107–108** (1983), 201–213.
10. R. Langevin and Y. Nikolayevsky, *Three viewpoints on the integral geometry of foliations*, Illinois Journal of Mathematics, **43**-2 (1999), 233–255.
11. R. Langevin and C. Possani, *Total curvature of foliations*, Illinois Journal of Mathematics, **37**-3 (1993), 508–524.
12. L.A. Santalò, *Integral geometry and geometric probability*, Encyclopedia of Mathematics and its Applications, Addison Wesley, 1976.
13. R. Thom, *Généralisation de la théorie de Morse aux variétés feuilletées*, Annales de l'Institut Fourier, **14**-1 (1964), 173–189.
14. Ph. Tondeur, Foliation on Riemannian Manifolds, Universitex Springer Verlag, 1988.

Received November 3, 2000.

Proceedings of
FOLIATIONS: GEOMETRY AND DYNAMICS
held in Warsaw, May 29–June 9, 2000
ed. by Paweł WALCZAK *et al.*
World Scientific, Singapore, 2002
pp. 75–125

FOLIATIONS AND CONTACT STRUCTURES ON 3-MANIFOLDS

YOSHIHIKO MITSUMATSU
Department of Mathematics, Chuo University,
1-23-27 Kasuga, Bunkyo-ku, Tokyo, 112-8551, Japan,
e-mail: yoshi@math.chuo-u.ac.jp

This article is a re-presentation of the minicourse in Warsaw which dealt with recent interactions between the theory of foliations and that of contact structures on 3-manifolds. In the conference, the author prepared and distributed a rather lengthy note on 3-dimensional contact topology, which contains introductory expositions of fundamentals in contact geometry as well as its recent progress. The present article focuses more on the main themes such as a brief introduction to the theory of confoliations, contact structures and foliations associated with Anosov flows, and their generalizations, and contains less about the contact topology itself. In the final section, some problems which were presented in the problem session during the conference are raised.

0 Introduction to contact geometry

0.1 Introduction

A maximally nonintegrable hyperplane field on an odd dimensional manifold is called a contact structure. The progress of symplectic topology in recent years, especially in dimension 4, has called attention also to 3-dimensional contact topology. While a symplectic structure is defined as a 2-form on an even dimensional manifold, a contact structure is defined by 1-form on odd dimensional manifolds. Therefore the method of its study can be much more topological. Especially, in 3-dimensional contact topology, which is now in a very rapid progress, the topological method investigating submanifolds *i.e.*, knots and surfaces has a great importance. On the other hand, we can also get strong results by applying global analytic big machineries, such as Seiberg-Witten theory and J-holomorphic curves, to a symplectic manifold which has the contact structure on its end or boundary.

In this note, we begin with and basically keep ourselves within the topological framework.

In 90's, the relation between the theory of codimension 1 foliations on 3-manifolds and 3-dimensional contact topology has come to attract attentions. The one is an integrable plane field and the other is even infinitesimally non integrable. As plane fields they differ by much, like water and oil. However, since long it has been recognized by several people at least vaguely that they have some similarities between their methods of study (*e.g.*, embedding a surface and tracing the vector field defined by their intersection), as well as their dominant philosophies (*e.g.*, *h*-principle). Especially, the works of Eliashberg, Gromov, and Thurston on both subjects have been suggesting that there must be something more. Now we know that it showed at least one of its tail as the theory of Confoliations due to Eliashberg and Thurston (Section 3).

This exposition aims at reporting the recent development of the interactions between the studies of foliations and contact structures. Especially, the generations of contact structures and even symplectic structures from foliations is one of the main topics. A general theory is given as a part of the theory of confoliations and some important examples are given starting from Anosov foliations.

However, as contact structures are not yet very familiar to people in the foliation theory, we begin with a review of the fundamentals of the contact geometry and topology.

0.2 Definitions and basic notions

Definition 0.2.1 A C^∞ hyper plane field ξ on a $(2n-1)$-dimensional manifold M is called a *contact structure* if ξ is locally defined as $\xi = \ker\alpha$, where α is a C^∞ nonsingular 1-form satisfying $\alpha\wedge(d\alpha)^{n-1}\neq 0$ everywhere (*i.e.* it gives a volume form on M). This condition implies that ξ has the least integrability among codimension one plane fields. The 1-form α is called a *contact 1-form.*

In this article, we assume the differentiability of contact plane fields to be of class C^∞ unless otherwise specified.

As examples of (local) contact 1-forms

$$\alpha_1 = dz + \frac{1}{2}\sum_{i=1}^{n-1}(x_i dy_i - y_i dx_i) \qquad \text{or} \qquad \alpha_2 = dz + \sum_{i=1}^{n-1} x_i dy_i$$

are standard ones on the (local) coordinate $(z, x_1, y_1, \cdots, x_{n-1}, y_{n-1})$.

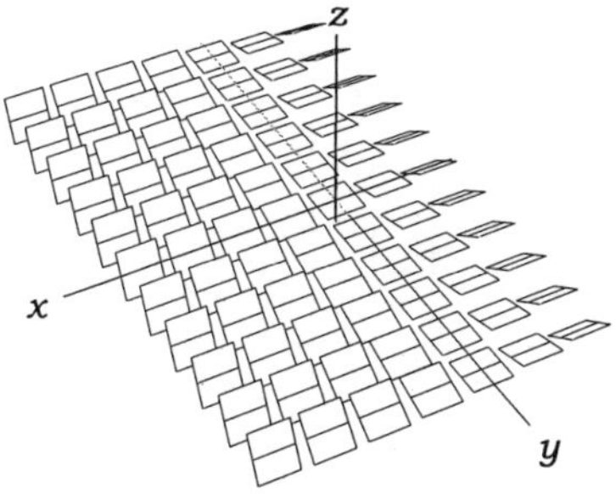

Figure 1. $\ker[\alpha_2 = dz + xdy]$

Remark 0.2.2 **1)** The property $\alpha \wedge (d\alpha)^{n-1} \neq 0$ is preserved even if α is replaced with $f\alpha$, where f is a nonzero smooth function on M. Furthermore, in the case of dimension $(4n-1)$ (especially in dimension 3), ξ itself defines an orientation of M by $\alpha \wedge (d\alpha)^{n-1}$, which is independent of the choice of the contact 1-form α. Hence if a manifold M of such dimensions admits a contact structure, then M is always orientable. When M is already oriented, ξ is called *positive* [*resp. negative*] if the orientation determined by ξ and that of M [*resp.* does not] coincide to each other. If a positive contact structure ξ is oriented as a hyper plane field, it naturally determines an orientation in the transverse direction. In this case ξ is called *co-oriented.* For a standard orientation $dx \wedge dy \wedge dz$ with respect to the coordinate (x, y, z), in the example above, we have $\alpha_1 \wedge d\alpha_1 = \alpha_2 \wedge d\alpha_2 = dx \wedge dy \wedge dz$, which shows that both $\ker \alpha_1$ and $\ker \alpha_2$ are positive contact structures.

2) The 2-form $d\alpha$ gives a symplectic structure on the contact hyperplane ξ_p for each point p, which is an area element of each ξ_p in 3-dimensional case.

Definition 0.2.3 In a 3-dimensional contact manifold, an immersed curve is called *Legendrian* if it is tangent to the contact plane field at any point. Remark that from the definition of the contact structure, there is no integral surface of the contact plane field.

Definition 0.2.4 **1)** Given a contact 1-form α, one obtains a vector field X_α which is uniquely determined by the conditions

$$\alpha(X_\alpha) = 1 \quad \text{and} \quad \iota_{X_\alpha} d\alpha = 0,$$

which are equivalent to "$\alpha(X_\alpha) = 1$ and $\mathcal{L}_{X_\alpha}\alpha = 0$". X_α is called the *Reeb vector field* associated with the contact 1-form α, and the flow generated by X_α is called the *Reeb flow* Sometimes it is also called the *contact flow.*

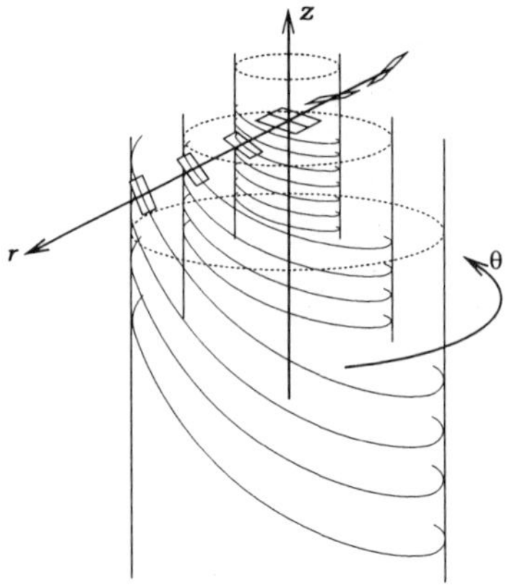

Figure 2.

2) If a vector field X preserves the contact plane field $\xi = \ker\alpha$, we call it a *contact vector field.* X may not preserve a given contact 1-form α. X is a contact vector field if and only if there exists a smooth function g such that $\mathcal{L}_X\alpha = g\alpha$.

In the above examples, $\frac{\partial}{\partial z}$ is the Reeb vector field for both α_1 and α_2. As stated below, a Reeb vector field is regarded as a time-independent Hamiltonian vector field restricted to a constant energy surface. The Reeb vector field X_α preserves not only ξ but also the contact 1-form α and $d\alpha$.

Definition 0.2.5 (Symplectization) For a $(2n-1)$-dimensional contact manifold (M, α), the 2-form $\omega = d(t\alpha)$ $(t \in (0, \infty))$ is an exact symplectic structure on the $2n$-dimensional manifold $W = (0, \infty) \times M$. (W, ω) is called the *symplectization* of (M, α) or of (M, ξ).

Then the Hamiltonian vector field X_t of the Hamiltonian t with respect to ω restricted to $M = \{t = 1\}$ is nothing but the Reeb vector field associated with α. W can be regarded as a connected component of the complement of the zero section of the real line bundle {1-form $\beta; \beta|_\xi = 0\} = (TM/\xi)^*$. The primitive $t\alpha$ of ω is nothing but the tautological 1-form β on W.

This leads us naturally to the Weinstein conjecture which is a contact version of the Arnol'd conjecture, which states the existence and an estimate of the number of closed orbits of a Hamiltonian vector field on a symplectic manifold. For the Arnol'd conjecture, see [19]. For the Weinstein conjecture, see also [39] and [72].

Conjecture 0.2.6 (The Weinstein conjecture) On a closed contact manifold (M, α), the Reeb vector field X_α admits a closed orbit.

In this note, unless otherwise stated, we assume every most of objects (manifolds, submanifolds and plane fields, tangentially and transversely, ...) are oriented or at least orientable.

0.3 Basic examples of contact structures

Let us look at some basic and important examples of contact structures. These contact structures have some good properties which are called 'fillablity' or 'tightness'. These properties will be explained later.

Example 0.3.1 (The most basic example) On the unit sphere S^{2n-1} in the standard complex euclidean space $\mathbb{C}^n$, the standard contact structure $(S^{2n-1}, \xi_0 = \ker\alpha_0)$ is given by $\alpha_0 = \frac{1}{2}\sum_{i=1}^n x_i dy_i - y_i dx_i$. This is also defined as the plane field which is perpendicular to the fibres of the Hopf fibration $S^{2n-1} \to \mathbf{CP}^{n-1}$. Here $z_i = x_i + \sqrt{-1}y_i$ $(i = 1, \ldots, n)$ denotes the standard coordinate on $\mathbb{C}^n$. It is easy to see

$$\alpha_0 \wedge (d\alpha_0)^{n-1} = \frac{(n-1)!}{2}\iota_R dx_1 \wedge dy_1 \wedge \cdots \wedge dx_n \wedge dy_n,$$

where $R = \sum_{i=1}^n (x_i \frac{\partial}{\partial x_i} + y_i \frac{\partial}{\partial y_i})$ is the radial vector field on $\mathbf{C}^n$, so that α_0 defines a contact structure.

This contact structure ξ_0 is regarded as the projectification of the standard symplectic structure $\omega_0 = \sum_{i=1}^n dx_i \wedge dy_i$ (or of its primitive $\lambda_0 = \frac{1}{2}\sum_{i=1}^n x_i dy_i - y_i dx_i$) and is again called *the standard contact structure.* Notice that each fibre of the Hopf fibration is a trajectory of the Reeb vector field X_{α_0}.

Example 0.3.2 (From classical mechanics) The canonical symplectic structure $\omega_0 = d\mathbf{p} \wedge d\mathbf{q}$ of the cotangent bundle $W = T^*V$ of arbitrary manifold V ($\dim V = n$) admits the canonical primitive $\lambda_0 = \mathbf{p}d\mathbf{q}$ ($\omega_0 = d\alpha_0$). Here $\mathbf{q} = (q_1, ..., q_n)$ denotes an arbitrary local coordinate around $q \in V$ and $\mathbf{p} = (p_1, ..., p_n)$ denotes the linear coordinate on T_q^*V which is defined from $\mathbf{q}$. As is well known, ω_0 and λ_0 are independent of the choice of local coordinate $\mathbf{q}$ on V. Then, for any Riemannian metric on V, the primitive α_0, restricted to the unit cotangent bundle $M = S^{n-1}(T^*V)$, gives us a contact 1-form α_0 which is called the *Liouville 1-form.*

Through the identification of the unit tangent bundle with the unit cotangent bundle by the Riemannian metric, the Reeb flow ϕ_t associated with the Liouville 1-form coincides with the geodesic flow of the Riemannian manifold V. It must be also noticed that each orbit of the geodesic flow is a solution to the equation of motion of a particle without potential in the classsical mechanics. Hence in this case, Weinstein's conjecture implies the existence of closed geodesics on closed Riemannian manifolds.

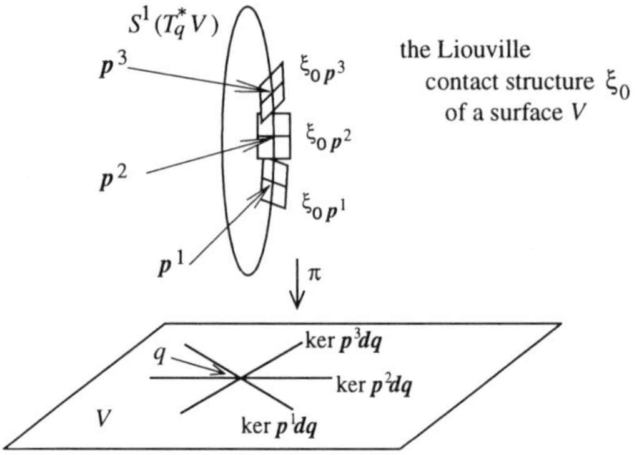

Figure 3.

Though the Liouville 1-form and the geodesic flow strongly depend on the Riemannian metric on V, the *Liouville contact structure* ξ_0 is independent of it. Identifying the unit cotangent bundle $M = S^{n-1}(T^*V)$ with the projectification $(T^*V \setminus \{0\})/\mathbb{R}_+$ of the cotangent bundle, on each point $p \in M$ on the fibre of $q \in V$ we have $\xi_{0p} = d\pi^{-1}(\ker[p : T_q^*V \to \mathbb{R}]\ )$, where π denotes the projection of the projectified cotangent bundle to V and $d\pi : T_p S^{n-1}(T^*V) \to T_q V$ denotes its differential. This construction shows that the Liouville contact structure ξ_0 is defined independently of the choice of Riemannian metric.

Each fibre of π is a Legendrian submanifold of ξ_0, *i.e.*, the fibres are tangent to ξ_0. In the case $\dim V = 2$, ξ_0 is negative with respect to the natural orientation of M as an S^1-bundle. This fact will have an importance in Section 4. Also in the succeeding examples the readers should be careful about the positivity and the negativity.

Example 0.3.3 (Quantum mechanical example) Let (N,ω) be a $(2n-2)$-dimensional symplectic manifold whose symplectic form represents an integral cohomology class $[\omega]$. Then N admits a complex line bundle $\mathcal{L}$ with a $U(1)$-connection ∇ whose curvature form is $2\pi\sqrt{-1}\omega$. Such $(\mathcal{L}, \nabla)$ is unique up to the action of the Gauge group. $\mathcal{L}$ is called the *pre-quantization* or *pre-quantum bundle* of (N, ω). The connection 1-form is restricted to the unit circle bundle M to be a connection 1-form α and defines a contact manifold (M, α). Each fibre is a closed orbit of the Reeb flow. In the case where $(N, \omega) = (\mathbb{CP}^{n-1}, \omega_0)$, the above construction gives us Example 0.3.1.

The complement of the zero section in the total space of $\mathcal{L}$ coincides with the symplectization of (M, α) (the correspondence of $\pm$ ends is opposite).

In the case of $n = 2$, the total space of the S^1-bundle with a non-trivial euler class on a closed surface Σ admits a contact structure given by the horizontal distribution coming from an S^1-connection with non-vanishing curvature. In this case, the S^1-bundle with *negative* euler class gives rise to a *positive* contact structure and vise versa.

Example 0.3.4 (Strict pseudo convexity in several complex variables) A complex manifold Ω is called a *Stein manifold* if Ω admits a strictly plurisubharmonic function ϕ which is proper and is bounded from below. For a regular value a of ϕ, the sub-level set $W = \phi^{-1}((-\infty, a])$ is a *strictly pseudo convex domain* (or a *Grauert domain*), and its boundary $M = \partial W$ is a *strictly pseudo convex boundary.* (Even when ϕ is defined only on a neighbourhood of ∂W, we call ∂W strictly pseudo convex.)

Then $\lambda = -J^*d\phi$ gives us a contact 1-form $\alpha = \lambda|_M$ on M, which define a contact structure $\xi = \ker\alpha$. ξ is also defined as $\xi = TM \cap J(TM)$. The strict pluri-subharmonicity of ϕ gives us a Kähler structure on W by $\omega = -dJ^*d\phi$, so that $\omega|_\xi$ is a positive symplectic structure on ξ. $\alpha \wedge (d\alpha)^{n-1}$ gives a positive orientation on M as the boundary of W. We will discuss further this in Section 2.2.

On $\Omega = \mathbb{C}^n$ if we take $\phi(z_1, ..., z_n) = 4^{-1} \sum_{1 \leq k \leq n} \|z_k\|^2$ and $a = 1/4$, we have Example 0.3.1. In a more algebraic manner, taking an intersection of a projective variety V embedded in $\mathbb{CP}^N$ with a tubular neighbourhood of the hyperplane $\mathbb{CP}^{N-1}$, we obtain a contact structure on its boundary $\partial(V \setminus \mathbb{CP}^{N-1})$. In this way we obtain a lot of examples of algebraic Stein manifolds and contact structures on their end.

Contact structures also relate to singularities. Taking the intersection of a complex surface with an isolated singularity and the ε-sphere centred at the singularity, we naturally obtain a 3-dimensional contact structure. For example, the intersection of the surface $W = \{(z_1, z_2, z_3) \in \mathbb{C}^3; z_1^2 + z_2^3 + z_3^5 = 0\}$ with a singularity at the origin and the sphere S^5_ε of radius ε gives us the *Poincaré homology 3-sphere* with a contact structure, which is given by dividing the left invariant standard contact plane field ξ_0 on the Lie group S^3 by a subgroup of order 120.

The strict pseudo convexity is one of important origins for the convexities of contact structures.

Example 0.3.5 (Invariant contact structures on 3-dimensional Lie groups) Simply connected 3-dimensional non-abelian Lie groups which admit a co-compact discrete subgroup Γ (equivalently to say, unimodular) are the following 5 types of groups: $S^3 = SU(2)$, $\widetilde{PSL}(2; \mathbb{R})$, *Solv* which is a solvable Lie group defined by the suspension of $\mathbb{R}^2$ by a hyperbolic element in $SL(2;\mathbb{Z})$, the 3-dimensional Heisenberg nilpotent Lie group Nil, and the universal covering of of the orientation preserving isometry group on the 2-dimensional euclidean space (2-dimensional group of euclidean motions) $\widetilde{Euc}(2)$.

It is not difficult to see that G is unimodular if and only if $[dvol] \neq 0 \in H^3(\mathfrak{g};\mathbb{R})$, where $\mathfrak{g}$ is the Lie algebra of G and $dvol$ is any nontrivial element in $\bigwedge^3 \mathfrak{g}^*$. This implies that for a unimodular 3-dimensional Lie algebra $\alpha \wedge d\beta = \beta \wedge d\alpha \in \bigwedge^3 \mathfrak{g}^*$ holds for any $\alpha, \beta \in \mathfrak{g}^*$.

On the dual $\mathfrak{g}^* = \{$ left invariant 1-forms on the Lie group G $\}$ of a 3-dimensional Lie algebra $\mathfrak{g}$, we can define a bilinear form

$$\nu : \mathfrak{g}^* \otimes \mathfrak{g}^* \to \bigwedge^3 \mathfrak{g}^* \cong \mathbb{R}, \quad (\alpha, \beta) \mapsto \alpha \wedge d\beta,$$

which measures the non-integrability of α and β in some sense. ν is symmetric for unimodular $\mathfrak{g}$. It is almost clear that ν is non-trivial if and only if $\mathfrak{g}$ is non-abelian. The symmetric bilinear form ν essentially coincides with the Killing form on $\mathfrak{g}$ in the case of $su(2)$ and $psl(2;\mathbb{R})$, but does not in general. Let (p, q, n) denote the type of these bilinear forms, *i.e.*, p [*resp.* q] $=$ max dim(positive [*resp.* negative] definite subspace), and $n =$ dim(annihilator). Then, for Lie algebras *solv* and *nil*, their Killing forms have the type $(1, 0, 2)$ and $(0, 0, 3)$ while ν has $(1, 1, 1)$ and $(1, 0, 2)$ respectively.

Regarding ν as a quadratic form on $\mathfrak{g}^*$, if $\nu(\alpha, \alpha) > 0$ [*resp.* < 0] then $\xi = \ker(\alpha)$ defines a positive [resp. negative] contact structure, and if $\nu(\alpha, \alpha) = 0$ then $\xi = \ker(\alpha)$ defines a foliation on G. These are induced naturally on $\Gamma\backslash G$ because of the left G-invariance. We calculate ν for the case of the Lie algebra $psl(2;\mathbb{R})$. Taking a basis

$$h = \begin{pmatrix} 1, & 0 \\ 0, & -1 \end{pmatrix}, \quad \ell = \begin{pmatrix} 0, & 1 \\ 1, & 0 \end{pmatrix}, \quad k = \begin{pmatrix} 0, & -1 \\ 1, & 0 \end{pmatrix}$$

of $psl(2;\mathbb{R})$, and let h^*, ℓ^*, k^* be its dual basis of $psl(2;\mathbb{R})^*$. From the bracket relation on the Lie algebra $[k, h] = \ell$, $[\ell, k] = h$, and $[\ell, h] = k$ we have $dh^* = -\ell^* \wedge k^*$, $d\ell^* = -k^* \wedge h^*$, and $dk^* = h^* \wedge \ell^*$, which imply

$\nu(\cdot,\cdot)$	h^*	ℓ^*	k^*
h^*	-1	0	0
ℓ^*	0	-1	0
k^*	0	0	1

where we are identifying $\bigwedge^3 psl(2;\mathbb{R})^* \cong \mathbb{R}$ by $dvol = h^* \wedge \ell^* \wedge k^*$. This table tells that the left-invariant 1-forms h^* and ℓ^* define negative contact structures and that k^* defines a positive one.

On the other hand, $\ell^* - k^*$ defines a foliation which is called an Anosov foliation associated with the geodesic flow of a surface of constant negative curvature. Here the geodesic flow is generated by h and is an Anosov flow(see Section 4). The figure below describes the quadratic form $\nu(\alpha, \alpha)$ for $psl(2;\mathbb{R})$ and for $solv$, in which surfaces drawn there present the light cone $\{\alpha \in \mathfrak{g}^*; \nu(\alpha,\alpha) = 0\}$. For the Lie algebra $solv$, we can choose a basis X, Y, T with the relations $[X,Y] = 0$, $[T,X] = X$, and $[T,Y] = -Y$, which also defines a system of coordinates $solv^* = \{xX^* + yY^* + tT^*\}$.

In Example 0.3.2, if we choose the sphere S^2, the torus T^2 or a hyperbolic surface as the base space, we obtain models for $G = S^3$, $\widetilde{Euc}(2)$, $\widetilde{PSL}(2;\mathbb{R})$ respectively. On the other hand, the quantum mechanical Example 0.3.3 for these base spaces gives us models for $G = S^3$, Nil, $\widetilde{PSL}(2;\mathbb{R})$.

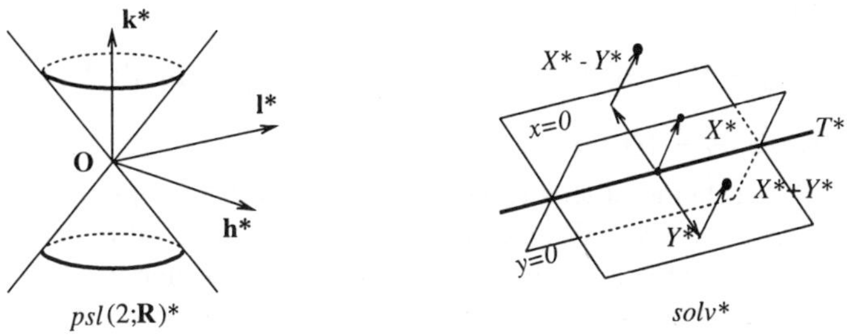

Figure 4.

As shown in the Figure 4, ν is indefinite only for the case of $\widetilde{PSL}(2;\mathbb{R})$ and $Solv$, in which we will observe an interesting phenomenon induced from these contact structures and from foliations defined by the null vectors. We will explain this in Section 4.

0.4 Rigidity of contact structures

Generally, it is a stable (*i.e.*, open) condition under C^1-topology for a hyperplane field ξ [*resp.* a 1-form α] on a $(2n-1)$-dimensional manifold to be a contact structure [*resp.* a contact form]. We can also say that it is rigid in the sense of following two theorems.

Theorem 0.4.1 (Darboux coordinate) *Any point p in a $(2n-1)$-dimensional contact manifold (M, α) has the following Darboux coordinate* *neighbourhood* $(U; x_1, y_1, \dots, x_{n-1}, y_{n-1}, z)$*:*

$$\alpha|_U = dz + \sum_{i=1}^{n-1} x_i dy_i, \qquad p = (0, \dots, 0).$$

The proof is done in the following way: We first apply the theorem of Darboux coordinate to the symplectic structure which is locally defined by $d\alpha$ on the quotient space w.r.t. the Reeb flow ϕ_t. Next, we regard a neighbourhood U of p in M as an open set of the total space of principal $\mathbb{R}$-bundles that admit a connection form α with the curvature form $d\alpha$. In the 3-dimensional case, as the Darboux theorem for 2-dimensional symplectic structure is almost trivial, the readers can complete the proof by themselves.

We can also prove this theorem by applying the symplectic Darboux theorem to the symplectization, which is higher in the dimension by 1.

Theorem 0.4.2 (Gray, [27]) *The 1-parameter family* ξ_t *of contact structures on a* $(2n-1)$*-dimensional closed manifold* M *is chased by the isotopy* Φ_t *defined by*

$$\dot{\Phi}_t = Y_t, \quad \alpha_t \wedge \dot{\alpha}_t \wedge (d\alpha_t)^{n-2} = \iota_{Y_t}(\alpha_t \wedge (d\alpha_t)^{n-1}).$$

In other words, $\xi_t = (\Phi_t)_* \xi_0$*.*

Proof. The scheme of this theorem is important for understanding contact structures locally. We particularly explain the 3-dimensional case. (The same proof applies in higher dimensions.) In principle, if we want to deform a contact plane ξ_{t_p} at the point p, there uniquely exists a direction in ξ_{t_p} (not T_pM) that realizes the alteration we need. Using the cylindrical coordinate, contact forms are presented as $\alpha = dz + r^2/2d\theta$ (which is equal to $dz + (xdy - ydx)/2$). Moving from the origin along a radius in the direction of θ, the plain field twists on the radius. If we define the isotopy Φ_t so as to pull back this twist, we should obtain the deformation we want. Indeed, if we write this condition infinitesimally w.r.t $\dot{\Phi}_t = Y_t$, we get $\alpha_t \wedge \dot{\alpha}_t = \iota_{Y_t}(\alpha_t \wedge d\alpha_t)$. □

Note that the definition of this vector field Y_t depends only on ξ_t and is independent of the choice of a contact form α_t. Describing this in a deformation theoretic fashion, an infinitesimal deformation is considered as a cross section of $\xi^* \otimes TM/\xi$. To give an identification $TM/\xi \cong \mathbb{R}$ and to fix a contact form α are equivalent. Then, the 2-form $d\alpha$ defines a symplectic structure on ξ, thus we obtain the isomorphism $d\alpha : \xi^* \to \xi$. This induces another isomorphism $\xi^* \otimes TM/\xi \to \xi$, which is more natural, and we find that infinitesimal deformations correspond to vector fields which are tangent to the contact plane field.

These results imply that contact forms do not have any local invariant. Each contact form induces the Reeb flow and its dynamical properties are important invariants. Particularly, neighbourhoods of closed orbits have various invariants. However, contact structures (contact plane fields) do not have even this kind of invariants. Thus, the study of contact structure is really a global problem. However, it is again important to study the topology of contact structure by means of fixing a contact form, like in studying topology of smooth manifolds by means of fixing a Riemannian metric.

The following lemma explains how a plane field on a 3-manifold deforms by a flow generated by a vector field which is tangent to the plane field. This argument has already appeared in the proof of Gray's theorem. This lemma plays an important role with relation to foliations later (Section 3, 4).

Lemma 0.4.3 *Let ξ be a plane field of class C^1 on a 3-dimensional manifold and Y be a nonsingular vector field that is tangent to the plane filed ξ.*

1) *The plane field ξ is invariant under the action of the flow generated by Y if and only if ξ is completely integrable. In this case, the plane field ξ defines a codimension-one foliation. If the plane field ξ is of class C^0, it is integrable, i.e., for each point there exits an integral surface passing through it. However, generally speaking,* **the integral surface is not unique!**

2) *Let us look at the flow from behind in such a way that each flow line reduces to a point. (We obtain a 2-dimensional view.) The plane field ξ (which is observed to be a line element on the plane) is a positive contact structure if and only if the plane field ξ rotates with a positive (anti-clockwise) angular velocity. It is negative if it rotates with a negative (clockwise) angular velocity.*

Proof. Take a local coordinate system $(x, y, z), 0 \leq x, y, z \leq 1$ satisfying the following two conditions.

(1) Every line which is parallel to the x-axis is tangent to ξ.

(2) On the plane $x = 0$, ξ is tangent to (x, y)-plane.

Under these conditions, we may assume that the plane field ξ is defined by the 1-form $\alpha = dz + f(x, y, z)dy$ (where $f(0, y, z) = 0$) and $Y = \frac{\partial}{\partial x}$. Then $\alpha \wedge d\alpha = \frac{\partial f}{\partial x} dx \wedge dy \wedge dz$. Thus the plane field ξ is a positive [*resp.* negative] contact plane field if $\frac{\partial f}{\partial x} > 0$ [*resp.* $\frac{\partial f}{\partial x} < 0$] . □

1 3D contact topology due to Bennequin

Bennequin introduced the notion of *tightness* which is A central and basic concept in 3-dimensional contact topology. In this section, after looking back basic constructions and some existence results in dimension 3 which were established before Bennequin, we explain the tightness as well as his solution to the problem which will be stated in the end of the first subsection. About the proof of Bennequin's main theorem, only a quite rough framework is described in 1.3 and in 1.4. About the detail the readers should refer to [6] or [50].

1.1 Existence and construction of 3D contact structures

Martinet showed the following theorem by using a Lickorish's theorem (any oriented closed 3-manifold can be obtained by the Dehn surgery along some link in S^3.)

Theorem 1.1.1 (Martinet, [44]) *There exist a positive contact structure on any oriented closed 3-manifold.*

Later Thurston-Winkelnkemper [70] proved this theorem by means of open book decomposition. Giroux made a more precise and beautiful argument in terms of 'f-essential surfaces' which is equivalent to open book decompositions ([22]). Any method of proving the existence of a contact structure directly from the Heegaard decomposition is not yet known. We will explain an operation, which is called Lutz twist or Lutz modification, by which we can create a new (?) contact structure from a given one. One of the basic idea is that if two contact structures with toral boundaries are both transverse to their boundaries and trace the same linear foliations on their boundaries, then we can paste their boundaries together to get a new contact manifold. This idea was used to prove the above existence theorems.

Theorem 1.1.2 (S^1-Darboux coordinate) *For any knot Γ which is transverse to a given contact structure ξ, there exist a tubular neighbourhood $U_\varepsilon(\Gamma)$ and a cylindrical coordinate system (r, θ, z) satisfying the following conditions.*

(1) $U_\varepsilon(\Gamma) = \{(r, \theta, z); r^2 < \varepsilon, z \in \mathbb{R}/2\pi\mathbb{Z}\} \supset \Gamma = \{r = 0\}$.

(2) $\xi|_{U_\varepsilon(\Gamma)} = \ker(\cos r^2 dz + \sin r^2 d\theta)$.

We call $U_\varepsilon(\Gamma)$ an *S^1-Darboux Tube.* Unlike in the case of the usual Darboux coordinate in Theorem 0.4.1, we can not specify the contact form in advance. Still we can easily prove this theorem in a similar way.

Definition 1.1.3 (Characteristic foliation) Here we prepare a useful and important terminology. An embedded (immersed) surface Σ in M^3 and a plane field ξ on M defines a (singular) foliation as $\xi|_\Sigma \cap T\Sigma$ on Σ. We call this the *characteristic foliation* of Σ and let ξ_Σ denote it.

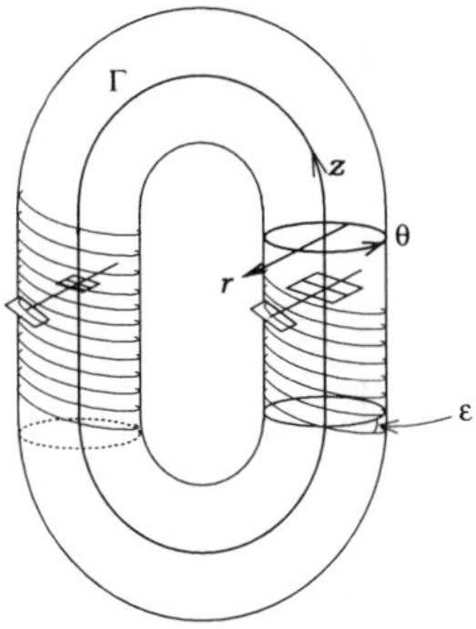

Figure 5. S^1-Darboux tube

The characteristic foliation of the toral boundary of S^1-Darboux tube $U_\varepsilon(\Gamma)$ is a linear foliation of slope $-\varepsilon = -r^2$ w.r.t. the coordinate (θ, z). Thus the slope of this characteristic foliation varies with the radius r of the tube. If the slope is rational, each leaf is compact and the foliation can be regarded as the trivial horizontal one with respect to some coordinate change.

Let V_R be a solid torus of radius r (where $R = r^2$) with the contact structure $\eta = \ker(\cos r^2 dz + \sin r^2 d\theta)$, which has the same presentation as in Theorem 1.1.2. Here the radius R can be arbitrarily large so that η rotates along the radius arbitrarily many times.

The characteristic foliation $\eta_{\partial V_R}$ on the boundary $\partial V_R \cong T^2$ is linear with slope $-\tan R$. Therefore we can regard the characteristic foliations on ∂V_{R_1} and ∂V_{R_2} the same if $R_1 \equiv R_2 \pmod{\pi}$. Of course this holds for the S^1-Darboux tube U_R.

Definition 1.1.4 (Lutz twist) For a knot Γ in a 3-dimensional manifold M which is transverse to the contact structure ξ, replace its S^1-Darboux tubular neighbourhood $U_\varepsilon(\Gamma)$ of with $V_{\varepsilon+2\pi}$, then we obtain a new contact structure $\xi_{FL}(\Gamma) = \xi|_{M \setminus U_\varepsilon} \cup \eta|_{V_{\varepsilon+2\pi}}$ without changing the manifold $M = (M \setminus U_\varepsilon(\Gamma)) \cup V_{\varepsilon+2\pi}$. This operation is called a *full Lutz twist* or a *Lutz modification.* It is possible to replace with $V_{\varepsilon+\pi}$ instead of $V_{\varepsilon+2\pi}$. This procedure is called a *half Lutz twist.* In this note, we call V_π (or $V_{2\pi}$) a *Lutz's tube.* These operations do not change the manifold itself.

On the manifold obtained by the Dehn surgery w.r.t. $\binom{a,b}{c,d} \in SL(2;\mathbb{Z})$ along a knot Γ that is transverse to the given contact structure ξ, we can construct a contact structure by replacing the tubular neighbourhood $U_\varepsilon(\Gamma)$ with V_R (where $R = (a\varepsilon + b)/(c\varepsilon + d)$).

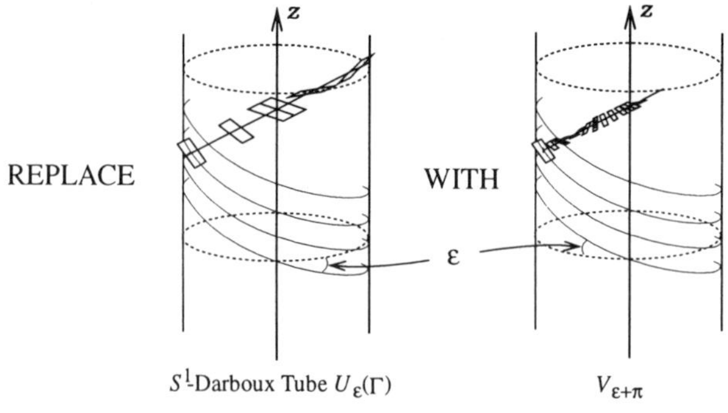

Figure 6. Half Lutz twist along Γ

Proof of Martinet's Theorem. Starting from the standard contact structure (S^3, ξ_0), construct a new manifold with a positive contact structure by means of the Dehn surgery along a transverse link. Thanks to the Lickorish's theorem, it is enough to show that any link on S^3 can be arranged to be transverse to the standard contact structure ξ_0. Two methods are known to achieve this. It is easily shown by using a classical Alexander's theorem: any oriented link on S^3 can be isotoped C^1-close to fibres of the Hopf fibration, i.e. any closed link can be realized as a closed braid. We can also use the following lemma, which is more general and well-known in the theory of contact geometry. This lemma holds because of the non-integrability of contact structures. □

Lemma 1.1.5 **1)** *Any oriented knot in a contact manifold is C^0-approximated by some Legendrian knot isotopically.*

2) *Any oriented Legendrian curve is C^1-approximated by some positively and negatively oriented transverse curve isotopically (See the proof of Lemma 0.4.3 and 1.3.5).*

Extending these techniques, Lutz also proved the following result.

Theorem 1.1.6 (Lutz, [42]) *For any given plane field on S^3, there exists a positive contact structure which is homotopic to the given plane field.*

Lutz twists do not change the topology of the manifold. What kind of change occurs to contact structures? How many contact structures are obtained from a given contact manifold by means of Lutz twists? In general, half Lutz twists change the homotopy class of plane field. This change is described by the 1-dimensional homology class represented by the axial knot (the axis of the twist) and the datum of its Seifert surface ([51]). How about the case of full Lutz twists?

Proposition 1.1.7 *Full Lutz twists do not change the homotopy class of plane field.*

Does full Lutz twist change the global topology of the contact structure? Especially the following problem is very natural to ask.

Problem 1.1.8 *A contact structure* (S^3, ξ_{0FL})*, which is obtained by operating a full Lutz twist on a standard contact structure* (S^3, ξ_0) *along some transverse knot, is diffeomorphic to* (S^3, ξ_0) *or not?*

Bennequin's answer in his famous thesis [6] to this problem is "**No!**".

1.2 Tight vs. overtwisted

Definition 1.2.1 ([6]) A contact structure ξ on a 3-dimensional manifold M is said to be *overtwisted* (OT for short) if there exists an embedded 2-disk $\mathcal{D}$ (which is called OT-*disk* henceforth) whose characteristic foliation $\xi_{\mathcal{D}}$ admits a periodic orbit. ξ is called *tight* if it is not OT.

Remark 1.2.2 Using the Elimination Lemma 1.3.8, OT-disk is always modified into a standard form, which has only one singular point of index 1 inside the periodic orbit.

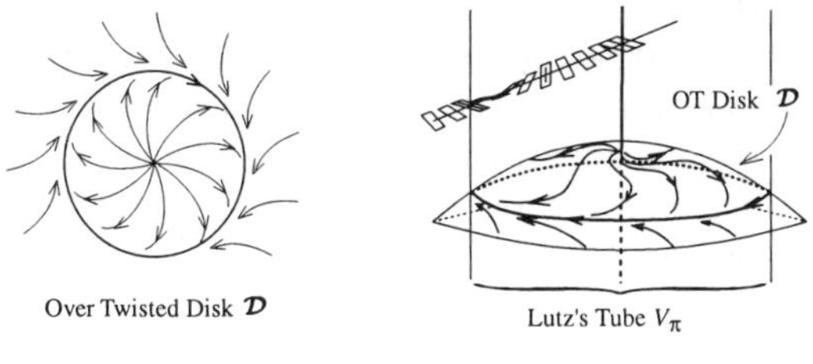

Figure 7.

Once Lutz twist is performed on a contact structure, a periodic orbit appears on the boundary of an almost horizontal disk in Lutz's tube as indicated in the above figure, and of course the new structure is OT. The operation of Lutz twist is similar to that of 'turbulization' on codimension 1 foliations which inserts a Reeb component into a codimension 1 foliation along a closed transeversal. Therefore it is plausible think that tight structures have less futility and are better structure.

Now, in a word, Bennequin's main theorem is stated as follows.

Theorem 1.2.3 (Bennequin's Main Theorem, [6]) (S^3, ξ_0) *is tight.*

On the other hand, about OT contact structures, there is a striking result due to Eliashberg. This asserts that the set of OT contact structures obeys the h-principle.

Theorem 1.2.4 (Eliashberg, [9]) 1) *For any closed 3-manifold M, the inclusion*

$$\{\text{OT contact plane fields on } M\} \hookrightarrow \{\text{plane fields on } M\}$$

is a weak homotopy equivalence.

2) *The isotopy classification of OT contact structures reduces to the homotopy classification of plane fields, i.e., to the calculation of $[M; S^2]$.*

Gray's Theorem 0.4.2 implies that 2) follows from 1). Moreover, taking Proposition 1.1.7 into account, a full Lutz twist ξ_{FL} of an OT contact structure ξ is again isotopic to ξ, *i.e.*, if ξ is OT, we can consider it already full Lutz twisted what million times! Therefore the idea that repeating the 'reverse' process of Lutz twist on a given OT contact structure would bring it finally to a tight one does not hold. Hence the important (unsolved ?) following problem comes out.

Problem 1.2.5 *On arbitrary oriented closed 3-dimensional manifolds do tight and positive contact structures exist?*

Recently Ko Honda and John Etnyre proved that there exists no positive tight contact structure on the Poincaré homology 3-sphere with reversed orientation. See [16].

Of course, a step further, one should ask classifying positive tight contact structures on each oriented closed 3-manifold, and further more, the homotopy type of the space of such structures. However at the moment, these problems seem still out of reach except in the case of very specific manifolds like 3-sphere.

1.3 Tightness and Bennequin's inequality

Definition 1.3.1 (Thurston-Bennequin's invariant) For an oriented Legendrian link Γ of (S^3, ξ_0) (or, more generally, a homologically trivial oriented Legendrian link of a contact 3-manifold (M, ξ)), let $\Gamma^\perp$ denote a link obtained by shifting Γ in the normal direction to ξ. Then the linking number $TB(\Gamma) = lk(\Gamma, \Gamma^\perp)$ is called *Thurston-Bennequin's invariant* of Γ.

Bennequin obtained his main theorem by proving what is called *Bennequin's inequality for a Legendrian knot* 1.3.2. This inequality estimates Thurston-Bennequin's invariant of a Legendrian knot in (S^3, ξ_0) in terms of the Euler number of its Seifert surface. This inequality is obtained from *Bennequin's inequality for a transverse knot* 1.3.3.

1.3.2 (Bennequin's inequality for Legendrian knots, [6], [7],[13]) *For an oriented Legendrian link Γ with a Seifert surface Σ, let $\mu(\Gamma)$ be the* *rotation number of $\dot{\Gamma}$ along Γ with respect to a trivialization of $\xi|_\Sigma$ as an $\mathbb{R}^2$-bundle. Then the following inequality holds:*

$$TB(\Gamma) \leq -\chi(\Sigma) \pm \mu(\Gamma).$$

1.3.3 (Bennequin's inequality for transverse knots, [6], [7]) *Suppose that an oriented link Γ with a Seifert surface Σ is positively transverse to the contact structure ξ. (The same goes well in the case where Γ is negatively transverse.) Let Γ^X be a link given by shifting Γ in the direction of a framing X which is obtained from the restriction of a trivialization of $\xi|_\Sigma$ to Γ, then the following inequality holds:*

$$lk(\Gamma, \Gamma^X) \leq -\chi(\Sigma).$$

Remark 1.3.4 We stated these inequalities for the standard contact structure. In general cases, if Γ is homologically trivial, we can (not claim but simply) state the same inequalities by taking the trivialization of ξ only over the Seifert surface Σ, because then $\mu(\Gamma)$ and Γ^X have exact meanings.

In fact, two inequalities still hold for any tight contact structures on closed 3-manifolds.

Lemma 1.3.5 **1)** *Suppose that the orientation of the contact plane field ξ coincides with the one induced from the trivialization $\langle v, \dot{\Gamma}\rangle$ of ξ restricted to an oriented Legendrian knot Γ. Let $\Gamma^\pm$ be a knot obtained by shifting Γ in the direction of $\pm v$ respectively, then they are transverse to ξ in the positive and negative direction respectively (refer to Lemma 1.1.5).*

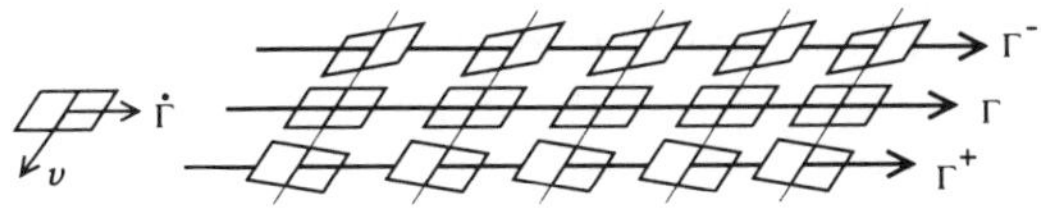

Figure 8.

2) *Suppose furthermore that Γ admits a Seifert surface Σ. Let X be a trivialization of ξ on Σ, then the following holds.*

$$TB(\Gamma) = lk(\Gamma, \Gamma^\perp) = lk(\Gamma, \Gamma^v) = lk(\Gamma^\pm, \Gamma^{\pm X}) \pm \mu(\Gamma)$$

This lemma enables us to easily deduce 1.3.2 from 1.3.3.

Theorem 1.3.6 (Bennequin's inequality $\Rightarrow$ tightness) *A contact 3-manifold (M, ξ) is tight if it satisfies the inequality 1.3.2.*

Proof. By definition, the limit cycle Γ of $\xi_\mathcal{D}$ on any OT-disk $\mathcal{D}$ is a Legendrian knot Γ with $TB(\Gamma) = 0$. Therefore if ξ is OT, then we have

$$0 = TB(\Gamma) \leq -\chi(\mathcal{D}) \pm \mu(\Gamma) = -1 \pm \mu(\Gamma),$$

which contradicts 1.3.2. □

As we shall explain briefly in 1.4, Bennequin showed these inequalities for (S^3, ξ_0). On the other hand, if we use the Elimination Lemma stated below, we obtain the converse of the above theorem, hence these inequalities turn out to be equivalent to the tightness.

Theorem 1.3.7 (Tight ⇒ Bennequin's inequality, [13]) *If a contact structure ξ on a closed 3-manifold M is tight, then 1.3.3 (, hence 1.3.2) holds for any homologically trivial oriented transverse link.*

The elimination lemma is extremely useful and important in the 3-dimensional contact topology.

Theorem 1.3.8 (Elimination Lemma, [12], [22]) *Let ξ_Σ be the vector field on an embedded oriented surface Σ obtained from the intersection of it with the contact plane field ξ on a contact 3-manifold. Suppose that there exist nondegenerate singular points P and Q of indices 1 and -1 respectively and an orbit γ of ξ_Σ connecting P to Q. If the orientation of tangency at P coincides with that at Q, then the two singular points P and Q can be eliminated from the singular vector field ξ_Σ without yielding new singular points by perturbing Σ by a small isotopy in C^0 in a sufficiently small neighbourhood of γ.*

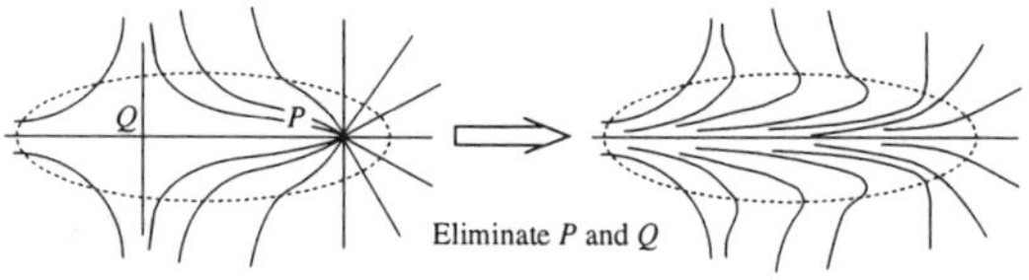

Figure 9.

Proof of Theorem 1.3.7. Assume that a positively transverse knot Γ in a tight contact manifold (M, ξ) is homologically trivial and has a Seifert surface Σ . As Γ is transverse to ξ, ξ_Σ is non-singular around the boundary and we assume that ξ_Σ has the natural orientation which is outward normal on the boundary. Let us introduce invariants $d_\pm$ for generically embedded such Seifert surfaces (as well as for closed surfaces). We assume that the singularities of ξ_Σ are non-degenerate so that their indices are ± 1 according to that they are elliptic or hyperbolic singularities. We also assume that ξ is co-oriented. Now $d_\pm$ are defined as,

$$e_+ = \sharp\{\text{positive elliptic point}\}, \qquad h_+ = \sharp\{\text{positive hyperbolic point}\},$$
$$e_- = \sharp\{\text{negative elliptic point}\}, \qquad h_- = \sharp\{\text{negative hyperbolic point}\},$$
$$d_+ = e_+ - h_+\,, \qquad d_- = e_- - h_-\,,$$

where the signs are given by the orientations of tangency of ξ to Σ at the singularities.

Since $\partial\Sigma = \Gamma$ is transverse to ξ, the relative euler class $e(\xi|_\Sigma) \in H^2(\Sigma, \partial\Sigma)$ is defined if we take $(\xi \cap T\Sigma)|_\Gamma$ as a boundary condition. Then

$$lk(\Gamma, \Gamma^X) = -\langle e(\xi|_\Sigma), [\Sigma, \partial\Sigma]\rangle$$

follows immediately from the definition. Using this, we can easily show

$$2d_+ = -lk(\Gamma, \Gamma^X) + \chi(\Sigma) \quad \text{and} \quad 2d_- = lk(\Gamma, \Gamma^X) + \chi(\Sigma)\,.$$

Therefore, $d_- \leq 0$ is equivalent to Bennequin's inequality for transverse knots.

We show $d_-(\Sigma) \leq 0$ by perturbing Σ so that the negative elliptic points are removed. First, we in advance remove every pair of a positive elliptic point and a positive hyperbolic point of ξ_Σ connected by an orbit by using Elimination Lemma. By a usual simple observation we can perturb Σ so that there is no connecting orbit between hyperbolic singular points.

Secondly we consider the attracting basin B of a negative elliptic point P which is the union of the orbits of ξ_Σ attracted by P, and its boundary. B is apart from the boundary of Σ by the assumption on the orientation. It is almost trivial that B is homeomorphic to a 2-dimensional open disk, however its boundary can be complicated to some extent.

If the boundary ∂B is a nonsingular Legendrian loop, then the closure of B is an OT-disk, *i.e.*, a contradiction. Therefore the boundary ∂B turns out to be a Legendrian loop with some singular points. There is no negative elliptic points on ∂B, because it is the boundary of an attracting basin.

Now from our assumptions and preparations, it is easy to see that on ∂B positive elliptic points and negative hyperbolic points appear alternately. Thus the negative elliptic point P at the centre of B and a negative hyperbolic point on ∂B can be eliminated. Applying this procedure for P to every negative elliptic points, we can eliminate them by a C^0-small perturbation without changing $d_-(\Sigma)$, hence we have $d_-(\Sigma) \leq 0$. □

As we shall explain in more detail, Bennequin's inequality for transverse knots (and its absolute version) is exactly alike to Thurston's inequality in the theory of codimension one foliations on 3-manifolds and even has the similar presentation to that of (generalized) Thom conjecture. To explain circumstances around them is also one of the main tasks in this article.

1.4 On the proof of Bennequin's inequality for (S^3, ξ_0)

In this subsection we only describe the framework of the proof (due to himself) of Bennequin's inequality for (S^3, ξ_0) and explain some features of his argument. For details, the readers should refer to his original paper [6]. The proof of Bennequin's inequality for a transverse knot consists of three steps.

First Step For any closed braid, show the existence of a special Seifert surface of minimal genus which is called Markov surface.

Second Step Show Bennequin's inequality for closed braids, which is stated below, by modifying further a given Markov surface of minimal genus.

Third Step Show that any positively transverse link can be deformed into a closed braid by an isotopy, preserving the transversality.

In a sense the argument does not relate to contact structure at all except for the third step. In particular, until then, Bennequin used not the standard contact structure but a Reeb foliation $\mathcal{F}_R$ as an approximation to it. We can assume that a closed braid is located in the central core of a Reeb component of $\mathcal{F}_R$ and is transverse to the leaves of $\mathcal{F}_R$.

If we replace *contact structures* by *foliations* in the setting of Bennequin's inequality for a transverse knot, we obtain an inequality which is known as *Thurston's inequality* 3.1.2 ([69]) for codimension one foliations on 3-manifolds. The following inequality which is proved at the second step is nothing else but Thurston's inequality for the Reeb foliation. In general, Thurston's inequality does not hold for foliations with Reeb components. The Reeb foliation, of course, consists of Reeb components, however since it is a good foliation without extras, we might be able to think that it is the reason why the equality still holds.

Theorem 1.4.1 (Bennequin's inequality for closed braids) *Let $c : B_n \to \mathbb{Z} \cong B_n/[B_n, B_n]$ be the abelianization of the n-braid group B_n and Σ a Seifert surface of a closed braid $\hat{\beta}$ representative of an n-braid β. Then the following inequality holds:*

$$|c(\beta)| \leq -\chi(\Sigma) + n.$$

The inequality 1.4.1 is proved by replacing a Markov surface with a better one. This is remarkably similar to the way to show 1.3.7 using Elimination Lemma. The manner of dealing with the characteristic number d_- by proving $d_- = \chi(\Sigma) - c - n$ made us anticipate what was achieved later in the argument of symplectic fillings.

2 Fillable contact structures and tightness

In this section, we introduce the notion of *fillability*, which is a generalization in the symplectic setting of the pseudo-convexity in complex analysis of several variables (see Example 0.3.4), induces convexity of contact structure, especially the tightness.

Differently from previous sections, discussions are made on 4-manifolds.

2.1 Tight contact structure

Generally, it is very difficult to show a contact structure to be tight. So far, roughly classifying, we know two or three distinct methods. Of course the first one is based on the work of Bennequin, which we explained in the previous section. The second one is the method of filling which is the main theme of this section. The third one is one of the advanced forms of the second, which was developed by Hofer [32].

We first review some elementary arguments to deduce the tightness in some cases directly from Bennequin's main theorem. Following proposition is trivial by the definition of tightness.

Proposition 2.1.1 **1)** *If (M,ξ) is tight then the restriction on any subset $N \subset M$ $(N, \xi|_N)$ is also tight.*

2) *If a covering $(\tilde{M},\tilde{\xi})$ of (M,ξ) is tight then (M,ξ) is also tight.*

3) *The standard contact structure $\xi_0 = \ker(\alpha_0 = dz + xdy)$ on $\mathbb{R}^3$ is isomorphic to (S^3,ξ_0) minus a point, therefor it is tight.*

We can regard the contact 1-form $\alpha_0 = dz + xdy$, which gives standard contact structure $(\mathbb{R}^3,\xi_0)$, as a connection form on a principal $\mathbb{R}$-bundle $\mathbb{R}^3 = \{(x,y,z)\} \to \mathbb{R}^2 = \{(x,y)\}$ and it gives the curvature form $= dx \wedge dy$. In this case, the contact plane field is nothing else than a horizontal distribution defined by the connection. Therefore quantum mechanical examples in Example 0.3.3 (some examples in 0.3.5), *i.e.*, the unit circle bundle of the pre-quantization of a surface, have $(\mathbb{R}^3,\xi_0)$ or (S^3,ξ_0) as their universal coverings. Hence they are tight by the above proposition.

Also in the case of classical mechanical Example 0.3.2, we set $\alpha = \alpha_0 + \varepsilon\beta$ which is perturbed from the Liouville form α_0 by a connection form β of a circle bundle. Then, if the perturbation is small, it is a contact plane field and is transverse to the fibres of the circle bundle. Moreover, thanks to Gray's Theorem 0.4.2, during the perturbation the structures stay in the same isotopy class of contact structure. Such contact structure is tight, because we can embed its universal covering to $(\mathbb{R}^3,\xi_0)$ or to (S^3,ξ_0), even though it is slightly more difficult than in the case of S^1-connection.

In fact these examples have a stronger property 'fillablity' than the tightness.

2.2 Fillable contact structure

Let us recall examples in 0.3.4. If a neighbourhood of the smooth boundary $M^3 = \partial W^4$ of a pre-compact domain W^4 in in a complex surface Ω admits a nice strictly pluri-subharmonic function ϕ with $M = \phi^{-1}(0)$ and $\phi^{-1}(-\infty, 0) \subset W$, W is called a strictly pseudo-convex domain or a Grauert domain, and the boundary $M = \partial W$ is called a strictly pseudo-convex boundary. Here, the function ϕ is strictly pluri-subharmonic if it satisfies $\Delta\phi > 0$ on any local holomorphic curve. In particular, if ϕ is defined on the whole W, then W is a Stein manifold (of finite type).

Then a symplectic structure $\omega = -dJ^*d\phi$ is given on the domain of ϕ and is compatible with original almost complex structure J of W, *i.e.*, $g(u, v) = \omega(u, Jv)$ gives a J-invariant Riemannian metric. This fact is equivalent to that ϕ is strictly pluri-subharmonic. Consequently we obtain a Kähler structure. Moreover, $\lambda = -J^*d\phi$ gives the contact form $\alpha = \lambda|_M$ on M. This contact structure ξ can be also presented as $\xi = TM \cap J(TM)$ using the almost complex structure J.

Definition 2.2.1 A Grauert domain W is called a *holomorphic filling* of the contact structure (M, ξ) of the boundary. Conversely such contact structure are said to be *holomorphically fillable.*

Now, the gradient vector field $Z = \nabla\phi$ with respect to the above Riemannian metric satisfies

2.2.2 Condition (C) : *$\mathcal{L}_Z\omega = \omega$, Z points outward at ∂W,*

and it also follows that $\iota_Z(\omega) = \lambda$ and $d\lambda = \omega$. For instance, in the case of$(S^3, \xi_0) \subset \Omega = \mathbb{C}^2$ we have

$$\phi(z_1, z_2) = \frac{1}{4}(\|z_1\|^2 + \|z_2\|^2 - 1), \quad Z = \frac{1}{2}\sum_{i=1,2}(x_i\frac{\partial}{\partial x_i} + y_i\frac{\partial}{\partial x_i}),$$

$$\omega = \sum_{i=1,2} dx_i \wedge df_i, \quad \alpha_0 = \frac{1}{2}\sum_{i=1,2}(x_i dy_i - y_i dx_i).$$

This situation is described only in terms of symplectic structures without using complex structures and naturally we get a positive contact 1-form $\alpha = \lambda|_M$ on the boundary $M = \partial W$.

Definition 2.2.3 ([14], [72], [7]) A compact symplectic manifold (W, ω) with boundary such that there exists a global vector field Z satisfying the condition (C) is called a *strong symplectic filling* of the contact structure $(M = \partial W, \xi = \ker\alpha)$ and this contact structure ξ is called *strongly symplectically fillable.* It is also said that W has a *contact type boundary* (M, ξ).

Z is called *Liouville vector field* and $\lambda = \iota_Z(\omega)$ is called *Liouville 1-form.* A symplectic structure on an open symplectic manifold W is called a *completely convex symplectic structure* if there exists *complete* vector field Z on an open symplectic manifold W such that it satisfies the Condition (C) and its any orbit intersects with a certain compact subset.

The notion of contact type boundary is originally proposed by Weinstein as a condition to formulate the Weinstein conjecture.

If we remove the boundary of a strong symplectic filling and extend the end infinitely using the conformal symmetry of the vector field Z of the symplectic form around a neighbourhood of the boundary, it becomes completely convex symplectic structure. (It reminds us that Grauert domain in a Stein manifold is again Stein.)

As to these fillings, it is important that a boundary has a convexity and (pseudo-)holomorphic curves can not be tangent to the boundary from inside. However, to assure this fact and for the discussions of pseudo-holomorphic curves to show the tightness, the following very weak fillability is sufficient.

Definition 2.2.4 ([10], [11], [14]) If there exists a compact symplectic manifold (W, ω) such that $\omega|_\xi$ belongs to the conformal class $\{f d\alpha; f > 0\}$ of the symplectic structure on ξ (which is defined by the contact structure $(M = \partial W, \xi)$), then the contact structure (M, ξ) is called *weakly symplectically fillable* or simply *fillable*, and (W, ω) is called a *symplectic filling* of (M, ξ).

Example 2.2.5 A contact structure ξ on the total space M of a S^1-bundle over a closed surface Σ which is transverse to each fibre is fillable. This is an easy consequence of the following.

First, we take the D^2-bundle over Σ associated with the S^1-bundle. Next we take a symplectic form of the base space Σ, i.e. an area form Ω. Note that we can choose a Thom form ϕ on the D^2-bundle so that it restricts to the area form on each fibre. Then if we take sufficiently large positive constant K, a weak filling for this contact structure is given on (W, ω) as $\omega = \phi + K\pi^*\Omega$.

Proposition 2.2.6 **1)** *Strongly symplectically fillable implies fillable.*

2) *There exists an ω-tame almost complex structure J on a symplectic filling (W, ω) such that it preserves ξ invariant (i.e. $\xi = TM \cap J(TM)$) and maps the Reeb vector field inward in W. (Such a boundary is called J-convex boundary).*

3) **(J-convexity)** *For such an almost complex structure J, no non-constant J-holomorphic curve $u : D \to W$ from the unit disk $D \subset \mathbb{C}$ can touch M at the interior of D.*

Unfortunately we do not have room in this article to develop the arguments on pseudo-holomorphic curves. Therefore we put up in this subsection with only introducing the following important theorem without proof.

Theorem 2.2.7 (Gromov-Eliashberg, [28],[11]) *A fillable contact structure is tight.*

3 Contact structures and foliations: focused on the theory of confoliations

People have been aware on some similar features between the theory of foliations and that of contact structures. In 90's, we have come to pay more attentions to such relationes. In this section, we first review some of these features which are common to both and next give a brief survey of the theory of confoliations due to Eliashberg-Thurston [15], through which we can see that there abundantly exist fillable contact structures. See [63], [26], and [8] for the basics of the foliation theory.

3.1 Common features between contact structures and foliations

It is an important topological method which is common to the theories of foliations and contact structures on 3-manifolds to chase the (singular) vector field, which we also call the characteristic foliation, defined as the intersection of an imbedded surface and the plane field. Especially in the foliation theory, the notion of holonomy is the core of the theory. However the situations around the singularities of such vector fields are rather different. The characteristic foliation ξ_Σ of an embedded surface Σ and a plane field $\xi = \ker\alpha$ and is given as the singular vector field X, which is defined by the equation $\iota_X dvol_\Sigma = \alpha|_\Sigma$. Here, $dvol_\Sigma$ denotes an area element of Σ. Since $div\, X \cdot dvol_\Sigma = d\alpha|_\Sigma$, then we have $div\, X = 0$ in the case of foliations, that is $\alpha \wedge d\alpha = 0$, and $div\, X \neq 0$ in the case of contact structures. The images around singularities of index ± 1 are indicated in the figure below.

Even around a non-singular point, the difference appears clearly by perturbing the imbedded surface. In the case of foliations, the picture does not change by small perturbations, but it does in the case of contact structures. See Lemma 0.4.3.

We also have various kinds of analogous notions and results between two theories, which are important backgrounds for the birth of the theory of confoliations.

Needless to say, we can add much more items to the next table. See [8], [15], [25], etc..., and try to do it by yourselves.

The relations between 3rd items and 4th ones in the table are featured as one of the main themes in this section.

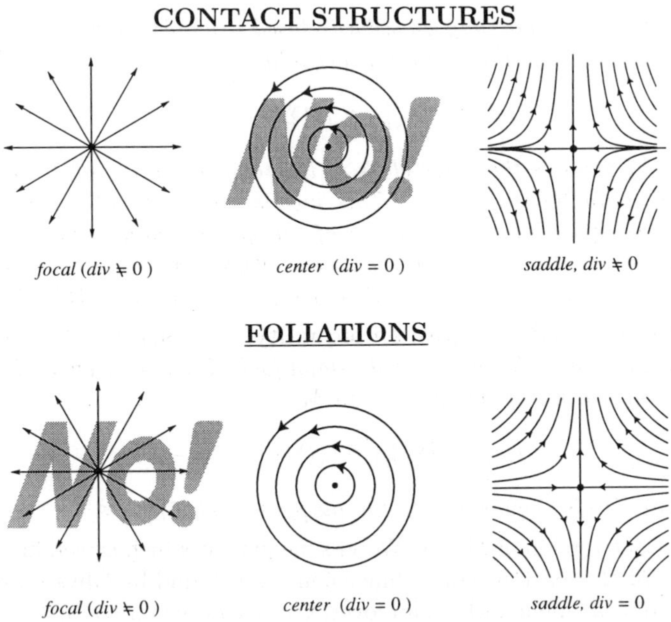

Figure 10.

Dictionary 3.1.1

THEORY OF CONTACT STRUCTURES		THEORY OF FOLIATIONS
limit periodic orbit of OT-disk	1	vanishing cycle
Lutz's tube	2	Reeb component
tight contact structure	3	foliation w/o Reeb comp.
semi-fillable contact structure	4	taut foliation
Bennequin's inequality	5	Thurston's inequality
Ghys-Giroux-Sato-Tsuboi's inequality	6	Milnor-Wood's inequality
a theorem of Giroux 3.1.7	7	Thurston's Thesis
Theorem 2.2.7 fillable $\Rightarrow$ tight	8	Barrett-Inaba's Theorem

First let us review 5th item, Thurston's and Bennequin's inequalities, whose similarity is already obvious, in the absolute version (*i.e.*, for closed surfaces).

Theorem 3.1.2 (Thurston's inequality, [69]) *Let M be a closed oriented 3-manifold and $\mathcal{F}$ an oriented codimension 1 foliation on M without*

Reeb components. For any embedded closed oriented surface Σ_g of genus $g > 0$ in M, we have the following inequality.

$$|\langle e(\tau\mathcal{F}), [\Sigma_g]\rangle| \leq |\chi(\Sigma_g)| = 2g - 2$$

Here $e(\tau\mathcal{F})$ denotes the euler class of the tangent bundle of $\mathcal{F}$ and $\chi(\Sigma_g)$ the euler characteristic of Σ_g. This means that compact leaves measure $e(\tau\mathcal{F})$ most effectively in the sense of the genus among embedded closed surfaces representing the same homology class. We may express this as $\|e(\tau\mathcal{F})\|_{Th} \leq 1$ by using Thurston's norm $\|\cdot\|_{Th}$ ([69]) on $H^2(M;\mathbb{Z})$.

Theorem 3.1.3 (Bennequin's inequality, [6],[13]) *Let ξ be a tight contact structure on a closed oriented 3-manifold M For any embedded closed oriented surface Σ_g of genus $g > 0$ in M,*

$$|\langle e(\tau\xi), [\Sigma_g]\rangle| \leq |\chi(\Sigma_g)| = 2g - 2$$

holds. This is also equivalent to the inequality $\|e(\tau\xi)\|_{Th} \leq 1$.

A contact version of Milnor-Wood's inequality, which is well known as a criterion for the existence of codimension 1 was found by Ghys-Giroux and Sato-Tsuboi independently. Moreover, Giroux proved a contact version of Thurston's thesis [68] minutely.

Theorem 3.1.4 (Milnor-Wood's inequality, [48], [73]) *Let $\rho : M \to \Sigma_g$ be an S^1-bundle over a closed oriented surface Σ_g of genus $g > 0$, whose euler number is $\chi(\rho) = \langle e(\rho), [\Sigma_g]\rangle$. Then there exists a codimension 1 foliations on M which is transverse to each fibre if and only if the following inequality holds.*

$$|\chi(\rho)| \leq |\chi(\Sigma_g)| = 2g - 2 \quad .$$

The same result is obtained for general Seifert fibrations. It applies also to the following G-G-S-T's inequality.

Theorem 3.1.5 (Ghys-Giroux-Sato-Tsuboi's inequality, [61], [25]) *The same $\rho : M \to \Sigma_g$ as above admits a positive contact structure which is transverse to each fibre if and only if the following inequality holds.*

$$\chi(\rho) \leq -\chi(\Sigma_g) = 2g - 2.$$

See the last note in Example 0.3.3 for directions of inequality signs. The reason why the inequality shifts from " $\cdots < 0$ " to " $\cdots < 2g-2$ " is that the horizontal distributions given by contact structures are not necessarily S^1-connections. Eliashberg-Thurston call such contact structures 'nonlinear pre-quantizations'.

Theorem 3.1.6 (Thurston's thesis, [68], [40]) *Let M be an S^1-bundle over a closed oriented surface and $\mathcal{F}$ a codimension 1 foliation on M. Suppose that $\mathcal{F}$ has no compact leaves. Then $\mathcal{F}$ can be isotoped so as to be transverse to each fibre .*

Theorem 3.1.7 (Giroux, [25]) *Suppose a contact structure ξ given on S^1-bundle M has no Legendrian knot both isotopic to the fibre and with the rotation number zero. Then ξ is isotoped to be transverse to each fibre.*

He reached these results by relying on his theory of convex surfaces [22].

We have already remarked that the tightness of contact structures was induced from the strict pseudo-convexity in the previous section. There is a result about foliations corresponding to this. This is the last part of the dictionary above.

Theorem 3.1.8 *[Gromov-Eliashberg, [28],[11]] A fillable contact structure is tight.*

Theorem 3.1.9 (Barrett-Inaba, [5]) *Let M be a compact Levi-flat real hypersurface of class C^∞ in a complex surface W and $\mathcal{F}$ the induced foliation. Then the foliation $\mathcal{F}$ has no Reeb components.*

Remark 3.1.10 **1)** It might be worth noticing that there are further differences between the theories of foliations and contact structures. In contact topology we have no analogue of Novikov's theorem, which claims that any foliation on S^3 (or in general on a closed 3-manifold with finite fundamental group) admits a Reeb component.

2) Related to Reeb components and vanishing cycles, Haefliger's theorem neither has its analogue in contact topology, which asserts that there are no analytic foliations on the same sort of manifolds. Because of the rigidity and the openness of being a contact structure, specifying the differentiability of contact structures makes no sense, while it has an essential importance in the theory of foliations.

3.2 Generation of contact structures through perturbations from foliations: theory of confoliations

It is the heart of the theory of confoliations to produce a contact structure from a foliation. It enables us to show the existence of fillable contact structures in certain cases. We explain perturbations of foliations into contact structures here and the mechanism by which fillable ones are obtained in 3.3 largely according to Confoliations [15].

Theorem 3.2.1 (Main Theorem of Confoliations, [15]) *Except for $(S^2 \times S^1, \zeta = \{S^2 \times \text{pt.}\})$, any codimension 1 oriented C^2 foliation $(M, \mathcal{F})$ on a closed oriented 3-manifold is approximated by positive (and of course* *negative) contact structures as plane fields by arbitrarily small perturbations in C^0-topology. is excluded as $(M, \mathcal{F})$.*

By the Reeb stability, if a foliation of a closed oriented 3-manifold has even a single leaf which is homeomorphic to S^2, it is diffeomorphic to 'ζ', which can never be perturbed into any other foliations nor even into any contact structures.

This theorem is proven through the following three steps of perturbations of tangent plane fields. It should be noticed that in each steps we can take perturbations of plane fields arbitrarily small in C^0-topology.

First Step (Adjusting foliations) Perturb a given foliation into one which has sufficiently many holonomies.

Second Step (Generating nonintegrability) Perturb the foliation into a positive contact structure near the support of the holonomy.

Third Step (Propagation of nonintegrability)

Propagate the nonintegrability to the whole manifold along leaves.

Let us deal with the Second and the Third Steps before the First, because we see easily what must be done in the First Step after the Second and Third.

Second Step (Generating nonintegrability) The following proposition is the key to construct a contact structure from a foliation.

Proposition 3.2.2 (Perturbation into confoliation by holonomy, [15]) 1) *If a leaf of a foliation $\mathcal{F}$ contains a simple loop γ with non-trivial linear holonomy, there exists a plane field ξ_1 obtained by slightly perturbing $\tau\mathcal{F}$ only in a sufficiently small tubular neighbourhood U of γ, which is a positive contact structure on U and coincides with $\tau\mathcal{F}$ outside U.*

2) *More generally, let us assume that a simple loop γ on a leaf of $\mathcal{F}$ has the holonomy which has contracting parts on both sizes and the supports of these contractions are sufficiently close to the leaf and on the region between the two contracting parts the holonomy is sufficiently close to the identity. Then there exists a small perturbation of $\tau\mathcal{F}$ into ξ_1, which is performed only in a sufficiently small tubular neighbourhood U of γ, and ξ_1 is a positive contact structure on U and coincides with $\tau\mathcal{F}$ outside U.*

Proof. **1)** Suppose γ has an expanding holonomy with non-trivial linear part. Then we can find a local coordinate (x, y, z), $(x \in S^1 = \mathbb{R}/\mathbb{Z})$ as follows on a sufficiently small tubular neighbourhood U of γ. The leaf L corresponds to (x, y)-plane and γ to x-axis. On U, the leaf looks almost like the graph of $z = c \exp ax$. Indeed, more exactly, we may suppose that the foliation $\mathcal{F}|_U$ is defined by a 1-form $\alpha = dz - f(x, z)dx$ (where $f = az + o(z)$, especially $\frac{\partial f}{\partial z} > 0$ on U). Then we have $d\alpha = df \wedge dx$. Take a bump function $g(y, z)$ such that

$$g|_{M \setminus U} = 0, \quad g|_U < 0, \quad f \cdot \frac{\partial g}{\partial z} \geq 0 ,$$

and put $\beta = gdy$. Define a deformation of α by $\alpha_t = \alpha + t\beta$ by using β above. By the straight forward calculation, we can see very easily that this α defines a positive contact structure on U when $t > 0$ and negative one when $t < 0$.

2) Let us outline the proof. Take a similar coordinate as above and assume that $\{\varepsilon < |z| \leq 1\}$ is contained in the contracting part.

Then we perform a similar perturbation on $U = S^1 \times (-1, 1) \times (-1, 1)$ as above to obtain new plane fields $\xi_t \ker\alpha_t$ for $|t| \ll 1$. ξ_t $0 < t \ll 1$ is a positive contact structure on $\{2\varepsilon < |z| < 1\}$ because of more or less the same reason as above. In the region $\{|z| \leq 2\varepsilon\}$ ξ_t can even be negative contact structure. However the negativity is small and bounded by the nature of the holonomy.

For not too much small $t > 0$, the curves which are tangent to ξ_t and $x \equiv$ const. go up and down so as to join the bad region $\{|z| \leq 2\varepsilon\}$ to the good region $\{2\varepsilon \leq |z| < 1\}$ where a good amount of positive nonintegrability is waiting for to compensate the badness in $\{|z| \leq 2\varepsilon\}$ from the point of view of the Third Step. □

The resulting plane field has both the part of foliations and that of positive contact structures.

Third Step (Propagation of nonintegrability)

Definition 3.2.3 (Confoliation, [15]) A plane field ξ of oriented 3-manifold M defined by a non-singular 1-form α is called a *confoliation* if it satisfies $\alpha \wedge d\alpha \geq 0$. For a given confoliation ξ, its contact part is denoted by $C(\xi) = \{x \in M; (\alpha \wedge d\alpha)_x > 0\}$, and the foliation part by $F(\xi) = \{x \in M; (\alpha \wedge d\alpha)_x = 0\}$. Furthermore, $\widetilde{C(\xi)}$ denotes the saturation $\{x \in M; \exists \gamma : [0, 1] \to M$ acurve along $\xi, \gamma(0) \in C(\xi), \gamma(1) = x\}$ of $C(\xi)$ by leaves of $F(\xi)$.

Generally, when a confoliation ξ is given, we can diffuse the nonintegrability on $C(\xi)$ to the outside along leaves of $F(\xi)$.

Proposition 3.2.4 (Propagation of nonintegrability, [15]) *Let B be a coordinate neighbourhood as in Lemma 0.4.3. Moreover, suppose the following condition.*

(3) $\frac{\partial f}{\partial x} > 0$ *on* $0 \leq x \leq 2\varepsilon$.

We can perturb the function f (so the plane field ξ) so that it satisfies the following conditions i), ii), and iii) by arbitrarily small perturbations on $B' = [\varepsilon, 1 - \varepsilon]^3$ ($=$ a sufficiently small neighbourhood of ∂B).

i) *On* $B'' = [2\varepsilon, 1-2\varepsilon]^3$, $\frac{\partial f}{\partial x} > 0$,
ii) *On* B, $\frac{\partial f}{\partial x} \geq 0$,
iii) *On* $B \setminus B'$, *it is same as the original one.*

Namely this means that the nonintegrability on $\{0 \leq x \leq 2\varepsilon\}$ *propagates to* B'' *along curves in leaves.*

Of course, the proof is easy. Then if a confoliation ξ of a compact manifold M satisfies $\widetilde{C(\xi)} = M$, we can perturb it until it becomes a positive contact structure on the whole of M by performing the operation in this proposition for finitely many times.

The propagation of nonintegrability along leaves was first achieved by Altschuler [1] in an analytic method. He used a new kind of heat equation. His work must have been one of good motivations for the theory of confoliations.

First Step (Adjusting foliations) If a foliation $(M, \mathcal{F})$ satisfies the following condition (F), we get a contact structure by applying the procedures in the last two Steps.

3.2.5 Condition (F): *There exist finitely many leaves with holonomies as in Proposition 3.2.2, and any other leaves (i.e., any minimal set of* $\mathcal{F}$*) meets the neighbourhoods of such leaves.*

From now on, we assume that foliations are of class C^2 because we need the qualitative theory of foliations. Now what we have to prove is the following, which is a good exercise for the experts on the qualitative theory of foliations.

Proposition 3.2.6 *Any* $\mathcal{F}(\neq \zeta)$ *is approximated by a sequence of foliations each of which satisfies (F).*

In general, minimal sets are classified into the following 3 types; (1) a compact leaf, (2) dense type, (3) an exceptional minimal set (EMS). Thanks to Sacksteder's theorem, EMS is very convenient.

Theorem 3.2.7 (Sacksteder, [60]) *In EMS, there is a leaf with a nontrivial linear holonomy.*

In the case of type (2), the situations are quite different according whether the holonomy is trivial or not. If there is a leaf with holonomy, we can easily find a subpseudogroup of Schottky type, *i.e.*, a pseudogroup of 1-dimensional local diffeomorphisms generated by γ_1 and γ_2, which look like $\gamma_1 : [0,1] \to [a,b]$, $\gamma_2 : [0,1] \to [c,d]$, where $0 \leq a < b < c < d \leq 1$. This subpseudogroup has an EMS. Then we can apply (the proof of) Sacksteder's theorem to this EMS to get a linear holonomy.

One the other hand, a foliation of which all leaves have no holonomy is a bundle foliation or is approximated by a sequence of bundle foliations (so called Tischler fibration). Therefore this case reduces to the case where infinitely many compact leaves pile on.

Since the genus of leaves are positive now, it is exactly easy to pick up some sheets of leaves discretely in the part of piling compact leaves and to perturb the foliation so that there are only finitely many compact leaves and they have linear holonomies or at least such holonomies as in 2) of Proposition 3.2.2.

In the case where $\mathcal{F}$ has infinitely many EMS's, Sacksteder's theorem prohibits them to accumulate to another EMS, therefore all but finite of them are trapped by compact leaves.

By these operations, any foliation except for ζ is perturbed so as to satisfy the condition (F).

3.3 Taut foliations and fillable contact structures

Now let us show the existence theorem of semi-fillable contact structures by using perturbations of foliations into contact structures. Gabai's theorem is the key from the side of the foliation theory.

Theorem 3.3.1 (Gabai, [20]) *Let M be an irreducible oriented closed 3-manifold with $H_2(M;\mathbb{Z}) \neq 0$. Given a non-trivial non-divisible element $c \in H_2(M;\mathbb{Z})$, there exists a C^∞ taut foliation $\mathcal{F}$ which has a compact leaf representing c.*

Proposition 3.3.2 ([15]) *If both positive and negative contact structures $\xi_\pm$ sufficiently approximate a taut foliation $(M, \mathcal{F})$, then $(M \sqcup (-M), \xi_+ \sqcup \xi_-)$ is fillable.*

Definition 3.3.3 ([15]) A contact structure (M, ξ) is called *semi-fillable*, if it consists of (some of) connected components of a fillable contact structure (N, η), *i.e.*, (M, ξ) is not necessarily the whole but a closed open subset of (N, η). Semi-fillable structures are of course tight by Proposition 2.1.1 and Theorem 2.2.7.

We have the following theorem as a direct corollary.

Theorem 3.3.4 (Existence of semi-fillable contact structures, [15]) *On any irreducible oriented closed 3-manifold M other than rational homotopy 3-spheres, there exists a semi-fillable contact structure.*

Let us prove Proposition 3.3.2 by using (3) of Sullivan's characterization of minimal foliations.

Theorem 3.3.5 (Sullivan's characterization of minimal foliations, [62]) *For transversely oriented codimension 1 foliation $(M, \mathcal{F})$ on a closed 3-manifold, the following four conditions are equivalent to each other.*

(1) *Every leaf intersects with a transverse knot, i.e., is taut.*
(2) *There is no dead end component.*
(3) *There is a closed 2-form Ω which is positive on each leaves.*
(4) *There is a Riemannian metric on M such that every leaf is a minimal surface.*

Remark 3.3.6 (3) implies if taut foliation has a compact leaf it represents a non-trivial homology class.

Proof of Proposition 3.3.2 Approximate a taut foliation $(M, \mathcal{F})$ $(\neq (S^2 \times S^1, \zeta))$ by positive and negative contact structures $\xi_\pm$. Let α be a 1-form which defines the foliation $\mathcal{F}$ and put $\omega = \Omega + d(t\alpha)$. Since $\omega^2 = 2dt \wedge \Omega \wedge \alpha$ as is easily calculated, ω defines a symplectic structure on $\mathbb{R} \times M$. Notice that $\omega|_{T\mathcal{F} \subset T(\{0\} \times M)} = \Omega|_{T\mathcal{F}} > 0$ by the construction. If you take sufficiently good approximations $\xi_\pm$ of $\mathcal{F}$ and sufficiently small $\varepsilon > 0$, a symplectic manifold $(W = [-\varepsilon, \varepsilon] \times M, \omega)$ gives a symplectic filling of the contact structures $\xi_\pm$ of the boundary $\partial W = \{\pm\varepsilon\} \times M$. □

3.4 Further results

Combining the results in this section with some topological constructions and some techniques to classify tight contact structures on noncompact manifolds, Eliashberg proved the following theorem.

Theorem 3.4.1 (Eliashberg, [15]) *Any foliation without Reeb component is approximated by tight contact structures.*

By this, we get a bridge from the right of 3rd item of the dictionary 3.1.1 at the top of this section to the left. One hand, we know that the standard contact structure on S^3 and its isotoped structures approximate the Reeb foliation. As the Reeb foliation satisfies the relative version of Thurston's inequality, there is no contradiction, while it is even approximated by OT contact structures.

Another important topic is 'finiteness'. Expressing Bennequin's and Thurston's inequalities by the Thurston norm, we see the following finiteness theorem.

Theorem 3.4.2 (Finiteness for Cohomology Classes) *Let M be a closed oriented 3-manifold. Then the number of cohomology classes in $H^2(M; \mathbb{Z})$ which can be realized as the euler class of a tight contact structure or the tangent bundle of a foliation without Reeb components is finite.*

Using the monopole equations, Kronheimer and Mrowka proved the following remarkable finiteness theorem on the homotopy classes as plane field.

Theorem 3.4.3 (Finiteness for homotopy classes, [38]) *For any oriented closed 3-manifold M, there exist only finitely many homotopy classes of plane fields which contain a fillable contact structure.*

So far, mainly the foliation theory served for the study of contact structures. Now a converse is established. Combining this result with the theory of confoliations, we get the following theorem as a direct corollary.

Theorem 3.4.4 *For any oriented closed 3-manifold M, there exist only finitely many homotopy classes of plane fields which contain the tangent bundle of a taut foliation.*

This is a really new result even in the foliation theory. We do not know whether the homotopy finiteness is true or not for tight contact structures nor for foliations without Reeb components.

4 Anosov flows and bi-contact structures

In the previous section, we introduced a general procedure of perturbing a foliation into a contact structure, under the name of confoliation. In the contrast with this, in this section we investigate some special situation where foliations deforms into contact structures and contact structures converges into foliations.

The most important class is that of foliations and contact structures associated with Anosov flows on 3-manifolds. Looking from the point of view of the hierarchy of convexities, this class shows the strongest convexities, *i.e.*, associated contact structures have strong symplectic fillings and Anosov foliations are highly mixing and taut. (We have already discussed on the weak fillablity and tightness until the previous section.)

The notion of Anosov flow is generalized into projectively Anosov flows, respecting the reason why foliations and contact structures are naturally associated with the flow.

This kind of flow exists on any oriented closed 3-manifolds so that it is expected to contribute to the study of convexity of structures. However, the associated integrable plane field does not have differentiability in general, and thus it does not define a foliation in a usual sense. This pathological phenomenon happens unavoidably. Moreover the resulting contact structure can be OT. On the other hand, if we assume the regularity on the weak Anosov splitting, we get their convexities as well as the classifying theorems 4.3.4 due to Noda.

As basic references for Anosov flows and foliations, we raise the very first reference [2] due to Anosov himself as well as [74], [31], and [37].

4.1 Anosov flows and strong filling

Question 4.1.1 (Calabi) Does there exist a completely convex (or strongly filling) symplectic 4-manifolds which does not come from a complex structure? More specifically do there exist those with disconnected end (or boundary)?

The first such example to this question was constructed by McDuff [46] as a convex symplectic structure on $[-1,1] \times S^1T^*\Sigma_g$. The two contact structures induced on the boundary components $\{\pm 1\} \times S^1T^*\Sigma_g$ were two really well known ones, classical and quantum mechanical examples 0.3.2 and 0.3.3, *i.e.*, the Liouville contact structure and the Riemannian connection.

E. Ghys and the author found a much simpler construction and description of this example in terms of the dual of the Lie algebra $sl(2;\mathbb{R})^*$ ([39], [49]) and it was soon clarified that the same construction applies to the solvable Lie group associated with a hyperbolic automorphism of the 2-torus T^2 (see Example 0.3.5 and Figure 4). As a matter of fact, these two are the Lie groups which admit a left invariant Anosov flow.

Remark 4.1.2 McDuff's construction was in 'some' sense essentially different from ours. According to hers the leaves of the Anosov foliation are symplectic and it is similar to the construction in §3.3. To the contrary, our construction makes the leaves Lagrangian. However, if we exchange stable and unstable (*i.e.*, reverse the flow), symplectic Lagrangian are also switched, and thus we can say they coincide finally. See Example 0.3.5 and Figure 4.

Definition 4.1.3 A nonsingular smooth flow ϕ_t on a closed 3-manifold M^3 is said to be an *Anosov flow* there exists a ϕ_t-invariant continuous splitting (Anosov splitting) $TM = T\phi \oplus E^{uu} \oplus E^{ss}$ of the tangent bundle which satisfies the following conditions for some Riemann metric; For some positive constant $c > 0$,

$$\forall v \in E^{uu}, \quad \forall t > 0, \qquad \|(\phi_t)_* v\| \geq \exp(ct)\|v\|,$$
$$\forall v \in E^{ss}, \quad \forall t < 0, \qquad \|(\phi_t)_* v\| \geq \exp(ct)\|v\|.$$

We set $E^s = T\phi \oplus E^{ss}$ ($E^u = T\phi \oplus E^{uu}$) and call $TM = T\phi \oplus E^u \oplus E^s$ the weak Anosov splitting. The C^r-section theorem ([31]) guarantees that E^u and E^s are plane fields of class C^1. Therefore Lemma 0.4.3 implies that they define two foliations $\mathcal{F}^u$ and $\mathcal{F}^s$ which are called unstable and stable Anosov foliations respectively. See Figure 11.

Theorem 4.1.4 ([49]) *Let ϕ_t be an Anosov flow on a closed oriented 3-manifold M. Then, the 4-manifold $W = [-1,1] \times M$ admits a convex* *symplectic structure, i.e., is a strong filling. On the other hand, W has the disconnected boundary so that it is not a Grauert domain.*

Remark 4.1.5 Stein manifold of $\dim_{\mathbb{C}} = n$ (*i.e.*, $\dim_{\mathbb{R}} = 2n$) carries a strictly pluri-subharmonic function whose Morse theory shows that the manifold has the homotopy type of at most n-dimensional CW complex. In particular, in the case where $\dim_{\mathbb{C}} \geq 2$ the boundary (end) is connected. This holds even for Grauert domains. See [58].

In order to visualize this theorem, let us introduce the notions of bi-contact structures and of linear perturbations.

Definition 4.1.6 (Linear perturbation) A 1-parameter family of plane fields $\{\xi_t\}$ is called a *linear perturbation* of the foliation ξ_0 if it is defined by a 1-parameter family of 1-forms $\{\alpha_t\}_{-\varepsilon<t<\varepsilon}$ which satisfies

$$\alpha_0 \wedge d\alpha_0 = 0 \qquad \text{and} \qquad \frac{d}{dt}(\alpha_t \wedge d\alpha_t)|_{t=0} > 0$$

everywhere on the manifold. If we set $\beta = \frac{d}{dt}(\alpha_t)|_{t=0}$ and $\alpha = \alpha_0$, the second condition (inequality) is equivalent to $\alpha \wedge d\beta + d\alpha \wedge \beta > 0$.

As an even more special case, we study linear perturbations such that $\xi_{t\neq 0}$ are transverse to the foliation, *i.e.*, $\alpha \wedge \beta \neq 0$. Anosov foliations are exactly in this case. Let us formulate what happens in the case of Anosov foliations looking from the side of contact structures.

Definition 4.1.7 (Bi-contact structure) A *bi-contact structure* on a 3-manifold M is a transverse pair (ξ, η) of contact structures on M with the opposite orientations .

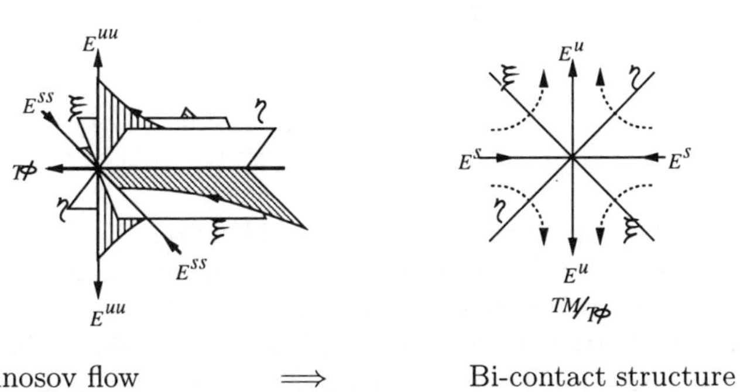

Anosov flow $\Longrightarrow$ Bi-contact structure

Figure 11.

As indicated in Figure 11, for an Anosov flow ϕ_t we define plane fields ξ and η to be those who contain the flow direction $T\phi$ and take the positions of $\pm 45°$ against E^u and E^s. Then neither ξ nor η are preserved by the flow ϕ_t. Actually they look twisted along the flow in the opposite directions. Therefore we obtain the following by Lemma 0.4.3.

Theorem 4.1.8 ([49]) *The pair of plane fields (ξ, η) thus associated with an Anosov flow is a bi-contact structure.*

Remark 4.1.9 We have a problem of the differentiability of the weak splitting, which has already caused a trouble in Theorem 4.1.8. Even in the case where E^{uu} or E^{ss} are not smooth, in this dimension the CR section theorem [31] guarantees that E^u and E^s are of class C^1. The definition for a plane field to be a contact structure has still its meaning even for C^1-plane fields and the condition is open. For Theorem 4.1.8, it suffices to look at the projectified normal bundle to the flow and to take a C^1-approximations of ξ and η by C^∞-plane fields as sections of that bundle.

Proof of Theorem 4.1.4 is outlined in the following.

The linear perturbation of the unstable foliation $\mathcal{F}^u$ by the stable foliation $\mathcal{F}^s$

$$\alpha_t = \alpha^u + t\alpha^s, \quad \ker\alpha^u = E^u, \ker\alpha^s = E^s,$$

gives rise to a convex symplectic structure $\omega = d\lambda$ on the 4-manifold $W = [-1,1] \times M$. Here λ is obtained by regarding the family of 1-forms $\{\alpha_t\}$ as a 1-form on $W = [-1,1] \times M$. On the boundary $\partial W = \{-1,1\} \times M$ we have contact forms $\alpha_{\pm 1} = \alpha^u \pm \alpha^s$ which are the restriction of λ and define ξ and η. (See also Figure 4.)

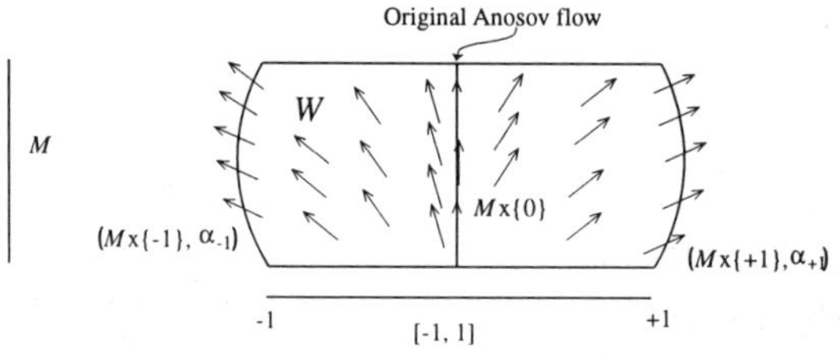

□

A series of works initiated by Handel-Thurston [29] allows us to obtain new Anosov flows by performing Dehn surgeries along closed orbits of Anosov flows. Therefore, there are so many Anosov flows other than algebraic ones. In Particular, most of graph manifolds admits Anosov flows. However, a theorem of Plante-Thurston [59] (see below) puts a constraint on the class of manifolds which admit Anosov flows.

Conversely, the class of foliations whose linear perturbation gives the primitive of a convex symplectic structure looks also quite restrictive and limited to Anosov foliations. A partial result is given by using a theorem of Inaba-Tsuchiya, which generalizes the theorem of Plante-Thurston to expansive foliations, *i.e.*, foliations who have the following property ; any small open transversal can be transformed into a transeversal of a uniformly fixed size by the holonomies along the leaves.

Theorem 4.1.10 1) (Inaba-Tsuchiya, [34]) *A manifold which admit an expansive foliation of codimension 1 has its fundamental group with exponential growth.*

2) (Plante-Thurston, [59]) *In particular, the fundamental group of a manifold which admits an Anosov flow grows exponentially.*

Theorem 4.1.11 ([52]) *Let* $\omega = d\lambda$ *be a convex symplectic structure on* $W^4 = [-1,1] \times M^3$*, which is obtained by a deformation* α_t *of a foliation* $\mathcal{F}$ *on* M *as* $\lambda = \alpha_t$*. Then, it is a linear perturbation,* $\beta = \dot{\alpha}|_{t=0}$ *is transverse to* α*, and the contact structures* $\ker \alpha_\varepsilon$ *and* $\ker \alpha_{-\varepsilon}$ *pair into a bi-contact structure. Moreover,* $\pi_1(M)$ *has an exponential growth.*

Outline of proof. If a linear perturbation $\lambda = \alpha + t\beta$ provides a primitive of a symplectic structure, *i.e.*, $\omega^2 = (d\lambda)^2 > 0$, it is not difficult to compute that the transverse metric expands exponentially along the lines defined by $\alpha \wedge \beta = 0$. The the statement follows from the theorem of Inaba-Tsuchiya.

4.2 Projectively Anosov flows and bi-contact structures

Conversely to Theorem 4.1.8, for any given bi-contact structure (ξ, η) do we always obtain an Anosov flow ϕ_t as the vector field given by their intersection $T\phi = \xi \cap \eta$?

The answer is No! and we should introduce the notion of projectively Anosov flows.

Definition 4.2.1 (Projectively Anosov Flow,[49]) A nonsingular smooth flow ϕ_t on a closed oriented 3-manifold M is a *Projectively Anosov* flow (or $\mathbb{P}$**A** flow for short) if the projectification $T\phi_t$ of the differential of the flow induced on the projectified normal bundle to the flow $S^1(TM/T\phi) = (TM/T\phi \setminus 0)/\mathbb{R}_+$ presents the same situation as that of Anosov flows. Precisely, it means the following.

(1) There are four mutually disjoint continuous sections $\mathcal{E}^u_\pm$ and $\mathcal{E}^s_\pm$ of $S^1(TM/T\phi)$ which are invariant under the induced flow $T\phi_t$ and are mutually anti-podal two by two.

(2) The orbit of any point which is not on these 4 cross sections is attracted to $\mathcal{E}^u_\pm$ when $t \to \infty$ and to $\mathcal{E}^s_\pm$ when $t \to -\infty$.

Remark that in this situation, $\mathcal{E}^u_\pm$ and $\mathcal{E}^s_\pm$ define two plane fields E^u and E^s on M which are invariant under the flow ϕ_t. We again call $TM = E^u + E^s$

the *weak Anosov splitting* of the $\mathbb{P}\mathbf{A}$ flow ϕ_t.

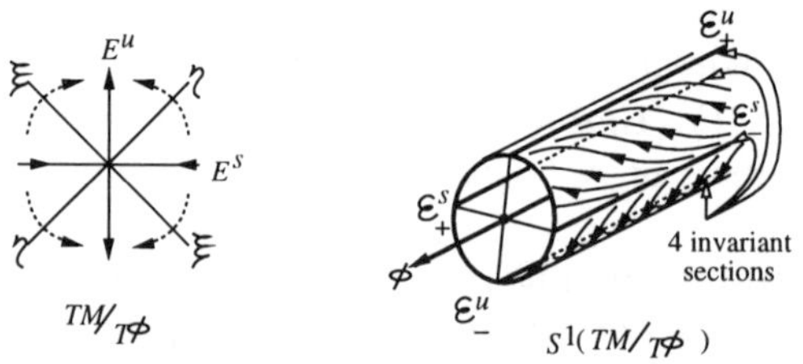

Bi-contact structure $\Longleftrightarrow$ Projectively Anosov flow

Figure 12.

The notions of $\mathbb{P}\mathbf{A}$ flows and bi-contact structures are equivalent to each other. That is to say, we have the following.

Theorem 4.2.2 ([49], [52]) 1) *Any $\mathbb{P}\mathbf{A}$ flow gives us a bi-contact structure in the same way as in the case of Anosov flows.*

2) *The flow ϕ_t defined from a bi-contact structure (ξ, η) as their intersection is a $\mathbb{P}\mathbf{A}$ flow.*

$\mathbb{P}\mathbf{A}$ flows exist much more abundantly than Anosov flows. The first example of non-Anosov $\mathbb{P}\mathbf{A}$ flow was constructed from an Anosov flow by modifying it in a neighbourhood of a closed orbit in such a way that the orbit comes to be totally attracting. This modification was pointed out by A. Zeghib in 1993, which was a rediscovery of an argument due to Franks-Willims [17]. Then a definitive example was found on T^3.

Example 4.2.3 (Propeller construction, [15], [49]) First prepare a pair of contact structures on T^3 which were introduced in Example 0.3.5.

$$\xi' = \xi_k = \ker[\alpha' = \cos(kz)dx - \sin(kz)dy], \quad (k \in \mathbb{N}),$$
$$\eta = \xi_{-l} = \ker[\beta = \cos(lz)dx + \sin(lz)dy], \quad (l \in \mathbb{N}).$$

As they are not transverse to each other, we add a bit of perturbation $\alpha = \alpha' + \varepsilon dz$ $(\varepsilon > 0)$ so as to arrange them transverse to each other and define $\xi = \ker[\alpha]$. Then, because $\frac{\partial}{\partial z} \notin \xi$ and $\frac{\partial}{\partial z} \in \eta$, ξ and η are transverse to each other and we get a bi- contact structure (ξ, η). In the case where $k = l$, indicated in the following figures, we obtain the resulting (un)stable foliations which are presented in very simple terms of trigonometric functions.

As the fundamental group $(\cong \mathbb{Z}^3)$ of T^3 grows of course as a polynomial of degree 3, it does not admit Anosov flows. What is even more characteristic here is that the two foliations have compact leaves and that they

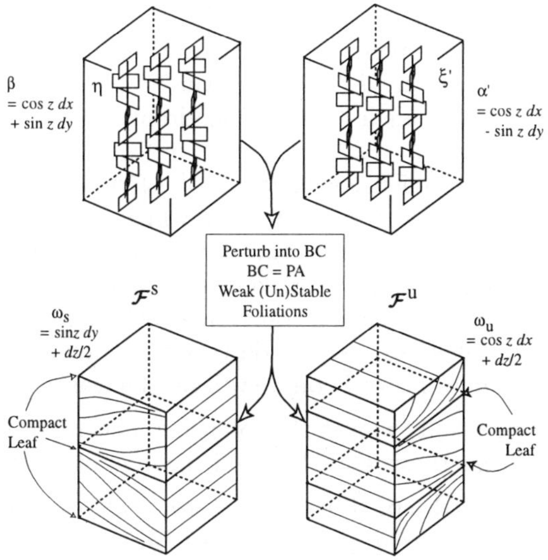

Figure 13. Propeller construction-1

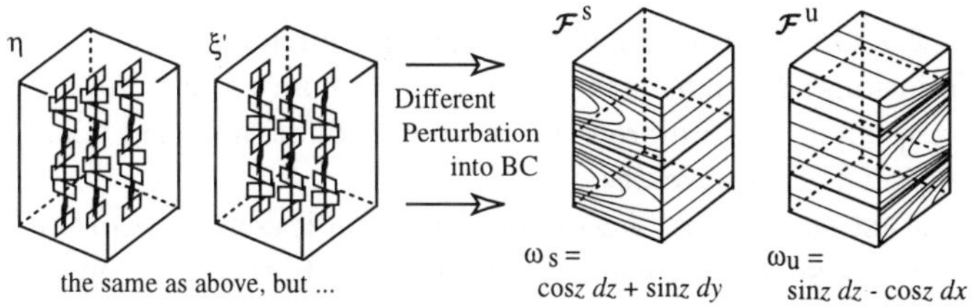

Figure 14. Propeller construction-2

coincide with the set of points on which before the perturbation ξ' and η are tangent to each other. Of course weak (un)stable foliation of an Anosov flow never admit a compact leaf.

To arrange ξ and η transverse to each other by perturbing ξ', it is enough to do it only on the neighbourhood of the points of tangency of ξ' and η (in the case of the above figure, four horizontal T^2's) by adding certain dz-component. Especially, we can choose different signs of the coefficients of dz on each connected component. What thus we get looks like in Figure 14. It is observed that both foliations $\mathcal{F}^u$ and $\mathcal{F}^s$ have two compact leaves which bound so called *dead end components*. That is to say, we are not allowed to expect to get taut foliations even assuming the differentiability of the weak splitting of a $\mathbb{P}\mathbf{A}$ flow.

Remark 4.2.4 **1)** The propeller construction is also applicable to produce bi-contact structures on any T^2-bundles over S^1 , if we respect the eigen directions or the twisting angles of the monodromy and adjust the rotation angle of propeller.

2) Even on S^3 an explicit construction of a $\mathbb{P}\mathbf{A}$ flow (and at the same time a bi-contact structure (ξ, η)) is known, which is due to H. Minakawa. In this example, the weak unstable and stable foliations are isomorphic to the standard Reeb foliation, however, so is the stable one only topologically. To arrange two Reeb foliations in a transverse position is not at all difficult, however, to get $\mathbb{P}\mathbf{A}$ flow we have to destroy the transverse differentiable structure of at least one of two. Soon we will see the reason (Theorem 4.3.1).

As an existence result, we get the following Theorem 4.2.6 as a direct corollary to Hardorp's thesis and the main theorem of confoliations.

Theorem 4.2.5 (Hardorp, [30]) *Any closed oriented 3-manifold admits a total foliation, i.e., a triad of foliations $(\mathcal{F}, \mathcal{G}, \mathcal{H})$ whose tangent planes $T\mathcal{F}$, $T\mathcal{G}$, and $T\mathcal{H}$ are linearly independent at each point of the manifold.*

Theorem 4.2.6 *Any closed oriented 3-manifold admits a bi-contact structure.*

Remark 4.2.7 On S^3, the situation seems quite confusing.

1) (Foliations) Novikov's theorem says that any foliation on S^3 admits a Reeb component. Therefore even if we get foliations from a $\mathbb{P}\mathbf{A}$ flow, we can not expect their convexity.

2) (Contact structures) According to the uniqueness theorem (see [12]) of Eliashberg, positive and negative tight contact structures on S^3 are both unique up to isotopy. Actually, they are given as the left and right invariant plane fields of the Lie group S^3 and they differ by 1 as homotopy class of plane fields, measured by the Hopf invariant. Therefore they can not pair into a single bi-contact structure because two contact structures of which a bi-contact structure consists belong to the same homotopy class of plane fields. Thus a bi-contact structure (ξ, η) on S^3 consists of at least one OT structure.

3) (Weak Anosov splitting) In general, we can not assume the differentiability on the weak Anosov splitting $E^u + E^s$ of $\mathbb{P}\mathbf{A}$ flows. Even though they are 'integrable' (*i.e.*, at any point there exists an integral submanifold) as Lemma 0.4.3 explains, the integral submanifolds can be branched. There exist many such examples. There also exist $\mathbb{P}\mathbf{A}$ flows whose weak stable and unstable foliations are without branching and still can not have differentiability. These are related to the above remarks 1) and 2).

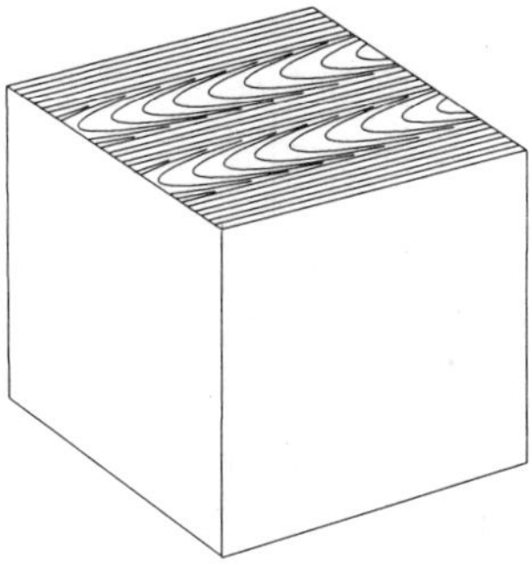

Figure 15. Irregular projectively Anosov flow

As in the figure on the previous page, prepare a 1-dimensional foliation on T^2 which has two 2-dimensional Reeb components and perform the propeller construction starting from it.

By the construction, the original 1-dimensional foliation on T^2 gives the flow lines of the resulting $\mathbb{P}\mathbf{A}$ flow and the T^2 must be a compact leaf.

Computing the holonomy along the boundary loop of 2-dimensional Reeb components as the Lyapunov exponents of the flow, we get some contradiction. This implies the resulting weak Anosov splitting is forced to have branching integral submanifolds. We will see a similar argument in the proof of Theorem 4.3.1.

4.3 Tightness of bi-contact structures

If we assume some regularity on the weak Anosov splitting of $\mathbb{P}\mathbf{A}$ flows, we get convexities both on foliations and contact structures.

Theorem 4.3.1 ([52]) *If a $\mathbb{P}\mathbf{A}$ flow has its weak Anosov splitting $E^u + E^s$ of class C^1, neither of the foliations $\mathcal{F}^u$ nor $\mathcal{F}^s$ have Reeb components.*

Proof. Proof relies on the Tamura-Sato classification ([64]) of the foliations which are transverse to a Reeb component.

Let $\mathcal{F}^u$ be the unstable foliation indicated in the figure on the right as a foliation drawn by solid lines. The classification tells us that as the stable foliation $\mathcal{F}^s$ (drawn with broken lines) is transverse to $\mathcal{F}^u$ there exists a *half Reeb component*, a region which looks like the right half of the Reeb component in the figure. Thus we can find two closed orbits γ_1 and γ_2 on the boundary(corner) of the half Reeb component.

We can compute the linear holonomy of $\mathcal{F}^u$ along the orbit of the $\mathbb{P}\mathbf{A}$ flow as the Lyapunov exponents of the (un-)stable direction. If we do this on γ_1 and γ_2 for both of $\mathcal{F}^u$ and $\mathcal{F}^s$, then we get a contradiction. □

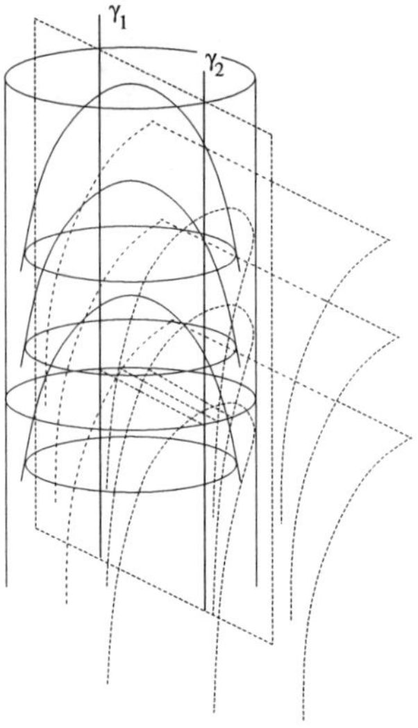

Definition 4.3.2 (Regular projectively Anosov flow, [54]) A $\mathbb{P}\mathbf{A}$ flow is *regular* if it satisfies the assumption of the above theorem. Hereafter in this note, this terminology implies the flow with C^∞-splitting unless otherwise specified.

Even in the case of regular $\mathbb{P}\mathbf{A}$ flows, as indicated in Figure 14 (Propeller Construction-2), $\mathcal{F}^u$ and $\mathcal{F}^s$ may have dead-end components. Therefore, it may be not reasonable to expect the fillability of contact structures. However, Theorem 4.3.1 and Eliashberg's theorem 3.4.1 suggests the following.

Conjecture 4.3.3 A bi-contact structure associated with a regular $\mathbb{P}\mathbf{A}$ flow consists both of tight contact structures.

In the case of regular $\mathbb{P}\mathbf{A}$ flows, sometimes not only the convexity but also even the classification is achieved. In the propeller construction 4.2.3 on T^3, if we take $k = l$, exceptionally we get real analytic foliations $\mathcal{F}^u$ and $\mathcal{F}^s$. Takeo Noda proved the following classification theorem concerning this phenomenon.

Theorem 4.3.4 (Noda, [53]) *Any regular $\mathbb{P}\mathbf{A}$ flow on T^3 is obtained by the propeller constructions starting from a linear foliation on T^2. The resulting bi-contact structure (ξ, η) consists both of tight contact structures and eventually they are ξ_k and ξ_{-k} for some $k \in \mathbb{N}$ in the classification theorem due to Giroux and Kanda ([24], [36]).*

His result is in fact far more precise. It classifies associated foliations completely. That is to say, such a foliation is a composition of the units of foliations in Figure 13 and 14.

As remarked before, the propeller constructions apply to any other T^2-bundles over S^1 than T^3. Noda extended his argument to such cases. Especially in the case of solvable manifolds which are obtained as the suspension of T^2 by hyperbolic elements in $SL(2;\mathbb{Z})$, he showed that the similar classification is possible under the presence of a compact leaf.

For those which have no compact leaves, Noda and Tsuboi further advanced their argument. They classified regular $\mathbb{P}\mathbf{A}$ flows on the solv-manifolds and the unit tangent bundle of hyperbolic surfaces, which are known to admit algebraic Anosov flows. Let us look around some backgrounds of this.

Let Σ_g be a closed oriented surface of genus $g \geq 2$ and $\mathcal{T}_g$ be its Teichmüler space, *i.e.*, the space of isotopy classes of hyperbolic structures on it. If we fix a hyperbolic structure $h \in \mathcal{T}_g$ on Σ_g, the unit tangent circle bundle $S^1_h T\Sigma_g$ is determined as a submanifold of the tangent bundle $T\Sigma_g$. However, since we can also regard this as the projectification of $T\Sigma_g$, we can take the same space as their unit tangent bundles $S^1 T\Sigma_g = S^1_{h_i} T\Sigma_g$ $(i = 1, 2)$ for distinct hyperbolic structures $h_1, h_2 \in \mathcal{T}_g$. For each hyperbolic structure $h \in \mathcal{T}_g \cong \mathbb{R}^{6g-6}$, the geodesic flow ϕ_h on $S^1 T\Sigma_g$ defines an algebraic Anosov flow and we obtain the weak (un)stable foliations $\mathcal{F}^s_h$ and $\mathcal{F}^u_h$ which are transverse to each other.

Now, if two hyperbolic structures $h_1, h_2 \in \mathcal{T}_g$ are close enough to each other, then $\mathcal{F}^s_{h_1}$ and $\mathcal{F}^u_{h_2}$ are again transverse, and moreover $\mathcal{F}^s_{h_1} \cap \mathcal{F}^u_{h_2}$ defines a 1-dimensional foliation which gives rise to (up to change of parameter) a flow ϕ_{h_1,h_2}. If h_1 and h_2 are still closer to each other, ϕ_{h_1,h_2} defines a $\mathbb{P}\mathbf{A}$ flow and if again closer defines an Anosov flow. E. Ghys called such a deformation of the geodesic Anosov flows *quasi-Fuchsian deformation* or *quasi-Fuchsian flow* ([21]).

Theorem 4.3.5 (Noda-Tsuboi, [54]) 1) *On a solv-manifold, only the (unique) algebraic Anosov flow can be a regular* $\mathbb{P}\mathbf{A}$ *flow without compact leaves.*

2) *On* $S^1 T\Sigma_g$, *only quasi Fuchsian flows can be regular* $\mathbb{P}\mathbf{A}$ *flows without compact leaves and eventually they are Anosov.*

What happens if we assume only the transversality of the intersection of two Anosov foliations (associated with possibly different Anosov flows) is studied by Matsumoto-Tsuboi.

Theorem 4.3.6 (Matsumoto-Tsuboi, [45]) 1) *On solv-manifolds, any transverse pair of foliations without compact leaves are obtained by a simul-*

taneous isotopy from the algebraic Anosov foliations.

2) *On $S^1T\Sigma_g$ there exists a transverse pair of foliations without compact leaves, which can not simultaneously isotoped to a quasi-Fuchsian pair of foliations. Eventually they do not define a* $\mathbb{P}\mathbf{A}$ *flow.*

As seen from the above discussions, the theory of regular $\mathbb{P}\mathbf{A}$ flows is now making progress in the framework of foliation theory. On the other hand, from the point of view of contact structures we have to investigate irregular ones. For instance, studies on bi-contact structure necessarily deal with irregular $\mathbb{P}\mathbf{A}$ flows. It is not difficult to construct bi-contact structures on S^3 which consist of an OT contact structure and a tight one as well as those which consist both of OT ones.

5 Problems

To close this article, let us raise some open problems which are related to the interaction of the theory of foliations and that of contact structures. Some might be nice to attack right now and some others are still in a vague form.

5.1 Finiteness problems

Problem 5.1.1 *Prove Theorem 3.4.4 without passing through symplectic geometry, Seiberg-Witten theory, nor J-curves.*

This might already have been achieved by Gabai. It seems plausible to the author that the method in which Eliashberg tried in [12] must work well in this situation.

Problem 5.1.2 *Does the same homotopy finiteness hold*

1) *for foliations without Reeb components?*

2) *for tight contact structures?*

By virtue of Theorem 3.4.1, 2) implies 1).

5.2 Anosov and projectively Anosov flows

We saw in §4 that an Anosov flow on a closed 3-manifold M induces a convex symplectic structure on $[-1,1] \times M$.

Problem 5.2.1 *Find an obstruction for M to the existence of convex symplectic structure on $W^4 = [-1,1] \times M$.*

McDuff's result in [46] tells us that M must not be spherical. Do there exist further obstructions?

If we assume further more that the symplectic structure on $W^4 = [-1,1] \times M$ comes from a linear perturbation from a foliation in the same way as in §4.1, we know that $\pi_1(M)$ has an exponential growth.

$\mathbb{P}\mathbf{A}$ flows in general are still in a chaotic situation. However, likely in the case of Anosov flows, if we assume the differentiability on the weak stable and unstable splitting, the situation is not bad as introduced in §4.3.

Problem 5.2.2 *In such a situation, is the foliation Anosov?*

Problem 5.2.3 *Are regular $\mathbb{P}\mathbf{A}$ flows without compact leaf in its weak stable foliation actually Anosov?*

Problem 5.2.4 *Are all regular projectively Anosov flow with compact leaves already listed on Noda's table? i.e., obtained by the propeller construction on T^3 or on solv-manifold?*

If these classification problems are affirmatively solved, the next (Conjecture 4.3.3) follows.

Problem 5.2.5 *Are contact structures associated with regular $\mathbb{P}\mathbf{A}$ flows always tight?*

For general $\mathbb{P}\mathbf{A}$ flows, so far not much is known.

Problem 5.2.6 *For a given 3-manifold M, which homotopy class of plane fields on M can be realized as a bi-contact structure? Especially how about in the case of S^3 ?*

The following problem, which is a bit reduced from above, is also natural to ask.

Problem 5.2.7 *Determine the homotopy classes of plane fields which can be realized as a transverse pair of foliations. How about for $S^3 H$.*

On S^3, so far only two homotopy classes (in some sense, minimal ones) are known to contain bi-contact structures.

Coming back to genuine Anosov flows, we have very vague but somewhat new problems as follows.

Problem 5.2.8 *In the convex symplectic structure associated with an Anosov flow, the weak unstable leaves are symplectic and the weak stable ones are Lagrangian. If we flip the direction of the Anosov flow, stable and unstable, i.e., symplectic and Lagrangian, are replaced with each other. Does this phenomenon have something to do with mirror symmetry?*

Problem 5.2.9 *Apply the convex symplectic structure associated with an Anosov flow to show the nonexistence of Anosov flows on certain 3-manifolds.*

5.3 2-dimensional foliations on 4-manifolds

As we have seen, the study of 3-dimensional contact structures is naturally related to the study of 4-dimensional symplectic manifolds or complex manifolds. It might be not too crazy to look for some relations between foliations and 4-manifolds. For example, if we apply the symplectic Thom conjecture to the weak filling constructed in §3.3, what we get is Thurston's inequality for taut foliations. The following problems arose from these circumstances.

Problem 5.3.1 *Find a topological obstructions or sufficient conditions for 2-dimensional foliation $\mathcal{F}$ on a closed oriented 4-manifold W to be a symplectic foliation, i.e., such there there exists a symplectic structure on W for which each leaf is symplectic submanifold.*

Symplectic foliations are minimal in the differential geometric sense. Therefore we want a stronger criterion than Sullivan's one for minimal foliations.

Symplectic foliations satisfies the same inequality as Thurston's one for foliations on 3-manifolds. This is due to so called the symplectic Thom conjecture ([57]).

Problem 5.3.2 *Do minimal 2-dimensional foliations on 4-manifolds satisfy Thurston's inequality?*

Problem 5.3.3 *Does there exist a (topological) condition for 2-dimensional foliations on 4-manifolds which interpolates between symplectic foliations and Thurston's inequality?*

These questions are tempting us to ask the following questions.

Problem 5.3.4 *Does there exist an analogue of Reeb component in the theory of 2-dimensional foliations on 4-manifolds?*

Problem 5.3.5 *Does there exist an analogue of Thurston's thesis 3.1.6 for S^2-bundles over closed surfaces?*

The last problem seems to have some intrinsic relation with problems on symplectic foliations raised above.

Acknowledgments

This article is based on the previous note distributed in the conference, which was translated from a text prepared in Japanese. Several people did not grudge so much effort to translate it into English. For this collaboration, the author would like to express his gratitude to Atsushi Sato, Hiroyuki Minakawa, Norikazu Hashiguchi, Ryoji Kasagawa, Hiroki Kodama, and Yasuharu Nakae.

Every figure in this article is taken from the booklet entitled "3-dimensional contact topology" by the same author which is mentioned above and is written in Japanese, and was published from the Mathematical Society of Japan.

References

1. S. Altschuler, *A geometric heat flow for one-forms on 3-dimensional manifolds*, Illinois J. Math., **39** (1995), 98–118.
2. D.V. Anosov, *Geodesic flows on closed Riemannian manifolds with negative curvature*, (English Transl.), Proc. Steklov Inst. Math., A. M. S., 1969.
3. V.I. Arnol'd, Mathematical methods of classical mechanics, Springer-Verlag, 1989.
4. V.I. Arnol'd and A. B. Givental', *Symplectic Geometry*, Dynamical Systems IV, Encyclopaedia of Mathematical Sciences, **4**, Springer-Verlag, 1990.
5. D.E. Barrett and T. Inaba, *On the topology of compact smooth three-dimensional Levi-flat hypersurfaces*, J. Geometric Analysis, **2**-6 (1992), 489–497.
6. D. Bennequin, *Entrelacements et équations de Pfaff*, Astérisque, **107-108** (1983), 83–161.
7. D. Bennequin, *Topologie symplectique, convexité holomorphe, et structures de contact, d'après Y.Eliashberg, D.McDuff, et al*, Séminaire BOURBAKI, n^o **725** (1989–90).
8. A. Candel and L. Conlon, Foliations I, Amer. Math. Soc., Graduate Studies in Mathematics, **23**, 1999.
9. Y. Eliashberg, *Classification of overtwisted contact structures on three manifolds*, Invent. Math., **98**-3 (1989), 623–637.
10. Y. Eliashberg, *Topological characterization of Stein manifolds of dimension > 2*, International J. Math., **1** (1990), 19–46.
11. Y. Eliashberg, *Filling by holomorphic discs and its applications*,in Geometry of Low-Dimensional Manifolds: 2, London Math. Soc. Lect. Note ser., **151**, Cambridge Univ. Press, 1990, 45–67.
12. Y. Eliashberg, *Contact 3-manifolds twenty years since J.Martinet's work*, Ann. Inst. Fourier, Grenoble, **42**-1-2 (1991), 165–192.
13. Y. Eliashberg, *Legendrian and transversal knots in tight contact 3-manifolds*, in Topological Method in Modern Mathematics, A Symposium in Honor of John Milnor's 60th Birthday, Publish or Perish Inc.,(1993, 171–195.
14. Y. Eliashberg and M. Gromov, *Convex symplectic manifolds*, Proc.

Sympo. Pure Math. A.M.S., **52**-2 (1991), 135–162.
15. Y. Eliashberg and W. Thurston, Confoliations, Amer. Math. Soc., University Lecture Series **13**, 1998.
16. J.B. Etnyre and K. Honda, *On the non-existence of tight contact structures*, Univ. Georgia Math. Preprint Ser., **8**-6 (2000).
17. J. Franks and R. F. Williams, *Anormalous Anosov flows*, in Global theory of dynamical systems, *ed.* by Z. Nitecki & C. Robinson, Springer Lecture Notes in Mathematics, **819**, 1980, 158–174.
18. K. Fukaya, Symplectic geometry, in Japanese, Iwanami Shoten, 1999.
19. K. Fukaya and K. Ono, *Arnold Conjecture and Gromov-Witten invariant*, Topology, **38** (1999), 933–1048.
20. D. Gabai, *Foliations and the topology of 3-manifolds*, J. Diff. Geom., **18** (1983), 445–503.
21. É. Ghys, *Déformations de flots d'Anosov et de groupes fuchsiens*, Ann. l'Inst. Fourier, **42**-1 & 2 (1992), 209–247. See also in Rigidité difféneréntiable des groupes fuchsiens, I.H.E.S. Publ. Math., **78** 1993.
22. E. Giroux, *Convexité en topologie de contact*, Comment. Math. Helvetici, **66** (1991), 637–677.
23. E. Giroux, *Topologie de contact en dimension 3, autour des travaux de Yakov Eliashberg*, Séminaire BOURBAKI, n^o **760** (1989–90).
24. E. Giroux, *Une infinité de structure de contact tendues sur une infinité de variétés*, preprint, (1998).
25. E. Giroux, *Structure de contact sur les variétés fibrés en cercles au-dessus d'une surface*, preprint, (1999), math.GT/9911235.
26. C. Godbillon, Feuilletages, Progress in Mathematics, **98**, Birkhäuser, 1991.
27. J.W. Gray, *Some global aspects of contact structures*, Ann. Math., **69** (1959), 421–450.
28. M. Gromov, *Pseudo-holomorphic curves in symplectic manifolds*, Invent. Math., **82** (1985), 307–347.
29. M. Handel and W.P. Thurston, *Anosov flows on new 3-manifolds*, Invent. Math., **59** (1980), 95–103.
30. D. Hardorp, All compact orientable three manifolds admit total foliations, Memoirs Amer. Math. Soc., **233**, 1980.
31. M. Hirsch, C. Pugh, and M. Shub, Invariant manifolds, Lecture Notes in Math., **583**, Springer Verlag, 1977.
32. H. Hofer, *Pseudoholomorphic curves in symplectizations with applications to the Weinstein conjecture in dimension three*, Invent. Math., **114** (1993), 515–563.
33. K. Honda, *On the classification of the tight contact structures* I: *Lens*

spaces, solid tori and $T^2 \times I$, II, *Torus bundles which fiber over the circle*, preprints, (1999).
34. T. Inaba and N. Tsuchiya, *Expansive foliations*, Hokkaido Math. J., **21** (1992), 39–49.
35. Y. Kanda, *On 3-dimensional contact topology*, in Japanese, 43rd Topology Symposium, (1996), 88–102.
36. Y. Kanda, *The classification of tight contact structures on the 3-torus*, Comm. Anal. Geom., **5**-3 (1997), 413–438.
37. A. Katok and B. Hasselblatt, Introduction to the modern theory of dynamical systems, Cambridge University Press, 1995.
38. P.B. Kronheimer and T.S. Mrowka, *Monopoles and contact structures*, Invent. Math., **130** (1997), 209–255.
39. F. Laudenbach, *Orbites périodiques et courbes pseudo-holomorphes, application à la conjecture de Weinstein en dimension 3, d'après H.Hofer, et al*, Séminaire BOURBAKI, n^o **786** (1993–94).
40. G. Levitt, *Feuilletages des variétés de dimension 3 qui sont des fibrés en cercles*, Comment. Math. Helvetici, **53** (1978), 572–594.
41. R. Lutz, Sur quelques propriétés des formes différentielles en dimension 3, Thèse, Strasbourg, 1971.
42. R. Lutz, *Structures de contact sur les fibrés principaux en cercle de dimension 3*, Ann. l'Inst. Fourier, **27**-3 (1977), 1–15.
43. L.A. Lyusternik and A.I. Fet, *Variational problems on closed manifolds*, Dokl. Akad. Nauk SSSR (NS), **81** (1951), 17–18, (in Russian). See also W. Klingenberg, Lectures on closed geodesics, Grundlehren der mathematischen Wissenschaften, **230**, Springer Verlag, 1978.
44. J. Martinet, *Formes de contact sur les variétés de dimension 3*, in Lecture Notes in Mathematics **209**, Springer Verlag, 1971, 142–163.
45. S. Matsumoto and T. Tsuboi, *Transverse intersections of foliations in three-manifolds*, preprint, (1999).
46. D. McDuff, *Symplectic manifolds with contact type boundaries*, Invent. Math., **103** (1991), 651–671.
47. D. McDuff and D. Salamon, Introduction to symplectic topology, Oxford Math. Monographs, Oxford U. P., 1995.
48. J. Milnor, *On the existence of a connection with curvature zero*, Comment. Math. Helvetici, **32** (1958), 215–223.
49. Y. Mitsumatsu, *Anosov flows and non-Stein symplectic manifolds*, Ann. Inst. Fourier, **45**-5 (1995), 1407–1421.
50. Y. Mitsumatsu, *Survey of recent 20 years of contact 3-manifolds, Part 1*, in Japanese, Proceedings of the symposium "Links of knot theory and various mathematics", Osaka City Univ., 1996, ed. by M. Sakuma,

(1997), 466–478.

51. Y. Mitsumatsu, *Lutz twist and homotopy class of plane field on 3-manifolds*, preprint in preparation.
52. Y. Mitsumatsu, *Projectively Anosov flows and bi-contact structures on 3-manifolds*, preprint in preparation.
53. T. Noda, *Projectively Anosov flows on T^3 whose stable and unstable foliations are differentiable*, preprint, (1998).
54. T. Noda and T. Tsuboi, *Regular projectively Anosov flows on solv-manifolds*, preprint, (1999).
55. K. Ono, *Existence problem of periodic solutions of Hamiltonian systems and J-holomorphic curves*, in Japanese, Abstracts of the symposium "Contact geometry and related topics", Hokkaido Univ., 1996, (1996), 1–22.
56. K. Ono, Symplectic topology and J-holomorphic curves, in Japanese, Suugaku, **51**-4, 1999.
57. P. Ozsváth and Z. Szabó, *The symplectic Thom conjecture*, Ann. Math., **151**-1 (2000), 93–124.
58. Th. Paternell, Pseudoconvexity, the Levi problem and vanishing theorems, Encyclopaedia of Mathematical Sciences, **74**, Several complex Variables VII, Chapter VIII, Springer-Verlag, Berlin, 1994.
59. J. Plante and W. Thurston, *Anosov flows and the fundamental group*, Topology, **11** (1972), 147–150.
60. R. Sacksteder, *Foliations and pseudogroups*, Amer. J. Math., **87** (1965), 79–102.
61. A. Sato and T. Tsuboi, *Contact structures of closed 3-manifolds fibred by the circle*, Mem. Inst. Sci. Tech. Meiji Univ., **33** (1994), 41–46.
62. D. Sullivan, *A homological characterization of foliations consisting of minimal surfaces*, Comment. Math. Helvetici, **54** (1979), 218–223.
63. I. Tamura, Topology of Foliations; an Introduction, Transl. by Kiki Hudson, Transl. Math. Monographs, **97**, Amer. Math. Soc., 1992.
64. I. Tamura and A. Sato, *On transverse foliations*, Publ. Math. I.H.E.S., **54** (1981), 205–235.
65. C.H. Taubes, *The Seiberg-Witten invariants and symplectic forms*, Math. Res. Lett., **1**-6 (1994), 809–822.
66. C.H. Taubes, *The Seiberg-Witten and Gromov invariants*, Math. Res. Lett., **2**-2 (1995), 221–238.
67. C.H. Taubes SW $\Rightarrow$ Gr:*from the Seiberg-Witten equations to pseudo-holomorphic curves*, J. Amer. Math. Soc., **9**-3 (1996), 845–918.
68. W.P. Thurston, Foliations of manifolds which are circle bundles, Thesis, University of California, Berkeley, 1972.

69. W.P. Thurston, *Norm for homology of 3-manifolds*, Memoirs Amer. Math. Soc., **339** (1986), 99–130.
70. W. Thurston and E. Winkelnkemper, *On the existence of contact forms*, Proc. Amer. Math. Soc., **52** (1975), 345–347.
71. I. Ustilovsky, PhD. thesis, Stanford University, 2000.
72. A. Weinstein, *On the hypotheses of Rabinowtz's periodic orbit theorems*, J. Diff. Eq., **33** (1979), 353–358.
73. J.W. Wood, *Foliations on 3-manifolds*, Ann. Math., (**2**), **89** (1991), 336–358.
74. K. Yano, Dynamical Systems 2, in Japanese, **32**, Iwanami-Shoten, 1998.

Received November 14, 2000.

Proceedings of
FOLIATIONS: GEOMETRY AND DYNAMICS
held in Warsaw, May 29–June 9, 2000
ed. by Paweł WALCZAK *et al.*
World Scientific, Singapore, 2002
pp. 127–155

OPERATOR ALGEBRAS AND THE INDEX THEOREM ON FOLIATED MANIFOLDS

HITOSHI MORIYOSHI
Department of Mathematics, Keio University,
Yokohama, 223-8522, Japan,
e-mail: moriyosi@math.keio.ac.jp

Introduction

We shall deliver a series of three lectures on Operator algebras and the Index Theorem on foliated manifolds. The topics dealt with in each lecture are as follows:

- **Topology of the leaf space $M/\mathcal{F}$ of foliated manifolds $(M, \mathcal{F})$.** We discuss topological method to study $M/\mathcal{F}$ such as: holonomy groupoid; groupoid for transformation groups; group C^*-algebras; the Morita equivalence; the K-theory; von Neumann algebras; the modular theory.

- **Analysis on the leaf space $M/\mathcal{F}$.** We develop Differential Geometry on the leaf space $M/\mathcal{F}$. We shall discuss notions that exploit analysis method such as: the index of longitudinal operators in K-group; cyclic cohomology; the pairing between K-theory and cyclic theory; invariant transverse measures and the Ruelle-Sullivan current; the Connes index theorem on foliated manifolds; the index theorem for Toeplitz operators on foliated manifolds;

- **Towards noncommutative geometry on foliated manifolds.** We discuss a few subjects that will be relevant to the future development of Noncommutative Geometry on foliated manifolds such as: the Godbillon-Vey class and the index theorem; the type II spectral flow and the Godbillon-Vey class; singular foliations on Poisson manifolds;

Woronowicz's quantum 3-sphere; the Atiyah-Patodi-Singer index theorem for singular foliations.

1 Topology of the leaf space $\mathbf{M}/\mathcal{F}$

1.1 Groupoid C^*-algebras

Let M be a smooth manifold of dimension $p+q$ and $\mathcal{F}$ a foliation of codimension q on M. To study the geometry of foliated manifolds $(M, \mathcal{F})$, the topology of the leaf spaces $M/\mathcal{F}$ would be the first concerns to us. In general, however, the spaces are not equipped with well-behaved topology; they are often non-Hausdorff spaces and it makes difficult to study them from topological viewpoint. To overcome the difficulties we shall introduce a holonomy groupoid, which in some sense resolves the topology of the leaf space $M/\mathcal{F}$.

Let L denote a leaf of $(M, \mathcal{F})$, and $\gamma,\ \gamma' : [0,1] \to L$ be continuous curves on L. We say that γ is equivalent to γ' if they have the same end points and induce the same holonomy. The holonomy groupoid G is then defined to be the set of equivalence classes of curves along leaves:

$$G = \{\gamma \mid \gamma \text{ is a curve along leaves }\}/\sim .$$

Put

$$s, r : G \to M, \qquad s(\gamma) = \gamma(0), \quad r(\gamma) = \gamma(1).$$

Then G is equipped with a structure of topological groupoid with s, r the source and target maps. The product is simply the composition of curves, and the inverse element is the same curve with the opposite parametrization:

$$\underset{x}{\bullet} \xleftarrow{\ \gamma\ } \underset{y}{\bullet} \xleftarrow{\ \gamma'\ } \underset{z}{\bullet} \qquad\qquad \underset{x}{\bullet} \xrightarrow{\ \gamma^{-1}\ } \underset{y}{\bullet}.$$

We can also verify that the holonomy groupoid G admits the system of coordinate neighbourhoods inherited from the foliation structure on M. Then G is a manifold of dimension $2p+q$; however G could be a non-Hausdorff space. Namely, the holonomy groupoid is a manifold but a non-Hausdorff space in general case.

Example 1.1 (Kronecker foliation) Let $T^2 = \mathbb{R}/\mathbb{Z} \times \mathbb{R}/\mathbb{Z}$ be the 2-dimensional torus and θ an irrational number. Take a vector field $X = \partial/\partial x + \theta\partial/\partial y$ on T^2. The Kronecker foliation $\mathcal{F}_\theta$ on T^2 is then defined by the flow of X. Here we consider the Kronecker foliation as a foliated bundle: $(T^2, \mathcal{F}_\theta)$ is diffeomorphic to the following space

$$M = (\mathbb{R} \times \mathbb{R}/\mathbb{Z})/\mathbb{Z}, \quad (x,t) \sim (x+1, t+\theta)$$

with leaves $\mathbb{R} \times \{t\}$ $(t \in \mathbb{R}/\mathbb{Z})$ in M. Since $\theta \in \mathbb{R} \setminus \mathbb{Q}$, each leave is diffeomorphic to $\mathbb{R}$ and curves along the leaves are uniquely determined by the end points up to homotopy. Thus the holonomy groupoid is given by the following space:

$$G = (\mathbb{R} \times \mathbb{R} \times \mathbb{R}/\mathbb{Z})/\mathbb{Z}, \quad (x, y, t) \sim (x + 1, y + 1, t + \theta).$$

Here the source and target maps $s,\ r : G \to M$ are:

$$s(x, y, t) = (y, t), \quad r(x, y, t) = (x, t).$$

On a foliated manifold $(M, \mathcal{F})$ we can choose a Riemannian metric along the leaves and obtain the volume form on each leaf. Given a point $x \in M$, we denote by L_x the leaf that contains x and set $\widetilde{L}_x = \{\gamma \in G \mid s(\gamma) = x\}$. It is called the holonomy covering space of L_x. In fact the target map r yields the covering projection $r : \widetilde{L}_x \to L_x$. Then we equip $\widetilde{L}_x$ with the volume form induced from L_x, and denote by $d\mu = \{d\mu_x\}_{x \in M}$ the family of resulting measures on $\{\widetilde{L}_x\}_{x \in M}$. Note that the family $d\mu$ is equivariant with respect to the right translation of an element $\gamma' \in G$: $\widetilde{L}_x \to \widetilde{L}_y$ with $s(\gamma') = y$, $r(\gamma') = x$. Let $C_c(G)$ denote the space of continuous functions on G with compact support. Here we make an observation. In general G is not a Hausdorff space. Thus some points in G are not distinguished by functions in $C_c(G)$. However each point admits a neighbourhood that is homeomorphic to an open set U in $\mathbb{R}^{2p+q}$. Let $C_c(U)$ denote the space of continuous functions on U with compact support. There exist a natural inclusion from $C_c(U)$ to the space $F(G)$ of functions on G. When G is not a Hausdorff space, we shall employ the subspace in $F(G)$ generated by $C_c(G)$ and the image of $C_c(U)$ assigned from non-Hausdorff points. The subspace is still denoted by $C_c(G)$ in the sequel.

We then introduce a convolution product on $C_c(G)$. Given $\varphi,\ \psi \in C_c(G)$, the convolution product is given by

$$(\varphi\psi)(\alpha) = \int_{\widetilde{L}_x} \varphi(\alpha\beta^{-1})\psi(\beta)d\mu_x(\beta) \qquad (\alpha,\ \beta \in \widetilde{L}_x).$$

We also define the involution * on $C_c(G)$ such that:

$$\varphi^*(\gamma) = \bar{\varphi}(\gamma^{-1}).$$

Let $\mathcal{H}_x$ denote the Hilbert space of L^2-functions on $\widetilde{L}_x$. Now we define a $*$-representation π_x on $\mathcal{H}_x$ such that:

$$[\pi_x(\varphi)\xi](\alpha) = \int_{\widetilde{L}_x} \varphi(\alpha\beta^{-1})\xi(\beta)d\mu_x(\beta) \qquad (\xi \in \mathcal{H}_x,\ \alpha, \beta \in \widetilde{L}_x).$$

The $*$-norm on $C_c(G)$ is then given by

$$\|\varphi\| = \sup_{x \in M} \|\pi_x(\varphi)\|,$$

where $\|\pi_x(\varphi)\|$ denotes the operator norm of $\pi_x(\varphi)$ on $\mathcal{H}_x$.

Definition 1.2 Let $(M, \mathcal{F})$ de a foliated manifold. The foliation C^*-algebra $C^*(M, \mathcal{F})$ is the C^*-completion of $C_c(G)$ with respect to the $*$-norm described above.

Remark. If we employ a family of half-densities along the leaves instead of $d\mu = \{d\mu_x\}_{x \in M}$, we obtain a C^*-algebra that is isomorphic to $C^*(M, \mathcal{F})$. Thus it yields a description of $C^*(M, \mathcal{F})$ which is free from the choice of a Riemannian metric along the leaves; see Baum-Connes [9].

Next we shall explain the crossed product algebra since for a foliated bundle it is closely related to the foliation C^*-algebra. Let Γ be a discrete group that acts on a locally compact Hausdorff space X from the right. Then a topological groupoid G associated to the Γ-action on X is defined as follows. Put $G = X \times \Gamma$, and set the source and target maps $s,\ r : G \to X$ to be

$$s(x, g) = xg, \quad r(x, g) = x.$$

We then define a product on G such that

$$(x, g)(y, h) = (x, gh), \quad (x, g),\ (y, h) \in G$$

if $s(x, g) = r(y, h)$. Furthermore we define the inverse of $(x, g) \in G$ by

$$(x, g)^{-1} = (xg, g^{-1}).$$

We denote the above groupoid by $X \rtimes \Gamma$ henceforth.

We also associate a C^*-algebra called a crossed product to the groupoid $X \rtimes \Gamma$. The construction is similar to that of the foliation C^*-algebra. In general, given a topological groupoid G with a system of Haar measures, there exists a groupoid C^*-algebra $C^*(G)$. Thus we can construct the crossed product as a groupoid C^*-algebra $C^*(X \rtimes \Gamma)$. We here describe the construction as follows.

We introduce a formal unitary element U_g corresponding to $g \in \Gamma$. The product is given by the formula $U_g U_h = U_{gh}$ $(g, h \in G)$ as usual. Let $C_c(X)$ be the space of continuous functions on X with compact support. We then denote by $C_c(X, \Gamma)$ the space of elements represented as a finite sum $\sum_{g \in G} a_g U_g$ $(a_g \in C_c(X))$. Given elements $a = \sum_{g \in G} a_g U_g$, $b =$

$\sum_{h\in G} b_h U_h$, the addition and multiplication are defined by

$$a + b = \sum_g (a_g + b_g) U_g,$$

$$ab = \sum_{g,h} a_g g(b_h) U_{gh}.$$

Here $g(b_h)$ denotes the function $b_h(xg)$, $x \in X$. Moreover we define the involution * on $C_c(X, \Gamma)$ such that:

$$\left(\sum_g a_g U_g \right)^* = \sum_g g^{-1}(\bar{a}_g) U_{g^{-1}}.$$

Put $G_x = \{(y, g) \in X \times G \mid yg = x\}$ for $x \in X$, and denote by $\mathcal{H}_x$ the Hilbert space of ℓ^2-sequences in G_x. Now we define a $*$-representation π_x on $\mathcal{H}_x$ such that:

$$(\pi_x(a)\xi)(y, h) = \sum_g a_g(y) \xi(yg, g^{-1}h), \quad (y, h) \in G_x.$$

The $*$-norm on $C_c(X, \Gamma)$ is then given by

$$\|a\| = \sup_{x\in X} \|\pi_x(a)\|,$$

where $\|\pi_x(a)\|$ denotes the operator norm of $\pi_x(a)$ on $\mathcal{H}_x$. The crossed product $C_o(X) \rtimes G$ is defined to be the C^*-algebra obtained by the C^*-completion of $C_c(X, \Gamma)$ with respect to the $*$-norm above.

We go back to the example of Kronecker foliation $(T^2, \mathcal{F}_\theta)$. Let N be the fibre $\mathbb{R}/\mathbb{Z}$ of the foliated bundle and Γ denote the global holonomy group acting on N. In this case the rotation by θ generates the holonomy group Γ. Here there exist two C^*-algebras $C^*(T^2, \mathcal{F}_\theta)$ and $C(N) \rtimes \Gamma$ arising from the same foliated bundle. It is then natural to ask the relationship between them. In fact it turns out that $C(N) \rtimes \Gamma$ is Morita equivalent to the foliation C^*-algebra $C^*(T^2, \mathcal{F}_\theta)$.

Definition 1.3 Let $\mathcal{A}$ and $\mathcal{B}$ be C^*-algebras. Let $\mathcal{K}$ denote the C^*-algebra of compact operators. We say that $\mathcal{A}$ is Morita equivalent to $\mathcal{B}$ if $\mathcal{A} \otimes \mathcal{K}$ is isomorphic to $\mathcal{B} \otimes \mathcal{K}$.

Remark. The above isomorphism is called a stable isomorphism. The notion of Morita equivalence of C^*-algebras is introduced by M. Rieffel, where it is called the strongly Morita equivalence. The original definition is concerned with the Morita equivalence bimodule; see Rieffel [38, 39]. However, it is known that two C^*-algebras are stable isomorphic if and only if they are strongly Morita equivalent. Therefore, we introduce the Morita equivalence here in the above manner.

The Morita equivalence plays an important role in the study of foliation C^*-algebras and topology of the leaf space. We shall explain it with the following example:

Example 1.4 Let M be a fibre bundle on N. Then M is foliated by the fibres and admits a foliation $\mathcal{F}$. In this case the foliation C^*-algebra $C^*(M, \mathcal{F})$ is isomorphic to $C_o(N) \otimes \mathcal{K}$, where $C_o(N)$ is the algebra of continuous functions on N vanishing at infinity. Thus $C^*(M, \mathcal{F})$ is Morita equivalent to $C_o(N)$. Here we note that the leaf space $M/\mathcal{F}$ is a well-behaved topological space. Actually it is homeomorphic to N. Generally it is difficult to study the topology of $M/\mathcal{F}$ in a direct way. However this example exhibits that the foliation C^*-algebra is considered as a substitute of the function algebra of $M/\mathcal{F}$ up to the Morita equivalence.

Example 1.5 (Hilsum and Skandalis, [27]) Let $(M, \mathcal{F})$ be a foliated manifold and denote by G the holonomy groupoid. Then we choose a open transversal N that intersects with all leaves of $\mathcal{F}$. Put

$$G_N^N = \{\gamma \mid s(\gamma),\ r(\gamma) \in N\}.$$

Then G_N^N is also a topological groupoid with the same source and target maps. It then turns out that the foliation C^*-algebra $C^*(M, \mathcal{F})$ is Morita equivalent to the groupoid C^*-algebra $C^*(G_N^N)$. The equivalence is realized quite geometrically by using free groupoid actions of G and G_N^N on $G^N = \{\gamma \mid s(\gamma) \in N\}$.

Here we sketch the idea by exploiting the following simple example. Let G be a Lie group, and take closed subgroups H and K of G. Let H and K act on G from the left and right respectively. We then obtain topological groupoids $H \ltimes G/K$ and $H\backslash G \rtimes K$. Exploiting the free action of H and K on G, we can construct the Morita equivalence bimodule over $C_o(G/K) \rtimes H$ and $C_o(H\backslash G) \rtimes K$. The bimodule is basically constructed from $C_c(G)$ with the action of H and K. It then follows that $C_o(G/K) \rtimes H$ is the to $C_o(H\backslash G) \rtimes K$.

Here we observe that the classifying space $B(H \ltimes G/K)$ is the universal space for a K-bundle whose G-extension admits an H-reduction. That is nothing but a G-bundle with both of H and K-reductions. In the same way $B(H\backslash G \rtimes K)$ also classifies a G-bundle with both of H and K-reductions. Therefore the spaces $B(H \ltimes G/K)$ and $B(H\backslash G \rtimes K)$ are homotopy equivalent to each other. That explains a geometrical significance of the Morita equivalence for topological groupoids.

1.2 K-theory for Banach algebras

As we observed that the leaf space $M/\mathcal{F}$ is not a well-behaved topological space in general. But there is a way out to study the topology of $M/\mathcal{F}$ exploiting the foliation C^*-algebras. In this section we shall explain the tool to investigate the topology of the leaf space $M/\mathcal{F}$ working on the C^*-algebras, that is K-theory.

Let $\mathcal{A}$ be a Banach algebra with unit and $M_k(\mathcal{A})$ denote the algebra of matrices with components in $\mathcal{A}$. There exists a natural inclusion into $M_{k+n}(\mathcal{A})$ such that:

$$a \longmapsto \begin{pmatrix} a & 0 \\ 0 & 0 \end{pmatrix}.$$

We then denote by $M_\infty(\mathcal{A})$ the direct limit of $M_k(\mathcal{A})$ with the above inclusions. Let $P(\mathcal{A})$ denote the set of idempotents $e \in M_\infty(\mathcal{A})$, i.e., $e^2 = e$. Then $P(\mathcal{A})$ admits the direct-sum operation given by

$$e_0 \oplus e_1 = \begin{pmatrix} e_0 & 0 \\ 0 & e_1 \end{pmatrix}.$$

We then introduce an equivalence relation on $P(\mathcal{A})$. Given $e_0,\ e_1 \in P(\mathcal{A})$, they are equivalent to each other if e_0 is connected with e_1 by a continuous path in $P(\mathcal{A})$. Note that direct sum is well defined on the quotient space $P(\mathcal{A})/\sim$.

Let $GL_k(\mathcal{A})$ denote the group of all invertible elements in $M_k(\mathcal{A})$. here exists a natural inclusion into $GL_{k+n}(\mathcal{A})$ such that:

$$g \longmapsto \begin{pmatrix} g & 0 \\ 0 & 1 \end{pmatrix}.$$

Here we denote the identity matrix by 1.

Let $GL_\infty(\mathcal{A})$ be the direct limit of $GL_k(\mathcal{A})$ with the above inclusions. We then introduce the direct-sum operation for $GL_\infty(\mathcal{A})$ by

$$g_0 \oplus g_1 = \begin{pmatrix} g_0 & 0 \\ 0 & g_1 \end{pmatrix}.$$

Given $g_0,\ g_1 \in GL_\infty(\mathcal{A})$, they are equivalent to each other if g_0 is connected with g_1 by a continuous path in $GL_\infty(\mathcal{A})$. Note also that direct sum is well defined on the quotient space $GL_\infty(\mathcal{A})/\sim$.

Definition 1.6 Let $\mathcal{A}$ be a C^*-algebra with unit.

1) The group $K_0(\mathcal{A})$ is the Grothendieck group of the semigroup $P(\mathcal{A})/\sim$. The addition is given by the direct-sum operation defined above.

2) The group $K_1(A)$ is the quotient space $GL_\infty(A)/\sim$. The addition is also given by the direct-sum operation defined above.

Remark. 1) Given C^*-algebras $\mathcal{A}$, $\mathcal{B}$ with unit and a homomorphism $\rho : \mathcal{A} \to \mathcal{B}$, it naturally defines a induced homomorphism $\rho_* : K_i(\mathcal{A}) \to K_i(\mathcal{B})$ $(i = 0, 1)$.

Let $C([0,1], \mathcal{A})$ be a continuous functions on the interval with values in $\mathcal{A}$. Given homomorphism ρ_0, $\rho_1 : \mathcal{A} \to \mathcal{B}$, ρ_0 is called homotope to ρ_1 if there exist a homomorphism $\rho : \mathcal{A} \otimes C[0,1] \to \mathcal{B}$ such that $\rho \circ \iota_i = \rho_i$ $(i = 0, 1)$. Here ι_0, $\iota_1 : C([0,1], \mathcal{A}) \to \mathcal{A}$ denote the restrictions to 0 and 1, respectively. Then the induced homomorphisms $(\rho_0)_*$, $(\rho_i)_*$ coincide with each other.

2) When $\mathcal{A}$ is not unital, we construct an algebra $\mathcal{A}^+$ by adjoining the unit. There exit a surjective homomorphism $\pi : \mathcal{A}^+ \to \mathcal{A}$. We then define the K-group $K_i(\mathcal{A})$ to be $\ker \pi_*[K_i(\mathcal{A}^+) \to K_i(\mathcal{A})]$.

Here we state a couple of important properties of the K-theory.

1) **The Morita equivalence** Let $\mathcal{A}$ and $\mathcal{B}$ be C^*-algebras. If $\mathcal{A}$ is Morita equivalent to $\mathcal{B}$, then $K_*(\mathcal{A})$ is naturally isomorphic to $K_*(\mathcal{B})$.

2) **The six-term exact sequence** Given a short exact sequence of C^*-algebras

$$0 \longrightarrow \mathcal{I} \xrightarrow[\iota]{} \mathcal{A} \xrightarrow[\pi]{} \mathcal{B} \longrightarrow 0,$$

there exists the six-term exact sequence

$$\begin{array}{ccccc} K_0(\mathcal{I}) & \xrightarrow{\iota_*} & K_0(\mathcal{A}) & \xrightarrow{\pi_*} & K_0(\mathcal{B}) \\ \partial \uparrow & & & & \downarrow \partial \\ K_1(\mathcal{B}) & \xleftarrow[\pi_*]{} & K_1(\mathcal{A}) & \xleftarrow[\iota_*]{} & K_1(\mathcal{I}) \end{array} .$$

3) **The Bott periodicity** Let $C_o(\mathbb{R}^n)$ be the algebra of continuous functions on $\mathbb{R}^n$ vanishing at infinity. For arbitrary C^*-algebra $\mathcal{A}$ the K-group $K_{i+n}(\mathcal{A})$ is isomorphic to $K_i(\mathcal{A} \otimes C_o(\mathbb{R}^n))$. Here we consider the degree of K-group modulo 2.

1.3 Relation to the topological K-theory

Let X be a compact Hausdorff space and $C(X)$ denote the C^*-algebra of continuous functions on X. We recall that the topological K-group $K^*(X)$ is defined as the Grothendieck group of isomorphism classes of complex vector bundles on X. Let E be a complex vector bundle on X. Since X is compact, we may assume that E is a subbundle of a trivial bundle $M \times \mathbb{C}^n$

for sufficiently large n. Chosen a metric on the trivial bundle, we can associate an idempotent $e \in M_n(C(X))$ to E taking the orthogonal projection onto E. Here we identify an element in $M_n(C(X))$ with a function on X with values in $M_n(\mathbb{C})$. Thus e is a family of the orthogonal projections onto the fibre E_x at each $x \in X$. We observe that the construction depends on the choice of the embedding into the trivial bundle. However, the corresponding class is independent of the choice when we pass over to the K-group. It is further known that the corresponding is an isomorphism between the K-group for $C(X)$ and the topological K-group.

Theorem 1.7 (Swan) *Let X be a compact Hausdorff space and $C(X)$ denote the algebra of continuous functions on X. There is a natural isomorphism between $K_*(C(X))$ and $K^*(X)$.*

The above theorem claims that $K_*(C(X))$ is identified with $K^*(M)$ and that the K-theory for C^*-algebras contains the theory of topological K-group. For foliated manifolds $(M, \mathcal{F})$, the topological K-group of the leaf space $M/\mathcal{F}$ is not defined directly. Instead we can exploit the K-group $K_*(C^*(M, \mathcal{F}))$ of C^*-algebra as the K-group for $M/\mathcal{F}$.

Let X be a locally compact Hausdorff space and G a compact Lie group that acts on X. Also in this case we can construct the crossed product $C_o(X) \rtimes G$ and consider the K-group of $C_o(X) \rtimes G$. On the other hand there exists a equivariant K-theory for the space with G-action. Due to P. Green [25], it is known that these theories are equivalent to each other:

Theorem 1.8 (P. Green) *Let X be a locally compact Hausdorff space and G a compact Lie group that acts on X. There exists a natural isomorphisms between $K_*(C_o(X) \rtimes G)$ and the equivariant K-group $K_G^*(X)$.*

Therefore the K-theory for C^*-algebras contains also the equivariant K-theory.

1.4 Foliation and the von Neumann algebras

In §1.1 we defined the foliation C^*-algebra for foliated manifold $(M, \mathcal{F})$. We can also define the von Neumann algebra out of the function algebra on the holonomy groupoid. Let $\mathcal{H} = (\mathcal{H}_x)_{x \in M}$ be the Hilbert bundle such that $\mathcal{H}_x = L^2(\widetilde{L}_x)$. We then take measurable sections $\xi = (\xi_x)$, $\eta = (\eta_x)$ of $\mathcal{H}$. Denote by L_x the leaf that contains $x \in M$. A family of operators $A = (A_L)_{L \in M/\mathcal{F}}$ is called a random operator if it is a measurable, that is, the function $M \ni x \mapsto \langle A_{L_x} \xi_x, \eta_x \rangle_x$ is Lebesque measurable for any ξ, η, where $\langle\ ,\ \rangle_x$ denotes the inner product on $\mathcal{H}_x$. Recall that a compactly supported continuous function φ on the holonomy groupoid G yields a family of operators $(\pi_x(\varphi))$ on $\mathcal{H}$. We can verify that it is a random operator also. Put

$$\|A\| = \operatorname*{ess.\,sup}_{x \in M} \|A_{L_x}\|,$$

where $\|A_{L_x}\|$ is the operator norm. The foliation von Neumann algebra $W^*(M, \mathcal{F})$ is then defined to be the weak completion of bounded random operators with respect to the above norm.

It is well known the von Neumann algebras are classified into three classes: type I, II and III. A von Neumann algebra $\mathcal{M}$ is called a factor if the center $Z(\mathcal{M})$ consists of the scalar multiplies of the identity operator. A factor plays an important role in the theory of von Neumann algebras. For instance it is known that von Neumann algebras on separable Hilbert spaces are decomposable into direct integrals of factors. With foliation von Neumann algebras $W^*(M, \mathcal{F})$ factors are characterized by quite geometrical properties.

Proposition 1.9 *Let $(M, \mathcal{F})$ be a foliated manifold. The foliation von Neumann algebra $W^*(M, \mathcal{F})$ is a factor if and only if all bounded measurable functions on the leaf space $M/\mathcal{F}$ are constant functions, that is, the foliation is ergodic.*

Any von Neumann algebra $\mathcal{M}$ is canonically decomposed to the direct sum $\mathcal{M}_I \oplus \mathcal{M}_{II} \oplus \mathcal{M}_{III}$ of von Neumann algebras, where $\mathcal{M}_I$, $\mathcal{M}_{II}$, $\mathcal{M}_{III}$ are of type I, II, III, respectively. The type classification is also translated into geometrical properties.

Theorem 1.10 *Let $(M, \mathcal{F})$ be a foliated manifold. The foliation von Neumann algebra $W^*(M, \mathcal{F})$ is of:*

1) Type I if and only if the leaf space $M/\mathcal{F}$ is isomorphic to the standard Borel measure space;

2) Type II if and only if there exists an invariant transverse measure and it is not of type I;

3) Type III if there exists no invariant transverse measure.

Example 1.11 The foliation von Neumann algebra of fibre foliation is of type I. Also with the Reeb foliation on T^2, it is of type I.

For the Kronecker foliation $\mathcal{F}_\theta$ ($\theta \in \mathbb{R} \setminus \mathbb{Q}$) on T^2 the foliation von Neumann algebra $W^*(T^2, \mathcal{F}_\theta)$ is of type II, and it is isomorphic to the unique hyperfinite factor of type II_∞.

Let X be a closed Riemann surface of genus ≥ 2 with hyperbolic metric. Let M denote the unit sphere bundle of TX. Then M admits a weakly stable foliation $\mathcal{F}$ of codimension 1 from the geodesic flow. Then the foliation von Neumann algebra $W^*(M, \mathcal{F})$ is of type III. In fact it is the unique hyperfinite factor of type III_1.

Given a von Neumann algebra $\mathcal{M}$, we set

$$\mathcal{M}_+ = \{x \in \mathcal{M} \mid x = y^*,\ y \in \mathcal{M}\}.$$

A mapping

$$w : \mathcal{M}_+ \to [0, +\infty]$$

is called a weight if it satisfies:

i) $w(x+y) = w(x) + w(y), \quad x, y \in \mathcal{M}_+;$

ii) $w(\lambda x) = \lambda w(x), \quad \lambda \geq 0.$

A weight τ is called a trace if it satisfies:

$$\tau(u^* x u) = \tau(x), \quad x \in \mathcal{M}_+$$

for any unitary element $u \in \mathcal{M}$.

When a foliated manifold $(M, \mathcal{F})$ admits a transverse invariant measure $d\nu$, we can verify that a trace τ on $W^*(M, \mathcal{F})$ is given by

$$\tau(\varphi) = \int_M \varphi|_M d\mu d\nu.$$

Here we consider $\varphi \in C_c(G)$ as a random operator as is explained previously. Also we identify a point $x \in M$ with the constant curve at x. Thus M is considered as a subset of the holonomy groupoid G and $\varphi|_M$ denotes the restriction to M.

Next we shall discuss the modular theory (the Tomita-Takesaki theory) of type III von Neumann algebras. When there is no invariant transverse measure on $(M, \mathcal{F})$, the foliation von Neumann algebra is of type III according to the above theorem. Given a type III von Neumann algebra $\mathcal{M}$ and a weight w on $\mathcal{M}$, it is known that there exists 1-parameter family of automorphisms on $\mathcal{M}$ called the modular automorphisms. That is a remarkable result with type III von Neumann algebras, which is a part of the Tomita-Takesaki theory. Here we do not discuss the deep result for general type III von Neumann algebras; however, with the foliated von Neumann algebras the theory can be understood rather clearly.

Let us explain the construction of the modular automorphisms for the foliated bundles for the sake of simplicity. It is not difficult to develop it to general foliations. Let $(M, \mathcal{F})$ be a foliated bundle over X with the fibre N. Thus there is a holonomy homomorphism $\rho : \pi_1(X) \to \mathrm{Diff}(N)$, and M is isomorphic to $\widetilde{X} \underset{\rho}{\times} N$. Here $\widetilde{X}$ is the universal covering space. Recall

Example 1.1 of the Kronecker foliation. The holonomy groupoid is then given by

$$G = (\widetilde{X} \times \widetilde{X} \times N)/\sim,$$

where $(x, y, t) \sim (xg^{-1}, yg^{-1}, \rho(g)t) \in \widetilde{X} \times \widetilde{X} \times N,\ g \in \pi_1(X)$. Then we choose a volume form $d\mu$ induced from X and an arbitrary volume form $d\nu$ on N. When the foliation von Neumann algebra is of type III, there is no invariant volume form (measure) on N with the action of ρ. On the other hand we choose a volume form on M and take the pullback to $\widetilde{X} \times N$ which we denoted by $d\lambda$. We then obtain a function

$$\psi = \frac{d\mu \times d\nu}{d\lambda}$$

on $\widetilde{X} \times N$. It is the Radon-Nykodim derivative of two measures. We then define the automorphisms $(\sigma_t)_{t\in\mathbb{R}}$ by

$$\sigma_t(\varphi) = \psi^{it}\varphi\psi^{-it}$$

for $\varphi \in C_c(G)$. Although ψ is not an element of $W^*(M, \mathcal{F})$, the element $\psi^{it}\varphi\psi^{-it}$ belongs to $W^*(M, \mathcal{F})$, and we can verify that the above formula defines the modular automorphisms on $W^*(M, \mathcal{F})$. In §3 we shall discuss the close relationship between the modular automorphisms and the Godbillon-Vey class.

Then we can consider the crossed product $W^*(M, \mathcal{F}) \rtimes_\sigma \mathbb{R}$. The Tomita-Takesaki theory tells us the property of the crossed product. It follows that $W^*(M, \mathcal{F}) \rtimes_\sigma \mathbb{R}$ is a type II von Neumann algebra and hence there exist a trace τ on it. On the other hand it is easy to verify that there is the dual action $\hat{\sigma}$ of $\mathbb{R}$ on $W^*(M, \mathcal{F}) \rtimes_\sigma \mathbb{R}$. The trace τ is not invariant with $\hat{\sigma}$, however, it satisfies the relative invariance under $\hat{\sigma}$:

$$\tau \circ \hat{\sigma}_t = e^{-t}\tau \quad (t \in \mathbb{R}).$$

We further construct the crossed product $W^*(M, \mathcal{F}) \rtimes_\sigma \mathbb{R} \rtimes_{\hat{\sigma}} \mathbb{R}$. Due to the Takesaki duality theorem it follows that $W^*(M, \mathcal{F}) \rtimes_\sigma \mathbb{R} \rtimes_{\hat{\sigma}} \mathbb{R}$ is isomorphic to $W^*(M, \mathcal{F}) \otimes \mathcal{L}$, where $\mathcal{L}$ denotes the algebra of all bounded operators on the Hilbert space. Therefore, we can reduce the investigation of type III von Neumann algebras to that of type II von Neumann algebras, which are easier to handle due to the existence of trace.

Thus the crossed product $W^*(M, \mathcal{F}) \rtimes_\sigma \mathbb{R}$ is the first clue to study type III von Neumann algebras. We claim here that it is still realized as a foliation von Neumann algebra. Let $TM/\mathcal{F}$ be the normal bundle of the foliation $\mathcal{F}$ and Q denote the principal $\mathbb{R}_+$-bundle of the determinant line bundle of $TM/\mathcal{F}$. It then follows that

$$Q = \widetilde{X} \times_{\rho} \bigwedge^{\dim N} TN_+,$$

where $\bigwedge^{\dim N} TN_+$ denotes the principal $\mathbb{R}_+$-bundle of $\bigwedge^{\dim N} TN$ This is also a foliated bundle and the foliation is denoted by $\mathcal{F}_o$. The codimension of $\mathcal{F}_o$ is $\dim N + 1$. The conclusion is that $W^*(M, \mathcal{F}) \rtimes_\sigma \mathbb{R}$ is isomorphic to the foliation von Neumann algebra $W^*(Q, \mathcal{F}_o)$. Furthermore the dual action $\hat{\sigma}$ on $W^*(Q, \mathcal{F}_o)$ is nothing but the principal action of $\mathbb{R}_+$ on Q. The trace τ on $W^*(M, \mathcal{F}) \rtimes_\sigma \mathbb{R}$ is also identified. Let $(t_1, t_2, \ldots, t_q)$ be a local coordinate of N, and t_{q+1} denote the coordinate in the fibre direction of the bundle $\bigwedge^{\dim N} TN_+$. Then the $(q+1)$-form

$$d\nu = d\left(\frac{dt_1 dt_2 \cdots dt_q}{t_{q+1}}\right)$$

on $\bigwedge^{\dim N} TN_+$ is invariant under the induced action of $\mathrm{Diff}(N)$. Thus we obtain a transverse invariant measure on $(Q, \mathcal{F}_o)$ and the trace τ on $W^*(Q, \mathcal{F}_o)$ from $d\nu$. The relative invariance of τ follows from straightforward computation.

2 Analysis on the leaf space $M/\mathcal{F}$

In this section we explain how to develop Differential Geometry on the leaf space. More precisely we discuss the Index Theorem of differential operators on the foliated manifold $(M, \mathcal{F})$. As mentioned in §1 the K-theory is suitably generalized on the leaf space. It is then reasonable to ask what are natural geometrical objects in the K-theory. In the following we explain that the index of longitudinal elliptic differential operator is realized as an element in the K-group of the foliation C^*-algebra.

2.1 The index in K-group

Let M be a compact manifold and T an elliptic differential operator on M. Then it is a Fredholm operator and the index is defined by

$$\mathrm{Ind}\, T = \dim \ker T - \dim \mathrm{coker} T.$$

Here we explain that the index is considered as an element of the K-group of the C^*-algebra $\mathcal{K}$ of compact operators. Choose a Riemannian metric on M and denote by $\mathcal{H}$ the Hilbert space of L^2-functions. Since M is complete, a formal selfadjoint differential operator on M extends uniquely to a selfadjoint operator D on $\mathcal{H}$. Due to the Stone Theorem we then obtain a 1-parameter family of unitary operators $\{e^{itD}\}_{t\in\mathbb{R}}$. The fundamental result is the following [40, p. 63][45]:

Proposition 2.1 *Let M be a closed manifold and D an elliptic differential operator on M.*
1) Given a rapidly decreasing function f on $\mathbb{R}$, the operator

$$f(D) = \int_{\mathbb{R}} f(t)e^{itD}dt$$

is a smoothing operator and admits a C^∞-kernel function $k : M \times M \to \mathbb{C}$;
2) The above construction extends to a homomorphism

$$\rho : C_o(\mathbb{R}) \to \mathcal{K}$$

so that $\rho(f) = f(D)$. Here $\mathcal{K}$ is the C^*-algebra of compact operators on $\mathcal{H}$.

Proposition 2.1 holds also for complete manifolds with bounded geometry. We refer the reader to [40, p. 63], [45] with the details.

Given an elliptic differential operator D and an involution ϵ such that $\epsilon D + D\epsilon = 0$. The Dirac operators on even-dimensional manifolds satisfies the condition. Let $\mathbb{Z}_2$ be the cyclic group of order 2 and $\mathbb{Z}_2$ act on $\mathbb{R}$ by the reflection $\epsilon(x) = -x$ $(x \in \mathbb{R})$. Let $a = f_0U_e + f_1U_\epsilon$ be an element of the crossed product $C_o(\mathbb{R}) \rtimes \mathbb{Z}_2$. Here U_e and U_ϵ are the formal unitary elements corresponding to the identity e and the generator $\epsilon \in \mathbb{Z}_2$. We often suppress the formal unitaries and denote them by 1 and ϵ. We then obtain a homomorphism:

$$\rho : C_o(\mathbb{R}) \rtimes \mathbb{Z}_2 \to \mathcal{K}$$

such that

$$\rho(f_0U_e + f_1U_\epsilon) = f_0(D) + f_1(D)\epsilon.$$

Definition 2.2 We call the homomorphisms

$$\rho : C_o(\mathbb{R}) \to \mathcal{K}$$

and

$$\rho : C_o(\mathbb{R}) \rtimes \mathbb{Z}_2 \to \mathcal{K}$$

the *index homomorphism* determined by the elliptic differential operator D.

Exploiting the index homomorphism ρ, we can generalize the index for longitudinal elliptic operator on foliated manifolds. But before proceeding further we explain the relation between the Fredholm index and the homomorphism ρ.

We first recall that $K_*(C_o(\mathbb{R}) \rtimes \mathbb{Z}_2)$ is isomorphic to the K-group of a point. In fact the generator of $K_0(C_o(\mathbb{R}) \rtimes \mathbb{Z}_2) = \mathbb{Z}$ is given by the formal difference of projections e_x and e_1:

$$e_x - e_1 \in K_0(C_o(\mathbb{R}) \rtimes \mathbb{Z}_2) = \mathbb{Z}, \quad e_x = \frac{(1+x)^2}{2(1+x^2)} + \frac{1-x^2}{2(1+x^2)}\epsilon, \quad e_1 = \frac{1-\epsilon}{2}.$$

Also recall that the generator of $K_1(C_o(\mathbb{R})) = \mathbb{Z}$ is given by :

$$u - 1 \in K_1(C_o(\mathbb{R})), \quad u = \frac{x-i}{x+i}.$$

Definition 2.3 1) For an elliptic differential operator D with an involution *epsilon* such that $\epsilon D + D\epsilon = 0$, the (even) index of D is:

$$\operatorname{Ind} D = \rho(e_x) - \rho(e_1) \in K_0(\mathcal{K}) = \mathbb{Z}. \tag{1}$$

2) For an elliptic differential operator D the (odd) index of D is:

$$\operatorname{Ind} D = \rho(u) - \rho(1) \in K_1(\mathcal{K}). \tag{2}$$

Remark. For $\mathcal{K}$ we have $K_1(\mathcal{K}) = 0$ and nothing is interesting with the odd index. However, we can consider various C^*-algebras in various geometric situations. For instance, if we consider a Γ-covering space $\widetilde{M}$, the C^*-algebra of Γ-invariant operators on $\widetilde{M}$ appears. As is explained later, if we consider a foliated manifold $(M, \mathcal{F})$, the foliation C^*-algebra appears. In each case, given a suitable differential operator D, it yields the index homomorphism form $C_o(\mathbb{R}) \rtimes \mathbb{Z}_2$ or $\mathbb{C}_o(\mathbb{R})$ to a suitable C^*-algebra $\mathcal{A}$. Accordingly the index of D is defined as an element of $K_0(\mathcal{A})$ or $K_1(\mathcal{A})$ following Definition 2.3. Since $K_i(\mathcal{A})$ could be nontrivial, the odd index is not necessarily vanish in general.

Now we examine Definition 2.3 is compatible with the Fredholm index for elliptic operators. First we introduce the following C^*-algebra:

$$\mathcal{A} = \left\{ \begin{pmatrix} a & b \\ c & d \end{pmatrix} \in C_o(\mathbb{R}) \otimes M_2(\mathbb{C}) : \ a,\ d\text{–even},\ \ c,\ d\text{–odd functions} \right\}.$$

We have an isomorphism $\sigma : C_o(\mathbb{R}) \rtimes \mathbb{Z}_2 \to \mathcal{A}$ given by

$$\sigma(f) = \begin{pmatrix} f^{ev} & f^{odd} \\ f^{odd} & f^{ev} \end{pmatrix}, \qquad \sigma(\epsilon) = \begin{pmatrix} 1 & 0 \\ 0 & -1 \end{pmatrix}.$$

Here $f^{ev}(x) = (f(x) + f(-x))/2$ and $f^{odd}(x) = (f(x) - f(-x))/2$ for $f \in C_o(\mathbb{R})$. We then choose continuous functions $\varphi,\ \psi$ on $\mathbb{R}$ such that:

i) φ is an even function and ψ is an odd function;

ii) $0 \leq \varphi(x) \leq 1$ $(x \in \mathbb{R})$, $\varphi(0) = 0$, $\varphi(\pm\infty) = 1$, $\psi^2 = \varphi(1-\varphi)$.

Via the isomorphism *sigma*, the generator in $K_0(\mathcal{A})$ corresponding to $e_x - e_1 \in K_0(C_o(\mathbb{R}) \rtimes \mathbb{Z}_2)$ is given by:

$$p_0 - p_1 \in K_0(\mathcal{A}), \qquad p_x = \begin{pmatrix} 1-\varphi & \psi \\ \psi & \varphi \end{pmatrix}, \quad p_1 = \begin{pmatrix} 0 & 0 \\ 0 & 1 \end{pmatrix}. \tag{3}$$

Thus we obtain $\operatorname{Ind} D = \rho(p_0) - \rho(p_1)$. Suppose that $\operatorname{supp}(1-\varphi)$ is contained in a small neighbourhood of $0 \in \mathbb{R}$. We then have

$$\rho(p_0) = \begin{pmatrix} 0 & 0 \\ 0 & 1 \end{pmatrix} = \rho(p_1)$$

on the eigenspaces of D with eigenvalues $\lambda \neq 0$, and

$$\rho(p_0) = \begin{pmatrix} 1 & 0 \\ 0 & 0 \end{pmatrix}, \quad \rho(p_1) = \begin{pmatrix} 0 & 0 \\ 0 & 1 \end{pmatrix}$$

on $\ker D$. Since $K_0(\mathcal{K})$ is identified with $\mathbb{Z}$ by the rank of projections, we obtain

$$\rho(p_0) - \rho(p_1) = \operatorname{Tr}[\rho(p_0)|\ker D - \rho(p_1)|\ker D] = \dim\ker D^+ - \dim\ker D^- .$$

Therefore the Fredholm index is recaptured from our definition of the index.

2.2 Cyclic cohomology group

Next we shall explain the cyclic cohomology theory, which plays the role of the de Rham theory to the ordinary manifolds. We then introduce the pairing between the K-group and the cyclic cohomology group. The pairing turns out to be a quite strong method to detect the index of operators in the K-group.

Let $\mathcal{A}$ be an algebra over $\mathbb{C}$ with unit. Define $C^n_\lambda(\mathcal{A})$ to be the space of $(n+1)$-multilinear maps τ on $\mathcal{A}$ into $\mathbb{C}$ satisfying the following cyclic condition:

$$\tau(a_0, a_1, \cdots, a_n) = (-1)^n \tau(a_n, a_0, \cdots, a_{n-1}) \qquad (a_i \in \mathcal{A}).$$

We have the coboundary map $b : C^{n-1}_\lambda(\mathcal{A}) \to C^n_\lambda(\mathcal{A})$ such that

$$b\tau(a_0, a_1, \cdots, a_n) = \sum_{k=0}^{n-1} (-1)^k \tau(a_0, \cdots, a_k a_{k+1}, \cdots, a_n) + (-1)^n \tau(a_n a_0, a_1, \cdots, a_{n-1}).$$

It is easy to see that $b^2 = 0$.

Definition 2.4 The cyclic cohomology $H^*_\lambda(\mathcal{A})$ is the cohomology group of the cochain complex $(C^*_\lambda(\mathcal{A}),\ b)$.

Example 2.5 Let $\tau : \mathcal{A} \to \mathbb{C}$ be a trace on $\mathcal{A}$. Since it satisfies the condition

$$b\tau(a,b) = \tau(ab) - \tau(ba) = 0,$$

it is a cyclic 0-cocycle on $\mathcal{A}$.

From τ we further obtain a cyclic $2m$-cocycle $\hat{\tau}$ on the matrix algebra $M_n(\mathcal{A})$ with components in $\mathcal{A}$. Let $a^{(k)} = (a^{(k)}_{ij}) \in M_n(\mathcal{A})$, where $a^{(k)}_{ij} \in \mathcal{A}$ denotes the (i,j)-component of the matrix $a^{(k)}$. We then define $\hat{\tau}$ to be

$$\hat{\tau}(a^{(0)}, a^{(1)}, \cdots, a^{(2m)}) = \sum_{i_0, i_1, \cdots, i_{2m}} \tau(a^{(0)}_{i_0 i_1}, a^{(1)}_{i_1 i_2}, \cdots, a^{(2m)}_{i_{2m} i_0}).$$

Example 2.6 We present a cyclic 2-cocycle on $C^\infty(T^2)$. Since T^2 is an abelian Lie group, it admits invariant vector fields X and Y with $[X,Y] = 0$, and the dual 1-forms dx and dy. The exterior differentiation is given by $d = dx\partial_X + dy\partial_Y$, where ∂_X and ∂_Y denote the differentiation along X and Y, respectively. We then have a cyclic 2-cocycle

$$\begin{aligned}\tau(a_0, a_1, a_2) &= \int_{T^2} a_0 da_1 da_2 \\ &= \int_{T^2} dxdy[a_0(\partial_X a_1)(\partial_Y a_2) - a_0(\partial_Y a_1)(\partial_X a_2)]\end{aligned}$$

on $C^\infty(T^2)$.

Example 2.7 Put $\mathcal{A} = C^\infty_c(\mathbb{R} \times \mathbb{R})$. Let $k \in \mathcal{A}$ act on $L^2(\mathbb{R})$ with k a kernel function. Let x and $\dfrac{d}{dx}$ act on $L^2(\mathbb{R})$ such that x is the multiplication by x and that $\dfrac{d}{dx}$ is the differentiation in x. We then define derivations ∂_1 and ∂_2 on $\mathcal{A}$ such that

$$\begin{aligned}(\partial_1 k)\xi &= [x, k]\xi = (xk - kx)\xi \\ (\partial_2 k)\xi &= [\frac{1}{i}\frac{d}{dx}, k]\xi = (\frac{1}{i}\frac{d}{dx}k - k\frac{1}{i}\frac{d}{dx})\xi,\end{aligned}$$

where $k \in \mathcal{A}$ and $\xi \in L^2(\mathbb{R})$. Note that ∂_1 and ∂_2 commute with each other. We then have a cyclic 2-cocycle

$$\tau(a_0, a_1, a_2) = \mathrm{Tr}\ (a_0(\partial_1 a_1)(\partial_2 a_2) - a_0(\partial_2 a_1)(\partial_1 a_2)).$$

Here Tr is given by

$$\mathrm{Tr}\,(k) = \int_{\mathbb{R}} k(x,x)dx.$$

2.3 The pairing between K-theory and cyclic cohomology

Let $\mathcal{A}$ be an algebra with unit. We shall discuss the pairing between $K_*(\mathcal{A})$ and $HC^*(\mathcal{A})$ obtained basically from the evaluation of elements in $\mathcal{A}$ with cyclic cocycles.

Definition 2.8 The pairing $\langle \ , \ \rangle$ between $K_i(\mathcal{A})$ and $HC^j(\mathcal{A})$ $(i \equiv j \bmod 2)$ is defined in such a way that:

i) Given $e \in K_0(\mathcal{A})$ and $\tau \in HC^{2m}(\mathcal{A})$, it is

$$\langle e, \tau \rangle = \tau(e, e, \cdots, e);$$

ii) Given $u \in K_1(\mathcal{A})$ and $\tau \in HC^{2m+1}(\mathcal{A})$, it is

$$\langle u, \tau \rangle = \tau(u^{-1} - 1, u - 1, \cdots, u^{-1} - 1, u - 1).$$

It is remarkable that the values of evaluation above depend only on the classes of e, u and τ.

Remark. In the above definition we omitted the normalizing constant appeared in Connes [16]. The constant will be important when we introduce the stabilizing operator S.

Example 2.9 Let $\mathcal{A}$ be $C^\infty(T^2)$ and τ the cyclic 2-cocycle on $M_k(\mathcal{A})$ such that

$$\tau(a_0, a_1, a_2) = \int_{T^2} dxdy \,\mathrm{Tr}\, (a_0(\partial_X a_1)(\partial_Y a_2) - a_0(\partial_Y a_1)(\partial_X a_2))$$

for $a_i \in M_k(\mathcal{A})$. Here we consider that a_i is a smooth function on T^2 with values in $M_k(\mathbb{C})$. The derivations ∂_X and ∂_Y are extended naturally to act on such elements, and Tr means taking the trace at each point of T^2. We also consider that a_i are operators acting on the space $C^\infty(T^2, \mathbb{C}^k)$ of all $\mathbb{C}^k$-valued smooth functions on T^2. The derivations are then given by the following form:

$$(\partial_X a)\xi = (Xa - aX)\xi, \quad (\partial_Y a)\xi = (Ya - aY)\xi$$

for $a \in M_k(\mathcal{A})$ and $\xi \in C^\infty(T^2, \mathbb{C}^k)$, where X and Y denote the differentiations considered as operators acting on $C^\infty(T^2, \mathbb{C}^k)$. In other words, we have

$$\partial_X a = [X, a], \quad \partial_Y a = [Y, a].$$

Let $e \in M_k(\mathcal{A})$ be an idempotent. It then follows that

$$\begin{aligned}\tau(e,e,e) &= \int_{T^2} dxdy \operatorname{Tr}(e[X,e][Y,e] - e[Y,e][X,e]) \\ &= \int_{T^2} dxdy \operatorname{Tr}(e(Xe-eX)(Ye-eY) - e(Ye-eY)(Xe-eX)) \\ &= \int_{T^2} dxdy \operatorname{Tr}([eXe, eYe] - e[X,Y]e).\end{aligned}$$

When we put $eZe = \nabla_Z$ for vector fields Z on T^2, we finally obtain

$$\begin{aligned}\tau(e,e,e) &= \int_{T^2} dxdy \operatorname{Tr}([\nabla_X, \nabla_Y] - \nabla_{[X,Y]}) \\ &= \int_{T^2} dxdy \operatorname{Tr}(R_{X,Y}),\end{aligned}$$

where R denotes the curvature tensor with respect to the covariant differentiation $\nabla = ede$ on the vector bundle E on T^2 which is given by the image of e. Recall that the first Chern class $c_1(E)$ is represented by $\operatorname{Tr}\left(\frac{i}{2\pi}R\right)$. Thus, the above equality amounts to

$$\frac{i}{2\pi}\langle e, \tau\rangle = \frac{i}{2\pi}\tau(e,e,e) = \int_{T^2} c_1(E).$$

Example 2.10 Let $\mathcal{A}$ and τ be the algebra and the cyclic cocycle in Example 2.7. Let e be a projection in $\mathcal{A}$ of finite rank. Recall that τ is given by

$$\tau(a_0, a_1, a_2) = \operatorname{Tr}\left(a_0[x, a_1][\frac{1}{i}\frac{d}{dx}, a_2] - a_0[\frac{1}{i}\frac{d}{dx}, a_1][x, a_2]\right).$$

By a similar computation to the previous example, we obtain

$$\tau(e,e,e) = \operatorname{Tr}\left([exe, e\frac{1}{i}\frac{d}{dx}e] - e[x, \frac{1}{i}\frac{d}{dx}]e\right).$$

Note that $\operatorname{Tr}\left([exe, e\frac{1}{i}\frac{d}{dx}e]\right) = 0$ since $exe, e\frac{1}{i}\frac{d}{dx}e \in A$. Observing the canonical commutation relation $[x, \frac{1}{i}\frac{d}{dx}] = i$, we thus obtain

$$\tau(e,e,e) = \operatorname{Tr}(-ie) = -i\operatorname{rk}(e).$$

If we pursue the analogy of the previous example, we can consider that $\tau(e,e,e)$ evaluates the *noncommutative curvature* with the projection e.

2.4 The Connes index theorem for longitudinal elliptic operators

Let M be a closed manifold and $\mathcal{F}$ a foliation on M of dimension p. Suppose that there exists an invariant transverse measure ν on $(M,\mathcal{F})$. Due to Ruelle-Sullivan [42] we obtain the current C_{RS} of degree p such that:

$$\int_{C_{RS}} \omega = \sum_i \int_{T_i} d\nu \int_{V_i\times\{t\}} \rho_i \omega.$$

for a differential p-form ω on M. Here we take the foliation chart $\mathcal{U} = \{U_i \cong V_i \times T_i\}_{i\in I}$ such that T_i is a transversal to $\mathcal{F}$, and choose the partition of unity $(\rho_i)_{i\in I}$ subordinate to $\mathcal{U}$. Then $\int_{V_i\times\{t\}}$ denotes the integration along the leaves $V_i \times \{t\}$ $(t \in T_i)$, and $\int_{T_i} d\nu$ is the integration with respect to ν. Since ν is holonomy invariant, the current C_{RS} is closed. Moreover C_{RS} induces a linear map:

$$C_{RS}\colon \Omega^p(\mathcal{F}) \to \mathbb{R}$$

for the complex $\Omega(\mathcal{F})$ of longitudinal differential forms. It also satisfies that $C_{RS}(d'\omega) = 0$, where d' denotes the exterior differential along the leaves.

Suppose that $(M,\mathcal{F})$ admits an invariant transverse measure ν. We then obtain the trace $\tau : C^*(M,\mathcal{F}) \to \mathbb{C}$ on $C^*(M,\mathcal{F})$ exploiting the Ruelle-Sullivan current C_{RS} and the leafwise measure $d\mu$:

$$\tau(\varphi) = \sum_i \int_{T_i} d\nu \int_{V_i\times\{t\}} \rho_i \varphi dv_{\mathcal{F}}.$$

Here φ is a continuous function on the holonomy groupoid G with compact support.

Example 2.11 Consider the Kronecker foliation $\mathcal{F}_\theta$ on $M = (\mathbb{R}\times\mathbb{R}/\mathbb{Z})/\mathbb{Z}$, $(x,t) \sim (x+1, t+\theta)$. Recall that the holonomy groupoid is given by

$$G = (\mathbb{R}\times\mathbb{R}\times\mathbb{R}/\mathbb{Z})/\mathbb{Z}.$$

Note that the standard volume form dt on $\mathbb{R}/\mathbb{Z}$ yields an invariant transverse measure for $(M,\mathcal{F}_\theta)$. We then identify $C^*(M,\mathcal{F}_\theta)$ with the C^*-completion of the algebra of kernel functions such that:

i) $k : \mathbb{R}\times\mathbb{R}\times\mathbb{R}/\mathbb{Z} \to \mathbb{C}$ is continuous;

ii) $k(x+1, y+1, t+\theta) = k(x,y,t), \quad (x,y,t) \in \mathbb{R}\times\mathbb{R}\times\mathbb{R}/\mathbb{Z}$ and

iii) k has compact support when it is considered as a function on $(\mathbb{R}\times\mathbb{R}\times\mathbb{R}/\mathbb{Z})/\mathbb{Z}$.

Then the trace is defined by

$$\tau(k) = \int_{\mathbb{R}/\mathbb{Z}} dt \int_0^1 k(x,x,t)dx.$$

Take a closed foliated manifold $(M,\mathcal{F})$ of even-dimensional leaves. Suppose that there exists a family of longitudinal Dirac operators on $(M,\mathcal{F})$. Let $D = (D_L)_{L\in M/\mathcal{F}}$ denote the family of longitudinal Dirac operators D_L that is lifted to the holonomy covering $\widetilde{L}$ from each leaf L. Then we can apply the construction in Proposition 2.1 to D to obtain the index homomorphism

$$\rho : C_o(\mathbb{R}) \rtimes \mathbb{Z}_2 \to C^*(M,\mathcal{F}).$$

Hence we define the index of the longitudinal Dirac operator D to be

$$\operatorname{Ind} D = \rho(e_x) - \rho(e_1) \in K_0(C^*(M,\mathcal{F})).$$

It is completely similar to Definition 2.3. Then the index theorem due to Connes for the longitudinal Dirac operator on $(M,\mathcal{F})$ is stated as follows:

Theorem 2.12 (Connes index theorem on foliated manifolds [15]) *Let M be a closed manifold and $\mathcal{F}$ a foliation on M. Suppose that there exists an invariant transverse measure ν on $(M,\mathcal{F})$. It then follows that:*

$$\langle \tau, \operatorname{Ind} D\rangle = \int_{C_{RS}} \widehat{A}\left(\frac{\sqrt{-1}}{2\pi}R_{\mathcal{F}}\right).$$

Here $\int_{C_{RS}}$ denotes the Ruelle-Sullivan current associated to ν, and $\widehat{A}\left(\frac{\sqrt{-1}}{2\pi}R_{\mathcal{F}}\right) \in \Omega(\mathcal{F})$ is the $\widehat{A}$-form associated to the Riemannian curvature form $R_{\mathcal{F}}$ along the leaves.

2.5 The Toeplitz index theorem for longitudinal elliptic operators

In the above theorem we assumed that the leave are of even dimension. There exists as well an index theorem on foliated manifolds of odd-dimensional leaves. In the following we shall describe the index theorem on the Kronecker foliation.

Let $\mathcal{F}_\theta$ be the Kronecker foliation on $M = (\mathbb{R}\times\mathbb{R}/\mathbb{Z})/\mathbb{Z}$. Here $\theta \in \mathbb{R}\backslash\mathbb{Q}$. Recall that $C^*(M,\mathcal{F}_\theta)$ is identified with the C^*-completion of the algebra of kernel functions $k : \mathbb{R}\times\mathbb{R}\times\mathbb{R}/\mathbb{Z} \to \mathbb{C}$ in Example 2.11. Note that with $t \in \mathbb{R}/\mathbb{Z}$ fixed such a function $k(x,y,t)$ is considered as the kernel function of a operator k_t that acts on the Hilbert space $L^2(\mathbb{R}\times\{t\})$. Let

$D = (D_t)_{t\in\mathbb{R}/\mathbb{Z}}$ be the longitudinal elliptic operator on $(T^2, \mathcal{F}_\theta)$ such that:

$$D_t = \frac{1}{i}\frac{d}{dx}$$

on the holonomy covering $\mathbb{R}\times\{t\}$. Furthermore, we define a family $F = (F_t)$ such that F_t is the Hilbert transformation on each $L^2(\mathbb{R}\times\{t\})$:

$$[F_t\xi](x) = \lim_{\epsilon\to 0}\frac{i}{\pi}\int\limits_{|x-y|>\epsilon}\frac{\xi(y)}{x-y}dy.$$

Thus F is constant in $t \in \mathbb{R}/\mathbb{Z}$. Let $P = \left(\dfrac{1+F_t}{2}\right)$ be a family of projections. Here we take a C^∞-function

$$\varphi : M = (\mathbb{R}\times\mathbb{R}/\mathbb{Z})/\mathbb{Z} \to \mathbb{C}\setminus\{0\}.$$

Then we make the pullback of φ to $\mathbb{R}\times\mathbb{R}/\mathbb{Z}$ and let φ act on $L^2(\mathbb{R}\times\{t\})$ Thus the operator $T_\varphi = \varphi^{-1}P\varphi$ is defined, which is called the Toeplitz operator given by φ. It then turns out that the operator

$$\varphi^{-1}P\varphi - P$$

yields an element of $K_0(C^*(M,\mathcal{F}_\theta))$. It is called the index of Toeplitz operator T_φ. We have the following theorem (see also [21]).

Theorem 2.13 *Let $(M, \mathcal{F}_\theta)$ be the Kronecker foliation. We take a C^∞-function*

$$\varphi : M \to \mathbb{C}\setminus\{0\}$$

and consider the index of Toeplitz operator T_φ

$$\varphi^{-1}P\varphi - P \in K_0(C^*(M,\mathcal{F}_\theta)).$$

It then follows that

$$\tau(\varphi^{-1}P\varphi - P) = \frac{1}{2\pi i}\int_{T^2}\varphi^{-1}d\varphi dt$$

where τ is the trace on $C^*(M,\mathcal{F}_\theta)$ defined previously.

We shall illustrate the theorem to calculate the index for $\varphi(x,t) = e^{2\pi i(\theta x - t)}$. We first observe $\varphi(x,t) = \varphi(x+1, t+\theta)$. On each $L^2(\mathbb{R}\times\{t\})$ we have

$$[(\varphi^{-1}P\varphi - P)\xi](x) = \lim_{\epsilon\to 0}\frac{i}{2\pi}\int\limits_{|x-y|>\epsilon}\frac{\varphi^{-1}(x,t)(\varphi(y,t)-\varphi(x,t))}{x-y}\xi(y,t)dy.$$

Thus the kernel function $k(x, y, t)$ of $\varphi^{-1}P\varphi - P$ is given by

$$k(x, y, t) = \frac{i}{2\pi} \frac{\varphi^{-1}(x, t)(\varphi(y, t) - \varphi(x, t))}{x - y}.$$

Since φ is a C^∞-function, $k(x, y, t)$ is continuous at the diagonal set $\{(x, x, t) \mid x \in \mathbb{R}\}$, and it follows that $k(x, x, t) = \theta$. Thus we obtain

$$\tau(\varphi^{-1}P\varphi - P) = \int_{\mathbb{R}/\mathbb{Z}} dt \int_0^1 \theta dx = \theta.$$

On the other hand we can verify

$$\frac{1}{2\pi i} \int_{T^2} \varphi^{-1} d\varphi dt = \theta.$$

Strictly speaking, the element $\varphi^{-1}P\varphi - P$ belongs to not $C^*(T^2, \mathcal{F}_\theta)$ but the foliation von Neumann algebra $W^*(T^2, \mathcal{F}_\theta)$. However, we can modify the argument and construct an element in $C^*(T^2, \mathcal{F}_\theta)$ that yields the index of Toeplitz operator T_φ.

3 Toward noncommutative geometry on foliated manifolds

In this section we discuss a few topics related to noncommutative geometry due to A. Connes.

3.1 The Godbillon-Vey cyclic cocycle and the modular automorphisms

Let X be a closed Riemann surface of genus ≥ 2. We consider X as the quotient of the universal covering $\widetilde{X}$ that is the hyperbolic space. Put $\Gamma = \pi_1(X)$. Then the group Γ acts on $\widetilde{X}$ isometrically. Let Γ also act on S^1 via orientation preserving diffeomorphisms. The diagonal action of Γ on $\widetilde{X} \times S^1$ then define a foliated bundle

$$M = (\widetilde{X} \times S^1)/\Gamma.$$

The holonomy groupoid is given by

$$G = (\widetilde{X} \times \widetilde{X} \times S^1)/\Gamma.$$

Recall that the foliation C^*-algebra $C^*(M, \mathcal{F})$ can be constructed from functions $k : \widetilde{X} \times \widetilde{X} \times S^1 \to \mathbb{C}$ such that:

i) $k : \widetilde{X} \times \widetilde{X} \times S^1 \to \mathbb{C}$ is continuous;

ii) $k(x\gamma, y\gamma, t\gamma) = k(x, y, t)$ for $\gamma \in \Gamma$;

iii) k has compact support when it is considered as a function on $(\widetilde{X} \times \widetilde{X} \times S^1)/\Gamma$.

Let $w : C^*(N, \mathcal{F}) \to \mathbb{C}$ a weight given by

$$w(k) = \int_{M(\Gamma)} k(x, x, t) dt dx.$$

Here $M(\Gamma)$ denotes the fundamental domain with respect to the Γ-action on $\widetilde{X} \times S^1$. Then we choose the hyperbolic volume form $d\mu$ on $\widetilde{X}$ and the standard volume form dt on S^1. We also choose a volume form on M and take the pullback to $\widetilde{X} \times N$. It is denoted by $d\lambda$. We then obtain a function

$$\psi = \frac{d\mu \times dt}{d\lambda}$$

on $\widetilde{X} \times N$. The modular automorphisms $(\sigma_t)_{t \in \mathbb{R}}$ is then defined by

$$\sigma_t(\varphi) = \psi^{it} \varphi \psi^{-it}$$

for $\varphi \in C_c(G)$. We can verify that the modular automorphisms $(\sigma_t)_{t \in \mathbb{R}}$ also preserves the foliation C^*-algebra $C^*(M, \mathcal{F})$.

We then define a cyclic 2-cocycle that is associated to he Godbillon-Vey class. For the details we refer to Moriyoshi-Natsume [34].

Definition 3.1 The Godbillon-Vey cyclic cocycle on $C^*(N, \mathcal{F})$ is

$$\tau(a_0, a_1, a_2) = \int_{M(\Gamma)} dx dt \; (a_0[\phi, a_1][\dot{\phi}, a_2] - a_0[\dot{\phi}, a_1][\phi, a_2]),$$

where $\phi = \log \psi$ and $\dot{\phi} = \partial \phi / \partial t$.

Let $D = (D_L)_{L \in M/\mathcal{F}}$ be the family of longitudinal Dirac operators lifted to $\widetilde{X} \times N$. The index $\operatorname{Ind} D \in K_0(C^*(M, \mathcal{F}))$ can be defined via the index homomorphism

$$\rho : C_o(\mathbb{R}) \rtimes \mathbb{Z}_2 \to C^*(M, \mathcal{F})$$

as is observed in §2. We then have the following:

Theorem 3.2 (Moriyoshi-Natsume [34]) *Let τ the Godbillon-Vey cyclic cocycle defined as above. We denote by* gv *the Godbillon-Vey class of the foliated bundle M. It then follows that*

$$\langle e, \tau \rangle = \frac{1}{\pi i} \int_X \mathrm{gv},$$

where $\langle e, \tau \rangle$ is the pairing between the K-theory and the cyclic cohomology group.

We observe the similarity of the form of the Godbillon-Vey cocycle and the usual curvature given by the covariant derivative. Thus the above theorem exhibits that the Godbillon-Vey number of the foliated bundle of M can be understood as the *noncommutative curvature* with respect to two derivations derived from the modular automorphisms on $C^*(M, \mathcal{F})$.

3.2 Noncommutative geometry for singular foliations

Let us review briefly the definition of quantum $SU(2)$ due to Woronowicz [46]. Recall that

$$SU(2) = \left\{ \begin{pmatrix} \alpha & -\bar{\gamma} \\ \gamma & \bar{\alpha} \end{pmatrix} : \alpha\bar{\alpha} + \gamma\bar{\gamma} = 1 \right\}.$$

The quantum $SU(2)$ is a deformation of the function algebra $C(SU(2))$. Precisely that is the universal C^*-algebra $C(S_\mu U(2))$ generated by two elements α and γ satisfying the following equations:

$$\begin{matrix} \alpha\alpha^* + \mu^2\gamma\gamma^* = 1 \\ \alpha^*\alpha + \gamma^*\gamma = 1 \end{matrix} \qquad \begin{matrix} \mu\gamma\alpha = \alpha\gamma \\ \mu\gamma^*\alpha = \alpha\gamma^* \\ \gamma^*\gamma = \gamma\gamma^* \end{matrix} \tag{4}$$

for $-1 \le \mu \le 1$. When $\mu = 1$, $C(S_\mu U(2))$ is exactly isomorphic to the function algebra $C(SU(2))$. There is another way to construct $C(S_\mu U(2))$. Let $C(T)$ be the function algebra of 1-dimensional torus T and $\mathcal{K}$ denote the C^*-algebra of compact operators. It is then obtained as an extension given by the following exact sequence:

$$0 \longrightarrow C(T) \otimes \mathcal{K} \longrightarrow C(S_\mu U(2)) \longrightarrow C(T) \longrightarrow 0. \tag{5}$$

According to Sheu [43] the C^*-algebra $C(S_\mu U(2))$ is considered as a deformation quantization whose infinitesimal deformation is the Poisson structure on $SU(2)$. He also understand the structure of $C(S_\mu U(2))$ from the viewpoint of singular foliation induced by the Poisson structure. Roughly speaking it is stated as follows. We first consider $SU(2)$ as S^3, and the join of S^1:

$$S^3 = (S^1 \times [0,1] \times S^1)/\sim,$$

where $(x, 0, y) \sim (x', 0, y')$ if $y = y'$, and $(x, 1, y) \sim (x', 1, y')$ if $x = x'$. There is no equivalence relations otherwise. Let D be the unit disk $\{z \in \mathbb{C} : |z| \le 1\}$. Then S^3 is also realized as the quotient of

$$(D \times S^1)/\sim,$$

where $(z,y) \sim (z',y')$ if $z = z' \in \partial D$. Thus S^3 is foliated with leaves $L_y = \{(z,y) : |z| < 1\}_{y \in S^1}$ outside the singular set $T = \{(z,y) : |z| = 1\}/\sim$. From the construction we easily obtain the following short exact seaquake of function algebras:

$$0 \longrightarrow C_o(\overset{\circ}{D} \times S^1) \longrightarrow C(S^3) \longrightarrow C(T) \longrightarrow 0, \tag{6}$$

where $\overset{\circ}{D} = \{z \in \mathbb{C} : |z| < 1\}$. Then we can consider that the short exact sequence (5) is obtained as a deformation of the above exact sequence. It is thus interesting to study $C(S_\mu U(2))$ as a singular foliation C^*- algebra. In fact, outside T we have nonsingular foliation on $S^3 \setminus T = \overset{\circ}{D} \times S^1$. Equip $S^3 \setminus T$ with leafwise complete metric, we can develop the index theorem for longitudinal operators. Even in the case of a single leaf, the index theorem turns out to be the Atiyah-Patodi-Singer index theorem [3] for manifolds with cylindrical ends. The eta invariant then appears as the evaluation of the index with a *relative cyclic cocycle* derived from the Heaviside function with cutoff on the boundary. The index theorem for general singular foliations is thus considered as a family version of the Atiyah-Patodi-Singer index theorem.

References

1. M. Atiyah, *Elliptic operators, discrete groups and von Neumann algebras*, Astérisque, **32** (1976), 43–72.
2. M. Atiyah, R. Bott and V.K. Patodi, *On the heat equation and the index theorem*, Invent. Math., **19** (1973), 279–330.
3. M. Atiyah, V.K. Patodi and I.M. Singer, *Spectral asymmetry and Riemannian geometry* I, Cambr. Phil. Soc., **77** (1975), 43–69; II, **78** (1975), 405–432; III, **79** (1976), 71–99.
4. M. Atiyah and W. Schmid, *A geometric construction of discrete series for semisimple Lie groups*, Invent. Math., **42** (1977), 1–62.
5. M. Atiyah and G. Segal, *The index of Elliptic operators* II, Ann. of Math., **87** (1968), 531–545.
6. M. Atiyah and I. M. Singer, *The index of elliptic operators on compact manifolds,* Bull. Amer. Math. Soc., **69** (1963), 422–433.
7. M. Atiyah and I. M. Singer, *The index of Elliptic operators* I, Ann. of Math., **87** (1968), 484–530; III, **87** (1968), 546–604; IV, **93** (1971), 119–138; V, **93** (1971), 139–149.
8. M. Atiyah and I. M. Singer, *Index theory for skew-adjoint Fredholm operators,* Publ. Math. IHES, **37** (1969), 305–326.

9. P. Baum and A. Connes, *Leafwise homotopy equivalence and rational Pontriagin classes,* in Foliations, (Tokyo 1983), Adv. Stud. Pure Math., **5**, North-Holland, Amsterdam, 1985, 1–14.
10. P. Baum and A. Connes, *Chern character for discrete groups,* in A Fête of Topology, Academic Press, Boston, 1988, 163–232.
11. P. Baum, A. Connes and N. Higson, *Classifying space for proper actions and K-theory of group C^*-algebras*, Contemporary Math., **167** (1994), 241–291.
12. P. Baum and R. Douglas, *K-homology and index theory*, Proc. Symp. Pure Math., **38** part 1 (1982), 521–628.
13. N. Berline, E. Getzler and M. Vergne, Heat kernels and Dirac operators, Springer-Verlag, New York-Berlin, 1992.
14. B. Blackadar, K-theory for operator algebras, Mathematical Sciences Research Institute Publications **5**, Springer-Verlag, New York-Berlin, 1986.
15. A. Connes, *A survey of foliations and operator algebras*, Proc. Symp. Pure Math., **38** (1982), 521–628.
16. A. Connes, *Noncommutative differential geometry* I, II, Publ. Math. IHES, **62** (1986), 257–360.
17. A. Connes, *Cyclic cohomology and the transversal fundamental class of a foliation*, in Geometric Method in Operator Algebras (H. Araki and E. G. Effros, eds.), Pitman Research Notes in Math. Series, **123** (1986), Longman Scientific and Technical, 52–144.
18. A. Connes, Noncommutative Geometry, Academic Press, 1994.
19. A. Connes and H. Moscovici, *The L^2-index theorem for homogeneous spaces of Lie groups*, Ann. of Math., **115** (1982), 291–330.
20. A. Connes and H. Moscovici, *Cyclic cohomology, Novikov conjecture and hyperbolic groups*, Topology, **29** (1990), 345–388.
21. R.G. Douglas, S.Hurder and J. Kaminker, *The longitudinal cocycle and the index of Toeplitz operators*, J. Func. Anal., **101** (1991), 120–144.
22. E. Getzler, *Pseudo-differential operators on supermanifolds and the Atiyah-Singer index theorem*, Comm. Math. Phys., **92** (1983), 163–178.
23. P.B. Gilkey, Invariance theory the heat equation, and the Atiyah-Singer index theorem, 2nd ed., CRC Press, Boca Raton-Ann Arbor-London-Tokyo, 1995.
24. I.Q. Gohberg and M.G. Krein, *The basic propositions on defect numbers, root numbers and indices of linear operators* , Amer. Math. Soc. Transl., **13** (1960), 185–264.
25. P. Green, *Equivariant K-theory and crossed products*, Proc. Symp.

Pure Math., **38** part 1, (1981), 337–338.
26. N. Higson, *An approach to* $\mathbb{Z}/k$*-index theory*, Inter. J. Math., **1** (1990), 189–210.
27. M. Hilsum and G. Skandalis, *Stabilité des* C^**-algèbres de feuilletages* , Ann. Inst. Fourier, (Grenoble) **33** no. 3, (1983), 201–208.
28. F. Hirzebruch, Topological methods in algebraic geometry, Springer, 1966.
29. S. Hurder, *Eta invariants and the odd index theorem for coverings*, Contemp. Math., **105** (1990), 47–82.
30. G.G. Kasparov, *The operator* K*-functor and extensions of* C^**-algebras,* Math. USSR Izv., **16** (1981), 513–572.
31. G.G. Kasparov, *Lorentz group: K-theory of unitary representations and crossed products*, Sov. Math. Dokl., **29** (1984), 252–260.
32. S. Kobayashi and K. Nomizu, Foundations of Differential Geometry I,II, Interscience Publ., 1963, 1969.
33. H.B. Lawson and M-L. Michelsohn, Spin Geometry, Princeton, 1989.
34. H. Moriyoshi and T. Natsume, *The Godbillon-Vey cyclic cocycle and longitudinal Dirac operators*, Pacific J. Math., **172** (1996), 483–539.
35. R.S. Palais, Seminar on the Atiyah-Singer index theorem, Ann. of Math. Studies **57**, Princeton, 1965.
36. V.K. Patodi, *An analytic proof of the Riemann-Roch-Hirzebruch theorem for K'ahler manifolds*, J. Diff. Geom., **5** (1971), 251–283.
37. G. Pedersen, C^*-algebras and their automorphism groups, Academic Press, 1979.
38. M. Rieffel, *Applications of strong Morita equivalence to transformation group* C^**-algebras. Operator* , Proc. Symp. Pure Math., **38** part 1, (1981), 299–310.
39. M. Rieffel, *Morita equivalence for operator algebras* , Proc. Symp. Pure Math., **38** part1, (1981), 285–298.
40. J. Roe, Elliptic operators, topology and asymptotic methods, Longman, 1988.
41. J. Roe, Coarse cohomology and index theory on complete Riemannian manifolds, Memoires of AMS **497**, 1993.
42. D. Ruelle and D. Sullivan, *Currents, flows and diffeomorphisms*, Topology, **14** (1975), 319–327.
43. A.J-L. Sheu, *Quantization of the Poison* $SU(2)$ *and its Poisson homogenous space - The 2-sphere*, Commun. Math. Phys., **135** (1991), 217–232.
44. I.M. Singer, *Some remarks on operator theory and index theory* , Lect. Notes in Math., **575** (1977), 128–138.

45. M. Taylor, Pseudo-differential operators, Princeton, 1982.
46. S.L. Woronowicz, *Twisted $SU(2)$ group: An example of noncommutative differential calculus*, Publ. R.I.M.S., **23** (1987), 117–181.

Received May 7, 2001.

RESEARCH PAPERS

Proceedings of
FOLIATIONS: GEOMETRY AND DYNAMICS
held in Warsaw, May 29–June 9, 2000
ed. by Paweł WALCZAK *et al.*
World Scientific, Singapore, 2002
pp. 159–183

DISTRIBUTIONAL BETTI NUMBERS OF TRANSITIVE FOLIATIONS OF CODIMENSION ONE

JESÚS A. ÁLVAREZ LÓPEZ

Departamento de Xeometría e Topoloxía, Facultade de Matemáticas, Universidade de Santiago de Compostela, 15706 Santiago de Compostela, Spain, e-mail: alvarez@zmat.usc.es

YURI A. KORDYUKOV

Department of Mathematics, Ufa State Aviation Technical University 12 K. Marx str., 450025 Ufa, Russia, e-mail: yurikor@math.ugatu.ac.ru

Let $\mathcal{F}$ be a transitive foliation of codimension one on a closed manifold M. This means that there is an infinitesimal transformation X of $(M, \mathcal{F})$ transverse to the leaves. The flow of X induces an $\mathbb{R}$-action on the reduced leafwise cohomology $\overline{H}(\mathcal{F})$. By using leafwise Hodge theory, the trace of this action on each $\overline{H}^i(\mathcal{F})$ can be defined as a distribution β^i_{dis} on $\mathbb{R}$, which is called distributional Betti number because it is kind of a finite measure of the "size" of $\overline{H}^i(\mathcal{F})$. So the corresponding distributional Euler characteristic, $\chi_{\mathrm{dis}}(\mathcal{F})$, is a distribution on $\mathbb{R}$ too. This is relevant because $\overline{H}(\mathcal{F})$ may be of infinite dimension, even when the leaves are dense, and its Euler characteristic makes no sense in general. The singularity at 0 of $\chi_{\mathrm{dis}}(\mathcal{F})$ is expressed in terms of the Connes' Λ-Euler characteristic, where Λ is the holonomy invariant transverse measure of $\mathcal{F}$ induced by the volume form dt on $\mathbb{R}$. Moreover the whole of $\chi_{\mathrm{dis}}(\mathcal{F})$ is computed by showing a dynamical Lefschetz formula.

1 Introduction

Let M be a closed manifold and $\mathcal{F}$ a smooth foliation on M of codimension one. As usual, let $\mathfrak{X}(M, \mathcal{F}) \subset \mathfrak{X}(M)$ denote the Lie subalgebra of infinitesimal transformations of $(M, \mathcal{F})$, and $\mathfrak{X}(\mathcal{F}) \subset \mathfrak{X}(M, \mathcal{F})$ the ideal of vector fields tangent to the leaves. For any $X \in \mathfrak{X}(M, \mathcal{F})$, the corresponding flow maps leaves to leaves, and will be denoted by $X_t : (M, \mathcal{F}) \to (M, \mathcal{F})$, $t \in \mathbb{R}$.

The foliation $\mathcal{F}$ is called *transitive* when

$$T_xM = \{X(x) \mid X \in \mathfrak{X}(M, \mathcal{F})\} .$$

Since

$$T_x\mathcal{F} = \{Y(x) \mid Y \in \mathfrak{X}(\mathcal{F})\} ,$$

we get that $\mathcal{F}$ is transitive if and only if there is some $X \in \mathfrak{X}(M, \mathcal{F})$ transverse to the leaves; i.e.,

$$T_xM = \mathbb{R}\, X(x) \oplus T_x\mathcal{F}$$

for all $x \in M$. Then the orbits of X_t, $t \in \mathbb{R}$, are non-singular and transverse to the leaves. Note that

$$\mathfrak{X}(M, \mathcal{F})/\mathfrak{X}(\mathcal{F}) \cong \mathbb{R} \tag{1}$$

if the leaves are dense, which is the most interesting case.

The leafwise de Rham complex of $\mathcal{F}$, $(\Omega(\mathcal{F}), d_{\mathcal{F}})$, is the the restriction of the de Rham complex of M to the leaves; i.e., it is given by the smooth sections of the exterior vector bundle $\bigwedge T\mathcal{F}^*$ over M. Its cohomology is called the *leafwise cohomology* of $\mathcal{F}$, and will be denoted by $H(\mathcal{F})$. Moreover $(\Omega(\mathcal{F}), d_{\mathcal{F}})$ is a topological complex with the C^∞ topology, and $H(\mathcal{F})$ is a topological vector space with the induced topology. It is well known that $H(\mathcal{F})$ may not be Hausdorff [14]. So it is interesting to consider its quotient over the closure of the trivial subspace, which is called the *reduced leafwise cohomology* of $\mathcal{F}$, and is denoted by $\overline{H}(\mathcal{F})$ in this paper.

Consider a Riemannian metric on M such that X is of norm one and orthogonal to the leaves. So all flow orbits are geodesics of speed one orthogonal to the leaves. This is what is called a *bundle-like* metric on M. Consider the induced Riemannian structure on the leaves, and let $\delta_{\mathcal{F}}, \Delta_{\mathcal{F}}$ be the leafwise coderivative and leafwise Laplacian on $\Omega(\mathcal{F})$, which are the restrictions to $\Omega(\mathcal{F})$ of the coderivative and Laplacian on the leaves. The kernel $\mathcal{H}(\mathcal{F})$ of $\Delta_{\mathcal{F}}$ is the space of harmonic forms on the leaves that are smooth on M. The L^2 inner product on M induces a Hilbert space structure in the space $L^2\Omega(\mathcal{F})$ of square integrable leafwise differential forms on M. Consider $\Delta_{\mathcal{F}}$ as an unbounded operator in $L^2\Omega(\mathcal{F})$ with domain $\Omega(\mathcal{F})$, and let $\overline{\Delta}_{\mathcal{F}}$ be its closure. It is well known that $\Delta_{\mathcal{F}}$ is symmetric on M when the metric is bundle-like (see, for instance, [5, 16]), so $\overline{\Delta}_{\mathcal{F}}$ is a self-adjoint operator. Let $\overline{\Pi}$ be the orthogonal projection $L^2\Omega(\mathcal{F}) \to \ker \overline{\Delta}_{\mathcal{F}}$. By [3], $\overline{\Pi}$ has the restriction $\Pi : \Omega(\mathcal{F}) \to \mathcal{H}(\mathcal{F})$, and there is an orthogonal decomposition

$$\Omega(\mathcal{F}) = \mathcal{H}(\mathcal{F}) \oplus \overline{\operatorname{im} d_{\mathcal{F}}} \oplus \overline{\operatorname{im} \delta_{\mathcal{F}}} ,$$

which can be called a leafwise Hodge decomposition. In particular, the inclusion $\mathcal{H}(\mathcal{F}) \subset \ker d_{\mathcal{F}}$ induces an isomorphism

$$\mathcal{H}(\mathcal{F}) \stackrel{\cong}{\to} \overline{H}(\mathcal{F}) , \tag{2}$$

whose inverse is induced by the orthogonal projection $\Pi : \ker d_{\mathcal{F}} \to \mathcal{H}(\mathcal{F})$.

For any function $f \in C_c^\infty(\mathbb{R})$, define an operator A_f on $\Omega(\mathcal{F})$ by the formula

$$A_f = \Pi \circ \int_{\mathbb{R}} X_t^* \cdot f(t)\, dt \circ \Pi ,$$

and let $A_f^{(i)}$ denote its restriction to $\Omega^i(\mathcal{F})$. Our first main result is the following.

Theorem 1.1 *For any function $f \in C_c^\infty(\mathbb{R})$, the operator A_f is of trace class, and the functional $f \mapsto \operatorname{Tr}\left(A_f^{(i)}\right)$ defines a distribution $\beta^i_{\text{dis}}(\mathcal{F})$ on $\mathbb{R}$ for each i.*

The distributions $\beta^i_{\text{dis}}(\mathcal{F})$ depend only on $\mathcal{F}$ and the class of X in $\mathfrak{X}(M,\mathcal{F})/\mathfrak{X}(\mathcal{F})$ (Lemma 2.3); thus, when the leaves are dense, they depend only on $\mathcal{F}$ up to linear isomorphisms of $\mathbb{R}$ by (1).

The usual dimension of the spaces $\overline{H}^i(\mathcal{F})$ can be infinite even when the leaves are dense [1, 2, 3]. So the Euler characteristic of $\overline{H}(\mathcal{F})$ can not be defined, and thus a leafwise Gauss-Bonnet theorem makes no sense in the usual way. This is surely a reason of the poor role played by the reduced leafwise cohomology in foliation theory, which should be similar to the important role played by de Rham cohomology of closed manifolds.

To have finite leafwise Betti numbers, they must be defined in another way, by using the another kind of dimension ("exotic dimension"). A solution was given by Connes for foliations with a holonomy invariant transverse measure Λ [8, 9]. In our case, Λ is the transverse Riemannian volume element, which corresponds to dt on $\mathbb{R}$. This Λ is used to make kind of an average on M of the "local dimension" of the space of square integrable harmonic forms on the leaves at each degree i, giving the finite Λ-Betti numbers $\beta^i_\Lambda(\mathcal{F})$, and thus a Λ-Euler characteristic $\chi_\Lambda(\mathcal{F})$. The technical difficulties of this idea are solved by using the noncommutative integration theory of Connes. But, if the leaves are not compact, the forms of our space $\mathcal{H}(\mathcal{F})$ are not square integrable on the leaves because they are smooth on M. So, a priori, the Λ-Betti numbers are not directly related with the reduced leafwise cohomology.

Now we give another "exotic" solution to the above problem. Observe that, for $f \in C_c^\infty(\mathbb{R})$ supported around 0, the operator A_f is kind of a diffusion of the orthogonal projection $\Pi : \Omega(\mathcal{F}) \to \mathcal{H}(\mathcal{F})$. So the germ of $\beta^i_{\mathrm{dis}}(\mathcal{F})$ at 0 can be considered as a finite measure of the size of $\mathcal{H}^i(\mathcal{F})$, and thus of $\overline{H}^i(\mathcal{F})$ as well. For this reason, the germs at 0 of the distributions $\beta^i_{\mathrm{dis}}(\mathcal{F})$ could be called distributional Betti numbers. But, for the sake of simplicity, the whole distributions $\beta^i_{\mathrm{dis}}(\mathcal{F})$ will be called the *distributional Betti numbers* of $\mathcal{F}$, even though they should be better considered as Lefschetz numbers away from 0. We also define the *distributional Euler characteristic* of $\mathcal{F}$ by the formula

$$\chi_{\mathrm{dis}}(\mathcal{F}) = \sum_i (-1)^i \beta^i_{\mathrm{dis}}(\mathcal{F}) .$$

The following theorem describes the singularity of $\chi_{\mathrm{dis}}(\mathcal{F})$ at 0 in terms of Connes' Λ-Euler characteristic $\chi_\Lambda(\mathcal{F})$. So Connes' Λ-Betti numbers are really strongly related with the reduced leafwise cohomology. The similar result was obtained in [18] when the flow is isometric.

Theorem 1.2 *In some neighbourhood of* 0 *in* $\mathbb{R}$, *we have*

$$\chi_{\mathrm{dis}}(\mathcal{F}) = \chi_\Lambda(\mathcal{F}) \cdot \delta_0 ,$$

where δ_0 *denotes the Dirac measure at* 0.

Recall that a closed orbit c of length l of the flow X_t on $(M, \mathcal{F})$ is called *simple* when

$$\det(\mathrm{id} - X^*_l : T_x\mathcal{F}^* \to T_x\mathcal{F}^*) \neq 0$$

for any $x \in c$. The following theorem proves, for this type of foliations, a conjecture stated by Deninger in [10]. Under some additional assumptions, it was proved in [11, 18].

Theorem 1.3 *Assume that all closed orbits of the flow* X_t *on* $(M, \mathcal{F})$ *are simple. Then we have*

$$\chi_{\mathrm{dis}}(\mathcal{F}) = \sum_c l(c) \sum_{k=1}^{\infty} \mathrm{sign}\,\det\left(\mathrm{id} - X^*_{l(c)} : T_x\mathcal{F}^* \to T_x\mathcal{F}^*\right) \cdot \delta_{kl(c)}$$

on $\mathbb{R}_+$, *where* c *runs over all primitive closed orbits of the flow* X_t, $l(c)$ *denotes the length of* c, *and* x *is an arbitrary point of* c.

Of course, in Theorem 1.3, a symmetric formula for $\chi_{\mathrm{dis}}(\mathcal{F})$ also holds in $\mathbb{R}_-$.

Observe that, if $\dim \mathcal{H}^i(\mathcal{F}) = \beta^i(\mathcal{F}) < \infty$, then $\beta^i_{\mathrm{dis}}(\mathcal{F})$ is a smooth measure whose value at 0 is $\beta^i(\mathcal{F})\,dt$. On the other hand, when $\mathcal{H}(\mathcal{F})$ is of finite dimension, its Euler characteristic can be defined :

$$\chi(\mathcal{F}) = \sum_i (-1)^i \beta^i(\mathcal{F}) .$$

But, by Theorems 1.2 and 1.3, the distributional Euler characteristic $\chi_{\rm dis}(\mathcal{F})$ is trivial if it is smooth, obtaining the following.

Corollary 1.4 *If* $\dim \mathcal{H}(\mathcal{F}) < \infty$, *then* $\chi_{\rm dis}(\mathcal{F})$, $\chi_\Lambda(\mathcal{F})$ *and* $\chi(\mathcal{F})$ *vanish.*

Theorem 1.3 also has the following consequence.

Corollary 1.5 *Assume that all closed orbits of the flow* X_t *on* $(M, \mathcal{F})$ *are simple. If* $\dim \mathcal{H}^{\cdot}(\mathcal{F}) < \infty$, *then, for any* $l \in \mathbb{R}$,

$$\sum_c \frac{1}{\mu(c)} \operatorname{sign} \det \left(\mathrm{id} - X_l^* : T_x\mathcal{F}^* \to T_x\mathcal{F}^*\right) = 0 ,$$

where c *runs over all closed orbits of the flow* X_t *of period* l, $\mu(c)$ *denotes the multiplicity of* c, *and* $x \in c$ *is an arbitrary point.*

When the dimension of $\mathcal{F}$ is two and the leaves are dense, it is possible to relate directly each distributional Betti number with the corresponding Λ-Betti number because we obviously have

$$\beta^0_\Lambda(\mathcal{F}) = \beta^2_\Lambda(\mathcal{F}) = 0 , \quad \dim \mathcal{H}^0(\mathcal{F}) = 1 , \quad \dim \mathcal{H}^2(\mathcal{F}) \le 1 .$$

So Theorem 1.2 has the following consequence.

Corollary 1.6 *Assume that* $\mathcal{F}$ *is of dimension two with dense leaves. Then the singular part of* $\beta^i_{\rm dis}(\mathcal{F})$ *around* 0 *is* $\beta^i_\Lambda(\mathcal{F}) \cdot \delta_0$ *for each degree* i.

It is possible that the statement of Corollary 1.6 holds in general. Indeed, a proof could be given by finding appropriate heat kernel estimates on the leaves. So we propose the following.

Question 1.7 *For each degree* i, *is it true that the singular part of* $\beta^i_{\rm dis}(\mathcal{F})$ *around* 0 *is* $\beta^i_\Lambda(\mathcal{F}) \cdot \delta_0$?

If this question has an affirmative answer, then $\dim \mathcal{H}^i(\mathcal{F}) = \infty$ whenever $\beta^i_\Lambda(\mathcal{F}) \neq 0$. This would mean that the existence of non-trivial square integrable harmonic i-forms on the leaves implies the existence of non-trivial harmonic i-forms on the leaves that are smooth on M. Similar results were shown in [1, 3], where integrable harmonic i-forms on the leaves are used instead of square integrable ones, which are much easier to find.

Let R_L be the curvature of the leafwise metric, and $\mathrm{Pf}(R_L/2\pi) \in \Omega^p(\mathcal{F})$ the leafwise Euler form, $p = \dim \mathcal{F}$. The product $\mathrm{Pf}(R_L/2\pi) \wedge \Lambda$ is a differential form of top degree on M. In particular,

$$\mathrm{Pf}(R_L/2\pi) \wedge \Lambda = \frac{1}{2\pi} K_{\mathcal{F}}(x)\, \omega_M(x)$$

if $\mathcal{F}$ is of dimension 2, where $K_{\mathcal{F}}$ is the Gauss curvature of the leaves and ω_M is the volume form on M. Then Theorem 1.2 and the foliation Gauss-Bonnet theorem from [8], which computes $\chi_\Lambda(\mathcal{F})$, have the following consequence.

Corollary 1.8 *We have*

$$\chi_{\mathrm{dis}}(\mathcal{F}) = \delta_0 \cdot \int_M \mathrm{Pf}(R_L/2\pi) \wedge \Lambda$$

around 0. *In particular, if $\mathcal{F}$ is of dimension two, then*

$$\chi_{\mathrm{dis}}(\mathcal{F}) = \delta_0 \cdot \frac{1}{2\pi} \int_M K_{\mathcal{F}}(x)\, \omega_M(x)$$

around 0.

Corollary 1.8 seems to be a powerful tool to produce examples of foliations with dense leaves on closed Riemannian manifolds with $\dim \mathcal{H}^i(\mathcal{F}) = \infty$; specially, if Question 1.7 has an affirmative answer.

There are obvious versions of these results with general coefficients, which were not considered here for the sake of simplicity.

This type of foliations are just Lie foliations of codimension one. So this is a particular case of our work on distributional Betti numbers for arbitrary Lie foliations [4]. It is worth to explain this particular case here because the arguments are much easier to understand, and moreover the codimension one case is relevant for Deninger's approach to Riemann Hypothesis [10, 11].

Finally, let us mention that our results are somehow related with the study of transversely elliptic operators for Lie group actions [6, 23, 20, 9, 15, 17].

2 Distributional Betti numbers

2.1 Leafwise homotopies

A C^∞ foliation map $f : (M, \mathcal{F}) \to (M, \mathcal{F})$ induces a homomorphism of topological complexes, $f^* : \Omega(\mathcal{F}) \to \Omega(\mathcal{F})$, by pulling-back differential forms. Then it also induces homomorphisms of graded topological vector spaces, $f^* : H(\mathcal{F}) \to H(\mathcal{F})$ and $f^* : \overline{H}(\mathcal{F}) \to \overline{H}(\mathcal{F})$.

Two maps C^∞ maps $f, f' : (M, \mathcal{F}) \to (M, \mathcal{F})$ are said to be leafwise homotopic if there is a C^∞ homotopy between them, $h_s : (M, \mathcal{F}) \to (M, \mathcal{F})$, $s \in I = [0, 1]$, such that each curve $s \mapsto h_s(x)$, $x \in M$, is contained in a leaf. Such a homotopy is called an *leafwise homotopy*, and the notation $f \simeq_{\mathcal{F}} f'$ will be used. Then the usual construction of an homotopy of de Rham complexes produces a linear continuous map $k : \Omega(\mathcal{F}) \to \Omega(\mathcal{F})$, homogeneous of degree -1, such that

$$f^* - f'^* = k \circ d_{\mathcal{F}} + d_{\mathcal{F}} \circ k. \tag{3}$$

Moreover k depends continuously on the homotopy h_s with respect to the C^∞ topology. We get

$$f \simeq_{\mathcal{F}} f' \Longrightarrow \begin{cases} f^* = f'^* : H(\mathcal{F}) \to H(\mathcal{F}) , \\ f^* = f'^* : \overline{H}(\mathcal{F}) \to \overline{H}(\mathcal{F}) , \end{cases}$$

and thus

$$f \simeq_{\mathcal{F}} f' \Longrightarrow \Pi \circ f^* = \Pi \circ f'^* : \mathcal{H}(\mathcal{F}) \to \mathcal{H}(\mathcal{F}) \tag{4}$$

by the isomorphism (2).

2.2 *Smoothing operators*

Let ω_M denote the Riemannian volume element of M, and $\omega_{\mathcal{F}}$ the Riemannian volume element of $\mathcal{F}$. A *smoothing operator* on $\Omega(\mathcal{F})$ is a linear map $P : \Omega(\mathcal{F}) \to \Omega(\mathcal{F})$, continuous with respect to the C^∞ topology, given by

$$(P\alpha)(x) = \int_M k(x,y)\, \alpha(y)\, \omega_M(y) , \quad \alpha \in \Omega(\mathcal{F}) ,$$

where $k \in C^\infty(\bigwedge T\mathcal{F}^* \boxtimes \bigwedge T\mathcal{F})$ is called the smoothing kernel of P. So

$$k(x,y) \in \bigwedge T\mathcal{F}^*_x \otimes \bigwedge T\mathcal{F}_y \equiv \operatorname{Hom}\left(\bigwedge T\mathcal{F}^*_y, \bigwedge T\mathcal{F}^*_x\right) , \quad (x,y) \in M \times M .$$

Any smoothing operator P is of trace class, and we have

$$\operatorname{Tr} P = \int_M \operatorname{tr} k(x,x)\, \omega_M(x) , \tag{5}$$

where k is its smoothing kernel.

Let $\Omega(\mathcal{F})'$ be the dual space of $\Omega(\mathcal{F})$; i.e., the space of continuous linear functionals $\Omega(\mathcal{F}) \to \mathbb{R}$, equipped with the weak dual topology (or topology of pointwise convergence). Let $\mathcal{L}(\Omega(\mathcal{F})', \Omega(\mathcal{F}))$ denote the space of continuous linear operators $\Omega(\mathcal{F})' \to \Omega(\mathcal{F})$, equipped with the topology of bounded convergence. Consider also the C^∞ topology on $C^\infty\left(\bigwedge T\mathcal{F}^* \boxtimes \bigwedge T\mathcal{F}\right)$. The following result is well known.

Lemma 2.1 *A continuous operator in $\Omega(\mathcal{F})$ is smoothing if and only if it extends to a bounded linear operator $\Omega(\mathcal{F})' \to \Omega(\mathcal{F})$. Furthermore the map*

$$\mathcal{L}\left(\Omega(\mathcal{F})', \Omega(\mathcal{F})\right) \to C^\infty\left(\bigwedge T\mathcal{F}^* \boxtimes \bigwedge T\mathcal{F}\right) ,$$

which assigns its kernel to each operator, is an isomorphism of topological vector spaces.

Of course, Lemma 2.1 can be stated in terms of Sobolev spaces $W^k\Omega(\mathcal{F})$ of leafwise differential forms; in particular, a continuous operator in $\Omega(\mathcal{F})$ is smoothing if and only if it extends to a bounded operator $W^k\Omega(\mathcal{F}) \to W^l\Omega(\mathcal{F})$ for all k, l.

A special type of smoothing operators on $\Omega(\mathcal{F})$ can be constructed as follows. A subspace $V \subset \mathfrak{X}(M)$ is called transitive if

$$T_xM = \{Z(x) \mid Z \in V\}$$

for all $x \in M$. Since M is compact, it easily follows that there exists a finite dimensional subspace $W \subset \mathfrak{X}(\mathcal{F})$ such that

$$T_x\mathcal{F} = \{Z(x) \mid Z \in W\} .$$

Then $V = W \oplus \mathbb{R}X \subset \mathfrak{X}(M, \mathcal{F})$ is a finite dimensional transitive subspace. Fix an Euclidean metric on V so that X has norm one and is orthogonal to W. Then the following result was shown in [22].

Lemma 2.2 (Sarkaria) *For any $f \in C_c^\infty(V)$, the operator*

$$P = \int_V Z_1^* \cdot f(Z)\, dZ$$

on $\Omega(\mathcal{F})$ is smoothing, and its smoothing kernel depends continuously on f (with respect to the C^∞ topologies).

2.3 Proof of Theorem 1.1

For each $Z \in W$ and $t \in \mathbb{R}$, the maps

$$((1-s)X)_t \circ (Z + stX)_1 : (M, \mathcal{F}) \to (M, \mathcal{F}) , \quad s \in I ,$$

define a leafwise homotopy between $X_t \circ Z_1 = (tX)_1 \circ Z_1$ and $(Z + tX)_1$. So

$$\Pi \circ Z_1^* \circ X_t^* = \Pi \circ (Z + tX)_1^* : \Omega(\mathcal{F}) \to \Omega(\mathcal{F})$$

by (4).

Now take any $f \in C_c^\infty(\mathbb{R})$ and any $g \in C_c^\infty(W)$ with $\int_W g(Z)\, dZ = 1$, and let

$$B_f = \int_V Z_1^* \cdot h(Z)\, dZ : \Omega(\mathcal{F}) \to \Omega(\mathcal{F}) ,$$

where $h \in C_c^\infty(V)$ is given by $h(Z+tX) = f(t) \cdot g(Z)$ for $Z \in W$ and $t \in \mathbb{R}$. By Lemma 2.2, such a B_f is a smoothing operator whose smoothing kernel depends continuously on h, and thus on f. We also have

$$A_f = \Pi \circ \int_W \int_{\mathbb{R}} Z_1^* \circ X_t^* \cdot f(t) \cdot g(Z)\, dt\, dZ \circ \Pi = \Pi \circ B_f \circ \Pi .$$

On the other hand, it was proved by the authors in [3] that $\Pi : \Omega(\mathcal{F}) \to \Omega(\mathcal{F})$ is continuous, and has an extension to a bounded linear operator on every Sobolev space of leafwise differential forms, and thus to $\Omega(\mathcal{F})'$. So, by Lemma 2.1, the operator $A_f = \Pi \circ B_f \circ \Pi$ is smoothing. Moreover its smoothing kernel depends continuously on A_f, and thus on B_f. In turn, B_f depends continuously on its smoothing kernel, and thus on f. So the smoothing kernel of A_f depends continuously on f. It follows that A_f is a trace class operator, as well as each $A_f^{(i)}$, and their traces depend continuously on f by (5). Therefore each β^i_{dis} is a distribution.

2.4 The dependence of the distributional Betti numbers

Lemma 2.3 *The distributional Betti numbers depend only on $\mathcal{F}$ and the class of X in $\mathfrak{X}(M,\mathcal{F})/\mathfrak{X}(\mathcal{F})$.*

Proof. Suppose that $Y \in \mathfrak{X}(M,\mathcal{F})$ defines the same class as X in $\mathfrak{X}(M,\mathcal{F})/\mathfrak{X}(\mathcal{F})$. Then, for all t,

$$(X + s(Y - X))_t : (M,\mathcal{F}) \to (M,\mathcal{F}) , \quad s \in I ,$$

is an leafwise homotopy between X_t and Y_t. So

$$\Pi \circ \int_{\mathbb{R}} X_s^* \cdot f(s)\, ds = \Pi \circ \int_{\mathbb{R}} Y_s^* \cdot f(s)\, ds \tag{6}$$

by (4).

Take another bundle-like metric on M so that Y is of norm one and orthogonal to the leaves, and let $\Pi' : \Omega(\mathcal{F}) \to \mathcal{H}'(\mathcal{F})$ be the corresponding orthogonal projection onto the corresponding leafwise harmonic forms. Then

$$\Pi' : \mathcal{H}(\mathcal{F}) \xrightarrow{\cong} \mathcal{H}'(\mathcal{F}) \tag{7}$$

by (2).

For any $f \in C_c^\infty(\mathbb{R})$, let

$$B_f = \Pi' \circ \int_{\mathbb{R}} Y_s^* \cdot f(s)\, ds \circ \Pi' : \Omega(\mathcal{F}) \to \Omega(\mathcal{F}) ,$$

and let $B_f^{(i)}$ denote its restriction to $\Omega^i(\mathcal{F})$. Then the distributional Betti numbers β'^i_{dis}, determined by $\mathcal{F}, Y$ and the new bundle-like metric, are given by $\langle \beta'^i_{\mathrm{dis}}, f\rangle = \mathrm{Tr}\left(B_f^{(i)}\right)$.

By (6), (7), and since Π, Π' are projections, it follows that

$$\begin{aligned}
\operatorname{Tr}\left(A_f^{(i)}\right) &= \operatorname{Tr}\left(\Pi \circ \int_{\mathbb{R}} X_s^* \cdot f(s)\, ds : \Omega^i(\mathcal{F}) \to \Omega^i(\mathcal{F})\right) \\
&= \operatorname{Tr}\left(\Pi \circ \int_{\mathbb{R}} Y_s^* \cdot f(s)\, ds : \Omega^i(\mathcal{F}) \to \Omega^i(\mathcal{F})\right) \\
&= \operatorname{Tr}\left(\Pi' \circ \int_{\mathbb{R}} Y_s^* \cdot f(s)\, ds : \Omega^i(\mathcal{F}) \to \Omega^i(\mathcal{F})\right) \\
&= \operatorname{Tr}\left(B_f^{(i)}\right) .
\end{aligned}$$

Therefore $\beta^i_{\mathrm{dis}} = \beta'^i_{\mathrm{dis}}$ as desired. □

3 The distributional Euler characteristic and functions of the leafwise Laplacian

3.1 A family of smoothing operators

Let $UB(\mathbb{R})$ be the space of uniformly bounded Borel functions on $\mathbb{R}$. Since the operator $D_{\mathcal{F}} = d_{\mathcal{F}} + \delta_{\mathcal{F}}$ is essentially self-adjoint in $L^2\Omega(\mathcal{F})$, the Spectral Theorem defines a "functional calculus map"

$$UB(\mathbb{R}) \to \operatorname{End}(L^2\Omega(\mathcal{F})) , \quad \phi \mapsto \phi(D_{\mathcal{F}}) ,$$

where $\operatorname{End}(L^2\Omega(\mathcal{F}))$ denotes the bounded linear endomorphisms of $L^2\Omega(\mathcal{F})$.

Let $\mathcal{A}$ be the set of functions $\phi : \mathbb{R} \to \mathbb{C}$ that extend to entire functions on $\mathbb{C}$ so that, for each compact subset $K \subset \mathbb{R}$, the set of functions $x \mapsto \phi(x+iy)$, $y \in K$, is bounded in the Schwartz space $\mathcal{S}(\mathbb{R})$. Such an $\mathcal{A}$ is a Fréchet algebra, and, in fact, a module over $\mathbb{C}[z]$. This algebra contains all functions with compactly supported Fourier transform, and functions $x \mapsto e^{-tx^2}$ with $t > 0$. By [21], the above functional calculus map, given by the Spectral Theorem, restricts to a "functional calculus map" $\mathcal{A} \to \operatorname{End}(\Omega(\mathcal{F}))$, which is a continuous homomorphism of $\mathbb{C}[z]$-modules and of algebras.

Let $x_1, \dots, x_p, y_1, \dots, y_q$ be foliation coordinates on a foliation patch U; i.e.,

$$(x_1, \dots, x_p, y_1, \dots, y_q) : U \to \mathbb{R}^p \times \mathbb{R}^q$$

is a diffeomorphism so that the slices $\mathbb{R}^p \times \{*\}$ correspond to the plaques of $\mathcal{F}$ in U. Then the differential forms

$$dx_I = dx_{i_1} \wedge \dots \wedge dx_{i_r} ,$$

for multiindices $I = (i_1, \dots, i_r)$ with $1 \le i_1 < \dots < i_r \le p$, form a base of $\Omega(\mathcal{F}|_U)$ as $C^\infty(U)$-module. An operator B in $\Omega(\mathcal{F})$ is local when, for any $\alpha \in \Omega(\mathcal{F})$ and any $x \in M$, the value $(B\alpha)(x)$ depends only on the germ of α at x, and thus B defines an operator $B|_U$ in $\Omega(\mathcal{F}|_U)$ for any open subset $U \subset M$. Recall that a leafwise differential operator B in $\Omega(\mathcal{F})$ is a local operator in $\Omega(\mathcal{F})$ such that, for arbitrary foliation coordinates $x_1, \ldots, x_p, y_1, \ldots, y_q$ on any foliation patch U, with respect to the $C^\infty(U)$-base dx_I of $\Omega(\mathcal{F}|_U)$, the restriction $B|_U$ is given by a matrix whose entries are linear combinations, with coefficients in $C^\infty(U)$, of the leafwise partial derivatives

$$\frac{\partial^k}{\partial x^K} = \frac{\partial^k}{\partial x_1^{k_1} \ldots \partial x_p^{k_p}},$$

for multiindices $K = (k_1, \ldots, k_p) \in \mathbb{N}^p$, where $k = k_1 + \ldots + k_p$. Now, a family $\{B_t \mid t \in \mathbb{R}\}$ of leafwise differential operators is called smooth when, for any foliation patch U with foliation coordinates $x_1, \ldots, x_p, y_1, \ldots, y_q$, in the corresponding expression of $B_t|_U$, the above coefficients of the partial derivatives $\partial^k/\partial x^K$ depend smoothly on t (they are C^∞ functions on $U \times \mathbb{R}$). The support of such a family is the closure in $\mathbb{R}$ of the set of points t with $B_t \neq 0$.

Proposition 3.1 *Let $\phi \in \mathcal{A}$ and let $\{B_t \mid t \in \mathbb{R}\}$ be a smooth compactly supported family of leafwise differential operators in $\Omega(\mathcal{F})$. Then*

$$B = \left(\int_{\mathbb{R}} X_t^* \circ B_t \, dt \right) \circ \phi(D_{\mathcal{F}})$$

is a smoothing operator in $\Omega(\mathcal{F})$ whose smoothing kernel depends continuously on ϕ and $\{B_t \mid t \in \mathbb{R}\}$.

Proposition 3.1 follows from Lemma 2.1 by showing that, for any k, l, there are some N and C such that

$$\left\| B : W^k\Omega(\mathcal{F}) \to W^l\Omega(\mathcal{F}) \right\| \leq \int_{\mathbb{R}} \left| (I - \partial_\xi^2)^N \hat{\phi}(\xi) \right| e^{C|\xi|} \, d\xi \,. \tag{8}$$

The proof of (8) is omitted here because it is rather technical. It is given in our work [3] with more generality.

Observe that Theorem 1.1 is a direct consequence of Proposition 3.1. In fact, by Proposition 3.1, for $f \in C_c^\infty(\mathbb{R})$ and $t > 0$, the operator

$$B_{t,f} = \int_{\mathbb{R}} X_s^* \cdot f(s) \, ds \circ e^{-t\Delta_{\mathcal{F}}}$$

on $\Omega(\mathcal{F})$ is smoothing and its smoothing kernel depends continuously on f. Hence $A_f = \Pi \circ B_{t,f} \circ \Pi$ satisfies the same properties. This also shows that the operator

$$A'_f = \int_{\mathbb{R}} X_t^* \cdot f(t)\,dt \circ \Pi = B_{t,f} \circ \Pi$$

is smoothing with smoothing kernel depending continuously on f, and we have $\mathrm{Tr}(A'_f) = \mathrm{Tr}(A_f)$ since Π is a projection.

3.2 The distributional Euler characteristic and the leafwise heat operator

Let $B_{t,f}^{(i)}$ denote the restriction of $B_{t,f}$ to $\Omega^i(\mathcal{F})$ for each degree i.

Lemma 3.2 *For any $f \in C_c^\infty(\mathbb{R})$, we have $\mathrm{Tr}B_{t,f}^{(i)} \to \mathrm{Tr}A_f^{(i)}$ as $t \to \infty$.*

Proof. Note that

$$\begin{aligned} B_{t,f} - A'_f &= \int_{\mathbb{R}} X_s^* \cdot f(s)\,ds \circ \left(e^{-t\Delta_{\mathcal{F}}} - \Pi\right) \\ &= \int_{\mathbb{R}} X_t^* \cdot f(t)\,dt \circ e^{-\Delta_{\mathcal{F}}} \circ \left(e^{-(t-1)\Delta_{\mathcal{F}}} - \Pi\right) . \end{aligned}$$

On the other hand, by [3], the operator $e^{-(t-1)\Delta_{\mathcal{F}}} - \Pi$ has a continuous extension $\Omega(\mathcal{F})' \to \Omega(\mathcal{F})'$, and converges to zero in $\mathcal{L}(\Omega(\mathcal{F})', \Omega(\mathcal{F})')$ (equipped with the topology of bounded convergence) as $t \to \infty$. Therefore $B_{t,f} - A'_f$ converges to zero in $\mathcal{L}(\Omega(\mathcal{F})', \Omega(\mathcal{F}))$ as $t \to \infty$, and thus its smoothing kernel converges to zero by Lemma 2.1, and the result follows. □

For the sake of simplicity, it is worthwhile to use the supertrace notation. Consider $\Omega(\mathcal{F})$ as a $\mathbb{Z}_2$-graded space:

$$\Omega(\mathcal{F}) = \Omega^+(\mathcal{F}) \oplus \Omega^-(\mathcal{F}) ,$$

where $\Omega^+(\mathcal{F}) = \Omega^{\mathrm{even}}(\mathcal{F})$ and $\Omega^-(\mathcal{F}) = \Omega^{\mathrm{odd}}(\mathcal{F})$. For any $\mathbb{Z}_2$-homogeneous operator P on $\Omega(\mathcal{F})$, let $P^\pm$ denote its restrictions to $\Omega^\pm(\mathcal{F})$. If moreover P is of trace class and $\mathbb{Z}_2$-degree zero, its *supertrace* is defined as

$$\mathrm{Tr}^{\mathrm{s}}(P) = \mathrm{Tr}(P^+) - \mathrm{Tr}(P^-) .$$

In particular,

$$\langle \chi_{\mathrm{dis}}(\mathcal{F}), f\rangle = \mathrm{Tr}^{\mathrm{s}}(A_f)$$

for all $f \in C_c^\infty(\mathbb{R})$.

Choose an even function in $\mathcal{A}$, which can be written as $x \mapsto \psi(x^2)$. Then, for $t > 0$ and $f \in C_c^\infty(\mathbb{R})$, let

$$C_{t,\psi,f} = \int_{\mathbb{R}} X_s^* \cdot f(s)\,ds \circ \psi(t\Delta_{\mathcal{F}})^2 : \Omega(\mathcal{F}) \to \Omega(\mathcal{F}) .$$

In particular, $B_{t,f} = C_{t,\psi,f}$ when $\psi(x^2) = e^{-x^2/2}$.

Lemma 3.3 $\mathrm{Tr}^{\mathrm{s}} C_{t,\psi,f}$ *is independent of* t.

Proof. It is similar to the proof of the corresponding result in the heat equation proof of the usual Lefschetz trace formula [7, 13]. We have

$$\begin{aligned}
\frac{d}{dt}\mathrm{Tr}^{\mathrm{s}} C_{t,\psi,f} &= 2\,\mathrm{Tr}^{\mathrm{s}}\left(\int_{\mathbb{R}} X_s^* \cdot f(s)\,ds \circ \Delta_{\mathcal{F}} \circ \psi'(t\Delta_{\mathcal{F}}) \circ \psi(t\Delta_{\mathcal{F}})\right) \\
&= 2\,\mathrm{Tr}\left(\int_{\mathbb{R}} X_s^* \cdot f(s)\,ds \circ d_{\mathcal{F}}^- \circ \delta_{\mathcal{F}}^+ \circ \psi'\left(t\Delta_{\mathcal{F}}^+\right) \circ \psi\left(t\Delta_{\mathcal{F}}^+\right)\right) \\
&\quad - 2\,\mathrm{Tr}\left(\int_{\mathbb{R}} X_s^* \cdot f(s)\,ds \circ d_{\mathcal{F}}^+ \circ \delta_{\mathcal{F}}^- \circ \psi'\left(t\Delta_{\mathcal{F}}^-\right) \circ \psi\left(t\Delta_{\mathcal{F}}^-\right)\right) \\
&\quad + 2\,\mathrm{Tr}\left(\int_{\mathbb{R}} X_s^* \cdot f(s)\,ds \circ \delta_{\mathcal{F}}^- \circ d_{\mathcal{F}}^+ \circ \psi'\left(t\Delta_{\mathcal{F}}^+\right) \circ \psi\left(t\Delta_{\mathcal{F}}^+\right)\right) \\
&\quad - 2\,\mathrm{Tr}\left(\int_{\mathbb{R}} X_s^* \cdot f(s)\,ds \circ \delta_{\mathcal{F}}^+ \circ d_{\mathcal{F}}^- \circ \psi'\left(t\Delta_{\mathcal{F}}^-\right) \circ \psi\left(t\Delta_{\mathcal{F}}^-\right)\right).
\end{aligned}$$

On the other hand, since the function $x \mapsto \psi'(x^2)$ is in $\mathcal{A}$, we have

$$\begin{aligned}
&\mathrm{Tr}\left(\int_{\mathbb{R}} X_s^* \cdot f(s)\,ds \circ d_{\mathcal{F}}^\mp \circ \delta_{\mathcal{F}}^\pm \circ \psi'\left(t\Delta_{\mathcal{F}}^\pm\right) \circ \psi\left(t\Delta_{\mathcal{F}}^\pm\right)\right) \\
&= \mathrm{Tr}\left(d_{\mathcal{F}}^\mp \circ \int_{\mathbb{R}} X_s^* \cdot f(s)\,ds \circ \psi'\left(t\Delta_{\mathcal{F}}^\pm\right) \circ \psi\left(t\Delta_{\mathcal{F}}^\pm\right) \circ \delta_{\mathcal{F}}^\pm\right) \\
&= \mathrm{Tr}\left(\psi\left(t\Delta_{\mathcal{F}}^\pm\right) \circ \delta_{\mathcal{F}}^\pm \circ d_{\mathcal{F}}^\mp \circ \int_{\mathbb{R}} X_s^* \cdot f(s)\,ds \circ \psi'\left(t\Delta_{\mathcal{F}}^\pm\right)\right) \\
&= \mathrm{Tr}\left(\int_{\mathbb{R}} X_s^* \cdot f(s)\,ds \circ \psi'\left(t\Delta_{\mathcal{F}}^\pm\right) \circ \psi\left(t\Delta_{\mathcal{F}}^\pm\right) \circ \delta_{\mathcal{F}}^\pm \circ d_{\mathcal{F}}^\mp\right) \\
&= \mathrm{Tr}\left(\int_{\mathbb{R}} X_s^* \cdot f(s)\,ds \circ \delta_{\mathcal{F}}^\pm \circ d_{\mathcal{F}}^\mp \circ \psi'\left(t\Delta_{\mathcal{F}}^\pm\right) \circ \psi\left(t\Delta_{\mathcal{F}}^\pm\right)\right),
\end{aligned}$$

where we have used the well known fact that, if A is a trace class operator and B is bounded, then AB and BA are trace class operators with the same trace. Therefore $\frac{d}{dt}\mathrm{Tr}^{\mathrm{s}} C_{t,\psi,f} = 0$ as desired. □

The following result follows directly from Lemmas 3.2 and 3.3.

Corollary 3.4 *We have*

$$\mathrm{Tr}^{\mathrm{s}} B_{t,f} = \langle \chi_{\mathrm{dis}}(\mathcal{F}), f \rangle$$

for any $t > 0$ *and* $f \in C_c^\infty(\mathbb{R})$.

Like in [21, p. 463], choose a sequence of smooth even functions $\phi_m \in \mathcal{A}$, which we write as $\phi_m(x) = \psi_m(x^2)$, with $\phi_m(0) = 1$, and whose Fourier transforms $\widehat{\phi_m}$ are compactly supported and tend to the function $\phi(x) = e^{-x^2/2}$ as $m \to \infty$ (in the topology of the Schwartz space $\mathcal{S}(\mathbb{R})$). Without loss of generality, we can also assume that, for each N and C,

$$\int \left|(\mathrm{id} - \partial_\xi^2)^N \left(\widehat{\phi_m}(\xi) - \hat{\phi}(\xi)\right)\right| e^{C|\xi|}\, d\xi \to 0 \tag{9}$$

as $m \to \infty$. Consider the operator

$$C_{t,m,f} = C_{t,\psi_m,f} = \int_G X_g^* \cdot f(g)\, dg \circ \psi_m(t\Delta_{\mathcal{F}})^2$$

on $\Omega(\mathcal{F})$.

Lemma 3.5 *For any $t > 0$ and $f \in C_c^\infty(\mathbb{R})$, we have*

$$\mathrm{Tr}^s C_{t,m,f} \to \mathrm{Tr}^s B_{t,f} = \langle \chi_{\mathrm{dis}}(\mathcal{F}), f\rangle$$

as $m \to \infty$.

Proof. Combining (9) and (8), we get that

$$C_{t,m,f} - B_{t,f} \to 0$$

in $\mathcal{L}(\Omega(\mathcal{F})', \Omega(\mathcal{F}))$ as $t \to \infty$. By Lemma 2.1, it follows that the smoothing kernel of $C_{t,m,f}$ converges uniformly to the smoothing kernel of $B_{t,f}$, and the result follows. □

3.3 Description of the smoothing kernels

According to the structure of Lie foliations [12, 19], the foliation $\mathcal{F}$ can be described as follows. There is a finitely generated subgroup $\Gamma \subset \mathbb{R}$ that acts on the right in some manifold L such that:

- M is diffeomorphic to the orbit space $L \times_\Gamma \mathbb{R}$ of the right Γ-action on $\widetilde{M} = L \times \mathbb{R}$ given by

$$(x, s) \cdot \gamma = (x \cdot \gamma, s + \gamma)\ , \quad (x, s) \in L \times \mathbb{R}\ , \quad \gamma \in \Gamma\ ;$$

  say $M = L \times_\Gamma \mathbb{R}$. Thus the canonical projection $\pi : \widetilde{M} \to M$ is a covering map.

- The leaves of the lifting $\widetilde{\mathcal{F}}$ of $\mathcal{F}$ to $\widetilde{M}$ are the fibres $L \times \{t\}$, $t \in \mathbb{R}$, of the second factor projection $D : \widetilde{M} \to \mathbb{R}$.

- The flow of the lifting $\widetilde{X}$ of X to $\widetilde{M}$ is given by $\widetilde{X}_t(x, s) = (x, t + s)$, $t \in \mathbb{R}$.

Let $\mathcal{G}, \widetilde{\mathcal{G}}$ denote the holonomy groupoids of $\mathcal{F}, \widetilde{\mathcal{F}}$ respectively. Since the leaves of $\mathcal{F}, \widetilde{\mathcal{F}}$ have trivial holonomy groups, we have

$$\mathcal{G} \equiv \{(x,y) \in M \mid x, y \text{ lie in the same leaf of } \mathcal{F}\} \,,$$

$$\widetilde{\mathcal{G}} \equiv \left\{(\tilde{x},\tilde{y}) \in \widetilde{M} \;\middle|\; \tilde{x},\tilde{y} \text{ lie in the same leaf of } \widetilde{\mathcal{F}} \text{ (fiber of } D)\right\} .$$

Thus $\mathcal{G}, \widetilde{\mathcal{G}}$ are C^∞ submanifolds of $M \times M$ and $\widetilde{M} \times \widetilde{M}$, respectively. Moreover $\pi \times \pi : \widetilde{\mathcal{G}} \to \mathcal{G}$ is a covering map whose group of deck transformations is

$$\operatorname{Aut}\left(\widetilde{\mathcal{G}} \to \mathcal{G}\right) \equiv \operatorname{Aut}(\pi) \equiv \Gamma \,,$$

where $\sigma \in \operatorname{Aut}(\pi)$ corresponds to $\sigma \times \sigma \in \operatorname{Aut}\left(\widetilde{\mathcal{G}} \to \mathcal{G}\right)$. Let $s, r : \mathcal{G} \to M$ the source and the range projections, which are the restrictions of the factor projections $M \times M \to M$.

Recall the definition of the global action of the convolution algebra

$$C^\infty_c\left(\mathcal{G}, r^* \bigwedge T\mathcal{F}^* \otimes s^* \bigwedge T\mathcal{F}\right)$$

in $\Omega(\mathcal{F})$. For any

$$k \in C^\infty_c\left(\mathcal{G}, r^* \bigwedge T\mathcal{F}^* \otimes s^* \bigwedge T\mathcal{F}\right) \,, \quad \alpha \in \Omega(\mathcal{F}) \,,$$

the element $k \cdot \alpha \in \Omega(\mathcal{F})$ is given by

$$(k \cdot \alpha)(x) = \int_{L_x} k(x,y)\, \alpha(y)\, \omega_{\mathcal{F}}(y) \,, \quad x \in M \,,$$

where L_x is the leaf of $\mathcal{F}$ through $x \in M$.

Consider the lifting of the fixed bundle-like metric on M to $\widetilde{M}$ and its restriction to the leaves of $\mathcal{F}$. Let $\omega_{\widetilde{M}}$ denote the Riemannian volume element of $\widetilde{M}$, and $\omega_{\widetilde{\mathcal{F}}}$ the Riemannian volume element of $\widetilde{\mathcal{F}}$. We also have a global action of any

$$\tilde{k} \in C^\infty\left(\widetilde{\mathcal{G}}, r^* \bigwedge T\widetilde{\mathcal{F}}^* \otimes s^* \bigwedge T\widetilde{\mathcal{F}}\right) \,,$$

supported in an R-neighbourhood of the diagonal $\widetilde{M} \equiv \widetilde{\Delta} \subset \widetilde{\mathcal{G}}$ for some $R > 0$, on the space $U\Omega(\widetilde{\mathcal{F}})$ of uniformly bounded differential forms in $\Omega(\widetilde{\mathcal{F}})$: For any $\alpha \in U\Omega(\widetilde{\mathcal{F}})$, the element $\tilde{k} \cdot \alpha \in U\Omega(\widetilde{\mathcal{F}})$ is given by

$$\left(\tilde{k} \cdot \alpha\right)(\tilde{x}) = \int_{\tilde{L}_{\tilde{x}}} \tilde{k}(\tilde{x},\tilde{y})\, \alpha(\tilde{y})\, \omega_{\widetilde{\mathcal{F}}}(\tilde{y}) \,, \quad \tilde{x} \in \widetilde{M} \,.$$

By [21], if h is a bounded Borel function on $\mathbb{R}$ such that its Fourier transform $\hat{h} \in C_c^\infty(\mathbb{R})$, then the operator $h(D_\mathcal{F})$ on $\Omega(\mathcal{F})$ is represented by some element of

$$C_c^\infty\left(\mathcal{G}, r^* \bigwedge T\mathcal{F}^* \otimes s^* \bigwedge T\mathcal{F}\right) .$$

Moreover, it follows from the proof of Assertion 1 in [21, p.461] that, for any function h in the Schwartz space $\mathcal{S}(\mathbb{R})$ with $\operatorname{supp}\hat{h} \subset [-R, R]$, the operator $h(D_\mathcal{F})$ on $\Omega(\mathcal{F})$ is represented by a leafwise smoothing kernel on $\mathcal{G}$ supported in the R-neighbourhood of the diagonal $M \equiv \Delta \subset \mathcal{G}$.

The map $\pi \times \pi : \widetilde{\mathcal{G}} \to \mathcal{G}$ restricts to a diffeomorphism $\pi \times \pi : \widetilde{L} \times \widetilde{L} \to L \times L$ for any leaf $\widetilde{L}$ of $\widetilde{\mathcal{F}}$ ($L = \pi(\widetilde{L})$). Hence, the lift of the leafwise smoothing kernel of $h(D_\mathcal{F})$ to $\widetilde{\mathcal{G}}$ is supported in the R-neighbourhood of the diagonal $\widetilde{M} \equiv \widetilde{\Delta} \subset \widetilde{\mathcal{G}}$, and thus defines an operator $h\left(D_{\widetilde{\mathcal{F}}}\right)$ on $U\Omega\left(\widetilde{\mathcal{F}}\right)$. It is clear that the diagram

$$\begin{array}{ccc} U\Omega\left(\widetilde{\mathcal{F}}\right) & \xrightarrow{h\left(D_{\widetilde{\mathcal{F}}}\right)} & U\Omega\left(\widetilde{\mathcal{F}}\right) \\ {\scriptstyle (\pi\times\pi)^*}\uparrow & & \uparrow{\scriptstyle (\pi\times\pi)^*} \\ \Omega(\mathcal{F}) & \xrightarrow{h(D_\mathcal{F})} & \Omega(\mathcal{F}) \end{array}$$

commutes.

Since $\widehat{\phi_m}$ is compactly supported, the operator $\psi_m(t\Delta_\mathcal{F})^2$ is represented by a leafwise smoothing kernel

$$k_{m,t} \in C_c^\infty\left(\mathcal{G}, r^* \bigwedge T\mathcal{F}^* \otimes s^* \bigwedge T\mathcal{F}\right) .$$

The action of $k_{m,t}$ on $\Omega(\mathcal{F})$ defines the operator $\psi_m(t\Delta_\mathcal{F})^2$ in $\Omega(\mathcal{F})$: For any $\alpha \in \Omega(\mathcal{F})$, we have

$$\left(\psi_m(t\Delta_\mathcal{F})^2\alpha\right)(x) = (k_{m,t} \cdot \alpha)(x) = \int_{L_x} k_{m,t}(x,y)\,\alpha(y)\,\omega_\mathcal{F}(y) , \quad x \in M .$$

This operator is equal to the operator $\psi_m(t\Delta_\mathcal{F})^2$ in $\Omega(\mathcal{F})$ defined by the Spectral Theorem.

Let $\tilde{k}_{m,t}$ be the lift of $k_{m,t}$ to $\widetilde{\mathcal{G}}$, which is also supported in the R-neighbourhood of the diagonal $\widetilde{M} \equiv \widetilde{\Delta} \subset \widetilde{\mathcal{G}}$ for some $R > 0$. For $\beta \in U\Omega\left(\widetilde{\mathcal{F}}\right)$, we have

$$\left(\psi_m\left(t\Delta_{\widetilde{\mathcal{F}}}\right)^2\beta\right)(\tilde{x}) = \int_{\widetilde{L}_{\tilde{x}}} \tilde{k}_{m,t}(\tilde{x},\tilde{y})\,\beta(\tilde{y})\,\omega_{\widetilde{\mathcal{F}}}(\tilde{y}) , \quad \tilde{x} \in \widetilde{M} .$$

For any $f \in C_c^\infty(\mathbb{R})$, consider the operator

$$\widetilde{C}_{t,m,f} = \int_{\mathbb{R}} \widetilde{X}_s^* \cdot f(s)\, ds \circ \psi_m(t\Delta_{\widetilde{\mathcal{F}}})^2 : U\Omega\left(\widetilde{\mathcal{F}}\right) \to U\Omega\left(\widetilde{\mathcal{F}}\right) .$$

Obviously, $\widetilde{C}_{t,m,f} \circ \pi^* = \pi^* \circ C_{t,m,f}$ on $\Omega(\mathcal{F})$.

Lemma 3.6 *For any $f \in C_c^\infty(\mathbb{R})$, $t > 0$ and m, the operator $\widetilde{C}_{t,m,f}$ is a smoothing operator whose smoothing kernel $\tilde{c}_{t,m,f}$ is supported in an R-neighbourhood of the diagonal in $\widetilde{M} \times \widetilde{M}$ for some $R > 0$, and*

$$\tilde{c}_{t,m,f}((x,u),(y,v)) = \widetilde{X}_{v-u}^* \circ \tilde{k}_{m,t}((x,v),(y,v)) \cdot f(v-u) \tag{10}$$

for $(x,u),(y,v) \in L \times \mathbb{R} = \widetilde{M}$.

Proof. For $\alpha \in U\Omega\left(\widetilde{\mathcal{F}}\right)$, we have

$$\begin{aligned}
&\left(\widetilde{C}_{t,m,f}\alpha\right)(x,u) \\
&= \left(\int_{\mathbb{R}} \widetilde{X}_s^* \cdot f(s)\, ds \circ \psi_m\left(t\Delta_{\widetilde{\mathcal{F}}}\right)^2 \alpha\right)(x,u) \\
&= \int_{\mathbb{R}} \widetilde{X}_s^* \left(\psi_m\left(t\Delta_{\widetilde{\mathcal{F}}}\right)^2 \alpha\right)(x,u+s) \cdot f(s)\, ds \\
&= \int_{\mathbb{R}} \widetilde{X}_{v-u}^* \left(\int_{L\times\{v\}} \tilde{k}_{m,t}((x,v),(y,v))\alpha(y,v)\, \omega_{\widetilde{\mathcal{F}}}(y,v)\right) \cdot f(v-u)\, dv \\
&= \int_{\widetilde{M}} \widetilde{X}_{v-u}^* \circ \tilde{k}_{m,t}((x,v),(y,v))\alpha(y,v) \cdot f(v-u)\, \omega_{\widetilde{M}}(y,v) ,
\end{aligned}$$

by using the change of variable $s = v - u$. Hence $\widetilde{C}_{t,m,f}$ is defined by the smoothing kernel given in (10), and the result follows. □

For each $(x,u) \in \widetilde{M}$, let $[x,u] = \pi(x,v)$. It is easy to see that the kernels $\tilde{c}_{t,m,f}$ and $c_{t,m,f}$ are related by the formula

$$c_{t,m,f}([x,u],[y,v]) \equiv \sum_{\gamma\in\Gamma} \tilde{c}_{t,m,f}((x,u),(y\cdot\gamma, v+\gamma)) , \tag{11}$$

where we use the identity

$$\bigwedge T_{[x,u]}\widetilde{\mathcal{F}}^* \otimes \bigwedge T_{[y,v]}\widetilde{\mathcal{F}} \equiv \bigwedge T_{(x,u)}\mathcal{F}^* \otimes \bigwedge T_{(y\cdot\gamma,v+\gamma)}\mathcal{F}$$

via the map π. The sum in (11) is finite because $\tilde{c}_{t,m,f}$ is supported in an R-neighbourhood of the diagonal in $\widetilde{M} \times \widetilde{M}$ for some $R > 0$.

4 Distributional Euler characteristic and Connes' Euler characteristic

The goal of this section is to prove Theorem 1.2.

Lemma 4.1 *Given $R > 0$, there is some neighbourhood U of 0 in $\mathbb{R}$ so that*

$$\pi : \{(y,v) \mid (y,u) \in B_{\widetilde{\mathcal{F}}}((x,u),R)\ ,\ v-u \in U\} \to M$$

is injective for any $(x,u) \in \widetilde{M}$, where $B_{\widetilde{\mathcal{F}}}((x,u),R)$ denotes the ball of radius R and centred at (x,u) in the leaf $L \times \{u\}$.

Proof. Since M is compact, there exists a compact subset $K \subset \widetilde{M}$ with $\pi(K) = M$. Note that, if the statement holds for $(x,u) \in K$, then it holds for all $(x,u) \in \widetilde{M}$.

Assume the result is false. Then there exist sequences $(x_i,u_i),(y_i,v_i) \in \widetilde{M}$, and a sequence $\gamma_i \in \Gamma$ with $(x_i,u_i) \in K$, $\gamma_i \neq 0$, and such that (y_i,v_i) and $(y_i \cdot \gamma_i, v_i + \gamma_i)$ approach $B_{\widetilde{\mathcal{F}}}((x_i,u_i),R)$ in the sense that the distance between the terms of this sequences to this set converges to zero.

Since K is compact, we can assume that there exists $\lim_i(x_i,u_i)_i = (x,u) \in \widetilde{M}$. Hence, (y_i,v_i) and $(y_i \cdot \gamma_i, v_i + \gamma_i)$ approach the relatively compact set $B_{\widetilde{\mathcal{F}}}((x,\ u),\ R)$. It follows that, for infinitely many i, the points $(y_i,\ v_i)$ and $(y_i\ \cdot\ \gamma_i,\ v_i\ +\ \gamma_i)$ lie in some compact neighbourhood Q of $B_{\widetilde{\mathcal{F}}}((x,u),R)$; thus $Q \cdot \gamma_i \bigcap Q \neq \emptyset$ for infinitely many i. This implies that there exists some $\gamma \in \Gamma$ such that $\gamma_i = \gamma$ for infinitely many i. In particular, $\gamma \neq 0$.

On the other hand, since (y_i,v_i) and $(y_i \cdot \gamma_i, v_i+\gamma_i)$ approach $B_{\widetilde{\mathcal{F}}}((x,u),R)$, which is relatively compact, we can assume that there exist

$$\lim_i(y_i,v_i)\ ,\quad \lim_i(y_i \cdot \gamma_i, v_i + \gamma_i)$$

in $\overline{B_{\widetilde{\mathcal{F}}}((x,u),R)}$. Therefore, if $\lim_i(y_i,v_i) = (y,v)$, then

$$(y,v) \in L \times \{u\}\ ,\quad (y \cdot \gamma, v + \gamma) = \lim_i(y_i \cdot \gamma_i, v_i + \gamma_i) \in L \times \{u\}\ ,$$

yielding $v = u = v + \gamma$, and thus $\gamma = 0$. The result follows from this contradiction. □

Lemma 4.2 *For each m, there is a neighbourhood U of 0 in $\mathbb{R}$ such that the map π is injective on the support of $\tilde{c}_{t,m,f}((x,u),\cdot)$ for all $(x,u) \in \widetilde{M}$ if t is small enough and the support of f is contained in U.*

Proof. For any fixed $R > 0$, choose some neighbourhood U of 0 in $\mathbb{R}$ satisfying the statement of Lemma 4.1. For any m, we have

$$\operatorname{supp}\left(F\left(x \mapsto \psi_m\left(tx^2\right)\right)\right) = \sqrt{t} \cdot \operatorname{supp}\left(F\left(x \mapsto \psi_m\left(x^2\right)\right)\right)\ ,$$

where F denotes the Fourier transform. So, since $\operatorname{supp}\left(F\left(x \mapsto \phi_m\left(x^2\right)\right)\right)$ is compact, it follows that

$$\operatorname{supp}\left(F\left(x \mapsto \psi_m\left(tx^2\right)\right)\right) \subset \left[-\frac{R}{2}, \frac{R}{2}\right]$$

if t is small enough, for each m. Thus the leafwise smoothing kernel of $\psi_m\left(t\Delta_{\widetilde{\mathcal{F}}}\right)$ is supported in the $\frac{R}{2}$-neighbourhood of the diagonal $\widetilde{\Delta} \subset \widetilde{\mathcal{G}}$, and the leafwise smoothing kernel of $\psi_m\left(t\Delta_{\widetilde{\mathcal{F}}}\right)^2$ is supported in the R-neighbourhood of the diagonal $\widetilde{\Delta} \subset \widetilde{\mathcal{G}}$. From Lemma 3.6, we get

$$\operatorname{supp}(\tilde{c}_{t,m,f}((x,u),\cdot)) \subset \left\{(y,v) \mid (y,u) \in B_{\widetilde{\mathcal{F}}}((x,u),R)\ ,\ v-u \in U\right\}$$

for any $(x,u) \in \widetilde{M}$ if t is small enough, and the result follows by Lemma 4.1. □

By noncommutative integration theory [8], the holonomy invariant transverse measure Λ defines a trace on the von Neumann algebra of $\mathcal{F}$, which can be shortly described as follows. The twisted convolution algebra

$$C_c^\infty\left(\mathcal{G}, r^* \bigwedge T\mathcal{F}^* \otimes s^* \bigwedge T\mathcal{F}\right)$$

is contained in the (twisted) von Neumann algebra $W^*\left(\mathcal{F}, \bigwedge T\mathcal{F}^*\right)$, and, for any

$$k \in C_c^\infty(\mathcal{G}, r^* \bigwedge T\mathcal{F}^* \otimes s^* \bigwedge T\mathcal{F})\ ,$$

the trace $\operatorname{Tr}_\Lambda(k)$ is finite and given by the formula

$$\operatorname{Tr}_\Lambda(k) = \int_M \operatorname{tr} k([x,u],[x,u])\,\omega_M([x,u])\ . \tag{12}$$

Now fix U as in Lemma 4.2, and let $f \in C_c^\infty(\mathbb{R})$ with $\operatorname{supp} f \subset U$.

Proposition 4.3 *For all $t > 0$, we have*

$$\operatorname{Tr}^{\mathrm{s}} C_{t,m,f} = f(0) \cdot \operatorname{Tr}^{\mathrm{s}}_\Lambda \psi_m(t\Delta_{\mathcal{F}})^2\ . \tag{13}$$

Proof. Recall that $\operatorname{Tr}^{\mathrm{s}} C_{t,m,f}$ is independent of t by Lemma 3.3, and $\operatorname{Tr}^{\mathrm{s}}_\Lambda \psi_m(t\Delta_{\mathcal{F}})^2$ is also independent of t by [21]. So we need only prove this statement for a single t.

If t is small enough, we have

$$\begin{aligned}\operatorname{Tr}^{\mathrm{s}} C_{t,m,f} &= \int_M \operatorname{tr}^{\mathrm{s}}(c_{t,m,f}([x,u],[x,u])\,\omega_M([x,u]) \\ &= \int_{\widetilde{M}} \operatorname{tr}^{\mathrm{s}}(\tilde{c}_{t,m,f}((x,u),(x,u)))\,\omega_{\widetilde{M}}(x,u)\end{aligned}$$

$$
\begin{aligned}
&= f(0) \cdot \int_{\widetilde{M}} \operatorname{tr}^{\mathrm{s}}(\tilde{k}_{m,t}((x,u),(x,u)))\, \omega_{\widetilde{M}}(x,u) \\
&= f(0) \cdot \int_{M} \operatorname{tr}^{\mathrm{s}}(k_{m,t}([x,u],[x,u]))\, \omega_{M}([x,u]) \\
&= f(0) \cdot \operatorname{Tr}^{\mathrm{s}}_{\Lambda} \psi_m \left(t\Delta_{\mathcal{F}}\right)^2
\end{aligned}
$$

by (11), Lemma 4.2, Lemma 3.6 and (12) since

$$
k_{m,t} \in C^{\infty}_{c}\left(\mathcal{G}, s^* \bigwedge T\mathcal{F}^* \otimes r^* \bigwedge T\mathcal{F}\right) .
$$

□

Now we recall some facts on Connes' Betti numbers. The family

$$
\{P_{i,L} \mid L \text{ is a leaf of } \mathcal{F}\} ,
$$

where each $P_{i,L}$ is the orthogonal projection onto the space of square integrable harmonic i-forms on L, defines a projection P_i in the twisted foliation von Neumann algebra $W^*\left(\mathcal{F}, \bigwedge T\mathcal{F}^*\right)$. As in [8], one can define Λ-Betti numbers $\beta^i_{\Lambda}(\mathcal{F})$ as

$$
\beta^i_{\Lambda}(\mathcal{F}) = \operatorname{Tr}_{\Lambda} P_i .
$$

Then the Λ-Euler characteristic of $\mathcal{F}$ is

$$
\chi_{\Lambda}(\mathcal{F}) = \sum_i (-1)^i \beta^i_{\Lambda}(\mathcal{F}) .
$$

Using the corresponding supertrace notion, this formula can be rewritten as

$$
\chi_{\Lambda}(\mathcal{F}) = \operatorname{Tr}^{\mathrm{s}}_{\Lambda} P ,
$$

where $P = \sum_i P_i$. By [21], we have that $\operatorname{Tr}^{\mathrm{s}}_{\Lambda}\left(\psi_m\left(t\Delta_{\mathcal{F}}\right)^2\right)$ is independent of t, and

$$
\operatorname{Tr}^{\mathrm{s}}_{\Lambda}\left(\psi_m\left(t\Delta_{\mathcal{F}}\right)^2\right) \to \chi_{\Lambda}(\mathcal{F}) \tag{14}
$$

as $m \to \infty$ for all $t > 0$ (independently of m!).

Proof of Theorem 1.2. Fix a neighbourhood U of 0 in $\mathbb{R}$ as in Lemma 4.2. Let $f \in C^{\infty}_{c}(\mathbb{R})$ with $\operatorname{supp} f \in U$. Combining Proposition 4.3, Lemma 3.3 and (14), we have

$$
\operatorname{Tr}^{\mathrm{s}} C_{t,m,f} = f(0) \cdot \operatorname{Tr}^{\mathrm{s}}_{\Lambda}\left(\psi_m\left(t\Delta_{\mathcal{F}}\right)^2\right) \to f(0) \cdot \chi_{\Lambda}(\mathcal{F}) \tag{15}
$$

as $m \to \infty$ for any $t > 0$.

Fix any $\varepsilon > 0$. From Lemmas 3.5 and 3.3, it follows that

$$|\mathrm{Tr}^{\mathrm{s}} C_{t,m,f} - \mathrm{Tr}^{\mathrm{s}} B_{t,f}| < \varepsilon$$

for any $t > 0$ if m is large enough. But

$$\mathrm{Tr}^{\mathrm{s}} B_{t,f} = \langle \chi_{\mathrm{dis}}(\mathcal{F}), f \rangle$$

for any $t > 0$ by Corollary 3.4. So

$$|\mathrm{Tr}^{\mathrm{s}} C_{t,m,f} - \langle \chi_{\mathrm{dis}}(\mathcal{F}), f \rangle| < \varepsilon \tag{16}$$

for any $t > 0$ if m is large enough.

From (15) and (16), it follows that

$$|\langle \chi_{\mathrm{dis}}(\mathcal{F}), f \rangle - f(0) \cdot \chi_{\Lambda}(\mathcal{F})| < \varepsilon$$

for any $\varepsilon > 0$, yielding $\chi_{\mathrm{dis}}(\mathcal{F}) = \chi_{\Lambda}(\mathcal{F}) \cdot \delta_0$ on U since $\varepsilon > 0$ and $f \in C_c^\infty(U)$ are arbitrary. □

5 Localization theorem

Theorem 5.1 *The distribution $\chi_{\mathrm{dis}}(\mathcal{F})$ is supported in the set of all $s \in \mathbb{R}$ such that X_s has a fixed point in M.*

Proof. Let V be an open subset in $\mathbb{R}$ such that X_s has no fixed points for all $s \in V$. We have to prove that $\chi_{\mathrm{dis}}(\mathcal{F}) = 0$ on V. Note that the fact that X_s has no fixed points for all $s \in V$ is equivalent to

$$\widetilde{X}_s(x, u) \neq (x \cdot \gamma, u + \gamma)$$

for any $s \in V$, $\gamma \in \Gamma$ and $(x, u) \in \widetilde{M}$.

One can prove an analogue of Lemma 4.2, asserting that there exists a neighbourhood U of 0 in $\mathbb{R}$ such that, for any $s_0 \in \mathbb{R}$ and for each m, if $f \in C_c^\infty(U + s_0)$, then π is injective on the support of $\tilde{c}_{t,m,f}((x, u), \cdot)$ for $(x, u) \in \widetilde{M}$ if t is small enough. ¿From this, it follows that, for each m, if t is small enough, then, for any $[x, u] \in M$, either

$$c_{t,m,f}([x, u], [x, u]) = 0$$

or

$$c_{t,m,f}([x, u], [x, u]) = \tilde{c}_{t,m,f}((x, u), (x \cdot \gamma, u + \gamma))$$

for some $\gamma \in \Gamma \cap (U + s_0)$. In the latter case, such a γ is uniquely determined by (x, u).

Take $s_0 \in V$ and some neighbourhood U of 0 as above, satisfying also $U + s_0 \subset V$. Since s_0 is an arbitrary point of V, it is enough to show that $\langle \chi_{\rm dis}, f\rangle = 0$ for any $f \in C^\infty(U+s_0)$. Then, by Lemma 3.5 and Corollary 3.4, it is enough to show that $\mathrm{Tr}^{\mathrm{s}} C_{t,m,f} = 0$ for each m and t small enough (depending on m).

We have

$$\begin{aligned}
\mathrm{Tr}^{\mathrm{s}} C_{t,m,f} &= \int_M \mathrm{tr}^{\mathrm{s}}(c_{t,m,f}([x,u],[x,u]))\,\omega_M([x,u]) \\
&= \int_{\widetilde{M}} \mathrm{tr}^{\mathrm{s}}\left(\tilde{c}_{t,m,f}((x,u),(x\cdot\gamma,u+\gamma))\right)\,\omega_{\widetilde{M}}(x,u) \\
&= \int_{\widetilde{M}} \mathrm{tr}^{\mathrm{s}}\left(\widetilde{X}^*_\gamma \circ \tilde{k}_{m,t}((x,u+\gamma),(x\cdot\gamma,u+\gamma))\right)\cdot f(\gamma)\,\omega_{\widetilde{M}}(x,u)\ ,
\end{aligned}$$

for the appropriate choice of $\gamma \in \Gamma$, where we use the identity

$$\bigwedge T_{(y,v)}\widetilde{\mathcal{F}}^* \equiv \bigwedge T_{(y\cdot\gamma,v+\gamma)}\widetilde{\mathcal{F}}^*$$

given by the diagonal action of γ on $\widetilde{M}$.

Since $\mathrm{supp} f \subset U + s_0$, we can consider only those $[x,u] \in M$ with $\gamma \in U + s_0$ for some choice of $\gamma \in \Gamma$ with $(x\cdot\gamma, u+\gamma)$ in the support of $\tilde{c}_{t,m,f}$. Thus $\gamma \in V$, yielding

$$(x\cdot\gamma,u+\gamma) \neq (x,u+\gamma)$$

by assumption. It follows that

$$\tilde{k}_{m,t}((x,u+\gamma),(x\cdot\gamma,u+\gamma)) \to 0$$

as $t \to 0$ uniformly on $(x,u) \in M$. Since $\mathrm{Tr}^{\mathrm{s}} C_{t,m,f}$ is independent of t, we get $\mathrm{Tr}^{\mathrm{s}} C_{t,m,f} = 0$ for each m, as desired. □

6 The Lefschetz trace formula

The goal of this section is to prove Theorem 1.3.

By Theorem 5.1, in order to evaluate

$$\mathrm{Tr}^{\mathrm{s}} C_{t,m,f} = \int_M \mathrm{tr}^{\mathrm{s}} c_{t,m,f}([x,u],[x,u])\,\omega_M([x,u])$$

asymptotically as $t \to 0$, it is enough to integrate over small neighbourhoods of closed orbits.

As in the proof of Theorem 5.1, take a neighbourhood U of 0 in $\mathbb{R}$ such that, for any $s_0 \in \mathbb{R}$ and for each m, if $f \in C^\infty_c(U+s_0)$, then π is injective in the support of $\tilde{c}_{t,m,f}((x,u),\cdot)$ for all $(x,u) \in \widetilde{M}$ if t is small enough.

Let s_0 be the period of some closed orbit of X. There exist finitely many closed orbits with the period in $U + s_0$. Hence, the neighbourhood

U can be chosen so that s_0 is the only period that belongs to $s_0 + U$, and thus only this period may be in supp$f$. Take a closed orbit of period s_0, and let c be the corresponding primitive closed orbit with length $l = l(c)$; thus $s_0 = kl$ for some integer $k > 0$. We also get that $l \in \Gamma$, and

$$(x, u + l) = (x \cdot l, u + l) \ , \quad (x, u + kl) = (x \cdot kl, u + kl)$$

if $[x, u]$ is in c. So x is a fixed point of the action of l on L, and there are no other fixed points of elements of $\Gamma \cap (s_0 + U)$ in some open neighbourhood W of x in L because all X-orbits are simple. Note also that

$$\pi(\{x\} \times [0, l]) = c$$

and

$$\pi : \{x\} \times (0, l) \to c$$

is a C^∞ embedding. Moreover $\pi(W \times [0, l])$ is an open neighbourhood of c where there are no other orbits of period in $s_0 + U$, and

$$\pi : W \times (0, l) \to M$$

is a C^∞ embedding.

Denote by

$$\tilde{k}_t \in C^\infty \left(\widetilde{\mathcal{G}}, r^* \bigwedge T\widetilde{\mathcal{F}}^* \otimes s^* \bigwedge T\widetilde{\mathcal{F}} \right)$$

the leafwise smoothing kernel of the leafwise heat operator $e^{-t\Delta_{\widetilde{\mathcal{F}}}}$. Then, since $\tilde{k}_{m,t}(x, \cdot)$ converges to $\tilde{k}_t(x, \cdot)$ as $m \to \infty$, by Lemma 3.6, it is enough to compute the asymptotics as $t \downarrow 0$ of

$$\int_{W \times [0,l]} \operatorname{tr}^{\mathrm{s}} \left(\widetilde{X}_\gamma^* \circ \tilde{k}_t \left((y, v + kl), (y \cdot kl, v + kl) \right) \right) \cdot f(kl) \, \omega_{\widetilde{M}}(y, v)$$

$$= f(kl) \cdot \int_0^l \int_W \operatorname{tr}^{\mathrm{s}} \left(\widetilde{X}_{kl}^* \circ \tilde{k}_t \left((y, v + kl), (y \cdot kl, v + kl) \right) \right) \omega_{W,v}(y) \, dv \ ,$$

where $\omega_{W,v}$ is the restriction of $\omega_{\widetilde{\mathcal{F}}}$ to $W \times \{v\} \equiv W$, and we use the identity

$$\bigwedge T_{(y,v)} \widetilde{\mathcal{F}}^* \equiv \bigwedge T_{(y \cdot kl, v + kl)} \widetilde{\mathcal{F}}^*$$

given by the diagonal action of kl on $\widetilde{M}$. But, by [7, 13], the integral

$$\int_W \operatorname{tr}^{\mathrm{s}} \left(\widetilde{X}_{kl}^* \circ \tilde{k}_t \left((y, v + kl), (y \cdot kl, v + kl) \right) \right) \omega_{W,v}(y)$$

converges as $t \downarrow 0$ to

$$\operatorname{sign} \det \left(\mathrm{id} - X_{kl*} : T_{(x,v)}(W \times \{v\}) \to T_{(x,v)}(W \times \{v\}) \right) \ ,$$

which is independent of v, and the proof is finished. □

References

1. J.A. Álvarez López and S. Hurder, *Pure-point spectrum for foliation geometric operators*, preprint, 1995.
2. J.A. Álvarez López and G. Hector, *The dimension of the leafwise reduced cohomology*, preprint, 2000.
3. J.A. Álvarez López and Y.A. Kordyukov, *Long time behaviour of leafwise heat flow for Riemannian foliations*, preprint dg-ga/9612010, 1996, to appear in *Compositio Math.*
4. J.A. Álvarez López and Y.A. Kordyukov, *Distributional Betti numbers for Lie foliations*, preprint, 2000.
5. J.A. Álvarez López and P. Tondeur, *Hodge decomposition along the leaves of a Riemannian foliation*, J.Funct. Anal., **99** (1991), 443–458.
6. M.F. Atiyah, *Elliptic operators and compact groups*, in Lecture Notes in Mathematics, **401**, 1–93. Springer Verlag, Berlin-Heidelberg-New York, 1974.
7. M.F. Atiyah and R. Bott, *A Lefschetz fixed point formula for elliptic complexes* I, Ann. of Math., **86** (1967), 374–407.
8. A. Connes, *Sur la théorie non-commutative de l'intégration*, in Algèbres d'opérateurs, Lecture Notes in Math. **725**, Springer Verlag, 1979, 19–143.
9. A. Connes, *Noncommutative differential geometry*, Publ. Math., **62** (1986), 41–144.
10. C. Deninger, *Some analogies between number theory and dynamical systems on foliated spaces*, Doc. Math. J. Extra Volume ICM98, **1** (1998), 163–186.
11. C. Deninger and W. Singhof, *A note on dynamical trace formulas*, manuscript, 2000.
12. E. Fedida, *Sur les feuilletages de Lie*, C. R. Acad. Sci. Paris. Ser. A-B, **272** (1971), A999–A1001.
13. P.B. Gilkey, Invariance Theory, the Heat Equation and the Atiyah-Singer Index Theorem, Publish or Perish Press, Berkeley, 1984.
14. A. Haefliger, *Some remarks on foliations with minimal leaves*, J. Diff. Geom., **15** (1980), 269–384.
15. Y.A. Kordyukov, *Transversally elliptic operators on G-manifolds of bounded geometry*, Russ. J. Math. Ph., **2** (1994), 175–198.
16. Y.A. Kordyukov, *Functional calculus for tangentially elliptic operators on foliated manifolds*, in Analysis and Geometry in Foliated Manifolds, Proceedings of the VII International Colloquium on Differential Geometry, Santiago de Compostela, 1994, 113–136, World Scientific, Singapore, 1995.

17. Y.A. Kordyukov, *Noncommutative spectral geometry of Riemannian foliations*, Manuscripta Math., **94** (1997), 45–73.
18. C. Lazarov, *Transverse index and periodic orbits*, Geom. and Funct. Anal., **10** (2000), 124–159.
19. P. Molino, *Géométrie globale des feuilletages riemanniens*, Proc. Nederl. Acad. A1, **85** (1982), 45–76.
20. A. Nestke and P. Zuckermann, *The index of transversally elliptic complexes*, Rend. Circ. Mat. Palermo, **34** Suppl. 9 (1985), 165–175.
21. J. Roe, *Finite propagation speed and Connes' foliation algebra*, Math. Proc. Cambridge Philos. Soc., **102** (1987), 459–466.
22. K.S. Sarkaria, *A finiteness theorem for foliated manifolds*, J. Math. Soc. Japan, **30** (1978), 687–696.
23. I.M. Singer, *Recent applications of index theory for elliptic operators*, in Proc. Symp. Pure Appl. Math. **23**, 11–31. Amer. Math. Soc., Providence, R. I., 1973.

Received November 3, 2000.

Proceedings of
FOLIATIONS: GEOMETRY AND DYNAMICS
held in Warsaw, May 29–June 9, 2000
ed. by Paweł WALCZAK *et al.*
World Scientific, Singapore, 2002
pp. 185–195

CIRCLE AT INFINITY INFLUENCES ON THE SMOOTHNESS OF SURFACE FLOWS

SAMUIL ARANSON

Department of Applied Mathematics, Nizhny Novgorod State Technical University
24 Minina Str., Nizhny Novgorod 603600, Russia,
e-mail: ara@uic.nnov.ru

EVGENY ZHUZHOMA

Department of Applied Mathematics, Nizhny Novgorod State Technical University
24 Minina Str., Nizhny Novgorod 603600, Russia,
e-mail: zhuzhoma@focus.nnov.ru

Let M be a closed hyperbolic surface of negative Euler characteristic and Δ the universal covering space of M. Let S_∞ be the circle at infinity of Δ. Then there exists a continual set $U \in S_\infty$ with the following property. Suppose a flow f^t on M has a semitrajectory $l^\pm$ such that a lift $\overline{l}^\pm$ of $l^\pm$ has an asymptotic direction defined by a point of U; then f^t is not analytic and has a continual set of fixed points. Moreover, f^t has neither nontrivially recurrent trajectories nor closed transversals nonhomotopic to zero. The set U is dense and has zero Lebesgue measure on S_∞. Given any point $\sigma \in U$, there exists a C^∞-flow on M and the corresponding covering flow $\overline{f}^t$ on Δ such that some semitrajectory of $\overline{f}^t$ has an asymptotic direction defined by the point σ.

Introduction

The purpose of this paper is to consider some aspects of Anosov-Weil's theory. Most generally by the *theory of Anosov-Weil* one understands the study of asymptotic properties of curves with no self-intersections lifted to the universal covering, and their 'deviation' from the lines of constant geodesic curvature that have the same asymptotic direction (see reviews [4] and [15], Chapter 10).

Throughout the paper M is a closed surface of negative Euler characteristic $\chi(M) < 0$. Let Δ be a hyperbolic (or Lobachevsky) plane, which is a simply connected complete Riemannian 2-manifold of constant negative curvature -1. For Δ, we use the Poincare disk model $|z| < 1$ of the complex z-plane endowed with the metric $ds = 2|dz|/(1-|z|^2)$. The circle $S_\infty = \partial\Delta = (|z| = 1)$ is called the *circle at infinity*. The geodesics of Δ are Euclidean circular arcs orthogonal to S_∞. We shall suppose that endpoints of any geodesic are ideal points belonging to the circle at infinity.

Any M can be thought of as the orbit space Δ/Γ, where Γ is a discrete group of isometries of Δ acting freely on Δ. The group Γ is isomorphic to the fundamental group of M. The natural projection $\pi : \Delta \to \Delta/\Gamma \cong M$ is a universal cover.

Suppose f^t is a flow on M and $\overline{f}^t$ is a covering flow on Δ. Let $l^\pm$ be a positive (negative) semitrajectory of f^t and $\overline{l}^\pm$ a semitrajectory of $\overline{f}^t$ which covers $l^\pm$. If $\overline{l}^\pm$ tends to a unique point of S_∞, say σ, as $t \to \pm\infty$, we shall say that $\overline{l}^\pm$ has an *asymptotic direction* defined by σ. Sometimes we shall say that $l^\pm$ has an *asymptotic direction* whenever some lift of $l^\pm$ (and so every lift) has an asymptotic direction. The point $\sigma \in S_\infty$ is called a *point achieved by* f^t. Denote by $A_{fl} \subset S_\infty$ the set of points achieved by all flows on M.

The main result of the paper is the following theorem.

Theorem 2.1 *Let M be a closed surface of negative Euler characteristic and S_∞ be the circle at infinity of the universal covering space Δ of M. Then there exists a continual set $U(I^-_{fol,h}(M)) \subset A_{fl}$ with the following properties. Suppose a flow f^t on M has a semitrajectory $l^\pm$ such that a lift $\overline{l}^{(\pm)}$ has an asymptotic direction defined by a point of $U(I^-_{fol,h}(M))$; then f^t is not analytic and has a continual set of fixed points. Furthermore, f^t has neither nontrivially recurrent semitrajectories nor closed transversals nonhomotopic to zero. The set $U(I^-_{fol,h}(M))$ is dense and has zero Lebesgue measure on S_∞.*

Note that due to the papers [2] and [10], given any point of $U_g = U(I^-_{fol,h}(M))$, there is a C^∞-flow f^t which satisfies theorem 2.1 (see also [3], where a C^∞-flow f^t with a preassigned smooth invariant measure was constructed).

1 Preliminaries

A foliation F on M is called *irrational* if the following conditions hold:

- F has a finite number of singularities.

- every one-dimensional leaf of F is dense on M.
- F has no fake saddles.

It follows from the definition that any singularity s of F is of saddle type i.e., some neighbourhood of s is a union of saddle sectors divided by separatrices. The index of s is equal to $ind\ s = 1 - \frac{\nu(s)}{2}$, where $\nu(s) \geq 1$ is a number of separatrices of s. Recall that s is called a *fake saddle* if $\nu(s) = 2$.

Denote by $I_{fol}(M)$ the set of irrational foliations on M. The following lemmas are needed for the sequel.

Lemma 1.1 *Let M_g^2 be a closed orientable surface of genus $g \geq 2$. There exists an irrational foliation $\mathcal{F}$ on M_g^2 such that any singularity of $\mathcal{F}$ has negative index and there are at least two singularities of half-integer index.*

Proof. According to the paper [8], there exists an irrational foliation $\mathcal{F}_{3,1}$ on the projective plane P^2 with four singularities: three thorns and one tripod. Let $p : S^2 \to P^2$ be a double covering, where S^2 is the 2-sphere. By the Path Lifting Theorem, there is a covering foliation $\mathcal{F}_{6,2}$ on S^2. It is easy to see that $\mathcal{F}_{6,2}$ has six thorns $n_1, \dots, n_6$ and two tripods t_1, t_2. Since $\mathcal{F}_{3,1}$ is irrational and p is a double covering, $\mathcal{F}_{6,2}$ is irrational as well.

Fix a natural number $g \geq 2$ and take a set $\Sigma_{2g-4} \subset S^2$ of $2g - 4$ points (if $g = 2$, then $\Sigma_{2g-4} = \emptyset$) so that these points belong to pairwise disjoint leaves and no points on separatrices of the foliation $\mathcal{F}_{6,2}$. Calling every point of Σ_{2g-4} a singularity (a fake saddle which has exactly two saddle sectors), we get the new foliation denoted by $\mathcal{F}'_{6,2}$. By the choice of Σ_{2g-4}, $\mathcal{F}'_{6,2}$ is a highly transitive foliation.

Let $q : M_g^2 \to S^2$ be a branched double cover having the branched set $\Sigma_{2g-4} \cup_{i=1}^{6} \{n_i\}$ with every point of branched index 2. Since $\Sigma_{2g-4} \cup_{i=1}^{6} \{n_i\}$ belongs to the set of singularities of $\mathcal{F}'_{6,2}$, it follows that there is a foliation $\mathcal{F}'$ on M_g^2 that is a lift of $\mathcal{F}'_{6,2}$ under q. The singularities of $\mathcal{F}'$ are six fake saddles $q^{-1}(n_1), \dots, q^{-1}(n_6)$, and four tripods belonging to $q^{-1}(t_1)$, $q^{-1}(t_2)$, and $2g - 4$ Morse's saddles each of the index -1. Note that the index of a tripod equals $-0,5$.

Since q is a branched double cover, $\mathcal{F}'$ is a highly transitive foliation. Deleting the fake saddles, we get the desired foliation $\mathcal{F}$ on M_g^2. □

Another way to obtain irrational foliations with singularities of half-integer negative index is to apply Theorem 2 [13] where irrational foliations are invariant foliations of pseudo-Anosov homeomorphisms. Denote by $I^-_{fol}(M_g^2)$ the set of irrational foliations on the surface M_g^2 which have only singularities of negative index. Denote also by $I^-_{fol,h}(M) \subset I^-_{fol}(M)$ the subset of foliations having singularities of half-integer index. By Lemma 1.1 $I^-_{fol,h}(M_g^2) \neq \emptyset$.

We recall that a *geodesic lamination* on M is a non-empty closed subset of M which is a disjoint union of simple (without self-intersections) geodesics. Given an irrational foliation $\mathcal{F} \in I^-_{fol}(M^2_g)$, one can constructs a special geodesic lamination called a geodesic framework.

Consider a leaf L of $\mathcal{F} \in I^-_{fol}(M^2_g)$ which is not a separatrix. Then both positive and negative semileaves of L have asymptotic directions [12]. Let $\overline{L}$ be any lift of L to Δ. Then the ideal endpoints $\alpha(\overline{L})$, $\omega(\overline{L})$ of $\overline{L}$ belongs to S_∞ and $\alpha(\overline{L}) \neq \omega(\overline{L})$ because the index of every singularity is negative. Let $\overline{g}(\overline{L})$ be the geodesic with the endpoints $\alpha(\overline{L})$, $\omega(\overline{L})$. This geodesic is called a *geodesic corresponding to* $\overline{L}$. The geodesic $\pi(\overline{g}(\overline{L})) \overset{\text{def}}{=} g(L)$ does not depend on a lift of L and is called a *geodesic corresponding to the leaf* L.

Since L is a simple curve (i.e., with no self-intersections), $g(L)$ is a simple geodesic. Hence the topological closure $clos[g(L)]$ of $g(L)$ is a geodesic lamination, Lemma 3.1 in [9]. This geodesic lamination is independent of the choice of L because any leaf of $\mathcal{F}$ is dense in M^2_g. The geodesic lamination

$$clos[g(L)] \overset{\text{def}}{=} G(\mathcal{F})$$

is called a *geodesic framework* of $\mathcal{F}$.

Lemma 1.2 *Let $\mathcal{F} \in I^-_{fol}(M^2_g)$ and let $G(\mathcal{F})$ be the geodesic framework of the foliation $\mathcal{F}$. Then $G(\mathcal{F})$ is a minimal geodesic lamination consisting of nontrivially recurrent geodesics, each dense in $G(\mathcal{F})$. Moreover, any ideal endpoint of a lift of every geodesic $g \in G(\mathcal{F})$ is irrational, i.e. not a fixed point of an element of the Fuchsian group Γ.*

Proof. Since any leaf of $\mathcal{F}$ is dense in M^2_g, it follows that any geodesic of $G(\mathcal{F})$ is nontrivially recurrent and dense in $G(\mathcal{F})$. Hence, $G(\mathcal{F})$ is a minimal geodesic lamination i.e., it has no proper non-empty sublaminations.

Due to [7], any ideal endpoint of a lift of every geodesic $g \in G(\mathcal{F})$ is irrational. □

Lemma 1.3 *Let $\mathcal{F} \in I^-_{fol}(M^2_g)$ and let $G(\mathcal{F})$ be the geodesic framework of the foliation $\mathcal{F}$. Given any geodesic $g \in G(\mathcal{F})$ and closed geodesic g_0, the intersection $g \cap g_0$ is nonempty.*

Proof. Due to [12], there is either a leaf L or generalized leaf L (denoted by the same letter) such that $g(L) = g$. Further, any component of $M^2_g - G(\mathcal{F})$ is simply connected because $\mathcal{F}$ is an irrational foliation (see also [9] Chapter 4). Combining this with the density of g in $G(\mathcal{F})$, we get the desired result. □

Recall that any component of $M^2_g - G(\mathcal{F})$ is an ideal polygon (see, e.g. [9]), Figure 1.

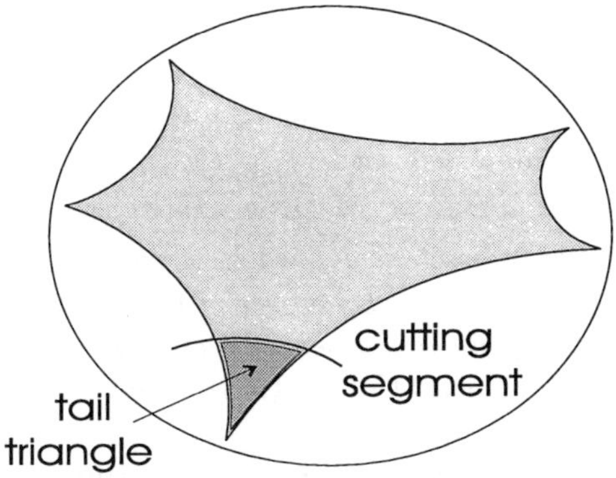

Figure 1. An ideal polygon.

Let g_1, g_2 be adjacent sides of an ideal polygon $P \subset M_g^2 - G(\mathcal{F})$. Take a cutting segment Σ through g_1, g_2 so that some subsegment $(a, b) \subset \Sigma \cap int\ P$ and geodesics rays $g_1^+ \subset g_1$, $g_2^+ \subset g_2$ with the initial points $a \in \Sigma \cap g_1$, $b \in \Sigma \cap g_2$ respectively form an ideal triangle denoted by $T(\Sigma, g_1, g_2)$, see Figure 1. This triangle is called a *tail triangle* of P formed by the adjacent sides g_1, g_2 and the cutting segment Σ.

Lemma 1.4 *Let $G(\mathcal{F})$ be the geodesic framework of a foliation $\mathcal{F} \in I_{fol}^-(M_g^2)$ and P an ideal polygon of $G(\mathcal{F})$. Then every tail triangle of P is dense in $G(\mathcal{F})$.*

Proof. Let $T(\Sigma, g_1, g_2)$ be a tail triangle of P formed by adjacent sides g_1, g_2 of P and a cutting segment Σ. Then $T(\Sigma, g_1, g_2)$ is bounded by geodesics rays $g_1^+ \subset g_1$, $g_2^+ \subset g_2$. It follows from Lemma 1.2 that both g_1^+ and g_2^+ are dense in $G(\mathcal{F})$. This concludes the proof. □

2 Proof of the main theorem

Let $\mathcal{F}$ be an irrational foliation on M that satisfies the conditions of Lemma 1.1 i.e., any singularity of $\mathcal{F}$ has negative index and there are at least two singularities of half-integer index. Denote now by $I_{fol,h}^-(M) \subset I_{fol}^-(M)$ the subset of such foliations on the surface M. Let $\overline{\mathcal{F}}$ be a lift of $\mathcal{F}$ on Δ. Then any semileaf $\overline{L}^{\pm}$ of $\overline{\mathcal{F}}$ that is not a separatrix goes to infinity and has an asymptotic direction $\sigma(\overline{L}^{\pm}) \in S_\infty$ [12]. Denote by $A(\mathcal{F}) \subset S_\infty$ the set of points that are achieved by semileaves of $\overline{\mathcal{F}}$. By definition, put

$$U(I_{fol,h}^-(M)) = \bigcup_{\mathcal{F}} A(\mathcal{F}), \qquad \text{where } \mathcal{F} \text{ ranges over } I_{fol,h}^-(M).$$

Due to [12], $U(I^-_{fol,h}(M)) \neq \emptyset$.

Theorem 2.1 *Let M be a closed surface of negative Euler characteristic and S_∞ be the circle at infinity of the universal covering space Δ of M. Then there exists a continual set $U(I^-_{fol,h}(M)) \subset A_{fl}$ with the following properties. Suppose a flow f^t on M has a semitrajectory $l^\pm$ such that a lift $\bar{l}^{(\pm)}$ has an asymptotic direction defined by a point of $U(I^-_{fol,h}(M))$; then f^t is not analytic and has a continual set of fixed points. Furthermore, f^t has neither nontrivially recurrent semitrajectories nor closed transversals nonhomotopic to zero. The set $U(I^-_{fol,h}(M))$ is dense and has zero Lebesgue measure on S_∞.*

Proof. Without loss of generality one can assume that M is a closed orientable surface M^2_g of genus $g \geq 2$, otherwise we pass to a double (non-branched) cover. To simplify matters denote $U(I^-_{fol,h}(M)) = U_g$. For the reader's convenience, we divide the proof into steps. The end of proof of a step will be denoted by $\diamond$. We keep the notation of Section 1.

Step 2.1 *Take $\mathcal{F} \in I^-_{fol,h}(M^2_g)$ and let $G(\mathcal{F})$ be a geodesic framework of the foliation $\mathcal{F}$. Then $G(\mathcal{F})$ has at least two ideal polygons with odd number of sides.*

Proof of Step 2.1 Since $\mathcal{F}$ has at least two singularities with a half-integer index, it immediately follows that there are at least two saddles, say s_1 and s_2, with odd number of separatrices each. Let $\bar{s}_1$ and $\bar{s}_2$ be lifts of s_1 and s_2 respectively. Then the corresponding lifts of separatrices define the ideal vertices of the ideal polygons with odd number of sides. $\diamond$

Step 2.2 *Let $G(\mathcal{F})$ be the geodesic framework of a foliation $\mathcal{F} \in I^-_{fol,h}(M^2_g)$ and Σ a geodesic segment which is transversal to $G(\mathcal{F})$, int $\Sigma \cap G(\mathcal{F}) \neq \emptyset$. We consider Σ to be a segment endowed with some normal orientation. Then given any geodesic $g \in G(\mathcal{F})$ endowed with a natural parametrization $\theta : \mathbb{R} \to g$, there is a sequence of parameters $t_i \in \mathbb{R}$ such that $t_i \to \infty$ as $i \to \infty$, $\theta(t_i) \cap \Sigma \neq \emptyset$, and the index of intersection $g \cap \Sigma$ at the point $\theta(t_i)$ equals $(-1)^i$, $i \in \mathbb{N}$.*

Proof of Step 2.2 By Step 2.1, there is an ideal polygon $P \subset G(\mathcal{F})$ with odd number of sides. By Lemma 1.2, any positive ray g^+ of g is dense in $G(\mathcal{F})$, and thus $\partial P \subset clos\ g = G(\mathcal{F})$. Taking into account that number of sides of P is odd, it follows that there is a cutting segment Σ_{12} intersecting two adjacent sides $g_1, g_2 \in \partial P$ of P at the points $m_1 \in g_1 \cap \Sigma_{12}$, $m_2 \in g_2 \cap \Sigma_{12}$ respectively such that

1. $int\ P \cap (m_1, m_2) = \emptyset$.
2. There is a sequence of parameters $t'_i \in \mathbb{R}$ with $\theta(t'_i) \to m_1 \cup m_2$ as

$i \to \infty$.

3. The index of intersection $g \cap \Sigma_{12}$ at the point $\theta(t_i')$ equals $(-1)^i$, $i \in \mathbb{N}$ (see Figure 2).

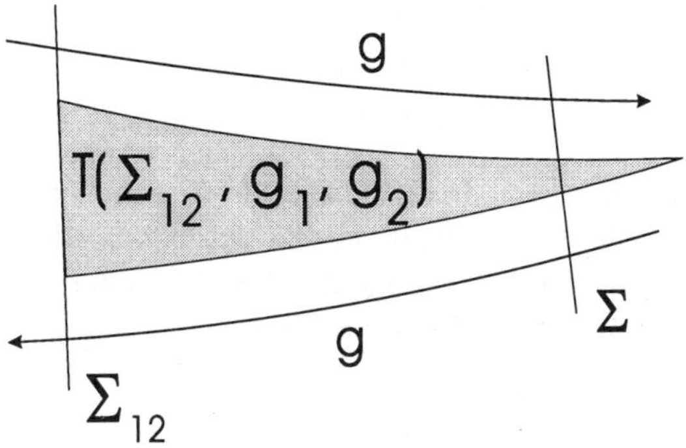

Figure 2. The tail triangle $T(\Sigma_{12}, g_1^+, g_2^+)$

Let $g_i^+ \subset g_i$ $(i = 1, 2)$ be the geodesic rays which form together with Σ_{12} the tail triangle $T(\Sigma_{12}, g_1^+, g_2^+)$ of P, Fig. 2. By Lemma 1.4, $T(\Sigma_{12}, g_1^+, g_2^+)$ is dense in $G(\mathcal{F})$. Hence, $T(\Sigma_{12}, g_1^+, g_2^+)$ intersects Σ. To conclude the proof, it remains to apply the theorem on the continuous dependence of geodesics of $G(\mathcal{F})$ on initial conditions. $\diamond$

Step 2.3 *Suppose a flow f^t on M_g^2 has a semitrajectory l^+ such that a lift $\overline{l}^+$ of l^+ has an asymptotic direction defined by a point $\sigma(\overline{l}^+) \in U_g$; then f^t has no closed transversals nonhomotopic to zero.*

Proof of Step 2.3 Assume the contrary. Then there is a closed simple transversal T of f^t which is nonhomotopic to zero [5]. It is well known that there exists a closed simple geodesic g_0 which is freely homotopic to T.

By the condition, there is a semileaf L^+ of some irrational foliation $\mathcal{F} \in I^-_{fol,h}(M)$ and the lift $\overline{L}^+$ of L^+ such that $\sigma(\overline{l}^+) = \sigma(\overline{L}^+) \overset{def}{=} \sigma$. If L^+ does not belong to an α-separatrix of a saddle, then L^+ belongs to a leaf denoted by L. If L^+ belongs to an α-separatrix of some saddle, say s, then we denote by the same letter L a generalized leaf containing L^+ (any of two generalized leaves of s which contains L^+). Let $G(\mathcal{F})$ be the geodesic framework of $\mathcal{F}$ and $g(L) \in G(\mathcal{F})$ the geodesic corresponding to L. Fix a natural parametrization $\theta : \mathbb{R} \to g(L)$.

Let $\overline{L}$ be the lift of L containing $\overline{L}^+$ and $\overline{g}(\overline{L})$ the geodesic on Δ corresponding to $\overline{L}$. Then $\sigma^+(\overline{g}(\overline{L})) = \sigma$. Note that due to Lemma 1.3, $g(L)$

intersects g_0. By Step 2.2, there is a sequence of parameters $t_i \in \mathbb{R}$ such that $t_i \to \infty$ as $i \to \infty$, $\theta(t_i) \cap g_0 \neq \emptyset$, and the index of intersection $g(L) \cap g_0$ at the point $\theta(t_i)$ equals $(-1)^i$, $i \in \mathbb{N}$. Hence there is a sequence of $\overline{g}_i$ of lifts g_0 which satisfy the following properties (see Figure 3):

1. $\overline{g}(\overline{L})$ intersects $\overline{g}_i$ at a unique point, say $\overline{m}_i$.
2. The topological limit of geodesics $\overline{g}_i$ equals σ.
3. The index of intersection $\overline{g}(\overline{L}) \cap \overline{g}_i$ at $\overline{m}_i$ equals $(-1)^i$.

Since the geodesic g_0 is freely homotopic to T, it follows that $\overline{g}_i$ is a corresponding geodesic of some lift $\overline{T}_i$ of T for every $i \in \mathbb{N}$, i.e., $\overline{g}_i$ and $\overline{T}_i$ have the same ideal endpoints on S_∞. By item 2, $\overline{l}^+$ intersects $\overline{T}_i$ beginning with $i \geq i_0$ for some i_0 sufficiently large, see Figure 3.

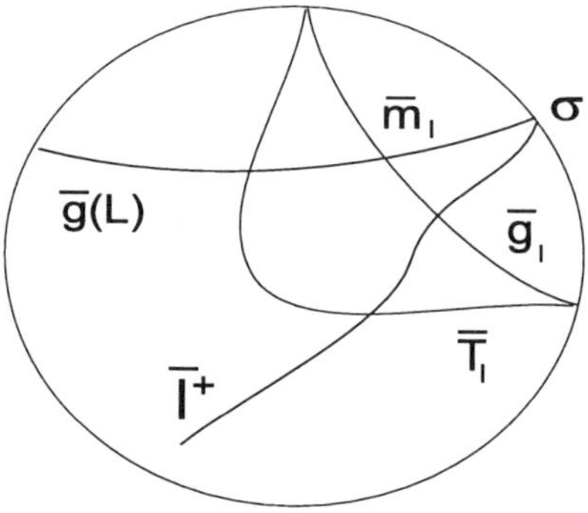

Figure 3. The sequence of $\overline{g}_i$.

By item 3, the index of intersection $\overline{l}^+ \cap \overline{T}_i$, $i \geq i_0$ is equal to $(-1)^i$. Hence, l^+ must intersect T in opposite directions, which is impossible. ◇

As a consequence of Step 2.3, f^t has no nontrivially recurrent semitrajectories and trajectories, otherwise f^t would have a closed simple transversals nonhomotopic to zero (Lemma 1.2.3 in [5]), which is impossible.

Step 2.4 *Suppose a flow f^t on M_g^2 has a semitrajectory l^+ such that a lift $\overline{l}^+$ of l^+ has an asymptotic direction defined by a point $\sigma(\overline{l}^+) \in U_g$; then f^t has a continual set of fixed points. Moreover, the ω-limit set of l^+ consists of fixed points.*

Proof of Step 2.4 Assume the contrary, i.e., the set of fixed points of f^t is not continual. The ω-limit set $\omega(l^+)$ of l^+ is not a unique fixed point because $\overline{l}^+$ has an asymptotic direction. Hence, $\omega(l^+)$ contains at least two points. It is well known (see, e.g., [5]) that an ω-limit set of any positive semitrajectory is a connected set. Combining this with our assumption, we get that $\omega(l^+)$ contains a one-dimensional trajectory, say l_1.

Let S_1 be a transversal segment through a point $a \in l_1$, $a \in int\ S_1$. Then l^+ intersects S_1 infinitely many times because $a \in \omega(l^+)$. As a consequence, there is a closed simple transversal T intersecting l^+, Corollary 1.1.2 [5]. Since $\overline{l}^+$ has an asymptotic direction, T is not homotopic to zero. This contradiction with Step 2.3 concludes the proof. ◇

Step 2.5 *Suppose a flow f^t on M_g^2 has a semitrajectory l^+ such that a lift $\overline{l}^+$ of l^+ has an asymptotic direction defined by a point $\sigma(\overline{l}^+) \in U_g$; then f^t is not analytic.*

Proof of Step 2.5 Assume the contrary. Denote by $Fix(f^t)$ the set of fixed points of f^t. According to [1], pp. 38–39, $Fix(f^t)$ contains a finite number of isolated points, a finite number of non-isolated points S, which are endpoints of curves formed by fixed points, and closed curves, and nonclosed curves with endpoints in S. Outside of S, these curves of $Fix(f^t)$ are analytic, pairwise disjoint, and have no self-intersections. As a consequence of this description, $Fix(f^t)$ is a compact set.

By Step 2.4, the ω-limit set of l^+ is a compact subset of $Fix(f^t)$. The description above of $Fix(f^t)$ shows that l^+ tends spirally to a closed contour nonhomotopic to zero because $\overline{l}^+$ has an asymptotic direction. Hence, $\sigma(\overline{l}^+)$ is a rational point. This contradicts Lemma 1.2. ◇

To conclude the proof it remains to show that U_g has zero Lebesgue measure. Following [11], we shall call a geodesic $\overline{g} \subset \Delta$ *transitive* if given any intervals U_1 and $U_2 \subset S_\infty$, there is a map $\gamma \in \Gamma$ such that one ideal endpoint of $\gamma(\overline{g})$ belongs to U_1 and the other ideal endpoint of $\gamma(\overline{g})$ belongs to U_2. Denote by $TR(\Gamma) \subset S_\infty$ the set of the points $a \in S_\infty$ with the following property: given any other distinct point $b \in S_\infty$, the geodesic joining a, b and directed from b to a is transitive. Myrberg [14] proved that the Lebesgue measure of $TR(\Gamma)$ is equal to the Lebesgue measure of S_∞. In particular, $S_\infty - TR(\Gamma)$ has zero Lebesgue measure.

Step 2.6 $U_g \subset S_\infty - TR(\Gamma)$. *In particular, U_g has zero Lebesgue measure.*

Proof. In fact, it was proved by Hedlund [11] that any geodesic with an endpoint in $TR(\Gamma)$ is transitive. So, if we assume that there is a point $\sigma \in U_g \cap TR(\Gamma)$, then σ is an endpoint of a lift $\overline{g}$ of some geodesic $g \in G(\mathcal{F})$, where $\mathcal{F} \in I^-_{fol,h}(M_g^2)$ is an irrational foliation. Obviously, g has self-intersections because $\overline{g}$ is transitive. This contradicts the fact that a geodesic lamination consists of geodesic with no self-intersections. □

Acknowledgments

The authors are grateful to D. Anosov, V. Grines, R. Plykin, A. Stepin and A. Zhirov for useful discussions. We thank a referee for useful remarks which helped us to correct many errors.

The research was partially supported by the INTAS grant 97-1843 and RFBR grant 99-01-00230.

References

1. D.V. Anosov, *On the behaviour of trajectories, on the Euclidian and Lobachevsky plane, covering trajectories of flows on closed surfaces, I,* Izvestia Acad. Nauk SSSR, Ser. Mat., **51** (1987), no 1, 16–43 (in Russian); *Transl. in:* Math. USSR, Izv., **30** (1988).
2. D.V. Anosov, *On the behaviour of trajectories, on the Euclidian and Lobachevsky plane, covering the trajectories of flows on closed surfaces, II,* Izvestia Acad. Nauk SSSR, Ser. Mat., **52** (1988), 451–478 (in Russian); *Transl. in:* Math. USSR, Izv., **32** (1989), no 3, 449–474.
3. D.V. Anosov, *On the behaviour of trajectories, in the Euclidian and Lobachevsky plane, covering the trajectories of flows on closed surfaces, III,* Izvestia Ross. Akad. Nauk, Ser. Mat., **59** (1995), no 2, 63–96 (in Russian).
4. D.V. Anosov, *Flows on closed surfaces and behaviour of trajectories lifted to the universal covering plane.* Jour. of Dyn. and Control Sys., **1** (1995), 125–138.
5. S. Aranson, G. Belitsky and E. Zhuzhoma, An Introduction to Qualitative Theory of Dynamical Systems on Surfaces, Amer. Math. Soc., Math. Monogr., Providence, 1996.
6. S. Aranson, I. Bronshtein, I. Nikolaev and E. Zhuzhoma, *Qualitative theory of foliations on closed surfaces,* J. Math. Sci., **90** (1998), no 3, 2111–2149.
7. S. Aranson and E. Zhuzhoma, *Maier's theorems and geodesic laminations of surface flows,* Jour. of Dyn. and Contr. Sys., **2** (1996), no 4, 557–582.
8. P. Arnoux and J.C. Yoccoz, *Construction de difféomorphismes pseudo-Anosov,* C. R. Acad. Sci., **292** (1981), 75–78.
9. A.J. Casson and S.A. Bleiler, Automorphisms of Surfaces after Nielsen and Thurston, London Math. Soc. Student Texts, Cambridge Univ. Press, 1988.
10. C. Gutierrez, *Smoothing continuous flows on 2-manifolds and recurrences,* Ergod. Th. and Dyn. Sys., **6** (1986), 17–44.

11. G. Hedlund, *Two-dimensional manifolds and transitivity,* Ann. Math., **37** (1936), no 3, 534–542.
12. G. Levitt, *Foliations and laminations on hyperbolic surfaces,* Topology, **22** (1983), no 2, 119–135.
13. H. Masur and J. Smillie, *Quadratic differentials with prescribed singularities and pseudo-Anosov diffeomorphisms,* Comment. Math. Helv., **68** (1993), 289–307.
14. P.J. Myrberg, *Ein Approximationssatz fur die Fuchsschen Gruppen,* Acta Math., **57** (1931), 389–409.
15. I. Nikolaev and E. Zhuzhoma, Flows on 2-dimensional manifolds, Lecture Notes in Math. **1705**, Springer Verlag, 1999.

Received June 7, 2000.

Proceedings of
FOLIATIONS: GEOMETRY AND DYNAMICS
held in Warsaw, May 29–June 9, 2000
ed. by Paweł WALCZAK *et al.*
World Scientific, Singapore, 2002
pp. 197–211

ENTROPIES OF HYPERBOLIC GROUPS AND SOME FOLIATED SPACES

ANDRZEJ BIŚ
Wydział Matematyki, Uniwersytet Łódzki,
ul. Banacha 22, 90-238 Łódź, Poland,
e-mail: andbis@math.uni.lodz.pl

PAWEŁ G. WALCZAK
Wydział Matematyki, Uniwersytet Łódzki,
ul. Banacha 22, 90-238 Łódź, Poland,
e-mail: pawelwal@math.uni.lodz.pl

1 Introduction

Hyperbolic groups in the Gromov's [9] sense play an important role in geometric group theory (see Grigorchuk and de la Harpe [8] and the references there). In particular, any non-elementary hyperbolic group has exponential growth and the compact boundary of positive finite Hausdorff dimension (see, Ghys and de la Harpe [6], pp. 126 and 157). Also, a hyperbolic group G (generated by a finite symmetric set S) acts on the boundary of its Cayley graph $X = C(G, S)$ *via* Lipschitz quasi-conformal maps (ibidem, p. 127). Roughly speaking, if (x_n) is a sequence of elements of G representing a point ξ of ∂X and $g \in G$, then $g(\xi)$ is a point of ∂X represented by the sequence $(g \cdot x_n)$. The dynamics of this action is of great interest. For instance, it has been shown (see Coornaert and Papadopoulos [3]) that this action is finitely presented, i.e. it is semiconjugate to a subshift of finite type in such a way that the fibres of the conjugating map are finite of bounded length and the equivalence relation determined by this map (two points are related whenever their images are equal) is another subshift of finite type. Also, one can consider the topological entropy $h(G, S)$ of this action in the sense of Ghys, Langevin and Walczak [7]. Let us recall at this point that the entropy of a group action, as well as the related geometric entropy of a foliation, generalizes the topological entropy of a continuous transformation in the sense of Bowen [1]. For readers' convenience, suitable definitions are given in Section 2.

Given a finitely generated group G and a finite symmetric generating set S, one can consider its rate of growth $\mathrm{gr}(G, S)$ defined as the rate of growth of the cardinality of the families of these elements of G which can be expressed as words of given length in the alphabet S (see Section 3 for more details). It seems that the numbers $h(G, S)$ and $\mathrm{gr}(G, S)$ should be closely related. In fact, we expect that they should be equal as it happens in the case of free groups (compare Section 5) when the Cayley graph is just a tree. The graph of an arbitrary hyperbolic group, even for a good generating set, is not a tree and has some "unwanted" links. These links provided technical difficulties which did not allow us to get the expected equality in general. However, their role becomes less and less significant when the scale is enlarging. This drove us to introduce a technically complicated notion of the rate of growth $\mathrm{gr}^{\mathrm{rel}}(G, S; \mu, \tau)$ relative to some bounds μ and τ depending on the geometry of G. For free groups, this relative rate of growth coincides with the standard one and this is in fact the reason for which the equality mentioned above holds in this case (see Section 5 again). The use of these bounds allowed us to overcome difficulties mentioned above and to prove the following.

Theorem 1 *The topological entropy $h(G, S)$ (with respect to a finite symmetric generating set S) of a hyperbolic group G acting on its ideal boundary lies between the exponential rate $\mathrm{gr}^{\mathrm{rel}}(G, S; \mu, \tau)$ of growth of G relative to suitable bounds μ and τ, and the exponential rate $\mathrm{gr}(G, S)$ of growth of G (with respect to the same generating set):*

$$\mathrm{gr}^{\mathrm{rel}}(G, S; \mu, \tau) \leq h(G, S; \partial G) \leq \mathrm{gr}(G, S). \tag{1}$$

The precise description of the bounds mentioned in the Theorem can be found in Section 5 which contains also the proof of the Theorem and some final remarks. As was mentioned before, the notion of the topological entropy of a group action is recalled in Section 2. The same Section contains also a definition of geometric entropy of foliations which is used in Section 6, where we apply the Theorem to some foliated spaces and prove the following.

Corollary 1 *Let M be a compact Riemannian manifold with a hyperbolic fundamental group $\pi_1(M) = G$. Let S be a finite symmetric set generating G, $\tilde{M}$ be the universal covering of M, $X = (\partial G \times \tilde{M})/G$ and $\mathcal{F}_G$ be the foliation of X obtained by suspending the natural action of G on $\partial G \times \tilde{M}$.*

Then, the entropy $h(\mathcal{F}_G)$ of $\mathcal{F}_G$ (with respect to the Riemannian metric along the leaves obtained by lifting the Riemannian structure of M) satisfies the inequalities

$$\frac{1}{a} \cdot \mathrm{gr}^{\mathrm{rel}}(G, S, \mu, \tau) \leq h(\mathcal{F}_G) \leq \frac{1}{b} \cdot \mathrm{gr}(G, S), \tag{2}$$

where a (resp., b) is the maximal (resp., minimal) length of free homotopy classes of curves corresponding to generating elements of $\pi_1(M)$, while μ and τ are the same as in (1).

Finally, we define the exponential rate of growth relative to given bounds in Section 3 and we provide a short review on hyperbolic groups and spaces in Section 4.

2 Entropy

Let G be a finitely generated group of homeomorphisms of a compact metric space (X, d) and S be a finite symmetric ($e \in S$, $S^{-1} = S$) set generating G. Equip G with the word metric d_S induced by S and let $B(n)$, $n \in \mathbb{N}$, denote the ball in G of radius n and centre e. Two points x and y of X are said to be (n, ϵ)-separated ($\epsilon > 0$, $n \in \mathbb{N}$) whenever

$$d(gx, gy) \geq \epsilon$$

for some $g \in B(n)$. Since X is compact, the maximal number $N(n, \epsilon)$ of pairwise (n, ϵ)-separated points of X is finite. Also, there exist finite (n, ϵ)-spanning subsets of X: A subset A of X is (n, ϵ)-spanning whenever for any $y \in X$ there exists $x \in A$ such that

$$d(gx, gy) < \epsilon$$

for all $g \in B(n)$. Let $N'(n, \epsilon)$ denote the minimal cardinality of an (n, ϵ)-spanning subset of X. Similarly to the case of classical dynamical systems (compare, Walters' book [12], p. 169), the families $N(n, \epsilon)$ and $N'(n, \epsilon)$ of functions have the same type of growth (see Egashira's paper [4]), more precisely, they have the same rate of exponential growth and the topological entropy $h(G, S, X)$ of G (w.r.t. S) can be defined by the formula

$$h(G, S, X) = \lim_{\epsilon \to 0} \limsup_{n \to \infty} \frac{1}{n} \log N(n, \epsilon) = \lim_{\epsilon \to 0} \limsup_{n \to \infty} \frac{1}{n} \log N'(n, \epsilon).$$

Note that If $G = \mathbb{Z}$ and $S = \{\pm 1\}$, then $h(G, S, X)$ coincides with the double of the Bowen's [1] topological entropy of the map $f : X \to X$ corresponding to the integer 1.

If all the maps of G are Lipschitz and X has finite Hausdorff dimension, then $h(G, S, X)$ is finite for any S (compare, Ghys *et al.* [7], Prop.2.7). Also, if $h(G, S, X) = 0$ for some S, then $h(G, S', X) = 0$ for any other generating set S'. Therefore, one can distinguish between groups of positive and vanishing entropy without referring to generating sets.

The same construction can be applied to finitely generated pseudogroups of local transformations of a metric space X.

Also, the entropy of a foliation $\mathcal{F}$ of any compact Riemannian manifold M can be defined. We shall follow Langevin *et al.* [11]. The definition there is slightly different than that of Ghys *e al.* [7] but provides us with the same notion of entropy (called *geometric*).

Let us consider a good covering $\mathcal{U}$ of M by charts distinguished by $\mathcal{F}$, and a corresponding complete transversal T (compare, for instance, Candel and Conlon's book [2]). Given $\epsilon > 0$ sufficiently small and $R > 0$, let us say that two points x and y of T are (R, ϵ)-separated whenever there exists a leaf curve $\gamma : [0, 1] \to L_x$ originated at one of them (here, x) of length $l(\gamma) \leq R$ and such that its orthogonal projection $\tilde{\gamma} : [0, 1] \to L_y$ to the other leaf (this time, L_y, the leaf through y) satisfies the inequality

$$d(\gamma(1), \tilde{\gamma}(1)) \geq \epsilon$$

whenever its origin $\tilde{\gamma}(0)$ belongs to the plaque through y of a chart of $\mathcal{U}$ which contains both x and y. (Here, d denotes the Riemannian distance on M.)

Denote by $N(R, \epsilon, \mathcal{F})$ the maximal cardinality of a subset A of T such that any two its points are (R, ϵ)-separated in the above sense. Let

$$h(\mathcal{F}) = \lim_{\epsilon \to 0} \limsup_{R \to \infty} N(R, \epsilon, \mathcal{F})$$

and call $h(\mathcal{F})$ the *geometric entropy* of $\mathcal{F}$. Note that $h(\mathcal{F})$ depends only on the Riemannian structure on M. The covering $\mathcal{U}$ involved in the definition plays only an auxilliary role. The value of $h(\mathcal{F})$ does not depend on the choice of $\mathcal{U}$.

The geometric entropy of a foliation $\mathcal{F}$ can be compared to entropies of its holonomy pseudogroups corresponding to all good coverings by distinguished charts. If $\mathcal{U}$ is such a covering, $\mathcal{H}_{\mathcal{U}}$ is the corresponding holonomy pseudogroup, and $S_{\mathcal{U}}$ is the set of all elementary holonomy maps corresponding to overlapping charts of $\mathcal{U}$, then $\mathcal{H}_{\mathcal{U}}$ acts on a suitable (built of connected transversals $T_U \subset U$, $U \in \mathcal{U}$) complete transversal $T_{\mathcal{U}}$ equipped with the Riemannian structure inherited from M, $T_{\mathcal{U}}$ carries the Riemannian distance, and the entropy $h(\mathcal{H}_{\mathcal{U}}, S_{\mathcal{U}}, T_{\mathcal{U}})$ can be considered. With this notation the equality

$$h(\mathcal{F}) = \sup_{\mathcal{U}} \frac{1}{\Delta(\mathcal{U})} \cdot h(\mathcal{H}_{\mathcal{U}}, S_{\mathcal{U}}, T_{\mathcal{U}}),$$

where $\mathcal{U}$ ranges over the family of all good coverings of M and $\Delta(\mathcal{U})$ denotes the maximal diameter of plaques of charts of $\mathcal{U}$, holds (see Ghys *et al.* [7]). Since the notion of holonomy extends directly to arbitrary foliated spaces, Candel and Conlon [2] proposed to use the last equality as the definition of a geometric entropy of any foliated space.

3 Growth

Let us keep the notation of the previous section and recall that the exponential *rate of growth* of G (with respect to S) is defined as

$$\mathrm{gr}(G, S) = \lim_{n\to\infty} \frac{1}{n} \log N(n) = \lim_{n\to\infty} \frac{1}{n} \log N_0(n),$$

where $N(n) = \#B(n)$ and $N_0(n) = \#S(n)$ is the cardinality of the sphere $S(n)$ of radius n and centre e. Also, if $\epsilon > 0$ (note that ϵ is hereafter rather large than small !) and $N_0(n;\epsilon)$ is the maximal cardinality of an ϵ-separated subset A of $S(n)$, then

$$\mathrm{gr}(G, S) = \lim_{n\to\infty} \frac{1}{n} \log N_0(n;\epsilon).$$

In fact, $N_0(n;\epsilon) \leq N_0(n)$ and, if A is such a subset of $S(n)$, then $\cup_{x\in A} B(x,\epsilon) \supset S(n)$ and, therefore, $N_0(n;\epsilon)N(\epsilon) \geq N_0(n)$ for any $n \in \mathbb{N}$.

Moreover, if $m \in \mathbb{N}$ and A_k, $k = 1, 2, \ldots$, are maximal ϵ-separated subsets of $S(mk)$, then for any $n \in \mathbb{N}$ and any $x \in A_n$ we can find a sequence $(x_0, x_1, \ldots x_n)$ of elements of the group G for which $x_k \in A_k$, $x_n = x$, $x_0 = e$ and $d(x_k, x_{k+1}) \leq m + \epsilon$ for all $k < n$. To construct such a sequence one can begin with $x_n = x$, join x to e by a geodesic segment γ_x, find the point x'_{n-1} of intersection of γ_x with the sphere $S((n-1)m)$ and a point $x_{n-1} \in A_{n-1} \cap B(x'_{n-1}, \epsilon)$, and continue by the induction. If y is another point of A_n, $(y_0, y_1, \ldots, y_n)$ is a corresponding sequence and k is the maximal natural number such that $x_k = y_k$, then $d(x_{k+1}, y_{k+1}) \geq \epsilon$.

Therefore, $N_0(mn,\epsilon)$ can be defined as the maximal cardinality of a subset A of the sphere $S(mn)$ which satisfies the condition

$(*)$ If x and y belong to A, then there exist sequences $(x_0, x_1, \ldots, x_n)$ and $(y_0, y_1, \ldots, y_n)$ of elements of G such that $x_k, y_k \in S(km)$, $x_0 = y_0 = e$, $x_n = x$, $y_n = y$,

$$d(x_j, x_{j+1}), d(y_j, y_{j+1}) \leq m + \epsilon,\ j = 0, \ldots, n-1, \tag{3}$$

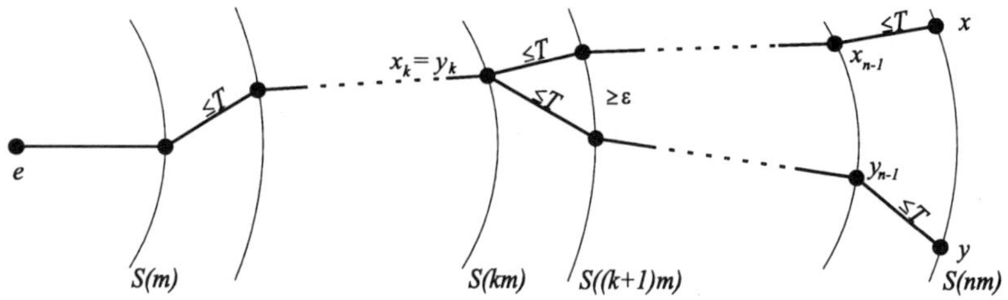

Figure 1.

and $d(x_{k+1}, y_{k+1}) \geq \epsilon$ when k is the maximal index for which $x_k = y_k$.

To define our relative growth we shall modify (*) by introducing an additional factor λ in (3):

Let us fix $m \in \mathbb{N}$, $\epsilon > 0$ and $\lambda \in]0, 1[$, and denote by $N_0(n; m, \epsilon, \lambda)$ the maximal cardinality of a subset A of $S(mn)$ satisfying the following condition:

(**) If x and y lie in A, then there exist sequences $(x_0, x_1, \dots x_n)$ and $(y_0, y_1, \dots, y_n)$ of elements of G such that $x_k,\ y_k \in S(km)$, $x_0 = y_0 = e$, $x_n = x$, $y_n = y$,

$$d(x_j, x_{j+1}), d(y_j, y_{j+1}) \leq T = m + \lambda\epsilon$$

for all j and $d(x_{k+1}, y_{k+1}) \geq \epsilon$ when k is the maximal index for which $x_k = y_k$ (Figure 1).

The number

$$\mathrm{gr}^{\mathrm{rel}}(G, S; m, \epsilon, \lambda) = \limsup_{n \to \infty} \frac{1}{mn} \log N_0(n; m, \epsilon, \lambda)$$

will be called the exponential *rate of growth* of G *relative to* m, ϵ and λ. Finally, if $\mu :]0, 1[\times \mathbb{R}_+ \to \mathbb{N}$ and $\tau :]0, 1[\to \mathbb{R}_+$ are arbitrary functions, then we define the *rate of growth* of G *relative to* μ and τ by

$$\mathrm{gr}^{\mathrm{rel}}(G, S; \mu, \tau) = \sup\{\mathrm{gr}^{\mathrm{rel}}(G, S; m, \epsilon, \lambda); m > \mu(\lambda, \epsilon), \epsilon > \tau(\lambda), \lambda \in]0, 1[\}.$$

Obviously,

$$\mathrm{gr}^{\mathrm{rel}}(G, S; \mu, \tau) \geq \mathrm{gr}^{\mathrm{rel}}(G, S; m, \epsilon, \lambda)$$

for any $\lambda \in]0, 1[$, ϵ and m sufficiently large.

Since $N_0(n; m, \epsilon, \lambda) \leq N_0(mn, \epsilon)$ for all m, n, ϵ and λ, we have

$$\mathrm{gr}^{\mathrm{rel}}(G, S; \mu, \tau) \leq \mathrm{gr}(G, S)$$

for all μ and τ as above. For the free group F_k generated by the set S_k of k free generators we have always

$$\mathrm{gr}^{\mathrm{rel}}(F_k, \tilde{S_k}; \mu, \tau) = \mathrm{gr}(F_k, \tilde{S_k}),$$

where $\tilde{S_k} = S_k \cup S_k^{-1} \cup \{e\}$. This is because F_k has no "dead ends" (see Grigorchuk and de la Harpe [8] for the definition and some information about some related problems) and in this case one can arrange ϵ-separated subsets A_n of the spheres $S(mn)$ in such a way that $\mathrm{dist}(x, A_n) = m$ for any $x \in A_{n+1}$. In general, one can expect that a relative rate of growth is strictly less than the "true" rate of growth. However, the above observation about free groups and inequality (1) imply that the boundary entropy of several hyperbolic groups is strictly positive (see Section 4).

4 Hyperbolic spaces and groups

Let (X, d) be a metric space. A curve $\gamma : [a, b] \to X$ is a *geodesic segment* when

$$d(\gamma(t), \gamma(s)) = |t - s|$$

for all $t, s \in [a, b]$. The space X is *geodesic* when any two points of X can be joined by a geodesic segment. For any finitely generated group G and any finite symmetric set S generating G, the Cayley graph $C(G, S)$ is geodesic.

Given three points x_0, y and z of a metric space X, the (based at x_0) *Gromov product* of y and z is given by

$$(y|z)_{x_0} = \frac{1}{2}\left(d(x_0, y) + d(x_0, z) - d(y, z)\right).$$

The space X is said to be *hyperbolic* (more precisely, δ-hyperbolic with $\delta \geq 0$) whenever the inequality

$$(x|z)_{x_0} \geq \min\{(x|y)_{x_0}, (y|z)_{x_0}\} - \delta$$

holds for arbitrary points x_0, x, y and z of X. Clearly, the Cayley graph of any free group F_k ($k = 1, 2, \ldots$) generated by the set S_k of k free generators is a tree, so becomes 0-hyperbolic. A finitely generated group is said to be *hyperbolic* whenever its Cayley graph with respect to some (equiv., any) generating set is hyperbolic. Free groups and fundamental groups of compact Riemannian manifolds of negative sectional curvature are hyperbolic.

Assume that X is geodesic, take three points x_1, x_2 and x_3 of X and connecting them geodesic segments γ_1, γ_2 and γ_3. The union

$$\Delta = \gamma_1 \cup \gamma_2 \cup \gamma_3$$

is a *geodesic triangle* with vertices x_i. The triangle Δ is η-*thin* ($\eta \geq 0$) when the canonical isometry f_Δ mapping Δ onto a tripod (i.e. the union of three segments with common origin) T_Δ satisfies the condition

$$d(x,y) \leq d(f_\Delta(x), f_\Delta(y)) + \eta$$

for all x and y of Δ.

In the proof of the Theorem we shall use the following.

Lemma 1 (Ghys and de la Harpe [6], p.41) *Let X be a geodesic metric space. If X is δ-hyperbolic, then all the geodesic triangles of X are 4δ-thin. Conversely, if all the geodesic triangles of X are η-thin, then X is 2η-hyperbolic.*

To construct the boundary ∂X of a hyperbolic space X let us fix a base point x_0 and say that a sequence (x_n) *diverges to infinity* whenever

$$\lim_{m,n\to\infty} (x_m|x_n) = \infty,$$

where $(\cdot|\cdot)$ denotes the Gromov product based at x_0. Two such sequences (x_n) and (y_m) are *equivalent* whenever

$$\lim_{m,n\to\infty} (x_m|y_n) = \infty.$$

The *boundary* ∂X of X consists of all the equivalence classes of sequences diverging to infinity. Note that ∂X can be described also in terms of equivalence classes of *geodesic rays* (i.e., maps $\gamma : [0,\infty) \to X$ such that $\gamma|[0,b]$ is a geodesic segment for any $b > 0$) or in terms of equivalence classes of *quasirays* (i.e. quasi-isometric maps of $[0,\infty)$ into X): Two such rays (or, quasirays) γ and σ are equivalent whenever their Hausdorff distance $d_H(\gamma,\sigma)$ is finite. The boundary point corresponding to the equivalence class of such γ is that determined by the sequence $x_n = \gamma(n)$, $n \in \mathbb{N}$. The equivalence of these constructions follows from the following fact which will be used later.

Lemma 2 (Ghys and de la Harpe [6], p.87) *Let X be a δ-hyperbolic geodesic metric space. For any $\alpha \geq 1$ and any $c \geq 0$ there exists $D = D(\delta,\alpha,c) \geq 0$ such that any (α,c)-quasigeodesic segment γ (i.e. any map $\gamma : [a,b] \to X$ (resp., $\gamma : [a,b] \cap \mathbb{Z} \to X$) such that the inequality $-c - \alpha|s-t| \leq d(\gamma(s),\gamma(t)) \leq \alpha|s-t| + c$ holds for all s and t of $[a,b]$ (resp., of $[a,b] \cap \mathbb{Z}$)) and any geodesic segment σ joining $\gamma(a)$ to $\gamma(b)$ satisfy the inequality*

$$d_H(\gamma,\sigma) \leq D.$$

The set ∂X can be equipped with the metric structure as follows. First, for any ξ and ζ of ∂X put

$$(\xi|\zeta) = \sup \liminf_{m,n\to\infty} (x_m|y_n),$$

where (x_m) and (y_n) run over the set of all diverging to infinity sequences representing, respectively, ξ and ζ. Note that if X is δ-hyperbolic, then

$$(\xi|\zeta) - 2\delta \leq \liminf_{m,n\to\infty} (x_m|y_n) \leq (\xi|\zeta)$$

for all sequences (x_m) and (y_m) representing ξ and ζ. Next, choose $\eta > 0$ and put

$$\rho_\eta(\xi,\zeta) = \exp(-\eta \cdot (\xi|\zeta)).$$

Finally, let

$$d_\eta(\xi,\zeta) = \inf\{\sum_{i=0}^{k} \rho_\eta(\xi_i,\xi_{i+1}); \xi_i \in \partial X, \xi_0 = \xi \text{ and } \xi_{k+1} = \zeta, k \in \mathbb{N}\}.$$

If $\eta > 0$ is small enough, then d_η is a distance function on ∂X and $(\partial X, d_\eta)$ becomes a compact metric space of finite Hausdorff dimension (Ghys and de la Harpe [6], pp. 122 - 126). Moreover, the inequalities

$$(1-2\eta')\rho_\eta(\xi,\zeta) \leq d_\eta(\xi,\zeta) \leq \rho_\eta(\xi,\zeta), \quad \xi,\zeta \in \partial X,$$

hold with

$$\eta' = \exp(\eta\delta) - 1.$$

Finally, let us recall that a hyperbolic group G is called *elementary* if either it is finite or contains a cyclic group of finite index. In this case, the boundary ∂G is either empty or finite (consisting of exactly two elements) and the entropy $h(G, S, \partial G)$ vanishes. If G is δ-hyperbolic and non-elementary, then ∂G is infinite and G itself contains a free subgroup H of two generators. If the inclusion $\iota : H \to G$ is (α, c)-quasi-isometric, then ι induces a map (denoted by ι again) of ∂H into ∂G which is Hölder with a constant $a = a(\delta, \alpha, c)$. It follows that the image $\iota(A)$ of any (n, ϵ)-saturated subset A of ∂H becomes $(kn, C\epsilon)$-separated in ∂G, where k is a natural number depending on the choices of generating sets S_H of H and S_G of G, while C is a positive constant depending on δ, α, c and η, a constant involved in the definition of the distance functions on ∂H and ∂G. This implies the inequalities

$$h(G, S_G, \partial G) \geq \frac{1}{k} h(H, S_H, \partial H) > 0,$$

where the last one follows from our Theorem and the final observation of Section 3.

5 Proof of the Theorem

Let again G be a finitely generated group, S - a finite symmetric set generating G and consider $C(G, S)$, the Cayley graph of G equipped with the distance function d satisfying

$$d(g_1, g_2) = |g_1^{-1} g_2|, \quad g_1, g_2 \in G,$$

where $|\cdot|$ is the length function on G determined by S, and making the edges of $C(G, S)$ isometric to Euclidean segments of length 1. Assume that $C(G, S)$ is δ-hyperbolic and let $\partial G = \partial C(G, S)$ be its boundary equipped, as in Section 4, with the distance function d_η, $\eta > 0$ being small enough.

The group G acts on $C(G, S)$ *via* isometries which, when restricted to $G \subset C(G, S)$, reduce to left translations L_g, $G \ni h \mapsto gh$. Therefore, each L_g extends to a Lipschitz homeomorphism, denoted by L_g again, of the boundary ∂G. Given $\theta > 0$ denote by $N(n, \theta; \partial G)$ the maximal number of points of ∂G pairwise (n, θ)-separated by this action of G. Also, let $N'(n, \theta; \partial G)$ be the minimal cardinality of an (n, θ)-spanning subset of ∂G and $h(G, S; \partial G)$ - the corresponding entropy.

Since G is hyperbolic, there exists a constant $c_0 > 1$ such that for any $g \in G$ there exists $g' \in G$ such that $d(g, g') \leq c_0$ and $|g'| = |g| + 1$ (see Grigorchuk and de la Harpe [8], p. 60). Let $D = D(\delta, c_0, c_0)$ be the corresponding constant such that any (c_0, c_0)-quasigeodesic segment lies at most D-apart (in the Hausdorff distance) from a true geodesic (compare Lemma 2). Let

$$\tau(\lambda) = \frac{4(D + \delta)}{1 - \lambda} \quad \text{and} \quad \mu(\lambda, \epsilon) = \frac{\lambda \epsilon}{c_0 - 1}.$$

We begin with the proof of the second inequality in (1).

Choose $\theta > 0$ and a natural number k for which the inequality

$$\exp(-\eta \cdot k) < \theta$$

holds. For any $n \in \mathbb{N}$ and any $g \in S(n + k)$ choose, if only possible, a point $\xi_g \in \partial G$ which can be connected to e by a geodesic ray, say γ_g, which passes through g. We claim that the set

$$A = \{\xi_g; g \in S(n + k)\}$$

is (n, ϵ)-spanning in ∂G. Indeed, if $\zeta \in \partial G$, $\sigma : [0, \infty) \to X$ is a geodesic ray connecting e to ζ, $h \in S(n)$ and $g = \sigma(n + k)$, then $g \in S(n + k)$, the corresponding ray γ_g exists and satisfies the conditions

$$d(x_j, y_i) \leq i + j - 2(n + k)$$

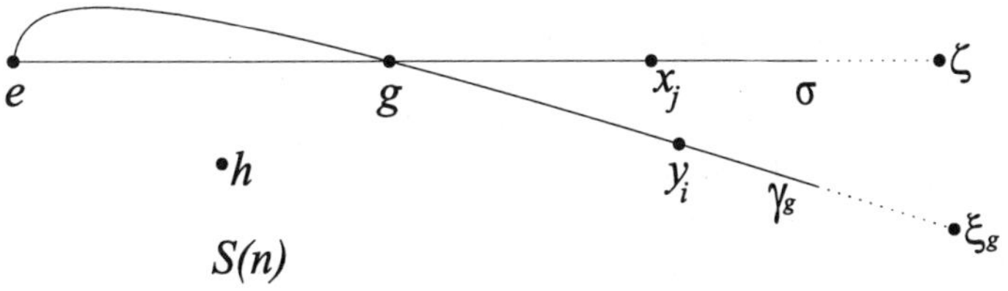

Figure 2.

and

$$2(h^{-1}y_i|h^{-1}x_j) \geq (i-n)+(j-n)-(i+j)+2(n+k) = 2k,$$

where $x_j = \sigma(j)$ and $y_i = \gamma(i)$ for all i and j sufficiently large (Figure 2).

Therefore,

$$d_\eta(L_{h^{-1}}\zeta, L_{h^{-1}}\xi_g) \leq \rho_\eta(L_{h^{-1}}\zeta, L_{h^{-1}}\xi_g) < e^{-k\eta} < \theta.$$

This shows the inequalities

$$N'(n,\theta;\partial G) \leq \#A \leq \#S(n+k) \leq N(n+k)$$

which imply immediately the required inequality in (1).

The proof of the first inequality in (1) is a bit more complicated.

Fix $\lambda \in]0,1[$, $\epsilon > \tau(\lambda)$ and $m > \mu(\lambda,\epsilon)$. Choose $n \in \mathbb{N}$ and a maximal subset A of $S(mn)$ satisfying condition (**) of Section 3. For any $x \in A$ set $x_n = x$, choose a point $x_{n-1} \in S(m(n-1))$ such that $d(x, x_{n-1}) \leq m + \lambda\epsilon$, then a point $x_{n-2} \in S(m(n-2))$ for which $d(x_{n-1}, x_{n-2}) \leq m + \lambda\epsilon$ and so on. Finally, put $x_0 = e$. The map

$$\{0, m, \ldots, mn\} \ni j \mapsto x_{j/m}$$

is c-quasi-isometric with $c = (m+\lambda\epsilon)/m < c_0$.

Each map considered above can be extended to a c_0-quasi-isometric map

$$\gamma_x : \mathbb{N} \ni j \mapsto x'_j \in G$$

such that $x'_{im} = x_i$ for $i = 0, 1, \ldots, n$ and the sequence (x_j) converges to a point ξ_x of ∂G. We are going to show that the set

$$\{\xi_x; x \in A\}$$

is (n,θ)-separated under the action of G for some θ independent of n.

To this end, let us take arbitrary points x and y of A, $x \neq y$, choose sequences $(x_0, x_1, \ldots, x_n)$ and $(y_0, y_1, \ldots, y_n)$ satisfying all the conditions of (**) and let k be the maximal element of $\{0, 1, \ldots, n\}$ for which $x_k = y_k$.

Denote this common value of x_k and y_k by g and consider the quasi-geodesic rays $\bar{\gamma}_x$ and $\bar{\gamma}_y$ obtained from γ_x and γ_y by restricting their domains to $\{mk, mk+1, \ldots\}$. Then, $\bar{\gamma}_x$ and $\bar{\gamma}_y$ originate at g and converge to ξ_x and ξ_y, respectively. By Lemma 2, there exist geodesic rays $\tilde{\gamma}_x$ and $\tilde{\gamma}_y$ originated at e and within the Hausdorff distance D from $L_{g^{-1}} \circ \bar{\gamma}_x$ and $L_{g^{-1}} \circ \bar{\gamma}_y$, respectively. Then, for any $j \in \mathbb{N}$ there exist positive real numbers s_j and t_j such that

$$d(\tilde{\gamma}_x(s_j), g^{-1}\bar{\gamma}_x(j)) \leq D \quad \text{and} \quad d(\tilde{\gamma}_y(t_j), g^{-1}\bar{\gamma}_y(j)) \leq D.$$

In particular,

$$d(\tilde{\gamma}_x(s_m), g^{-1}x_{k+1}) \leq D \quad \text{and} \quad d(\tilde{\gamma}_y(t_m), g^{-1}y_{k+1}) \leq D.$$

Clearly,

$$m - D \leq s_m, \quad t_m \leq m + D + \lambda\epsilon, \quad |s_m - t_m| \leq 2D + \lambda\epsilon$$

and

$$d(\tilde{\gamma}_x(s_m), \tilde{\gamma}_y(t_m)) \geq \epsilon - 2D.$$

Assume that $(\tilde{\gamma}_x(s_j)|\tilde{\gamma}_y(t_j)) > m + D + \lambda\epsilon$ for some $j \in \mathbb{N}$. The isometry f_Δ corresponding to the geodesic triangle Δ with vertices $e, \tilde{\gamma}_x(s_j), \tilde{\gamma}_y(t_j)$ (compare Lemma 1) maps the points $\tilde{\gamma}_x(s_m)$ and $\tilde{\gamma}_y(t_m)$ onto some points of the originated at e edge of the tripod $f_\Delta(\Delta)$ and, therefore, satisfies the condition

$$\begin{aligned}(\tilde{\gamma}_x(s_m), \tilde{\gamma}_y(t_m)) &\leq d(f_\Delta(\tilde{\gamma}_x(s_m)), f_\Delta(\tilde{\gamma}_y(t_m))) + 4\delta \\ &= |s_m - t_m| + 4\delta \leq 2D + \lambda\epsilon + 4\delta < \epsilon - 2D.\end{aligned}$$

Comparing the inequalities above we obtain a contradiction which shows that

$$(\tilde{\gamma}_x(s_j)|\tilde{\gamma}_y(t_j)) \leq m + D + \epsilon\lambda$$

for all $j \in \mathbb{N}$. This inequality proves that

$$\begin{aligned}d_\eta(g^{-1}\xi_x, g^{-1}\xi_y) &\geq (1 - 2\eta')\rho_\eta(g^{-1}\xi_x, g^{-1}\xi_y) \\ &\geq (1 - 2\eta')\exp(-\eta(m + D + \lambda\epsilon + 2\delta)),\end{aligned}$$

i.e. that the points ξ_x and ξ_y are (nm, θ)-separated with

$$\theta = (1 - 2\eta')\exp(-\eta(m + D + \lambda\epsilon + 2\delta)).$$

The above argument implies the inequality

$$N(nm, \theta; \partial G) \geq \#A = N_0(G, S; m, \epsilon, \lambda)$$

which holds for all n. Passing to suitable limits when $n \to \infty$ yields the required inequality in (1).

Friedland [5] defined the *minimal entropy* $h_{\min}(G)$ of a finitely generated group G of homeomorphisms of a compact metric space X:

$$h_{\min}(G) = \inf_S h(G, S),$$

where S ranges over all finite symmetric sets generating G. Similarly, the *minimal rate of growth* $\mathrm{gr}_{\min}(G)$ of any finitely generated group G can be defined as follows (compare Grigorchuk and de la Harpe [8]):

$$\mathrm{gr}_{\min}(G) = \inf_S \mathrm{gr}(G, S).$$

The reader can define the minimal relative rate of growth $\mathrm{gr}^{\mathrm{rel}}_{\min}(G)$ appropriately.

If G is hyperbolic, then id_G induces a Hölder homeomorphism of boundaries of G obtained from different generating sets (Ghys and de la Harpe [6], page 128). Therefore, the boundary entropy of such G (w.r.t. a given finite symmetric generating set S) does not depend on the choice of a generating set used in the construction of ∂G and our Theorem implies immediately the following.

Corollary 2 *For any hyperbolic group G the equalities*

$$\mathrm{gr}^{\mathrm{rel}}_{\min}(G) \le h_{\min}(G, \partial G) \le \mathrm{gr}_{\min}(G)$$

hold.

This answers partially the following question asked by Friedland [5]: Find a geometric interpretation of the minimal entropy of a Kleinian group acting on the ideal boundary of a hyperbolic space H^n.

For the free group F_k with k generators the above discussion and an argument of Gromov *et al.* [10] (p. 70) imply the equality

$$h_{\min}(F_k, \partial F_k) = \mathrm{gr}_{\min}(F_k) = \mathrm{gr}^{\mathrm{rel}}_{\min}(F_k) = \log(2k-1).$$

In fact, if S is any finite symmetric set generating F_k, then the elements of S represent members of a set S' generating $\mathbb{Z}^k$, the abelianization of F_k. S' contains a symmetric set R' such that $\#R' = 2k$ and the subgroup of $\mathbb{Z}^k$ generated by R' has finite index. The corresponding subset R of S consists also of $2k$ elements and generates the free group isomorphic to F_k. Therefore,

$$\begin{aligned} h(F_k, S; \partial F_k) &\ge h(F_k, R; \partial F_k) = h(F_k, S_k; \partial F_k) \\ &\ge \mathrm{gr}^{\mathrm{rel}}(F_k, S^k) = \mathrm{gr}(F_k, S_k) = \log(2k-1). \end{aligned}$$

The opposite inequality is obvious.

The other natural case, that of the fundamental group Γ_g of a closed oriented surface of genus $g > 1$ is more complicated: The minimal rate of growth of Γ_g is still unknown even if some estimates exist: $\mathrm{gr}_{\min}(\Gamma_g) \geq 4g - 3$, $\mathrm{gr}_{\min}(\Gamma_g) \leq \mathrm{gr}(\Gamma_g, S_g) \approx 4g - 1 - \epsilon_g$, where S_g is the canonical set of generators of Γ_g and ϵ_g is a small constant found numerically. In particular, $5 \leq \mathrm{gr}_{\min}(\Gamma_2) \leq 6.9798$. The calculation or estimation of the value of the minimal relative rate of growth of Γ_g is yet more difficult.

6 Suspensions

Let now G be a hyperbolic group and M a compact Riemannian manifold with the fundamental group $\pi_1(M) = G$. Let S be a finite symmetric set generating set and a (resp., b) be the maximum (resp., minimum) of lengths of free homotopy classes of curves homotopic to members of S. Let also $\tilde{M}$ be the universal covering of M and

$$X = (\partial G \times \tilde{M})/G.$$

X carries a natural structure of a foliated space $(X, \mathcal{F}_G)$ (compare Candel and Conlon [2], Chapter 11) whose leaves arise as $\pi(\{\xi\} \times \tilde{M})$, where $\xi \in \partial G$ and $\pi : \partial G \times \tilde{M} \to X$ is a canonical projection. Also, X fibres over M with fibres homeomorphic to ∂G. Moreover, the distance function d_η on ∂G (Section 3) and the Riemannian metric g_M on M lifted *via* π to the leaves of $\mathcal{F}_G$ determine a natural distance function d on X which coincides with d_η along the fibres and with the natural Riemannian distance along the leaves. Recall that (Section 2) the notion of geometric entropy introduced by Ghys *et. al.* [7] can be extended to foliated spaces as in the Candel and Conlon's book [2] (Section 13.3.B). Similarly to the situation considered in Example 4.3 of Ghys *et al.* [7] one can get easily the inequality

$$\frac{1}{a} \cdot h(G, S; \partial G) \leq h(\mathcal{F}_G) \leq \frac{1}{b} \cdot h(G, S; \partial G). \tag{4}$$

Comparing (1) and (4) yields (2) and proves our Corollary 1.

Acknowledgments

The second author was supported by the KBN grants P03A 066 10 and P03A 033 18.

References

1. R. Bowen, *Entropy for group endomorphisms and homogeneous spaces*, Trans. Amer. Math. Soc., **153** (1971), 401–414.
2. A. Candel and L. Conlon, Foliations I, Amer. Math. Soc., Providence, 2000.
3. M. Coornaert and A. Papadopoulos, Symbolic Dynamics and Hyperbolic Groups, Springer Verlag, Berlin-Heidelberg, 1993.
4. S. Egashira, *Expansion growth of foliations*, Ann. Fac. Sci. Univ. Toulouse, **2** (1993), 15–52.
5. S. Friedland, *Entropy of graphs, semigroups and groups*, in Ergodic theory of $\mathbb{Z}^d$ actions, eds. M. Policott and K. Schmidt London Math. Soc., London, 1996, 319 – 343.
6. E. Ghys and P. de la Harpe, Sur les Groupes Hyperboliques d'apres Mikhael Gromov, Birkhäuser, Boston-Basel-Berlin, 1990.
7. E. Ghys, R. Langevin and P. Walczak, *Entropie géométrique des feuilletages*, Acta Math., **160** (1988), 105–142.
8. R. Grigorchuk and P. de la Harpe, *On problems related to growth, entropy and spectrum in group theory*, J. Dyn. Control Sys., **3** (1997), 51–89.
9. M. Gromov, *Hyperbolic groups*, in Essays in group theory, ed. S. M. Gersten, Springer Verlag, Berlin-Heidelberg-New York, 1987, 75–263.
10. M. Gromov, J. Lafontaine and P. Pansu, Structures métriques pur les variétés riemanniennes, Cedic/F. Nathan, Paris, 1981.
11. R. Langevin and P. Walczak, *Entropy, transverse entropy and partitions of unity*, Ergodic Th. and Dynam. Sys., **14** (1994), 551–563.
12. P. Walters, An introduction to Ergodic Theory, Springer Verlag, New York-Heidelberg-Berlin, 1982.

Received June 7, 2000, revised November 22, 2000 and September 25, 2001.

Proceedings of
FOLIATIONS: GEOMETRY AND DYNAMICS
held in Warsaw, May 29–June 9, 2000
ed. by Paweł WALCZAK *et al.*
World Scientific, Singapore, 2002
pp. 213–224

TAUTLY FOLIATED 3-MANIFOLDS WITH NO R-COVERED FOLIATIONS

MARK BRITTENHAM
Department of Mathematics and Statistics,
University of Nebraska, Lincoln, NE 68588-0323,
e-mail: mbritten@math.unl.edu

We provide a family of examples of graph manifolds which admit taut foliations, but no **R**-covered foliations. We also show that, with very few exceptions, **R** - covered foliations are taut.

1 Introduction

A foliation $\mathcal{F}$ of a closed 3-manifold M can be lifted to a foliation $\widetilde{\mathcal{F}}$ of the universal cover $\widetilde{M}$ of M; if the foliation $\mathcal{F}$ has no Reeb components, the leaves of this lifted foliation are all planes, and Palmeira [19] has shown that $\widetilde{M}$ is homeomorphic to R^3. Palmeira also showed that $\widetilde{\mathcal{F}}$ is homeomorphic to (a foliation of R^2 by lines)$\times R$. The space of leaves of $\widetilde{\mathcal{F}}$, the quotient space obtained by crushing each leaf to a point, is homeomorphic to the space of leaves of the foliation of R^2, and is a (typically non-Hausdorff) simply-connected 1-manifold.

If the space of leaves is Hausdorff (and therefore homeomorphic to **R**), we say that the foliation $\mathcal{F}$ is **R**-*covered*. Examples of **R**-covered foliations abound, starting with surface bundles over the circle; the foliation by fibres is **R**-covered. Thurston's notion of 'slitherings' [20] also provide a large collection of examples. A great deal has been learned in recent years about **R**-covered foliations and the manifolds that support them (see, e.g., [1], [8], [11]), especially in the case when the underlying manifold M contains no incompressible tori.

The purpose of this note is to provide examples of 3-manifolds which admit taut foliations, but which do not admit any **R**-covered foliations. All of our examples are drawn from graph manifolds, and so all contain an incompressible torus, and, in fact, a separating one. The technology does not exist at present to identify examples which are atoroidal, and it is perhaps not clear that such examples should be expected to exist. That we can find the examples we seek among graph manifolds relies on two facts: (a) we have a good understanding [3] of how taut and Reebless foliations can meet an incompressible torus, and (b) we understand [4], [9], [17] which Seifert-fibered spaces can contain taut or Reebless foliations. These two facts have been used previously [2] to find graph manifolds which admit foliations with various properties, but no *"stronger"* properties; the examples we provide here are in fact the exact same examples used in [2] to illustrate its results. We will simply look at them from a somewhat different perspective.

This paper can, in fact, be thought of as a further illustration of this 'you can get this much, but no more' point of view towards foliating manifolds. Namely, you can get taut, but not **R**-covered. **R**-covered is, it turns out, almost (but not quite) always a stronger condition: in all but a very small handful of instances, an **R**-covered foliation must be taut, as we show in the next section. This result seems to have been implicit in much of the literature on **R**-covered foliations; a different proof of this result, along somewhat different lines, can be found in [14].

2 R-covered almost implies taut

In this section we show that in all but a very few instances, an **R**-covered foliation must be taut. We divide the proof into two parts; first we show that a Reebless **R**-covered foliation is taut, and then describe the manifolds that admit **R**-covered foliations with Reeb components.

Recall that a *Reeb component* is a solid torus whose interior is foliated by planes transverse to the core of the solid torus, each leaf limiting on the boundary torus, which is also a leaf. (There is a non-orientable version of a Reeb component, foliating a solid Klein bottle, which we will largely ignore in this discussion. It can be dealt with by taking a suitable double cover of our 3-manifold.) We follow standard practice and refer to both the solid torus and its foliation as a Reeb component. A foliation that has no Reeb components is called *Reebless*. A foliation is *taut* if for every leaf there is a loop transverse to the foliation which passes through that leaf. Taut foliations are Reebless.

Lemma 1 *If a closed, irreducible 3-manifold M admits a Reebless,* **R**-*covered foliation $\mathcal{F}$ containing a compact leaf F, then every component of $M|F$, the manifold obtained by splitting M open along F, is an I-bundle* *over a compact surface. In particular, M is either a surface bundle over the circle with fibre F, or the union of two twisted I-bundles glued along their common boundary F.*

Proof. Because $\mathcal{F}$ is Reebless, the surface F is π_1-injective [18], and so lifts to a collection of planes in $\widetilde{M}$. Their image C in the space of leaves of $\widetilde{\mathcal{F}}$ is a discrete set of points in $\mathbf{R}$, since it is closed, and any sequence of distinct points in C with a limit point can be used to find a sequence of points in F limiting on F in the transverse direction, contradicting the compactness of F. The complementary regions of F in M lift to the complementary regions of the lifts of F in $\widetilde{M}$; in the space of leaves they correspond to the intervals between successive points of C. Each is bounded by two points of C, and so every component $\widetilde{X}$ of the inverse image of a component X of $M|F$ has boundary equal to two lifts of F. Because $\widetilde{M}$ is simply-connected, as are the ∂-components of $\widetilde{X}$, $\widetilde{X}$ is simply-connected, and so $\widetilde{X}$ is the universal cover of X.

Because F π_1-injects into M, it π_1-injects into $M|F$, and hence into X. The index of $\pi_1(F)$ in $\pi_1(X)$ is equal to the number of connected components of the inverse image of F in the universal cover of X. To see this, choose a basepoint x_0 for X lying in F, and suppose that γ is a loop based at x_0 which is not in the image of $\pi_1(X)$. Then the lift $\widetilde{\gamma}$ of γ to $\widetilde{X}$ must have endpoints on distinct lifts of F, for otherwise the endpoints can be joined by an arc $\widetilde{\alpha}$ in the lift of F, whose projection, since $\widetilde{X}$ is simply connected (so $\widetilde{\gamma}*\widetilde{\alpha}$ is null-homotopic) is a null-homotopic loop $\gamma * \alpha$ in $\pi_1(X)$. This implies that $[\gamma] = [\bar{\alpha}] \in \pi_1(F)$, a contradiction. Choosing representatives from each coset of $\pi_1(F)$ in $\pi_1(X)$, and lifting each to arcs with initial points a fixed lift $\widetilde{x_0}$ of x_0, we find that their terminal points must therefore lie on distinct lifts of F.

But since $\widetilde{X}$ has only two boundary components, this means that $\pi_1(F)$ has index at most two in $\pi_1(X)$. Since X is irreducible (because F is incompressible and M is irreducible), ([15], Theorem 10.6) implies that X is an I-bundle over a closed surface. The resulting description of M follows. □

The foliation by fibres of a bundle over the circle is always taut. In the other case, when F separates, we understand [5] the structure of the foliation $\mathcal{F}$ on each of the two I-bundles, since their boundaries are leaves. If the induced foliations can be made transverse to the I-fibres of each bundle, then by taking a pair of I-fibres, one from each bundle, and deforming them so that they share endpoints on F, we can obtain a loop transverse to the leaves of $\mathcal{F}$, so $\mathcal{F}$ is taut. If the induced foliations cannot be made transverse to the I-fibres, then F is a torus, and (after possibly passing to a finite cover) the foliation contains a pair of parallel tori with a Reeb annulus in between. It is then straightforward to see that the resulting lifted foliation $\widetilde{\mathcal{F}}$ cannot be $\mathbf{R}$-covered, since this torus$\times I$ will lift to $R^2 \times I$ whose induced foliation has space of leaves $\mathbf{R}$ *together with* two points (the two boundary components) that are *both* the limit of the positive (say) ray of the line. In particular, the space of leaves of $\widetilde{\mathcal{F}}$ would not be Hausdorff. Therefore

Corollary 2 *A Reebless, $\mathbf{R}$-covered foliation is taut.*

We now turn our attention to $\mathbf{R}$-covered foliations with Reeb components. Such foliations do exist, for example, the foliation of $S^2 \times S^1$ as a pair of Reeb components glued along their boundaries; the lift to $S^2 \times R$ consists of a pair of solid cylinders, each having space of leaves a closed half-line. Gluing the solid cylinders together results in gluing the two half-lines together, giving space of leaves $\mathbf{R}$. We show, however, that, in some sense, this is the only such example. Recall that the Poincaré associate $P(M)$ of M consists of the connected sum of the non-simply-connected components of the prime decomposition of M, i.e., $M = P(M)\#$ (a counterexample to the Poincaré Conjecture).

Lemma 3 *If $\mathcal{F}$ is an $\mathbf{R}$-covered foliation of the orientable 3-manifold M, which has a Reeb component, then $P(M) = S^2 \times S^1$.*

Proof. The core loop γ of the Reeb component must have infinite order in the fundamental group of M, otherwise the Reeb component lifts to a Reeb component of $\widetilde{\mathcal{F}}$; but since the interior of a Reeb component has space of leaves S^1, this would imply that S^1 embeds in $\mathbf{R}$, a contradiction.

The Reeb solid torus therefore lifts to a family of infinite solid cylinders in $\widetilde{M}$, foliated by planes. The induced foliation of each closed solid cylinder has space of leaves a closed half-line properly embedded in the space of leaves of $\widetilde{\mathcal{F}}$. Each such half-line is disjoint from the others; but since $\mathbf{R}$ has only two ends, this implies that the Reeb component has at most two lifts to $\widetilde{M}$. This means that the inverse image of the core loop γ of the Reeb solid torus, in the universal cover $\widetilde{M}$, consists of at most two lines, and so the (infinite) cyclic group generated by γ has index at most 2 in $\pi_1(M)$. Because M is orientable, it's fundamental group is torsion-free, and so by [15], Theorem 10.7, $\pi_1(M)$ is free, hence isomorphic to $\mathbb{Z}$, and so ([15], Exercise 5.3) $P(M)$ is an S^2-bundle over S^1. Since M is orientable, this gives the conclusion. □

Note that the space of leaves in the universal cover does not change by passing to finite covers (there is only one universal cover), and so we can lose the orientability hypothesis by weakening the conclusion slightly. Putting the lemmas together, we get

Corollary 4 *If a 3-manifold M admits an* **R***-covered foliation $\mathcal{F}$, then either $\mathcal{F}$ is taut or $P(M)$ is double-covered by $S^2 \times S^1$.*

3 Taut but not R-covered

In [2], Theorem D, the authors exhibit a family of 3-manifolds which admit $C^{(0)}$-foliations with no compact leaves, but no $C^{(2)}$-foliations without compact leaves. Each of the examples is obtained from two copies M_1,M_2 of (a once-punctured torus)$\times S^1$, glued together along their boundary tori by a homeomorphism A. What we will show is that for essentially the same choices of A, the resulting manifolds admit taut foliations, but no **R**-covered ones.

A is determined by its induced isomorphism on first homology $\mathbb{Z}^2$ of the boundary torus, and so we will think of it as a 2$\times$2 integer matrix with determinant ±1. We choose as basis for the homology of each torus the pair ($*\times S^1$,$\partial F\times *$), where F denotes the once-punctured torus. (Technically, we should orient these curves, but because all of the conditions we will encounter will be symmetric with respect to sign, the orientations will make no difference, and so we won't bother.)

Each M_i is a Seifert-fibered space (fibered by $*\times S^1$); the manifold M_A resulting from gluing via A is a Seifert-fibered space iff A glues fibre (1,0) to fibre (1,0), i.e., A is upper triangular. We will assume that this is not the case.

Let T denote the incompressible torus $\partial M_1 = \partial M_2$ in M_A. By [10], any horizontal foliation of M_i, i.e., a foliation everywhere transverse to the Seifert fibering of M_i, must meet ∂M_i in a foliation with slope in the interval ($-$1,1). [Note that this disagrees with the statement in [2], where the result was quoted incorrectly.]

If $\mathcal{F}$ is an **R**-covered, hence Reebless, foliation of M, then by [3], we can isotope $\mathcal{F}$ so that either it is transverse to T, and the restrictions $\mathcal{F}_i$ of $\mathcal{F}$ to M_i, $i = 1, 2$, have no Reeb or half-Reeb components, or $\mathcal{F}$ contains a cylindrical component, and therefore a compact (toral) leaf. In the second case, the torus leaf must hit the torus T, and is split into a collection of non-∂-parallel annuli; these (essential) annuli must be vertical in the Seifert-fibering of each M_i, since $F \times S^1$ contains no horizontal annuli. But this implies that the gluing map A glues fibre to fibre, a contradiction. Therefore, we may assume that $\mathcal{F}$ restricts to Reebless foliations on each of the manifolds M_i. Note that [3] requires that we allow a finite amount of splitting along leaves to reach this conclusion; but since a splitting of an **R**-covered foliation is still **R**-covered (it amounts, in the space of leaves, to replacing points with closed intervals), this will not affect our argument.

By [6], each of the induced foliations $\mathcal{F}_i$ of M_i has either a vertical or horizontal sublamination. Every horizontal lamination in M_i can be extended to a foliation transverse to the fibres of M_i, and so meets ∂M_i in curves whose slope lies in $(-1,1)$. If $\mathcal{F}_i$ has a vertical sublamination, it either meets ∂M_i in curves of slope ∞ (i.e., in curves homologous to (1,0)) or is disjoint from the boundary.

It is this last possibility, a vertical sublamination disjoint from T, that we wish to require, and so we will now impose conditions on the gluing map A to rule out the other possibilities. If *both* $\mathcal{F}_1$ and $\mathcal{F}_2$ have either a horizontal sublamination or a vertical sublamination meeting T, then for both M_1 and M_2, the induced foliations meet T in curves with slope in $(-1,1)\cup\{\infty\}$. Therefore, the gluing map

$$A = \begin{pmatrix} a & b \\ c & d \end{pmatrix}, \qquad \text{which as a function on slopes, is} \quad A(x) = \frac{ax+b}{cx+d}$$

must have $A\left((-1,1)\cup\{\infty\}\right)\cap\left((-1,1)\cup\{\infty\}\right) \neq \emptyset$. But since

$$A(\infty) = \frac{a}{c}, \quad A\left(\frac{-d}{c}\right) = \infty, \quad A\left(\frac{1}{1}\right) = \frac{a+b}{c+d}$$

and A is increasing if $ad - bc = 1$ (take the derivative!), we can force $A\left((-1,1)\cup\{\infty\}\right)$ to be disjoint from $(-1,1)\cup\{\infty\}$ by setting

$$|a| > |c|, \ |d| > |c|, \text{ and } \frac{a+b}{c+d} < -1.$$

For example, we may choose $a = -n$, $b = -nm-1$, $c = 1$, and $d = m$, with $n, m \geq 2$, so that

$$\frac{a+b}{c+d} = -n - \frac{1}{m+1} < -1.$$

As the figure below shows, the conditions $-a > c$, $d > c$ and $ad - bc = 1$ are also sufficient; what is needed, essentially, is that neither the graph of A nor either of its asymptotes pass through the square $[-1,1]\times[-1,1]$.

For such a gluing map A and resulting manifold M_A, either $\mathcal{F}_1$ or $\mathcal{F}_2$ (without loss of generality, $\mathcal{F}_1$) must contain a vertical sublamination $\mathcal{L}$ disjoint from $T = \partial M_1$. $\mathcal{L}$ is the saturation, by circle fibres, of a 1-dimensional lamination λ in the punctured torus F. This lamination λ cannot contain a closed loop, since then $\mathcal{F}_1$, and therefore $\mathcal{F}$, would contain a torus leaf L missing T. Lemma 1 would then imply that $M|L$ is an I-bundle, a contradiction, since it contains M_2. λ is therefore a lamination by lines.

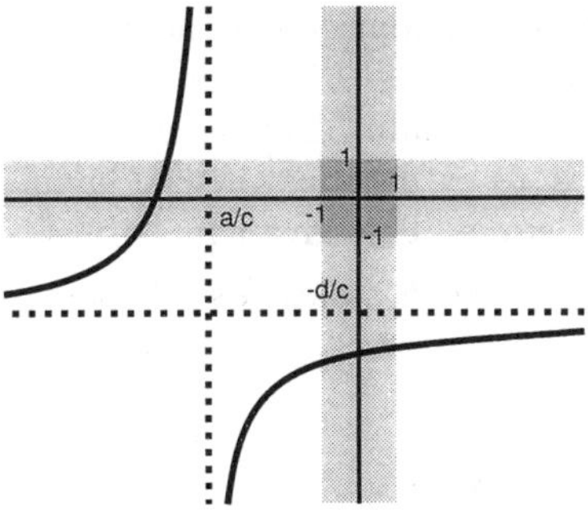

Figure 1.

By Euler characteristic considerations, the complementary regions of λ, thought as in a torus, are products, and so the complementary region of λ in F which meets ∂F is topologically a (∂-parallel) annulus, with a pair of points removed from the 'inner' boundary. Therefore, the component N of $M_A|\mathcal{L}$ which contains T is homeomorphic to M_2 with a pair of parallel loops removed from $\partial M_2 = T$.

Now we will assume that $\mathcal{F}$ is **R**-covered, and argue as in the proof of Lemma 1, to arrive at a contradiction. $\mathcal{L}$ lifts to a lamination in R^3 by planes, whose image in the space of leaves **R** of $\widetilde{\mathcal{F}}$ is a closed set. The two boundary leaves of $\mathcal{L}$ in ∂N are both annuli; they are the complements of the two parallel loops in ∂M_2 in the above description of $M_A|\mathcal{L}$. A lift of this complementary region to $\widetilde{M}$ has boundary consisting of lifts of the two annuli. Since the lift is a closed set in $\widetilde{M}$ its image in the space of leaves **R** is a connected, closed set, and therefore an interval. This implies that the lift of $M_A|\mathcal{L}$ has (at most) two boundary components, implying that the inverse image of each of the annulus leaves is a single (planar) leaf of $\widetilde{\mathcal{F}}$. This implies, as in the proof of Lemma 1, that the image in $\pi_1(M_A|\mathcal{L})$ $= \pi_1(M_2)$ of the fundamental group of each annulus has index at most 1, and so $\pi_1(M_2) = F_2 \times \mathbb{Z}$ is cyclic, a contradiction. Therefore, $\mathcal{F}$ is not **R**-covered. This implies

Theorem 5 *With gluing map A given as above, M_A admits no* **R***-covered foliations.*

On the other hand, every manifold M_A built out of the pieces we have used admits taut foliations and, in fact, foliations with no compact leaves. We simply choose a vertical lamination with no compact leaves in each of the Seifert-fibered pieces M_i, missing the gluing torus T. The complement of this lamination is homeomorphic to $T \times I$, with a pair of parallel loops removed from each of the boundary components. Treating this as a sutured manifold, and thinking of this as $(S^1 \times I) \times S^1$, we can foliate it, transverse to the sutures, by parallel annuli. Then, as in [12], we can spin the annular leaves near the sutures, to extend our vertical laminations to a foliation of M_A with no compact leaves. This give us

Proclaim 6 *There exist graph manifolds admitting taut foliations, but no* $\mathbf{R}$*-covered foliations.*

4 R-covered finite covers

Work of Luecke and Wu [16] implies that (nearly) every connected graph manifold is *finitely covered* by a graph manifold that admits an $\mathbf{R}$-covered foliation. In particular, for any graph manifold M whose Seifert fibered pieces all have base surfaces having negative (orbifold) Euler characteristic, they find a finite cover M' (which is also a graph manifold) admitting a foliation $\mathcal{F}$ transverse to the circle fibres of each Seifert fibered piece of M', and which restricts on each piece to a fibration over the circle. Note that this implies that every leaf of $\mathcal{F}$ meets every torus which splits M' into Seifert-fibered pieces.

Even more, every leaf of the lift, to the universal cover of M, of $\mathcal{F}$ meets every lift P_1, P_2 of the splitting tori. This can be verified by induction on the number of lifts of the tori that we must pass through to get from a lift we know the leaf hits, to our chosen target lift. The initial step follows by picking a path $\widetilde{\gamma}$ between two 'adjacent' lifts P_1 and P_2, whose interior misses every lift of the splitting tori, and projecting down to M; this gives a path γ in a single Seifert fibered piece of M'. This path can be made piecewise vertical (in fibres) and horizontal (in leaves of $\mathcal{F}$), missing, without loss of generality, the multiple fibres of M' (just do this locally, in a foliation chart for $\mathcal{F}$; the Seifert fibering can be used as the vertical direction for the chart). Each vertical piece can then be dragged to the boundary tori, since the saturation by fibres of an edgemost horizontal piece of γ is a (singular) annulus with induced foliation by horizontal line segments; see Figure 2. [This is where the fact that $\mathcal{F}$ is everywhere transverse to the fibers is really used.] The end result of this process is a loop γ', homotopic rel endpoints to γ, which consists of two paths each lying in a circle fibre in the boundary tori, with a single path in a leaf of $\mathcal{F}$ lying in between. This lifts to a path homotopic rel endpoints to $\widetilde{\gamma}$, consisting of paths in the two lifted tori, and a path in some lifted leaf. This middle path demonstrates that *some* lifted leaf L hits both P_1 and P_2.

By choosing a point where any other lifted leaf L' hits a lift P of a

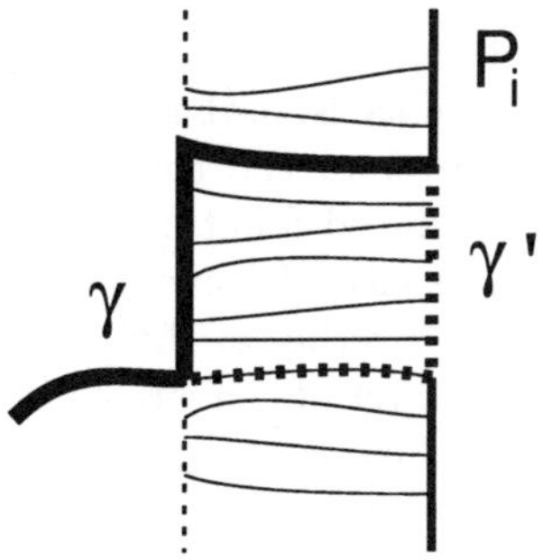

Figure 2.

splitting torus, and joining it by a path α to a point where L hits P_1 or P_2, we can apply the same straightening procedure as above (see Figure 3), to show that α is homotopic rel endpoints to paths, one of which lies in a lifted fibre and then lies totally in L', and the other of which lies totally in L, and then in a lifted fibre. This in particular implies that L' also hits P_1 and P_2. Therefore, every lifted leaf hits both P_1 and P_2, as desired.

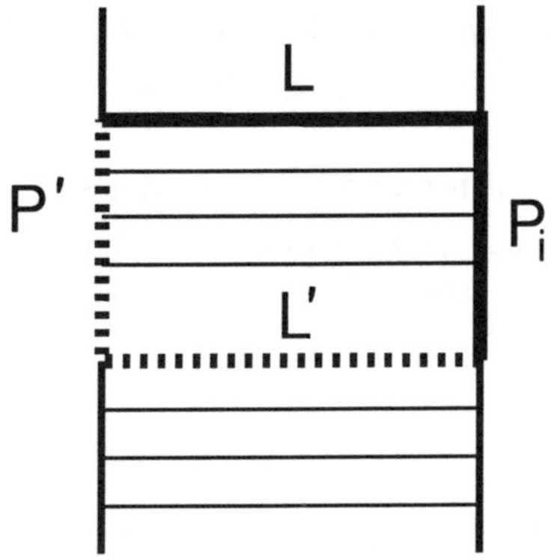

Figure 3.

The inductive step is nearly identical; assuming our two leaves L_1 and L_2 both hit $P_1, \ldots, P_{n-1}$, and P_n can be reached from P_{n-1} without passing through any other lift of a splitting torus, the above argument implies that two leaves, in the lift of the relevant Seifert fibered piece, and contained in L_1 and L_2, hit both L_{n-1} and L_n, implying that L_1 and L_2 also both hit P_n.

But this in turn implies that the lifted foliation $\widetilde{\mathcal{F}}$ has space of leaves **R**. This is because the foliation induced by $\widetilde{\mathcal{F}}$ on any lift P of a splitting torus is a foliation transverse to (either of the) foliations by lifts of circle fibres, and so has space of leaves **R**, which can be identified with one of the lifts of a circle fibre. [This is probably most easily seen in stages: first pass to a cylindrical cover of the torus, for which the circle fibers lift homeomorpically. The induced foliation from $\mathcal{F}$ is by lines transverse to this fibering, and so has space of leaves one of the circle fibers. The universal cover P is a cyclic covering of this, whose induced foliation has space of leaves the universal cover of the circle fiber.] The argument above implies that every leaf of $\widetilde{\mathcal{F}}$ hits P at least once. But no leaf of $\widetilde{\mathcal{F}}$ can hit a lift of a circle fibre *more* than once; by standard arguments, using transverse orientability, a path in the leaf joining two such points could be used to build a (null-homotopic) loop transverse to $\widetilde{\mathcal{F}}$, contradicting tautness of $\mathcal{F}$, via Novikov's Theorem [18]. We therefore have a one-to-one correspondence between the leaves of $\widetilde{\mathcal{F}}$ and (any!) lift of a circle fibre in any of the Seifert-fibered pieces, giving our conclusion

Proposition 7 *Any foliation of a graph manifold M, which restricts to a foliation transverse to the fibres of every Seifert-fibered piece of M, is* **R***-covered.*

Combining this with the result of Luecke and Wu, we obtain

Corollary 8 *Every graph manifold, whose Seifert-fibered pieces all have hyperbolic base orbifold, is finitely covered by a manifold admitting an* **R***-covered foliation.*

Combining the proposition with our main result, we obtain:

Corollary 9 *There exist graph manifolds, admitting no* **R***-covered foliations, which are finitely covered by manifolds admitting* **R***-covered foliations.*

5 Concluding remarks

Being finitely covered by a manifold admitting an **R**-covered foliation is nearly as good as having an **R**-covered foliation yourself. Any property that could be verified in the presence of an **R**-covered foliation, which remains 'virtually' true (e.g., virtually Haken, or having residually finite fundamental group), would then be true of the original manifold. It would then be of interest to know

Question 1 *Does every 3-manifold with universal cover R^3 have a finite cover admitting an* **R***-covered foliation?*

Or, even stronger

Question 2 *Does every irreducible 3-manifold with infinite fundamental group have a finite cover admitting an* **R***-covered foliation?*

Weaker, but still interesting

Question 3 *Does every tautly foliated 3-manifold have a finite cover admitting an* **R**-*covered foliation?*

The first two questions could be broken down into Question 3 and

Question 4 *Does every 3-manifold (in the appropriate class) have a tautly foliated finite cover?*

Note that showing that every irreducible 3-manifold with infinite fundamental group has a tautly foliated finite cover would settle the

Conjecture 1 *Every irreducible 3-manifold with infinite fundamental group has universal cover* R^3.

Questions 1 and 2 can be thought of as weaker versions of the (still unanswered) question, due to Thurston, of whether or not every hyperbolic 3-manifold is finitely covered by a bundle over the circle; the foliation by bundle fibres is **R**-covered. Gabai [13] has noted that there are Seifert-fibered spaces for which the answer to Thurston's question is 'No', although an observation of Luecke and Wu [16] implies that, via the results [10], the answer to our Question 1 is 'Yes', for Seifert-fibered spaces, since [7] a transverse foliation of a Seifert-fibered space is **R**-covered. [Note that the arguments of the previous section can be modified to give a different proof of this; look at how lifted leaves meet lifts of a single regular fiber, instead of lifts of the splitting tori.] Question 4, with its conclusion replaced by 'have a taut foliation', has as answer 'No'; examples were first found among Seifert-fibered spaces [4], [9]; there are no known examples among hyperbolic manifolds.

Question 5 *Do there exist hyperbolic 3-manifolds admitting no taut foliations?*

Finally, the result we have established here for graph manifolds is still unknown for hyperbolic 3-manifolds:

Question 6 *Do there exist hyperbolic 3-manifolds which admit taut foliations, but no* **R**-*covered foliations?*

References

1. T. Barbot, *Caractérisation des flots d'Anosov en dimension 3 par leurs feuilletages faibles*, Ergodic Th. & Dynam. Sys., **15** (1995), 247–270.
2. M. Brittenham, R. Naimi and R. Roberts, *Graph manifolds and taut foliations,* J. Diff. Geom., **45** (1997), 446–470.
3. M. Brittenham and R. Roberts, *When incompressible tori meet essential laminations,* Pacific J. Math., **190** (1999), 21–40.

4. M. Brittenham, *Essential laminations in Seifert-fibered spaces,* Topology, **32** (1993), 61–85.
5. M. Brittenham, *Essential laminations in* $\mathbb{I}$*-bundles,* Trans. Amer. Math. Soc,, **349** (1997), 1463–1485.
6. M. Brittenham, *Essential laminations in Seifert-fibered spaces: boundary behaviour,* Topology Appl., **95** (1999), 47–62.
7. M. Brittenham, *Exceptional Seifert-fibered spaces and Dehn surgery on 2-bridge knots,* Topology, **37** (1998), 665–672.
8. D. Calegari, Foliations and the geometry of 3-manifolds, Thesis, University of California at Berkeley, 2000.
9. W. Claus, Essential laminations in closed Seifert-fibered spaces, Thesis, Univ. of Texas at Austin, 1991.
10. D. Eisenbud, U. Hirsch and W. Neumann, *Transverse Foliations on Seifert Bundles and Self-homeomorphisms of the Circle,* Comment. Math. Helv., **56** (1981), 638–660.
11. S. Fenley, *Anosov flows in 3-manifolds,* Ann. of Math., **139** (1994), 79–115.
12. D. Gabai, *Foliations and the topology of 3-manifolds,* J. Diff. Geom., **18** (1983), 445–503.
13. D. Gabai, *On 3-manifolds finitely covered by surface bundles,* in Low-dimensional topology and Kleinian groups, (Coventry/Durham, 1984), Cambridge Univ. Press, 1986, 145–155.
14. S. Goodman and S. Shields, *A condition for the stability of* $\mathbf{R}$*-covered on foliations of 3-manifolds,* Trans. Amer. Math. Soc., **352** (2000), 4051–4065.
15. J. Hempel, 3-Manifolds, Ann. of Math. Studies No. 86, Princeton University Press, 1976.
16. J. Luecke and Y.-Q. Wu, *Relative Euler number and finite covers of graph manifolds,* in Geometric topology (Athens, GA, 1993), Amer. Math. Soc., 1997, 80–103.
17. R. Naimi, *Foliations transverse to fibres of Seifert manifolds,* Comment. Math. Helv., **69** (1994), 155–162.
18. S. Novikov, *Topology of foliations,* Trans. Moscow Math. Soc., **14** (1965), 268–305.
19. C. Palmeira, *Open manifolds foliated by planes,* Ann. of Math., **107** (1978), 109–121.
20. W. Thurston, *Three-manifolds, Foliations and Circles I,* preprint.

Received October 31, 2000.

Proceedings of
FOLIATIONS: GEOMETRY AND DYNAMICS
held in Warsaw, May 29–June 9, 2000
ed. by Paweł WALCZAK *et al.*
World Scientific, Singapore, 2002
pp. 225–261

ENDSETS OF EXCEPTIONAL LEAVES; A THEOREM OF G. DUMINY

JOHN CANTWELL
Department of Mathematics, St. Louis University
St. Louis, MO 63103,
e-mail: cantwelljc@slu.edu

LAWRENCE CONLON
Department of Mathematics, Washington University,
St. Louis, MO 63130,
e-mail: lc@math.wustl.edu

In 1977 Gerard Duminy proved that, if F is a semiproper leaf of a C^2 codimension–one foliation $\mathcal{F}$ of a compact n–manifold M and X is an exceptional local minimal set of $\mathcal{F}$, then the set $\mathcal{E}^X(F)$ of ends of F asymptotic to X, if nonempty, is homeomorphic to a Cantor set. No proof of this remarkable result has ever appeared, even in preprint form. Here, we offer a proof of our own.

1 Duminy's theorems

Let $(M, \mathcal{F})$ be a transversely oriented, C^2–foliated manifold of codimension one, with M compact and oriented. We assume each component of ∂M, if any, is a leaf of $\mathcal{F}$. Remark that each leaf of $\mathcal{F}$ is oriented.

Let $U \subseteq M$ be an open, $\mathcal{F}$–saturated subset and let $X \subset U$ be an exceptional minimal set of $\mathcal{F}|U$. As in [2] and elsewhere, we say that X is an exceptional LMS (local minimal set) of $\mathcal{F}$.

If F is a leaf of $\mathcal{F}$, let $\mathcal{E}(F)$ denote the space of ends of F and let $\mathcal{E}^X(F)$ denote the subspace of ends that are asymptotic to X.

Theorem 1.1 (Duminy) *If the leaf F of $\mathcal{F}$ is semiproper and if it accumulates on the exceptional LMS X, then $\mathcal{E}^X(F)$ is homeomorphic to the Cantor set.*

Here, *semiproper* is taken to mean that at least one side of F is proper, so we include the case in which F is a proper leaf.

Corollary 1.2 *If X is an exceptional minimal set of $\mathcal{F}$ and if $F \subset X$ is a semiproper leaf, then $\mathcal{E}(F)$ is homeomorphic to the Cantor set.*

Indeed, since X is minimal, $\mathcal{E}^X(F) = \mathcal{E}(F)$.

Theorem 1.1 is a corollary of another more directly useful result. For this, let L be any semiproper leaf of $\mathcal{F}$ that is approached by a possibly different semiproper leaf F. By the transverse orientation, L has a positive and a negative side, and we assume (reversing the transverse orientation, if necessary) that the positive side is a side on which F accumulates. Let $0 \in L$ and let $[0, \varepsilon]$ be a parametrized, $\mathcal{F}$–transverse arc issuing from 0 on the positive side. If σ is a loop on L based at 0, then h_σ will denote the holonomy in $[0, \varepsilon]$ defined by σ. If $H \subset L$ is a compact, connected, nonseparating, oriented submanifold of codimension one (a "handle"), then the homological intersection number of the loop σ with H will be denoted by $\sigma \frown H$.

Let L' be L cut apart along the handle H. Assume that $0 \in H$ and let $0^+ \in H^+$ and $0^- \in H^-$ be the two copies of 0 thus obtained. Let $\tilde{\alpha}$ be a curve from 0^+ to 0^- and α the curve obtained from $\tilde{\alpha}$ by identifying 0^+ to 0^-. Assume the definition of $\pm$ has been chosen so that $\alpha \frown H = +1$. In Section 7 we show that there exists an $\varepsilon > 0$, a contraction f of $[0, \varepsilon)$ to 0, a handle H, and curve α as above with $\alpha \frown H = +1$ and $f = h_\alpha$.

Theorem 1.3 *If $\varepsilon > 0$, f, H, and α are as above and (a, b) is a component of $(0, \varepsilon) \setminus \overline{F}$ then there is an integer $N \geq 0$ such that if σ is a loop on L' based at 0^+ then $h_\sigma(f^k(b)) = f^k(b)$ all $k \geq N$.*

Remarks. The integer N depends on the gap $[a, b]$ but ε, f, H, and α do not (see Proposition 7.16). The final choice of ε, f, H, and α is made just prior to Lemma 7.15 while the choice of the integer N is made in the proof of Lemma 7.15. The proof of Theorem 1.3 occurs just after the proof of Lemma 7.15. Note that the theorem remains true if we replace b with a in its statement. A careful proof of Theorem 1.3 will occupy Sections 3 through 7. In Section 2 we will deduce Theorem 1.1 from Theorem 1.3 and we will also deduce the following theorem that was the basis of Duminy's original proof of Theorem 1.1.

Theorem 1.4 (Duminy) *If $e \in \mathcal{E}^X(F)$, then every neighbourhood of e in F contains a complete submanifold that spirals on L with juncture H. Consequently, e is a cluster point of $\mathcal{E}^X(F)$ and this set is a Cantor set.*

It is natural to conjecture that, if $F \subset X$, the requirement in Theorem 1.1 that F be semiproper can be dropped. This is true for *Markov* LMS [4] but is unknown in general. A weaker conjecture would be that the generic leaf of an exceptional minimal set X have a Cantor set of ends. Here generic can be taken either in the sense of [5] or [7]. It is not hard to show that the generic leaf has one end or a Cantor set of ends.

In Section 8 we ask some other questions. Section 9 gives a general construction of examples of Markov minimal sets. The construction is general in the sense that it provides examples of any subshift of finite type occurring as holonomy of a codimension-one foliation.

Remark. Duminy deserves full credit for Theorem 1.1. The proof given here is essentially one that we worked out about 15 years ago and we take responsibility for whatever shortcomings it may have.

2 The proof of Theorem 1.1 and Theorem 1.4

The foliation $\mathfrak{F}$ of M^n is described by a finite, C^2 atlas $\Phi = \{(U_i, x_i, y_i)\}_{i=1}^r$ of foliated charts. Here,

$$x_i : U_i \to \mathbb{R}^{n-1},$$
$$y_i : U_i \to \mathbb{R}$$

and, on overlaps $U_i \cap U_j$, the coordinate changes are of the form

$$x_i = x_i(x_j, y_j), \qquad (*)$$
$$y_i = y_i(y_j). \qquad (**)$$

We can and do assume that the images $y_i(U_i)$ are bounded intervals in $\mathbb{R}$ with disjoint closures. As is standard, the set of transverse coordinate changes $(**)$ is interpreted as a symmetric generating set $\{h_1, \ldots, h_m\}$ of a pseudogroup Γ of local C^2 diffeomorphisms in $\mathbb{R}$, called the "holonomy pseudogroup of $\mathfrak{F}$". To relate this pseudogroup to the geometry of $\mathfrak{F}$, it is useful to fix imbeddings of the intervals $y_i(U_i) \subset U_i$ as disjoint transverse curves to $\mathfrak{F}$. The union W of these transversals is then a complete transversal and the holonomy pseudogroup maps open subsets of W onto open subsets of W simply by sliding along the leaves.

One also requires a certain "regularity" property of the foliated atlas, namely that each (U_i, x_i, y_i) is actually a subchart of a foliated chart (V_i, x_i, y_i) and that the closure $\overline{U}_i$ is a compact subset of V_i. Again, it is useful to fix imbeddings of the intervals $y_i(V_i) \subset V_i$ as disjoint transverse curves to $\mathfrak{F}$. The union $T \supset W$ of these transversals is then a complete transversal and the holonomy pseudogroup maps open subsets of T onto open subsets of T simply by sliding along the leaves. Again, the set of transverse coordinate changes $(**)$ is interpreted as a symmetric generating set $\{\tilde{h}_1, \ldots, \tilde{h}_m\}$ of a pseudogroup $\tilde{\Gamma}$, of local C^2 diffeomorphisms in $\mathbb{R}$.

Thus, the domain of h_i is an open interval $D(h_i)$ with compact closure $\overline{D}(h_i)$ and h_i is the restriction of a C^2 imbedding $\overline{h}_i : \overline{D}(h_i) \to \mathbb{R}$, $1 \leq i \leq m$. Thus, every element of Γ extends C^2–smoothly to the closure of its domain. Abusing notation we denote $\tilde{h}_i$ by h_i, $1 \leq i \leq m$, and $\tilde{\Gamma}$ by Γ.

In this section we assume the truth of Theorem 1.3 and deduce Theorem 1.1 and Theorem 1.4. Let $X \subset M$ be an exceptional LMS of $\mathcal{F}$, let F be a semiproper leaf that accumulates on X, L a semiproper leaf lying in X. We assume we have chosen our atlas $\Phi = \{(U_i, x_i, y_i)\}_{i=1}^r$ of foliated charts so that boundary plaques of $\overline{\mathcal{U}}_i$ do not lie in $\overline{F}$, $1 \leq i \leq r$.

Lemma 2.1 *The leaf L lies at finite level and the closure $\overline{F}$ of F in M accumulates on F from at most one side.*

Proof. No leaf at infinite level can have a proper side [1, Lemma 8.3.23], so F and L each lie at finite level. This means that F lies in a unique LMS Q and Q must either reduce to F itself (and F is a *proper* leaf) or it must be exceptional. In either case, $\overline{F}$ is a compact, nowhere dense, $\mathcal{F}$–saturated set containing X. Further, by [1, Corollary 8.3.12], $\overline{F}$ is a finite union of LMS, each of which is either a proper leaf or an exceptional LMS and none of which, except for Q, can accumulate on F. If Q reduces to F itself, $\overline{F}$ does not accumulate on F. Alternatively, Q is exceptional and accumulates on F only from the nonproper side of that leaf. □

By Lemma 2.1, each open, $\mathcal{F}$–transverse arc meets $\overline{F}$, if at all, in a relatively closed, nowhere dense set and each point in which F meets the arc is the endpoint of a gap of this set. Since X is exceptional, the closed subset $\overline{X}$ of $\overline{F}$ meets each open, $\mathcal{F}$–transverse arc, if at all, in an open subset of a Cantor set. The set $Y = \overline{F} \cap W$ is a compact, nowhere dense, Γ–invariant set in $\mathbb{R}$, as is $Z = \overline{X} \cap W$. By the above observations, Z is a Cantor set, but Y may contain isolated points.

The leaf F accumulates on the semiproper leaf $L \subset X$. Since L accumulates on itself, F accumulates on L from the nonproper side. If $F \not\subset X$, it is possible that F also accumulates on L from the proper side, but this is immaterial for our purposes.

We assume Theorem 1.3 and adopt the notation in the paragraph prior to Theorem 1.3. By Lemma 2.1, the Γ–orbit $F \cap W$ consists of some of the endpoints of gaps of the set Y. For definiteness, we suppose that $F \cap W$ consists of upper endpoints of gaps and indicate, where pertinent, the minor adjustments needed for the other case. Let $[a, b] \subset (0, \varepsilon)$ be a gap of Y with $b \in F \cap (0, \varepsilon)$.

Let $y \in L'$ and let $\tau(t)$, $0 \leq t \leq 1$, be a path from 0^+ to y in L', missing $\partial L'$ except for $t = 0$ and, perhaps, $t = 1$. Subdivide τ into segments, each lying in an $\mathcal{F}$–plaque, obtaining thereby a chain $P_0, P_1, \ldots, P_m$ of $\mathcal{F}$–plaques such that $0 \in P_0$ and $P_j \cap P_{j+1} \neq \emptyset$, $0 \leq j < m$. This defines a holonomy chain h_τ at 0. While this has involved choices, another such holonomy chain $\tilde{h}_\tau$ will necessarily be germinally equivalent to h_τ. Germinal independence of choices will not be enough for our purposes, but what we do need is given in the next lemma.

Let f and $N \geq 0$ be as in Theorem 1.3, $x_i = f^{N+i}(b)$ and $z_i = f^{N+i}(a)$, $\forall i \geq 0$. If τ is a path from 0^+ to $y \in L'$, then $[0, x_0] \subset \mathcal{D}(h_\tau)$. In fact, by Theorem 1.3, since $\sigma = \tau^{-1} \circ \tau$ is a loop at 0^+, h_σ is defined on $[0, x_0]$.

Lemma 2.2 *Let τ_1 and τ_2 be paths in L' from 0^+ to the same point y. Then, for any corresponding holonomy chains at 0, chosen as above, $h_{\tau_1}(x_i) = h_{\tau_2}(x_i)$ and $h_{\tau_1}(z_i) = h_{\tau_2}(z_i)$.*

Proof. If σ is a loop on L', based at 0^+, then Theorem 1.3 implies that both $h_\sigma(x_i) = x_i$ and $h_\sigma(z_i) = z_i$. If σ is the loop on L' formed by following τ_1 by τ_2^{-1}, then $h_\sigma = h_{\tau_2}^{-1} \circ h_{\tau_1}$ and the assertion follows. □

If $y \in L'$ and τ in L' is a path from 0^+ to y, then $h_\tau(0)$ can be interpreted as an $\mathcal{F}$–plaque P' containing y and $h_\tau(x_0)$ is defined and can be interpreted as another $\mathcal{F}$–plaque $P_0 \subset F$. Let $\pi_0 : P' \to P_0$ be the projection along plaques of $\mathcal{F}^\perp$. In particular, $\pi_0(y) \in F$. If the points of $F \cap W$ are *lower* endpoints of gaps of the set Y, we replace $h_\tau(x_0)$ with $h_\tau(z_0)$ in the definition of π_0.

Lemma 2.3 *A finite covering map $\pi_0 : L' \to F$, with image a complete, connected submanifold $B_0 \subseteq F$, is well–defined by the above procedure. By replacing x_0 with x_i, $i \geq 0$, we obtain finite covering maps π_i of L' onto complete, connected submanifolds B_i.*

Proof. By Lemma 2.2, π_i is well defined, $\forall i \geq 0$. It is a covering map by definition. Since H^+ is compact and $\pi_i | H^+ : H^+ \to H_i$ is a covering map, it follows that π_i is a finite covering map. The remaining assertions are elementary consequences of this. □

Let $H_i = \pi_i(H), i \geq 0$. It is clear that $H_i \neq H_j$ whenever $i \neq j$. It is also clear that $\partial B_i = H_i \cup H_{i+1}, i \geq 0$.

Lemma 2.4 *If $i \leq j$ and $B_i \cap B_j \neq \emptyset$, then either $j = i$ or $j = i + 1$ and $B_i \cap B_{i+1} = H_{i+1}$.*

Proof. It will be enough to show that, if $i \neq j$, then $\operatorname{int} B_i \cap \operatorname{int} B_j = \emptyset$. It is an easy consequence of the definition of π_i and π_j that this intersection is relatively open and closed in each of $\operatorname{int} B_i$ and $\operatorname{int} B_j$. By connectivity, either $\operatorname{int} B_i = \operatorname{int} B_j$ or these sets are disjoint. In the first case, it follows easily that $H_i = H_j$, hence that $i = j$. □

Definition 2.5 For each integer $j \geq 0$, $V_j = \bigcup_{i=j}^\infty B_i$.

Corollary 2.6 *The family $\{V_j\}_{j=0}^\infty$ is a fundamental neighbourhood system* *of an end $e^* \in \mathcal{E}^X(F)$ and e^* is a cluster point of $\mathcal{E}^X(F)$.*

Proof. It is obvious that $\{V_j\}_{j=0}^\infty$ is a fundamental neighbourhood system of an end $e^* \in \mathcal{E}(F)$. Since *every* V_j is asymptotic to L and L is asymptotic to X, it follows that $e^* \in \mathcal{E}^X(F)$. Finally, each B_j is asymptotic to everything to which L is asymptotic. Indeed, if $s \in M$, if I is the leaf of $\mathcal{F}^\perp$ through s, and if $\{s_i\}_{i=1}^\infty \subset L \cap I$ clusters at s, then, for fixed j, $\lim_{i\to\infty} \|[s_i, \pi_j(s_i)]\| = 0$ (by [1, p. 170] or Lemma 4.6), so $\{\pi_j(s_i)\}_{i=1}^\infty$ also clusters at s. Thus, each B_j is asymptotic to X and must be a neighbourhood of at least one end $e_j \in \mathcal{E}^X(F)$. By Lemma 2.4, $e_j \neq e_k$ for $j \neq k$. since V_j is a neighbourhood of $\{e_k\}_{k=j}^\infty$, it follows that e^* is a cluster point of $\{e_j\}_{j=0}^\infty$. □

Proof.[Proof of Theorem 1.1] Let $e \in \mathcal{E}^X(F)$. Any neighbourhood V of e in F is a connected component of the complement in F of a compact, connected subset $E \subset F$. The set E is also a compact subset of M. Let C be a compact submanifold of L such that the handle H and the curve α with $\alpha \frown H = +1$ lies in the interior of C. A suitable normal neighbourhood $\nu(C)$ of C in M will not meet E. In the above discussion, we choose ε so that $[0, \varepsilon)$ is a fibre of $\nu(C)$. We take $b \in V \cap (0, \varepsilon)$ such that all positive iterates α^k lift to paths on F that start at b and stay in $\nu(C)$. Since these paths cannot meet E, they stay in V, so each $x_i \in V$. But $H_i \subset \nu(C)$ cannot meet E and $x_i \in H_i \cap V$, so each $H_i \subset V$. By Lemma 2.4 and the connectedness of E, $B_i \cap E = \emptyset$ for all $i \geq 0$. It follows that $V_j \subset V$, if j is sufficiently large. By Corollary 2.6, V is a neighbourhood of a cluster point e^* of $\mathcal{E}^X(F)$. This cluster point may or may not be e but, in any case, the fact that V is an arbitrary neighbourhood of e will imply that e is a cluster point of $\mathcal{E}^X(F)$. Since $e \in \mathcal{E}^X(F)$ is arbitrary, it follows that $\mathcal{E}^X(F)$ is a Cantor set. □

Definition 2.7 A complete, connected submanifold $V \subset F$ is said to *spiral* (see [1, §8.4]) onto L with *juncture* H if

(1) for each $x \in V$, there is a choice $[p(x), x]$ of a compact subarc of the leaf of $\mathcal{F}^\perp$ through x such that $p(x) = y \in L$ and $p : V \to L$ is a surjection that is locally a homeomorphism;

(2) for each $y = p(x) \in L$, $p^{-1}(y) \cap [y, x] = \{x_i\}_{i=0}^\infty$ is a sequence such that $x_i \downarrow y$ and $x_0 = x$;

(3) $p^{-1}(H)$ falls into connected components $H_i \cong H$, $i = 0, 1, 2, \ldots$, such that $H_0 = \partial B$, and these copies of H partition B into a sequence $\{B_i\}_{i=0}^\infty$ of complete submanifolds, $\partial B_i = H_i \cup H_{i+1}$, such that each B_i is homeomorphic to L' via p.

Remark. We have actually defined the notion of V spiralling onto L *from above.* With obvious changes, one obtains a similar notion of spiralling *from below.* The two kinds of spiralling are interchanged by a change of transverse orientation, so no generality is lost in restricting attention to one type.

Proposition 2.8 *For each $i \geq 0$, $\pi_i : L' \to B_i$ is a homeomorphism.*

Proof. We deal explicitly with the case $i = 0$, the general case being no different. It will be enough to show that π_0 is one–one. Since π_0 is a finite covering map, it is enough to show that $\pi_0|H^+$ is one–one. Suppose not. Then there exists $y_0 = 0 \neq y_1 \in H^+$ so that $\pi_0(y_1) = \pi_0(y_0) = x_0$. Let $\tilde{\sigma}$ be a path in H^+ from y_0 to y_1 and let σ be the loop at x_0 in H_0 which is the projection of $\tilde{\sigma}$ under π_0. By reversing the direction of σ if necessary and rechoosing y_1, we can assume that $h_\sigma(0) = y_1 \in [0, x_0]$. Then the set of points $y_j = h_\sigma^j \in [0, x_0]$, $j \geq 0$, satisfies $y_0 < y_1 \dots$ in $[0, x_0]$ and thus is a set of distinct points in $\pi_0^{-1}(x_0)$. This contradiction the fact that π_0 is a finite cover. □

Proof.[Proof of Theorem 1.4] Given the end e, find an end e^* as in the proof of Theorem 1.1. It is obvious that V_j spirals on L for all $j \geq 0$. □

3 Unbounded holonomy

In this section we use some notation and Lemma 4.6 from Section 4. Lemma 4.6 is the "key lemma" of [1, p. 170].

Let L and F be semiproper leaves of $\mathcal{F}$ and assume that F accumulates on L. We emphasize that the important case in which $L = F$ is allowed. In that case L is exceptional. Fix a choice of transverse orientation and of basepoint $0 \in L$ so that the leaf of $\mathcal{F}^\perp$ through 0 has compact subarcs $J_0 = [y_0, 0]$ and $I_0 = [0, x_0]$, with J_0 contained in a gap of the set $Z = \overline{L} \cap W$ and with $\lambda \cdot |I_0|/|J_0| \leq 1$ (λ as in [1, p. 170] or Lemma 4.6).

Definition 3.1 *The holonomy of L is* unbounded *in I_0 (respectively, in J_0) if there does not exist a sequence $\{x_n\}_{n=1}^\infty \subset I_0$ (respectively, J_0) such that x_n converges monotonically to 0 and, $\forall n \geq 1$, $\Gamma_L(x_n)$ is the singleton $\{x_n\}$. Here Γ_L is the holonomy pseudogroup on transverse, open arcs containing 0 consisting of $g \in \Gamma$ such that $g_j(0) \in L$, $1 \leq j \leq p$, $g_p(0) = 0$.*

Our goal in this section is to prove the following easy but important step in the proof of Theorem 1.3.

Theorem 3.2 *The holonomy of L is unbounded on any side that is approached by F.*

Remark. By the semi-stability theorem of Dippolito [6, Theorem 3], unboundedness on a *proper* side of L will fail if and only if there is a collar $C \cong [0,1] \times L$ of $L = \{0\} \times L$ on that side in M and a sequence $\{t_n\}_{n=1}^{\infty} \subset [0,1]$, clustering at 0, such that $\{L_n = \{t_n\} \times L\}_{n=1}^{\infty}$ is a sequence of proper leaves of $\mathcal{F}$. In this case, no leaf can accumulate on L on the side in question. The proof of Theorem 3.2, therefore, is reduced to the case in which L is exceptional and the side in question is nonproper. In this case, L can always play the role of F. We fix these hypotheses.

We suppose that the sequence $\{x_n\}_{n=1}^{\infty} \subset I_0$, as in the above definition exists. We take n sufficiently large, let L_n denote the leaf of $\mathcal{F}$ through x_n, and proceed very much as in the previous section, to construct a local homeomorphism π_n of L onto L_n such that $[y, \pi_n(y)]$ is a subarc of a leaf of $\mathcal{F}^{\perp}$, $\forall y \in L$. Indeed, since Γ_L fixes x_n, the construction of $\pi_i : L' \to B_i$ in Section 2 can be mimicked exactly, there being no need to cut L along a handle H, to produce a local homeomorphism $\pi_n : L \to L_n$ which is also surjective. By discarding finitely many of the points x_n and renumbering the sequence, we assume the construction works for all $n \geq 1$.

Lemma 3.3 *For each integer $n \geq 1$, the points of $L \cap (y, \pi_n(y))$ range over all of L as y ranges over L.*

Proof. By the definition of π_n, the set $L \cap (y, \pi_n(y))$ is the union of plaques, hence is an open subset of L. This set is also complete (hence closed) in L, so it exhausts L. □

Proof.[Proof of Theorem 3.2] By an application of Lemma 4.6, we can choose n so large that the interval $[y, \pi_n(y)]$ has length less than that of J_0, $\forall y \in L$. In fact, appealing to the octopus decomposition (see Subsection 4.2), one sees that there are at most finitely many J_p with $|J_0| < |J_p|$. If J_p is one of these, one might have to increase n to guarantee that, for $y = y_p$, $|[y, \pi_n(y)]| < |J_0|$. If J_p is not one of these, then for $y = y_p$, $|[y, \pi_n(y)]| \leq |I_p| < |J_p| \leq |J_0|$. But Lemma 4.6 implies that $J_0 \subset [y, \pi_n(y)]$, some $y \in L$. This contradiction establishes Theorem 3.2. □

4 Some derivative estimates

4.1 Basic inequalities

Recall that $\{h_1, \ldots, h_m\}$ is a symmetric generating set for the holonomy group Γ.

Definition 4.1 A reduced word $g = h_{i_p} \circ \cdots \circ h_{i_1}$ is called a *chain* of length $|g| = p$. We write $g_j = h_{i_j} \circ \cdots \circ h_{i_1}$, $1 \leq j \leq p$, and $g_0 = \text{id}$.

Convention 1 Whenever g is a chain of length p and $Y \subseteq D(g)$, we set

$$Y_j = g_j(Y), \quad 0 \leq j \leq p.$$

In particular if $Y = \{y_0\}$, then one has $y_0, \dots, y_p$.

Definition 4.2 If $y_0 \in D(g)$, $\mathcal{C}$ is a subset of the Γ-orbit of y_0, and $y_0, \dots, y_p \in \mathcal{C}$, the chain g is said to be a *chain at* y_0 *in* $\mathcal{C}$. This is a *simple chain* at y_0 if $y_j \neq y_k$, whenever $j \neq k$. If $|g| = p$, if g_{p-1} is a simple chain at y_0, and if $g(y_p) = y_0$, then g is called a *simple loop* at y_0. Finally, if $g = h^{-1} \circ l \circ h$, where h is a simple chain at y_0 and l is a simple loop at $h(y_0)$, then g is called a *basic loop* at y_0.

Definition 4.3 If $Y \subseteq D(g)$ is a nondegenerate interval, the chain g is said to be a *chain at* Y. This is a *simple chain* at Y if $\operatorname{int} Y_j \cap \operatorname{int} Y_k = \emptyset$, whenever $j \neq k$. If $|g| = p$, if g_{p-1} is a simple chain at Y, and if $g(Y) = Y$, then g is called a *simple loop* at Y. Finally, if $g = h^{-1} \circ l \circ h$, where h is a simple chain at Y and l is a simple loop at $h(Y)$, then g is called a *basic loop* at Y.

Choose constants $A > 0$, $B > 0$, $\lambda > 0$, and $\theta > 1$ such that

1. $h_i' > A$, $1 \leq i \leq m$,
2. $|h_i''| \leq B$, $1 \leq i \leq m$,
3. $\theta = A/B$,
4. $\lambda = \exp(6\theta|W|)$ (where $|Y|$ denotes the Lebesgue measure of a set Y).

Lemma 4.4 *Let $g \in \Gamma$ be a chain of length p, and let $u_0, v_0 \in D(g)$. Then*

$$\frac{g'(u_0)}{g'(v_0)} \leq \exp\left(\theta \cdot \sum_{j=0}^{p-1} |u_j - v_j|\right).$$

For this fundamental estimate see [10] or [1, p. 168]. Here is an easy consequence.

Lemma 4.5 *Let $g \in \Gamma$ and $J \subset D(g)$ be a compact, nondegenerate interval. Suppose g is a simple chain of length p at J. Then*

$$\frac{g'(u)}{g'(v)} \leq \exp(\theta|W|), \quad \forall u, v \in J.$$

However, we need the following more delicate version of Lemma 4.5. It is the "key lemma" of [1, p. 170].

Lemma 4.6 *Let $g \in \Gamma$ and let J_0 and I_0 be compact subintervals of $D(g)$ such that*

- *J_0 is non-degenerate, I_0 is possibly degenerate,*
- *$J_0 \cap I_0 = \{y_0\}$ is a single point of ∂J_0,*
- *$|I_0|/|J_0| \leq 1/\lambda$,*

If g is either a simple chain or a basic loop at J_0 then

1. $|g(I_0)| < |g(J_0)|$,
2. $\dfrac{g'(u)}{g'(v)} < \lambda, \forall u, v \in J_0 \cup I_0$.

This lemma will give us significant control of holonomy on the nonproper side of a semiproper leaf.

4.2 The octopus decomposition

Fix a choice of a smooth, one-dimensional foliation $\mathcal{F}^\perp$ that is everywhere transverse to $\mathcal{F}$. We assume the atlas Φ from Subsection 4.1 is biregular relative to $\mathcal{F}$ and $\mathcal{F}^\perp$ in the sense defined in [1, §5.1]. Each $U_i = P_i \times I_i$, $1 \le i \le m$. In standard fashion, the arc I_i is to be identified with a set of $\mathcal{F}$–plaques by $t \leftrightarrow P_i \times \{t\}$ and with the set $y_i(U_i)$, $1 \le i \le m$.

In the theory of foliated manifolds of codimension one, the open, connected, saturated sets U play a fundamental role [6, 8, 2, 1]. One considers the "transverse completion" $\widehat{U}$, obtained by taking the metric completion of U relative to any Riemannian metric on U relativized from a Riemannian metric on M. If U is connected, this simply attaches to U finitely many boundary components, completing $\mathcal{F}|U$ to a foliation $\widehat{\mathcal{F}}$ of $\widehat{U}$ tangent to the boundary. We emphasize that $\widehat{U}$ is generally noncompact.

The inclusion map $i : U \hookrightarrow M$ induces a natural immersion $\widehat{i} : \widehat{U} \hookrightarrow M$ that may identify some pairs of boundary leaves to a single leaf of $\mathcal{F}$. The leaves of $\widehat{i}(\partial\widehat{U})$ are called the "border leaves" of U. If U is connected, there is an "octopus decomposition"

$$\widehat{U} = K \cup A_1 \cup \cdots \cup A_r,$$

where K is a compact, connected manifold with boundary and corners and each A_i is a foliated interval bundle, also with corners. The corners separate the subset $\partial_\pitchfork K$ of ∂K (respectively, the subset $\partial_\pitchfork A_i$ of ∂A_i) transverse to $\widehat{F}$ from the subset $\partial_\tau K$ (respectively, $\partial_\tau A_i$) tangent to $\widehat{\mathcal{F}}$. The manifold K is called a *nucleus* of $\widehat{U}$ and each A_i is an *arm*. Each arm has $\partial_\pitchfork A_i$ connected and attaches to K so that $\partial_\pitchfork A_i$ is identified with one component of $\partial_\pitchfork K$. The nucleus can be chosen as large as desired, generally at a cost of increasing the number of arms. For more details, see [1, Chapter 5].

If L is one of the boundary leaves of $\widehat{U}$ (generally L is taken to be a boundary leaf on the positive side of $\widehat{U}$), we let $C = K \cap L$ be the *core of* L.

For $1 \le i \le m$, the components of $U \cap U_i$ will be of the form $P_i \times (a, b)$. If both $P_i \times \{a\}$ and $P_i \times \{b\}$ are subsets of U_i, then $P_i \times (a, b)$ will be called an *ordinary product.* Otherwise $P_i \times (a, b)$ will be called a *special products.* By the definition of nucleus [1, Chapter 5] every special product is contained in the nucleus. An ordinary product that meets both the nucleus and an arm is called a *boundary product.*

Let $\mathcal{U} = U \cap W$. The components of $\mathcal{U}$ are the intervals (a, b) from the previous paragraph. Define $\widehat{\mathcal{U}}$ to be the disjoint union of the closure in T of these intervals. That is, one defines the *components* of $\widehat{\mathcal{U}}$ to be the intervals $[a, b]$.

If J is a component of $\widehat{\mathcal{U}}$, then J is a *component of an arm*, if the product $P_i \times J$ is contained in an arm for some $1 \leq i \leq m$. It is a *nuclear component* if the product is contained in the nucleus. It is a *boundary component* if the product meets both an arm and the nucleus. We let $\mathcal{J}$ be the set of all components in an arm. Let $|\mathcal{J}|$ be the sum of the lengths of components in an arm.

Remark. By choosing the nucleus larger one can make $|\mathcal{J}|$ as small as desired.

We let L be one of the boundary leaves of $\widehat{U}$ (without loss assume L is a boundary leaf on the positive side of $\widehat{U}$). Let $\mathcal{S} = L \cap T$. Let $\mathcal{A} \subset \mathcal{S}$ denote the set of upper endpoints of components in arms, $\mathcal{C} \subset \mathcal{S}$ the set of upper endpoints of nuclear components whose upper endpoints lie in $\mathcal{S}$, and $\partial\mathcal{C}$ the set of upper endpoints of boundary components. The set $\mathcal{C}$ will be called the *core* of $\mathcal{S}$ and the set $\partial\mathcal{C}$ will be called the *boundary of the core.* Then $\mathcal{S} = \mathcal{A} \cup \mathcal{C}$ and $\partial\mathcal{C} = \mathcal{A} \cap \mathcal{C}$. If $y_0 \in \mathcal{S}$, then $\mathcal{S}$ is the Γ-orbit of y_0.

4.3 Holonomy in the arms

If $X \subset M$ is a compact, $\mathcal{F}$-saturated set with empty interior, it is possible to select the regular foliated atlas so that $Z = X \cap W$ is a compact, Γ-invariant set that meets each of the open intervals $D(h_i)$ in a compact set. Assume this and choose $\mu > 0$ so small that any point in W that is μ-close to a point of the compact set $Z \cap D(h_i)$ lies in $D(h_i)$, $1 \leq i \leq m$. Also assume any component of an arm has length less than μ.

Let U be a component of the complement of X. Let K be a nucleus and let $J_0 = [x_0, y_0]$ be a component of an arm. Then $x_0, y_0 \in Z$.

Remark. If $g = h_{i_p} \circ \cdots \circ h_{i_1}$ is a chain at y_0 in $\mathcal{A}$, then g is a chain at J_0. It will be a simple chain, simple loop or basic loop, respectively, at J_0 if and only if it is such at y_0.

Let $I_0 \subset W$ be a compact interval meeting J_0 in exactly one of the endpoints, say y_0. In order to use Lemma 4.6, choose I_0 so that $|I_0|/|J_0| < 1/\lambda$.

Lemma 4.7 *If g is a simple chain or basic loop at y_0 in $\mathcal{A}$, then $I_0 \cup J_0 \subset D(g)$ and*

(1) $g'(u)/g'(v) \leq \exp(6\theta|\mathfrak{J}|)$ *if* $u, v \in I_0 \cup J_0$*;*

(2) if g is a basic loop, then $\exp(-6\theta|\mathfrak{J}|) \leq g'(u) \leq \exp(6\theta|\mathfrak{J}|)$, $\forall u \in I_0 \cup J_0$.

Proof. By the remark, $J_0 \subset D(g)$. For $0 \leq j \leq p$, we prove that I_j is defined.

For $j = 0$, there is nothing to prove. Suppose that, for some $i < p$ and $0 \leq j \leq i$, I_j is defined. Using Lemma 4.6, we obtain the inequality

$$\frac{|I_j|}{|J_j|} < 1, \quad 0 \leq j \leq i.$$

In particular, $J_i \in \mathfrak{J}$, so $|I_i| < |J_i| < \mu$. Therefore, $I_i \subset D(h_{i+1})$ and I_{i+1} is defined. By finite induction, I_p is defined and $I_0 \cup J_0 \subset D(g)$.

Let $u, v \in I_0 \cup J_0$. By Lemma 4.4,

$$\frac{g'(u)}{g'(v)} \leq \exp\left(\theta \cdot \sum_{j=0}^{p-1} |u_j - v_j|\right).$$

Furthermore, $|u_j - v_j| \leq |I_j| + |J_j| \leq 2|J_j|$ and, as in the proof of Lemma 4.6 (see [1, p. 170]), it follows that

$$\sum_{j=0}^{p-1} |u_j - v_j| \leq 6|\mathfrak{J}|.$$

This proves (1). Finally, if g is a basic loop, find $v \in J_0$ with $g'(v) = 1$ and apply (1) to prove (2). □

Corollary 4.8 *Let $0 < r < 1 < s$. If the octopus decomposition of $\widehat{U}$ has large enough nucleus, if $I_0 \cup J_0$ is as in Lemma 4.7, and if g is a simple chain or basic loop at y_0 in $\mathcal{A}$, then $r < g'(u)/g'(v) < s$, $\forall u, v \in I_0 \cup J_0$. If g is a basic loop, then $r < g'(u) < s$, $\forall u \in I_0 \cup J_0$.*

Proof. By making K large enough, we guarantee that

$$\exp(6\theta|\mathfrak{J}|) < \min\{s, 1/r\}.$$

An application of Lemma 4.7 completes the proof. □

Fix a boundary orbit $\mathcal{S} \subset X$ on the positive side of $\mathfrak{U}$, as above, together with an exhaustion

$$K = K_0 \subset K_1 \subset \cdots \subset K_k \subset \dots$$

of $\widehat{U}$ by nucleii of octopus decompositions. For a fixed $k = 0, 1, 2, \dots$, define $\mathcal{A}_k$, $\partial\mathcal{C}_k$, and $\mathcal{C}_k$ as in Subsection 4.2. In particular, each $\mathcal{C}_k$ is Γ-connected.

Fix a basepoint in $\mathcal{C}_0$ and, by recoordinatizing, assume that this basepoint is $0 \in T \subset \mathbb{R}$. Let $\operatorname{int} \mathcal{C}_k = \mathcal{C}_k \setminus \partial\mathcal{C}_k$.

The orbit $\mathcal{S}$ corresponds to a semiproper leaf $L \subset X$ and any choice of J_0, as above, lies on a proper side. If the leaf is not proper, then I_0 lies on the nonproper side and $\mathcal{S} \cap I_0$ clusters on 0. It is our delicate task to analyze the holonomy along paths in L on this nonproper side.

4.4 The set $\Gamma_\# \subset \Gamma_L$

The set $\Gamma_\#$ will serve as a set of generators for the germinal holonomy of L at 0, there exists an $\varepsilon_0 > 0$ such that every element of $\Gamma_\#$ is defined on the interval $[-\varepsilon_0, \varepsilon_0]$, and we have very delicate control of the maps γ for every $\gamma \in \Gamma_\#$.

Definition 4.9 The set $\Gamma_0 \subset \Gamma_L$ consists of basic loops at 0 in $\mathcal{C}_0$. If $\Gamma_k \subset \Gamma_L$ has been defined, some $k \geq 0$, then $\Gamma_L \supset \Gamma_{k+1} \supset \Gamma_k$ is constructed as follows. The complement $\Gamma_{k+1} \setminus \Gamma_k$ is to be the set of basic loops at 0 of the form $h^{-1} \circ g \circ h$, where g is a basic loop in $\mathcal{C}_{k+1} \setminus \operatorname{int} \mathcal{C}_k$ at $h(0) \in \partial\mathcal{C}_k$ and where h is a simple chain at 0 in $\mathcal{C}_k$ of length (say) p such that for $1 \leq j \leq k$, $\exists 1 < i_1 < \dots i_j < \cdots < i_k = p$, satisfying:

$$h_i(0) \in \operatorname{int} \mathcal{C}_j, i < i_j; h_{i_j}(0) \in \partial\mathcal{C}_j; h_i(0) \notin \operatorname{int} \mathcal{C}_j, i > i_j$$

Finally $\Gamma_\# = \bigcup_{k=0}^{\infty} \Gamma_k$. The product $\gamma = h^{-1} \circ g \circ h$ is called the *k-representation* of γ.

Lemma 4.10 *Let L be semiproper. If the holonomy of L is unbounded on a given side of L, then $\Gamma_\#$ has no fixed points arbitrarily close to L on that side.*

Indeed, in a standard fashion, each plaque–loop σ on L at 0 is "factored" into a product of loops, each of which corresponds to an element of $\Gamma_\#$. This requires that σ be modified by inserting finitely many plaque–chains of the form $\tau \star \tau^{-1}$ (where $\star$ denotes the adjunction of plaque chains). It follows that, if $\Gamma_\#$ has fixed points $x_n \downarrow 0$, then $h_\sigma(x_n) = x_n$, $\forall n$. Details are left to the reader.

The following is an application of Lemma 4.7.

Lemma 4.11 *For some $\varepsilon_0 > 0$, $[-\varepsilon_0, \varepsilon_0] \subset D(\gamma)$, $\forall \gamma \in \Gamma_\#$. Further, for any such ε_0, one can choose $0 < \varepsilon_1 < \varepsilon_0$ such that $\gamma[-\varepsilon_1, \varepsilon_1] \subset [-\varepsilon_0, \varepsilon_0]$, $\forall \gamma \in \Gamma_\#$.*

Proof. Choose δ sufficiently small that $[-\delta, \delta] \subset D(\gamma)$ for the finitely many basic loops $\gamma \in \Gamma_0$. We can also assume that the finitely many simple chains in $\mathcal{C}_0$ from 0 to points of $\partial\mathcal{C}_0$ are defined on $[-\delta, \delta]$. Every $\gamma \in \Gamma_\# \setminus \Gamma_0$ is of the form $\gamma = h^{-1} \circ g \circ h$ where g is a basic loop in

$\mathcal{S} \setminus \mathcal{C}_0$ at a point $y_0 \in \partial\mathcal{C}_0$ and h is a simple chain in $\mathcal{C}_0$ from 0 to y_0. Here $y_0 = I_0 \cap J_0$ where I_0 and J_0 are chosen as in Lemma 4.7 and, by that lemma, $J_0 \cup I_0 \subset D(g)$. Thus, choosing δ smaller, if necessary, we can assume that $g \circ h$ is defined on $[-\delta, \delta]$, for all choices of g and h as above. We must choose $0 < \varepsilon_0 < \delta$ so that h^{-1} is defined on $g \circ h[-\varepsilon_0, \varepsilon_0]$.

By the mean value theorem, there is a point $\xi \in h[-\delta, \delta]$ such that

$$|g[h(-\delta), h(\delta)]| = |[h(-\delta), h(\delta)]| \cdot |g'(\xi)|.$$

Again appealing to Lemma 4.7, we note that the derivative $g'(\xi)$ is uniformly bounded away from 0, for all basic loops g as above. Noting that g^{-1} varies over all such basic loops as g does and that h varies over finitely many simple chains, one easily chooses ε_0 so that $h[-\varepsilon_0, \varepsilon_0] \subset g^{-1} \circ h[-\delta, \delta]$, for all such g and h.

Further, the previous argument yields $\gamma[-\varepsilon_0, \varepsilon_0] \subset [-\delta, \delta]$, $\forall \gamma \in \Gamma_\#$. Replacing δ by ε_0 in the previous argument yields an $\varepsilon_1 < \varepsilon_0$ so that $\gamma[-\varepsilon_1, \varepsilon_1] \subset [-\varepsilon_0, \varepsilon_0]$, $\forall \gamma \in \Gamma_\#$. □

Every $g \in \Gamma_\#$ is a basic loop at 0. Choose ε_0 as in Lemma 4.11. Let $J_0 \subset [-\varepsilon_0, 0]$ have upper endpoint 0. Then g is a basic loop at J_0. If one chooses $I_0 \subset [0, \varepsilon_0]$ with $|I_0|/|J_0| \leq 1/\lambda$ and $I_0 \cap J_0 = \{0\}$, an application of Lemma 4.6 yields $|I_i \cup J_i| \leq 2|J_i|$.

Definition 4.12 Let $\ell = |I_0 \cup J_0|$ and $\Theta = (6\theta|W|/\ell) \cdot \exp(6\theta|W|)$. The constant Θ is fixed throughout the rest of the paper.

Lemma 4.13 *Let $g \in \Gamma_\#$ and I_0 and J_0 be as above. Then*

$$\frac{g'(u)}{g'(v)} \leq \exp(\Theta|u - v|), \quad \forall u, v \in I_0 \cup J_0.$$

Proof. Set $\Lambda = I_0 \cup J_0$. Then $\ell = |\Lambda|$. As usual, set $\Lambda_j = g_j(\Lambda)$, $J_j = g_j(J_0)$, $1 \leq j \leq p$. Then

$$\frac{|u_j - v_j|}{|\Lambda_j|} = \frac{g_j'(\xi)}{g_j'(\zeta)} \cdot \frac{|u - v|}{|\Lambda|} \leq \frac{|u - v|}{|\Lambda|} \cdot \exp(6\theta|W|)$$

by Lemma 4.6. Since I_0 is chosen with $|I_0|/|J_0| \leq 1/\lambda$ and g is a basic loop at J_0

$$\sum_{j=0}^{p-1} |\Lambda_j| \leq 2 \sum_{j=0}^{p-1} |J_j| \leq 6|W|.$$

Thus

$$\sum_{j=0}^{p-1} |u_j - v_j| \leq \frac{|u - v|}{|\Lambda|} \exp(6\theta|W|) \sum_{j=0}^{p-1} |\Lambda_j| \leq \frac{|u - v|}{|\Lambda|} \exp(6\theta|W|) \cdot 6|W|.$$

By Lemma 4.4,

$$\frac{g'(u)}{g'(v)} \le \exp\left(\theta \sum_{j=0}^{p-1} |u_j - v_j|\right) \le \exp(\Theta|u - v|).$$

□

Definition 4.14 A pseudogroup element $\gamma = g_m \circ g_{m-1} \circ \cdots \circ g_1$, where $g_i \in \Gamma_\#$, $1 \le i \le m$ is called a $\Gamma_\#$–chain of length m. Let γ be such a chain and let $\Lambda \subseteq D(\gamma)$ be a compact interval. For $1 \le p \le m$, set $\gamma_p = g_p \circ \cdots \circ g_1$ and let $g_0 = \text{id}$. Set $\Lambda_p = \gamma_p(\Lambda)$, $0 \le p \le m$. If $\text{int}\,\Lambda_0, \text{int}\,\Lambda_1, \dots, \text{int}\,\Lambda_m$ are disjoint, then γ is said to be a *simple* $\Gamma_\#$–chain at Λ. *Simple* $\Gamma_\#$–loops and *basic* $\Gamma_\#$–loops at Λ could be defined, but they will not be needed.

The following lemma is an analog of Lemma 4.5. The difference is that simple chains are constructed from the finite generating set $\{h_1, \dots, h_m\}$ of the pseudogroup Γ, while the simple $\Gamma_\#$–chains are constructed from the generally infinite set $\Gamma_\#$ which provides a set of generators for the germinal holonomy group at 0.

Lemma 4.15 *Let $\Lambda \subset I_0 \cup J_0$ be a compact interval and $\gamma = g_m \circ \cdots \circ g_1$ be a simple $\Gamma_\#$–chain of length m at Λ. Then*

$$\frac{\gamma'(u)}{\gamma'(v)} \le \exp(\Theta|W|), \quad \forall u, v \in \Lambda.$$

Proof. Let $u_j = \gamma_j(u)$ and $v_j = \gamma_j(v)$, $0 \le j \le m-1$. Then

$$\frac{\gamma'(u)}{\gamma'(v)} = \frac{g_m'(u_{m-1})}{g_m'(v_{m-1})} \cdots \frac{g_1'(u_0)}{g_1'(v_0)}.$$

Applying Lemma 4.13 to each g_j gives

$$\frac{\gamma'(u)}{\gamma'(v)} \le \exp\left(\Theta \sum_{j=0}^{m-1} |u_j - v_j|\right) \le \exp(\Theta|W|).$$

□

Remark. In Lemma 4.15, it is sufficient that γ_{m-1} be a simple $\Gamma_\#$–chain at Λ.

The following will be useful in Section 7.

Corollary 4.16 *Let $f \in \Gamma_\#$ be a contraction of $[0, \varepsilon) \subset I_0 \cup J_0$ to 0. Let $x_0 \in [0, \varepsilon)$ and $x_k = f^k(x_0)$, $\forall k \ge 0$. Then, for all integers p, $k \ge 1$ and all $u, v \in [x_p, x_0]$, $(f^k)'(u)/(f^k)'(v) \le \exp(p\Theta|W|)$.*

Proof. Choose $u = u_0, u_1, \dots, u_p = v$ in $[x_p, x_0]$ so that, for suitable choices of $z_i \in [x_p, x_0]$, $\{u_i, u_{i-1}\} \subset \Lambda_i = [f(z_i), z_i]$, $1 \le i \le p$. This is clearly possible. Then $\gamma = f^k$ and $\Lambda = \Lambda_i$ satisfy the hypotheses of Lemma 4.15, $1 \le i \le p$, so we see that $\gamma'(u_{i-1})/\gamma'(u_i) \le \exp(\Theta|W|)$, $1 \le i \le p$. The desired conclusion follows. □

Remark. In Theorem 6.1 we will show that there does exist a contraction $f \in \Gamma_\#$.

Lemma 4.17 *Let Λ and γ be as in Lemma 4.15 and let $g \in \Gamma$ be such that $\gamma(I_0 \cup J_0) \subset D(g)$. Let $Q = [a, b] \subset \Lambda$. Choose $r < 1$ so close to 1 that*

$$1 - r < \exp(-\Theta|W|) \cdot |Q|/|\Lambda|.$$

If there exists $x \in \gamma(\Lambda)$ such that

(1) $g(x) = x < \gamma(b)$ (respectively, $g(x) = x > \gamma(a)$),

(2) $g(\gamma(b)) \le \gamma(a)$ (respectively, $g(\gamma(a)) \ge \gamma(b)$),

then there exists $t \in \gamma(\Lambda)$ such that $g'(t) < r$.

Proof. We treat the case $x < \gamma(b)$ and $g(\gamma(b)) \le \gamma(a)$. The alternative case is obtained by an entirely similar argument.

By the mean value theorem and Lemma 4.15,

$$\frac{|\gamma(\Lambda)|}{|\gamma(Q)|} = \frac{\gamma'(u)}{\gamma'(v)} \cdot \frac{|\Lambda|}{|Q|} \le \exp(\Theta|W|) \cdot \frac{|\Lambda|}{|Q|},$$

for suitable values $u \in \Lambda$ and $v \in Q$. Thus,

$$\begin{aligned} 1 - r < \exp(-\Theta|W|) \cdot \frac{|Q|}{|\Lambda|} &\le \frac{|\gamma(Q)|}{|\gamma(\Lambda)|} \\ &\le \frac{\gamma(b) - \gamma(a)}{\gamma(b) - x} \le \frac{\gamma(b) - g(\gamma(b))}{\gamma(b) - x}. \end{aligned}$$

Therefore

$$r > 1 - \frac{\gamma(b) - g(\gamma(b))}{\gamma(b) - x} = \frac{g(\gamma(b)) - x}{\gamma(b) - x} = \frac{g(\gamma(b)) - g(x)}{\gamma(b) - x}.$$

By the mean value theorem, there exists $t \in \gamma(\Lambda)$ such that $g'(t) < r$. □

Remark. A simpler version of Lemma 4.17 can be given with Θ replaced by θ. In the proof of the simple version, the use of Lemma 4.15 is replaced by a use of Lemma 4.5. Otherwise the proof is identical. We need the stronger lemma.

4.5 Better estimates

Every $\gamma \in \Gamma_\#$ is defined on $I_0 \cup J_0$. Choose ε_1 in Lemma 4.11 such that $[-\varepsilon_1, \varepsilon_1] \cup \gamma[-\varepsilon_1, \varepsilon_1] \subset I_0 \cup J_0$, $\forall \gamma \in \Gamma_\#$. In the rest of the paper we work in an interval $(-\varepsilon, \varepsilon) \subset [-\varepsilon_1, \varepsilon_1]$ and take ε smaller as needed.

Lemma 4.18 *Let $0 < r < 1 < s$. Then $k \geq 0$ can be chosen so large and $\varepsilon > 0$ so small that for all $\gamma \in \Gamma_\# \smallsetminus \Gamma_k$, the inequalities $r < \gamma'(u) < s$ hold for $-\varepsilon \leq u \leq \varepsilon$. Furthermore, if $\{\gamma_1, \gamma_2, \ldots, \gamma_n\} \subseteq \Gamma_k$ is a subset such that each γ_i fails to be germinally equivalent to a contraction to 0, either in $[0, \varepsilon]$ or in $[-\varepsilon, 0]$, then the choice of ε can be made so small that $r < \gamma_i'(u) < s$, $1 \leq i \leq n$, $-\varepsilon \leq u \leq \varepsilon$.*

Proof. By Corollary 4.8, we choose $k \geq 0$ such that, $\forall \gamma \in \Gamma_\# \smallsetminus \Gamma_k$ with k–representation $\gamma = h^{-1} \circ g \circ h$, we have

$$\sqrt{r} < g'(h(u)) < \sqrt{s}, \quad -\varepsilon \leq u \leq \varepsilon.$$

By elementary calculus,

$$\gamma'(u) = \frac{h'(u)}{h'(\gamma(u))} \cdot g'(h(u)), \quad -\varepsilon \leq u \leq \varepsilon,$$

and each of the finitely many functions h' is continuous. By making ε small enough, we guarantee that

$$\sqrt{r} < \frac{h'(u)}{h'(\gamma(u))} < \sqrt{s}, \quad -\varepsilon \leq u \leq \varepsilon,$$

hence

$$r < \gamma'(u) < s, \quad -\varepsilon \leq u \leq \varepsilon.$$

There remain the elements $\gamma_i \in \Gamma_k$, $1 \leq i \leq n$. By the assumption on these, we see that $\gamma_i'(0) = 1$, $1 \leq i \leq n$. Thus, making ε possibly smaller, we complete the proof. □

Lemma 4.19 *Let ε_1 be as above and $0 < \beta \leq \varepsilon_1$. Then $k \geq 0$ can be chosen so large that for $\gamma \in \Gamma_\# \smallsetminus \Gamma_k$, the inequalities $1/(2\lambda) < \gamma'(u) < 2\lambda$ hold for $-\varepsilon_1 \leq u \leq \varepsilon_1$. Further for this choice of k, $[0, \beta/(2\lambda)] \subset \gamma[0, \beta] \subset [0, 2\lambda\beta]$, $\forall \gamma \in \Gamma_\# \smallsetminus \Gamma_k$.*

Proof. By Corollary 4.8, we choose $k \geq 0$ such that, $\forall \gamma \in \Gamma_\# \smallsetminus \Gamma_k$ with k–representation $\gamma = h^{-1} \circ g \circ h$, we have

$$1/2 < g'(h(u)) < 2, \quad -\varepsilon_1 \leq u \leq \varepsilon_1.$$

By elementary calculus,

$$\gamma'(u) = \frac{h'(u)}{h'(\gamma(u))} \cdot g'(h(u)), \quad -\varepsilon_1 \leq u \leq \varepsilon_1.$$

By Lemma 4.6 and Lemma 4.11,

$$1/\lambda < \frac{h'(u)}{h'(\gamma(u))} < \lambda, \quad -\varepsilon_1 \le u \le \varepsilon_1,$$

hence

$$1/(2\lambda) < \gamma'(u) < 2\lambda, \quad -\varepsilon_1 \le u \le \varepsilon_1.$$

By the mean-value theorem, $|[0,\beta/(2\lambda)]| \le |\gamma[0,\beta]| = |[0,\beta]| \cdot \gamma'(\xi) \le |[0,2\lambda\beta]|$ □

5 In the absence of a contraction

Let L be a semiproper leaf of $\mathcal{F}$. Fix a choice of basepoint $0 \in L$ and the arc $[-\varepsilon_1, \varepsilon_1]$ of the leaf of $\mathcal{F}^\perp$ through 0. Let $\Gamma_\#$ be as in Definition 4.9. By Lemma 4.11, every element of $\Gamma_\#$ is defined on all of $[-\varepsilon_1, \varepsilon_1]$. Further, $[-\varepsilon_1, \varepsilon_1] \subset I_0 \cup J_0$ with I_0 and J_0 as in Definition 4.12. This allows us to use Lemma 4.15 and Lemma 4.17 with Θ as defined in Definition 4.12. We fix the following assumptions.

(i) No $\gamma \in \Gamma_\#$ restricts to a contraction to 0 on any subarc $[0,\eta) \subset [0,\varepsilon]$.

(ii) The leaf L has unbounded holonomy on the positive side.

In the next section, we will see that these assumptions lead to a contradiction. Here, we simply produce some technical consequences of the assumptions that will be used in Section 6.

It will not matter whether the positive side is a proper side of L. By Lemma 4.18, we can choose the value of $0 < \varepsilon < \varepsilon_1$ so small that

$$\frac{9}{10} < \gamma'(x) < \frac{11}{10}, \qquad \forall \gamma \in \Gamma_\#, \qquad -\varepsilon < x < \varepsilon \tag{\#}$$

Remark. In the proofs that follow, it is enough that the constants $\frac{9}{10} < 1 < \frac{11}{10}$ be sufficiently close to 1.

Lemma 5.1 *There is a sequence $\{y_1, y_2, x_1, y_3, x_2, y_4, \dots, y_{i+1}, x_i, \dots\}$ in $(0,\varepsilon)$, converging monotonically to 0, and a sequence $\{g_i\}_{i=1}^\infty \subset \Gamma_\#$ such that*

(a) $g_i|[x_i, y_i]$ is a contraction to x_i, $1 \le i < \infty$;

(b) $g_i(y_i) = x_{i-1}$, $2 \le i < \infty$.

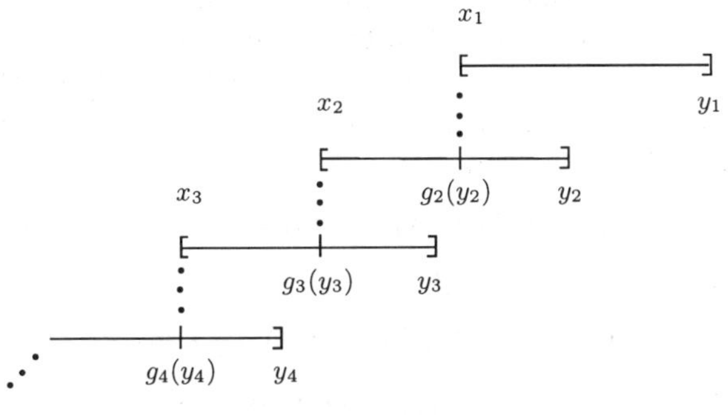

Figure 1.

Proof. In the rather technical argument, the reader may find the visual schema in Figure 1 helpful. By (ii) and Lemma 4.10, there is an $\alpha > 0$ so small that the elements of $\Gamma_{\#}|(0, \alpha)$ have no common fixed point. On the other hand, by (i), none of these are germinally contractions to 0. Thus, given $z \in (0, \alpha) \subset (0, \varepsilon)$, choose $g_z \in \Gamma_{\#}$ and $[x_z, b_z) \subset (0, \alpha)$ such that $z \in (x_z, b_z)$ and $g_z|[x_z, b_z)$ is a contraction to x_z. We consider two cases.

Case 1. Assume that $\alpha > 0$ can be chosen so that there is a set of points $z \in [0, \alpha)$, clustering at 0, for each of which one can choose $b_z = \alpha$. Choose a sequence $\{g_i, [x_i, \alpha)\}_{i=1}^{\infty}$ with $g_i = g_{z_i}$, $x_i = x_{z_i}$, and $x_i \downarrow 0$. Choose x_1 so that $2\lambda x_1 < \alpha$. In general the choice of x_{i+1} is to be made so that $2\lambda x_{i+1} < x_i$. If $i \neq j$, then it is clear that $g_i \neq g_j$. If i is sufficiently large, Lemma 4.19 implies that $g_i^{-1}(x_1) < \alpha$, so renumber g_i as g_2, x_i as x_2, and set $y_1 = \alpha$ and $y_2 = g_2^{-1}(x_1)$. Again, for j sufficiently large, $g_j^{-1}(x_2) < x_1$, so renumber g_j as g_3, x_j as x_3, and set $y_3 = g_3^{-1}(x_2)$. Iteration of this procedure yields a sequence $\{g_i, [x_i, y_i)\}_{i=1}^{\infty}$ with all the desired properties.

Case 2. Alternatively, $\alpha > 0$ cannot be chosen as in Case 1. Thus, for some $(0, \eta)$ and every $z \in (0, \eta)$, g_z and $[x_z, b_z)$ can be chosen as above with the additional property that $g_z(b_z) = b_z < \alpha$.

By the local compactness of $(0, \eta/2]$, choose a sequence

$$\{g_i = g_{z_i}, [x_i, b_i] = [x_{z_i}, b_{z_i}]\}_{i=1}^{\infty}$$

such that $\mathcal{O} = \{(x_i, b_i)\}_{i=1}^{\infty}$ covers $(0, \eta/2]$ and $x_i \downarrow 0$. We claim that $\lim_{i \to \infty} b_i = 0$. Otherwise, $\{b_i\}_{i=1}^{\infty}$ would have a cluster point $\delta > 0$ and we could rechoose $\alpha = \delta/2$, contradicting the hypothesis that no choice of $\alpha > 0$ places us in Case 1. Therefore, $\mathcal{O}$ is a *locally finite* cover of $(0, \eta/2]$. Also, *wlog*, we can assume that every component of $\mathcal{O}$ meets $(0, \eta/2]$. In particular, each component of $\mathcal{O}$ meets at least one other.

We seek a subcover $\mathcal{O}^* \subset \mathcal{O}$ without triple intersections. Set $\Lambda_i = [x_i, b_i]$, $i \geq 1$. If $\operatorname{int}\Lambda_1 \subset \bigcup_{i\neq 1} \operatorname{int}\Lambda_i$, let $\mathcal{O}^1 = \mathcal{O} \smallsetminus \{\operatorname{int}\Lambda_1\}$. Otherwise, set $\mathcal{O}^1 = \mathcal{O}$. Inductively, suppose that $\mathcal{O}^i \subseteq \mathcal{O}^{i-1} \subseteq \cdots \subseteq \mathcal{O}^1 \subseteq \mathcal{O}$, where $\mathcal{O}^i$ covers $(0, \eta/2]$ and, for $1 \leq j \leq i$,

$$\operatorname{int}\Lambda_j \subset \bigcup\{\operatorname{int}\Lambda \mid \operatorname{int}\Lambda_j \neq \operatorname{int}\Lambda \in \mathcal{O}^i\} \quad \Longrightarrow \quad \operatorname{int}\Lambda_j \notin \mathcal{O}^i.$$

Define $\mathcal{O}^{i+1}$ as follows. If

$$\operatorname{int}\Lambda_{i+1} \subset \bigcup\{\operatorname{int}\Lambda \mid \operatorname{int}\Lambda_{i+1} \neq \operatorname{int}\Lambda \in \mathcal{O}^i\},$$

then set $\mathcal{O}^{i+1} = \mathcal{O}^i \smallsetminus \{\operatorname{int}\Lambda_{i+1}\}$. Otherwise set $\mathcal{O}^{i+1} = \mathcal{O}^i$. Let $\mathcal{O}^* = \bigcap_{i\geq 1} \mathcal{O}^i$.

We claim that $\mathcal{O}^*$ covers $(0, \eta/2]$. Indeed, if $z \in (0, \eta/2]$, then the local finiteness of $\mathcal{O}$ implies that $z \notin \operatorname{int}\Lambda_j$ for sufficiently large values of j. Since every $\mathcal{O}^i$ is a cover, some $\operatorname{int}\Lambda_j$ containing z is never discarded in the above process.

Suppose that $\operatorname{int}\Lambda_i, \operatorname{int}\Lambda_j$, and $\operatorname{int}\Lambda_k$ are distinct elements of $\mathcal{O}^*$ and that $\operatorname{int}\Lambda_i \cap \operatorname{int}\Lambda_j \cap \operatorname{int}\Lambda_k \neq \emptyset$. Since these are intervals, we can assume that $\operatorname{int}\Lambda_i \subset \operatorname{int}\Lambda_j \cup \operatorname{int}\Lambda_k$, in which case $\operatorname{int}\Lambda_i \notin \mathcal{O}^i \supseteq \mathcal{O}^*$, a contradiction. It

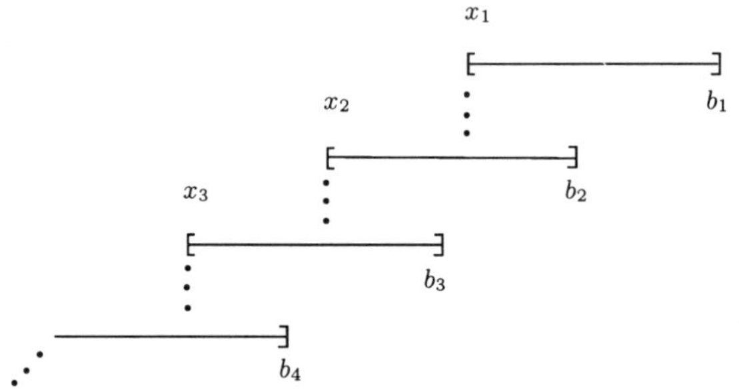

Figure 2.

follows that we can renumber $\mathcal{O}^* = \{(x_i, b_i)\}_{i=1}^{\infty}$ so that $(x_i, b_i) \cap (x_{i+1}, b_{i+1})$ is nonempty, $\forall i \geq 1$, and conclude that $b_1 > b_2 > x_1 > b_3 > x_2 > b_4 > x_3 > \ldots$ (see Figure 2). Set $y_1 = b_1$ and, if $i \geq 2$, set $y_i = g_i^{-1}(x_{i-1})$. It follows that this choice of $\{x_i, y_i, g_i\}_{i=1}^{\infty}$ satisfies the requirements. □

Lemma 5.2 *If F is a leaf and $F \cap (x_1, y_1) \neq \emptyset$, then we can assume that $b \in F \cap (x_1, y_1)$ is such that $[g_1(b), b] \subset (y_2, y_1)$.*

Proof. We must consider the two cases in the proof of Lemma 5.1. In Case 1, choose x_1 so that

$$2\lambda x_1 < b/(2\lambda).$$

The application of Lemma 4.19 can be made to ensure that g_1 and g_2 satisfy

$$y_2 = g_2^{-1}(x_1) < 2\lambda x_1 < b/(2\lambda) < g_1(b) < b < y_1.$$

In Case 2, iterated applications of g_1^{-1} to a point of $F \cap (x_1, y_1)$ produces a point b as desired. $\square$

Definition 5.3 $\Delta_k = (x_k, y_{k+1}]$, $\forall k \geq 1$.

Fix a choice of $\Lambda = [g_1(b), b] \subset (y_2, y_1)$ as in Lemma 5.2. Let $\Lambda' = [g_1(b), b)$. We will consider elements $\gamma \in \Gamma_L$ of the form $\gamma = g_{i_p} \circ g_{i_{p-1}} \circ \cdots \circ g_{i_1}$ where $i_1 = 1$ and either $i_j = i_{j-1}$ or $i_j = 1 + i_{j-1}$, $2 \leq j \leq p$, and such that $\Lambda \subset D(\gamma)$. Set

$$\gamma_0 = \text{id}$$
$$\gamma_j = g_{i_j} \circ g_{i_{j-1}} \circ \cdots \circ g_{i_1}$$
$$\Lambda_j = \gamma_j(\Lambda), \quad 0 \leq j \leq p.$$

Definition 5.4 An element $\gamma \in \Gamma_L$ as above is *admissible* if, whenever $i_j = 1 + i_{j-1}$ then

$$\Lambda_{j-1} \cap \Delta_{i_{j-1}} \neq \emptyset.$$

Also, $\gamma = \gamma_0 = \text{id}$ will be called *admissible*.

Lemma 5.5 *Let $x \in (0, b)$. Then there exists an admissible $\gamma \in \Gamma_L$ such that $x \in \gamma(\Lambda')$.*

Proof. If $x \in (x_1, b)$, we take γ to be a suitable iterate g_1^k, $k \geq 0$. If $x \in (0, x_1]$, there is a unique integer $n \geq 2$ such that $x \in (x_n, x_{n-1}]$. A suitable iterate of g_n^{-1} moves x to $g_n^{-k}(x) \in (x_{n-1}, y_n] = \Delta_{n-1}$. If $n = 2$, $\Delta_{n-1} \subset (x_1, b)$, while, if $n \geq 3$, then $\Delta_{n-1} \subset (x_{n-1}, x_{n-2}]$. Finite iteration of this process leads to $\lambda^{-1}(x) \in (x_1, b)$, hence to $g_1^{-r} \circ \lambda^{-1}(x) \in \Lambda'$. Then $\gamma = \lambda \circ g_1^r$ is admissible and $x \in \gamma(\Lambda')$. $\square$

Lemma 5.6 *If $\gamma = g_{i_p} \circ \cdots \circ g_{i_1}$ is admissible, then $\Lambda_0, \Lambda_1, \ldots, \Lambda_p$ have disjoint interiors.*

Proof. For $0 \leq i, j \leq p$ write $\Lambda_i < \Lambda_j$, if $x \in \operatorname{int} \Lambda_i$ and $y \in \operatorname{int} \Lambda_j$ implies $x < y$. For $0 < n \leq p$, we show inductively that $\Lambda_n < \Lambda_{n-1}$. This will prove the lemma.

For $n = 1, \Lambda_0 = [g_1(b), b]$ and $\Lambda_1 = g_1(\Lambda_0) = [g_1^2(b), g_1(b)]$, so $\Lambda_1 < \Lambda_0$. Inductively, assume that, for some $0 < n < p$, $\Lambda_n < \Lambda_{n-1}$. There are three cases to consider.

Case 1. Suppose that $i_{n+1} = i_n$. This together with the inductive hypothesis, gives

$$\Lambda_{n+1} = g_{i_n}^2(\Lambda_{n-1}) = g_{i_n}(\Lambda_n) < g_{i_n}(\Lambda_{n-1}) = \Lambda_n.$$

Case 2. Suppose that $i_{n+1} = 1 + i_n$ and $\Lambda_n \subset \Delta_{i_n}$. Then

$$\Lambda_{n+1} = g_{1+i_n}(\Lambda_n) \subset g_{1+i_n}(\Delta_{i_n}) < \Delta_{i_n},$$

hence $\Lambda_{n+1} < \Lambda_n$.

Case 3. Suppose that $i_{n+1} = 1 + i_n$ and $\Lambda_n \not\subseteq \Delta_{i_n}$. Set $i_n = k$, $\Lambda_n = [u, v]$, and consider the overlapping of Δ_k and Λ_n as indicated in Figure 3. By the inductive hypothesis, $[u, v] \le g_k^{-1}[u, v]$, so $v \le g_k^{-1}(u)$.

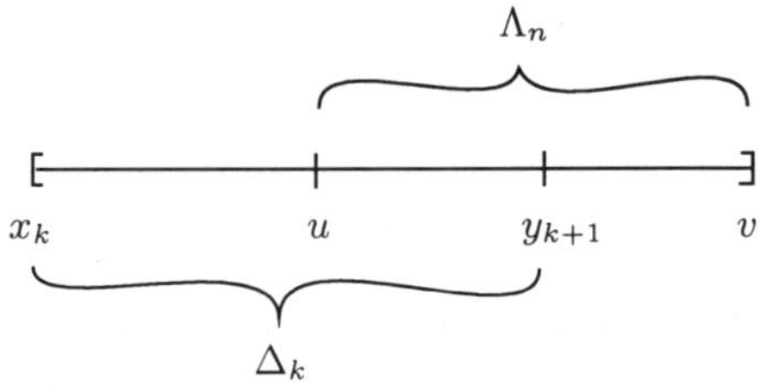

Figure 3.

Thus, using (#), we obtain

$$\begin{aligned} 0 < \frac{v - x_k}{y_{k+1} - x_k} &\le \frac{g_k^{-1}(u) - g_k^{-1}(x_k)}{y_{k+1} - x_k} \\ &\le \frac{g_k^{-1}(y_{k+1}) - g_k^{-1}(x_k)}{y_{k+1} - x_k} = (g_k^{-1})'(\zeta) < \frac{11}{10}. \end{aligned}$$

That is, $v - x_k < (11/10) \cdot |\Delta_k|$, hence

$$0 < v - y_{k+1} < \frac{1}{10} \cdot |\Delta_k| \tag{$*$}$$

Similarly

$$\begin{aligned} \frac{u - x_k}{y_{k+1} - x_k} \ge \frac{g_k(v) - g_k(x_k)}{y_{k+1} - x_k} &> \frac{g_k(y_{k+1}) - g_k(x_k)}{y_{k+1} - x_k} \\ &= g_k'(\xi) > \frac{9}{10}. \end{aligned}$$

It follows that $u - x_k > (9/10) \cdot |\Delta_k|$, hence that

$$0 \le y_{k+1} - u < \frac{1}{10} \cdot |\Delta_k|. \tag{$**$}$$

Putting together $(*)$ and $(**)$, we obtain

$$0 < v - u < \frac{1}{5} \cdot |\Delta_k|. \qquad (***)$$

Also,

$$0 < \frac{g_{k+1}(v) - x_k}{v - y_{k+1}} = \frac{g_{k+1}(v) - g_{k+1}(y_{k+1})}{v - y_{k+1}} = g'_{k+1}(\chi) < \frac{11}{10},$$

so $(*)$ implies that

$$0 < g_{k+1}(v) - x_k < \frac{11}{10} \cdot (v - y_{k+1}) < \frac{11}{100} \cdot |\Delta_k|.$$

That is,

$$g_{k+1}(v) < x_k + \frac{11}{100} \cdot |\Delta_k|.$$

We want to show that $u \geq g_{k+1}(v)$. If not, we have

$$u < x_k + \frac{11}{100} \cdot |\Delta_k|$$

and so

$$\begin{aligned} v - u &> v - x_k - \frac{11}{100} \cdot |\Delta_k| > y_{k+1} - x_k - \frac{11}{100} \cdot |\Delta_k| \\ &= \frac{89}{100} \cdot |\Delta_k| > \frac{1}{5} \cdot |\Delta_k|, \end{aligned}$$

contradicting $(***)$ □

Corollary 5.7 *Given $k \geq 1$, there exists an admissible γ such that $x_k \in \gamma(\Lambda')$ and $\gamma(b) \in (x_k, y_k)$.*

Proof. By lemma 5.5, choose γ so that $x_k \in \gamma(\Lambda')$. From the proof of Lemma 5.5, it is evident that $\gamma = g_{k+1} \circ g_k \circ \tilde{\gamma}$ and that $[u, v] = g_k \circ \tilde{\gamma}(\Lambda)$ is as in Case 3 of the proof of Lemma 5.6. Thus,

$$\gamma(b) = g_{k+1}(v) > g_{k+1}(y_{k+1}) = x_k$$

and

$$\gamma(b) = g_{k+1}(v) < x_k + \frac{11}{100} \cdot |\Delta_k| < y_{k+1} < y_k.$$

The proof is complete. □

6 Existence of the contraction

We make the same hypotheses as in Section 3 and use the same notation and conventions as we used there. In particular, the semiproper leaf F is asymptotic to L on the positive side of L.

Theorem 6.1 *There is $f \in \Gamma_\#$ and $\eta > 0$ such that $[0,\eta) \subset D(f)$ and $f : [0,\eta) \to [0,\eta)$ is a contraction to* 0.

Proof. Assume that there is no such contraction. Then, by Theorem 3.2, (i) and (ii) of Section 5 hold. We work in $[0,\varepsilon)$ chosen as in Section 5. We assume, *wlog*, that F is proper on its negative side. By Lemma 5.5, $F \cap (x_1, y_1) \neq \emptyset$, hence Lemma 5.2 allows us to choose $b \in F \cap (x_1, y_1)$ such that $\Lambda = [g_1(b), b] \subset (y_2, y_1)$. As before, let $\Lambda' = [g_1(b), b)$. If $Q = [a, b]$ is the component of $(0,\varepsilon) \setminus \overline{F}$ with upper endpoint b, then $Q \subset \Lambda$. Let $J = [0, b]$ and Θ be as in Subsection 4.4. Choose $r < 1$ so close to 1 that

$$1 - r < \exp(-\Theta|W|)|Q|/|\Lambda|.$$

By Lemma 4.18, all but finitely many $g \in \Gamma_\#$ satisfy $g'(x) > r$, $\forall x \in [0,\varepsilon)$.

Case 1. In Lemma 5.1, assume that infinitely many distinct elements of $\Gamma_\#$ occur in the list $\{g_i\}_{i=1}^\infty$. Select g_k such that $g_k'(x) > r$, $0 \leq x < \varepsilon$. By Corollary 5.7, let $\gamma \in \Gamma_L$ be admissible such that $x_k \in \gamma(\Lambda')$ and $\gamma(b) \in (x_k, y_k)$. Thus $g_k(x_k) = x_k$ and $x_k < g_k(\gamma(b)) < \gamma(b)$. Since $[a, b]$ is a gap of $\overline{F} \cap W$, it follows that $g_k(\gamma(b)) \leq \gamma(a)$. By Lemma 4.17, there is $t \in [x_k, \gamma(b)] \subset [0,\varepsilon)$ such that $r > g_k'(t)$, a contradiction.

Case 2. Assume that only finitely many distinct elements of $\Gamma_\#$ occur in the list $\{g_i\}_{i=1}^\infty$. Let $g \in \Gamma_\#$ be an element that occurs infinitely often in the list. By assumption, there is no value $\eta \in (0,\varepsilon)$ such that $g|[0,\eta)$ is a contraction to 0, so $g'(0) = 1$. It follows that there is an integer $k \geq 1$ such that $g = g_k$ and $g'|[0, y_k) > r$. It is now possible to argue exactly as in Case 1 so as to obtain the contradiction that $r > g'(t)$, some $t \in \gamma(\Lambda) \subset [0, y_k)$. □

7 A compactly supported cohomology class

In this section, we use the the contraction f of Theorem 6.1 to define a compactly supported cohomology class on L. We will use Poincaré duality to show that this class yields a handle on L with the properties in Theorem 1.3.

Unless otherwise specified, we keep the notation and assumptions of the previous section. By Theorem 6.1, we can assume that $f \in \Gamma_\#$ and $[0,\varepsilon) \subset D(f)$ are such that $f : [0,\varepsilon) \to [0,\varepsilon)$ is a contraction to 0. Let

(a,b) be a component of $(0,\varepsilon) \setminus \overline{F}$ and set $a_k = f^k(a)$, $b_k = f^k(b)$, $k \geq 0$. Finally, let Θ be as defined in Definition 4.12.

Lemma 7.1 *Let $h \in \Gamma_L$ and suppose that neither h nor h^{-1} is germinally equivalent to a contraction to 0 in $[0,\varepsilon)$. Then there exists an $N \geq 0$ such that for all $k \geq N$, $h(b_k) = b_k$. Furthermore, all but finitely many elements of $\Gamma_\#$ fix every b_k, $k \geq N$.*

Proof. Choose $0 < r < 1$ such that

$$1 - r < \exp(-\Theta|W|) \cdot \frac{b_0 - a_0}{b_0 - b_1}.$$

By assumption, $h'(0) = 1$. Thus, we can choose an integer $N \geq 0$ such that $h'(t) > r$, $0 \leq t \leq b_N$. If there is an integer $n \geq N$ for which $h(b_n) \neq b_n$, then such n can be found so that h has a fixed point $x \in [b_{n+1}, b_n)$. We fix the choice of integer n and assume *wlog* that $h(b_n) < b_n$. It follows that $h(b_n) \leq a_n$. We are set up to apply Lemma 4.17 with $\Lambda = [b_1, b_0]$, $Q = [a_0, b_0]$, and $\gamma = f^n$. This gives the contradiction that, for some $t_0 \in f^n[b_1, b_0] \subset [0, b_N]$, $h'(t_0) < r$.

By Lemma 4.18, $\gamma'(t) > r$, $0 \leq t \leq \varepsilon$, for all but finitely many $\gamma \in \Gamma_\#$. Therefore, by the result of the previous paragraph, for all but these finitely many $\gamma \in \Gamma_\#$, $\gamma(b_k) = b_k$, $\forall k \geq N$. □

Since all but finitely many $\gamma \in \Gamma_\#$ fix b_k, $\forall k \geq N$, there can be at most finitely many elements in $\Gamma_\#$ that are germinally equivalent to contractions to 0 in $[0,\varepsilon)$. Let $\{g_0, g_1, \cdots, g_q\} \subset \Gamma_\#$ be the subset of these elements. By choosing $\varepsilon > 0$ smaller, we can assume that $\{g_0, g_1, \cdots, g_q\}$ are all contractions to 0 in $[0,\varepsilon)$. Rechoose (a,b) to be a component of $(0,\varepsilon) \setminus \overline{F}$.

Remark. Lemma 7.1 holds with the contraction f equal any $g \in \{g_0, g_1, \cdots, g_q\}$, (a,b) any component of $(0,\varepsilon) \setminus \overline{F}$, and $b_k = g^k(b)$.

Let D be the smallest of the numbers

$$\exp(-\Theta|W|) \cdot \frac{b_0 - a_0}{b_0 - g_i(b_0)}, \quad 0 \leq i \leq q.$$

For notational simplicity, let $g \in \{g_0, g_1, \cdots, g_q\}$ and $a_k = g^k(a_0)$, $b_k = g^k(b_0)$. Then

Lemma 7.2 *Whenever $h \in \Gamma_L$ and $h(b_k) < b_k$, some $k \geq 0$, then*

$$\frac{b_k - h(b_k)}{b_k - g(b_k)} \geq D.$$

Proof. Since $h(b_k) < b_k$, we have $h(b_k) \leq a_k$, so

$$\begin{aligned}\frac{b_k - h(b_k)}{b_k - g(b_k)} &\geq \frac{b_k - a_k}{b_k - b_{k+1}} = \frac{g^k(b_0) - g^k(a_0)}{g^k(b_0) - g^k(b_1)} \\ &= \frac{(g^k)'(\xi)}{(g^k)'(\zeta)} \cdot \left(\frac{b_0 - a_0}{b_0 - b_1}\right) \geq \exp(-\Theta|W|) \cdot \left(\frac{b_0 - a_0}{b_0 - b_1}\right) \geq D,\end{aligned}$$

where the second last inequality follows from

$$\frac{(g^k)'(\zeta)}{(g^k)'(\xi)} \leq \exp(\Theta|W|)$$

(see Lemma 4.15). □

Definition 7.3 If $h \in \Gamma_L$, the germ of $h|[0,\varepsilon)$ at 0 will be denoted $[h]$. The group of all such germs will be denoted $\mathcal{H}^+(L)$.

Theorem 7.4 *There is a homomorphism $\nu : \mathcal{H}^+(L) \to \mathbb{R}$ with infinite cyclic image. Furthermore, if $\chi : \pi_1(L,0) \to \mathcal{H}^+(L)$ is the canonical surjection, then the cohomolgy class $[\nu \circ \chi] \in H^1(L;\mathbb{R})$ is compactly supported.*

The proof will be achieved in a series of lemmas. There are two cases to treat. In the harder case, $g_j'(0) = 1$, $0 \leq j \leq q$. We treat the easier case first.

Case 1. There is $g \in \{g_0, g_1, \dots, g_q\}$ such that $g'(0) < 1$. As in the proof of Lemma 7.2, $b_k = g^k(b_0)$, $a_k = g^k(a_0)$.

The following lemma will be needed later in this section but it is convenient to prove it here.

Lemma 7.5 *Let $h \in \Gamma_L$ be such that $h(b_k) \neq b_k$, for infinitely many values of k. Then, $h'(0) \neq 1$.*

Proof. Suppose that $h'(0) = 1$ and, *wlog*, assume that $h(b_k) < b_k$ for infinitely many values of k. Write

$$\begin{aligned} g(t) &= (\rho + r(t))t \\ h(t) &= (1 + s(t))t, \end{aligned}$$

where $\rho \neq 1$ and $r(0) = 0 = s(0)$. Then

$$\frac{t - h(t)}{t - g(t)} = \frac{-s(t)}{1 - \rho - r(t)}.$$

It follows that

$$\lim_{t \to 0+} \frac{t - h(t)}{t - g(t)} = 0,$$

contradicting Lemma 7.2. □

Lemma 7.6 *There is $\tau \in (0,1)$ such that, whenever $h \in \Gamma_L$ and $h'(0) < 1$, then $h'(0) \le 1 - \tau$.*

Proof. If this were false, then, $\forall \tau \in (0,1)$, there exists $h \in \Gamma_L$ with $1 - \tau < h'(0) = \rho < 1$. Set $\sigma = g'(0)$. Choose such τ and h with

$$\frac{4\tau}{1-\sigma} < D.$$

Express

$$\begin{aligned} g(t) &= (\sigma + s(t))t \\ h(t) &= (\rho + r(t))t, \end{aligned}$$

where $s(0) = 0$ and $r(0) = 0$. Then, for k sufficiently large,

$$\begin{aligned} 1 - \sigma - s(b_k) &> \frac{1-\sigma}{2} \\ 1 - \rho - r(b_k) &< 2(1-\rho). \end{aligned}$$

Consequently,

$$\frac{b_k - h(b_k)}{b_k - g(b_k)} = \frac{1 - \rho - r(b_k)}{1 - \sigma - s(b_k)} < \frac{4(1-\rho)}{1-\sigma} < \frac{4\tau}{1-\sigma} < D,$$

which contradicts Lemma 7.2. □

Lemma 7.7 completes the proof of Case 1 of Theorem 7.4.

Lemma 7.7 *Let the homomorphism $\nu : \mathcal{H}^+(L) \to \mathbb{R}$ be defined by the formula $\nu([h]) = -\log(h'(0))$. Then ν has infinite cyclic image and the cohomology class $[\nu \circ \chi] \in H^1(L; \mathbb{R})$ is compactly supported.*

Proof. Since $g'(0) \neq 1$, ν is nontrivial. By Lemma 7.6, the image of ν does not cluster at 0. Furthermore, $\Gamma_\#$ corresponds to a set of loops on L that generate $\pi_1(L, 0)$, so Lemma 7.1 implies that $[\nu \circ \chi]$ is compactly supported. □

Case 2. We assume that $g_i'(0) = 1$, $0 \le i \le q$, and we fix choices of elements $g, h \in \{g_0, g_1, \dots, g_q\}$. For $g = g_i$, we let $b_k = g^k(b_0)$. We define $l(n)$ to be the least positive integer such that $[h(b_n), b_n] \subseteq [b_{n+l(n)}, b_n]$, $\forall n \ge 0$.

Lemma 7.8 *There is a strictly increasing sequence $\{n_k\}_{k=1}^{\infty}$ of positive integers on which l is bounded.*

Proof. (1) Suppose the assertion false. Then, $\lim_{n\to\infty} l(n) = \infty$. Consider

$$\begin{aligned}\frac{b_n - h(b_n)}{b_n - g(b_n)} &\geq \frac{b_n - b_{n+l(n)-1}}{b_n - b_{n+1}} \\ &= \frac{b_n - b_{n+1}}{b_n - b_{n+1}} + \frac{b_{n+1} - b_{n+2}}{b_n - b_{n+1}} + \cdots + \frac{b_{n+l(n)-2} - b_{n+l(n)-1}}{b_n - b_{n+1}} \\ &= 1 + \frac{g(b_n) - g(b_{n+1})}{b_n - b_{n+1}} + \cdots + \frac{g^{l(n)-2}(b_n) - g^{l(n)-2}(b_{n+1})}{b_n - b_{n+1}} \\ &= 1 + g'(\xi_1) + (g^2)'(\xi_2) + \cdots + (g^{l(n)-2})'(\xi_{l(n)-2}),\end{aligned}$$

where all $\xi_i \in (b_{n+1}, b_n)$.

Given $r \in (0,1)$, take n so large that $g'(\xi) > r$, $\forall \xi \in [0, b_n]$. This is possible since $g'(0) = 1$. Then

$$(g^j)'(\xi) = g'(g^{j-1}(\xi)) \cdot g'(g^{j-2}(\xi)) \cdots g'(\xi) > r^j.$$

Therefore,

$$\frac{b_n - h(b_n)}{b_n - g(b_n)} > 1 + r + r^2 + \cdots + r^{l(n)-2} = \frac{1 - r^{l(n)-1}}{1 - r}.$$

Since $l(n) \to \infty$, we can choose n so large that

$$\frac{b_n - h(b_n)}{b_n - g(b_n)} \geq \frac{1}{2(1-r)}.$$

But the number r can be taken as close to 1 as desired, forcing n to be chosen sufficiently large. Therefore, given an arbitrarily large positive number R, we can choose an integer $N \geq 0$ such that

$$\frac{b_n - h(b_n)}{b_n - g(b_n)} \geq R, \quad \forall n \geq N.$$

(2) let $R > 0$ be given and choose N as in (1). Let $n \geq N$ and let $x \in [b_{n+1}, b_n]$. Let $y \in [b_1, b_0]$ be the point with $x = g^n(y)$ and let

$$M = \max_{b_1 \leq y \leq b_0} (y - g(y)).$$

Then

$$\frac{x - g(x)}{b_n - g(b_n)} = \frac{g^n(y) - g^n(g(y))}{g^n(b_0) - g^n(b_1)} = \frac{(g^n)'(\xi)}{(g^n)'(\eta)} \cdot \frac{y - g(y)}{b_0 - b_1},$$

for suitable $\xi, \eta \in [b_2, b_0]$. By Lemma 4.16,

$$\frac{x - g(x)}{b_n - g(b_n)} \leq \frac{\exp(2\Theta|W|) \cdot M}{b_0 - b_1}.$$

It follows that, for arbitrary $x \in [b_{n+1}, b_n]$,

$$\begin{aligned} x - h(x) &\geq b_{n+1} - h(b_n) = b_{n+1} - b_n + b_n - h(b_n) \\ &\geq (R-1)(b_n - g(b_n)) \\ &\geq \frac{(R-1)\exp(-2\Theta|W|)(b_0 - b_1)}{M} \cdot (x - g(x)). \end{aligned}$$

As N becomes arbitrarily large, so does R, hence $(x - g(x))/(x - h(x))$ becomes arbitrarily small, uniformly for $x \in (0, b_N)$. This contradicts Lemma 7.2. □

Lemma 7.9 *The sequence* $\{l(n)\}_{n=1}^{\infty}$ *is bounded.*

Proof. By Lemma 7.8, there is a strictly increasing sequence $\{n_k\}_{k=1}^{\infty}$ of integers and a positive integer λ such that $l(n_k) < \lambda$, $\forall k \geq 1$. Thus, $g^{-\lambda} \circ h(b_{n_k}) > b_{n_k}$. Suppose that $l(n)$ is not bounded. Then there are arbitrarily large values of n for which $g^{-\lambda} \circ h(b_n) < b_n$. Let $\gamma = g^{-\lambda} \circ h$. By the above remarks, there exist arbitrarily large integers m such that

$$\begin{aligned} \gamma(b_m) &< b_m \\ \gamma(b_{m+1}) &\geq b_{m+1}. \end{aligned}$$

Thus we can choose a γ–fixed point $z \in [b_{m+1}, b_m]$.

Recall that $g = g_i$, $b_k = g^k(b_0)$, and set $a_k = g^k(a_0)$. In Lemma 4.15, take $\Lambda = [b_1, b_0]$ and $Q = [a_0, b_0]$. Choose m as above so large that

$$1 - \gamma'(x) < \exp(-\Theta|W|) \cdot |Q|/|\Lambda|, \quad \forall x \in (0, b_m].$$

Then, for suitable $\xi \in [z.b_m]$,

$$\begin{aligned} 1 - \gamma'(\xi) &= 1 - \frac{\gamma(b_m) - \gamma(z)}{b_m - z} = \frac{b_m - \gamma(b_m)}{b_m - z} \geq \frac{b_m - a_m}{b_m - z} \\ &\geq \frac{|g^m(Q)|}{|g^m(\Lambda)|} = \frac{(g^m)'(\zeta)}{(g^m)'(\eta)} \cdot \frac{|Q|}{|\Lambda|} \geq \exp(-\Theta|W|) \cdot \frac{|Q|}{|\Lambda|}, \end{aligned}$$

a contradiction. □

Lemma 7.10 *There is a positive number* p *such that*

$$\frac{1}{p} \leq \frac{x - g_j(x)}{x - g_i(x)} \leq p, \quad 0 < x \leq \varepsilon, \quad 0 \leq i, j \leq q.$$

Proof. Set $g = g_i$ and $h = g_j$. Let $b_k = g^k(b_0)$ as above. Let $x \in [b_{n+1}, b_n]$, $x = g^n(y)$, $y \in [b_1, b_0]$. Let $\mu = \min_{b_1 \leq y \leq b_0}(y - g(y))$. By

Lemma 7.9, there exists an integer $\lambda > 0$, independent of n, such that

$$\frac{x-h(x)}{x-g(x)} \leq \frac{b_n - g^{\lambda+1}(b_n)}{x-g(x)} = \frac{g^n(b_0)-g^n(b_{\lambda+1})}{g^n(y)-g^n(g(y))}$$
$$= \frac{(g^n)'(\zeta)}{(g^n)'(\eta)} \cdot \frac{b_0 - b_{\lambda+1}}{y-g(y)}, \leq \exp((\lambda+1)\Theta|W|)\left(\frac{b_0-b_{\lambda+1}}{\mu}\right)$$
$$\zeta, \eta \in [b_{\lambda+1}, b_0],$$

the final inequality being by Lemma 4.16. Thus, we take p to be the largest of the many bounds obtained for the finitely many choices of $g = g_i$ and $h = g_j$, $0 \leq i, j \leq q$. □

Lemma 7.11 completes the proof of Case 2 of Theorem 7.4.

Lemma 7.11 *There is a homomorphism $\nu : \mathcal{H}^+(L) \to \mathbb{R}$ having the properities in Theorem 7.4.*

Proof. Since every element of $\Gamma_\# \setminus \{g_0^{\pm 1}, \dots, g_q^{\pm 1}\}$ fixes b_k, for all k sufficiently large, and since $\Gamma_\#$ defines a generating set $G_\# \subset \mathcal{H}^+(L)$, we see that $h'(0) = 1$, $\forall\ [h] \in \mathcal{H}^+(L)$.

We will use the method of Thurston's stability theorem to obtain the homomorphism ν. By Lemma 7.10,

$$\frac{1}{p} \leq \frac{x - g_i(x)}{x - g_0(x)} \leq p, \quad 0 \leq i \leq q, \quad 0 < x \leq \varepsilon.$$

Let $b_k = g_0^k(b_0)$. There is a subsequence $\{b_{n_k}\}_{k=1}^\infty$such that

$$A_i = \lim_{k\to\infty} \frac{b_{n_k} - g_i(b_{n_k})}{b_{n_k} - g_0(b_{n_k})}$$

exists, $0 \leq i \leq q$. Remark that $A_0 = 1$. But, $\forall h \in \Gamma_\# \setminus \{g_0^{\pm 1}, \dots, g_q^{\pm 1}\}$, $h(b_{n_k}) = b_{n_k}$ so we also have

$$0 = \lim_{k\to\infty} \frac{b_{n_k} - h(b_{n_k})}{b_{n_k} - g_0(b_{n_k})}.$$

Given $\phi, \theta \in \Gamma_L$, suppose that it has been shown that the limits

$$A = \lim_{k\to\infty} \frac{b_{n_k} - \phi(b_{n_k})}{b_{n_k} - g_0(b_{n_k})}$$
$$B = \lim_{k\to\infty} \frac{b_{n_k} - \theta(b_{n_k})}{b_{n_k} - g_0(b_{n_k})}$$

exist. Let $\eta = \phi \circ \theta$. For $h \in \Gamma_L$, write $h(x) = x + \tilde{h}(x)$, where, as remarked above, $\tilde{h}'(0) = 0$. Then,

$$\eta(x) = x + \tilde{\eta}(x) = \phi(\theta(x)) = x + \tilde{\theta}(x) + \tilde{\phi}(x + \tilde{\theta}(x))$$
$$= x + \tilde{\theta}(x) + \tilde{\phi}'(\xi_x)\tilde{\theta}(x) + \tilde{\phi}(x)$$

for some ξ_x lying between x and $x+\tilde{\theta}(x)$. That is, $\tilde{\eta}(x) = \tilde{\theta}(x)+\tilde{\phi}'(\xi_x)\tilde{\theta}(x)+\tilde{\phi}(x)$. Therefore

$$\begin{aligned}\lim_{k\to\infty} \frac{b_{n_k} - \eta(b_{n_k})}{b_{n_k} - g_0(b_{n_k})} &= \lim_{k\to\infty} \frac{\tilde{\eta}(b_{n_k})}{\tilde{g}_0(b_{n_k})} \\ &= \lim_{k\to\infty} \left(\frac{\tilde{\theta}(b_{n_k})}{\tilde{g}_0(b_{n_k})} + \tilde{\phi}'(\xi_{b_{n_k}}) \cdot \frac{\tilde{\theta}(b_{n_k})}{\tilde{g}_0(b_{n_k})} + \frac{\tilde{\phi}(b_{n_k})}{\tilde{g}_0(b_{n_k})} \right) \\ &= B + A\end{aligned}$$

Given $\phi \in \Gamma_L$, suppose that it has been shown that

$$A = \lim_{k\to\infty} \frac{b_{n_k} - \phi(b_{n_k})}{b_{n_k} - g_0(b_{n_k})} = \lim_{k\to\infty} \frac{\tilde{\phi}(b_{n_k})}{\tilde{g}_0(b_{n_k})}.$$

Set $\gamma = \phi^{-1}$. Then

$$x = \gamma(\phi(x)) = x + \tilde{\phi}(x) + \tilde{\gamma}(x + \tilde{\phi}(x)) = x + \tilde{\phi}(x) + \tilde{\gamma}'(\xi_x)\tilde{\phi}(x) + \tilde{\gamma}(x)$$

for some ξ_x lying between x and $x + \tilde{\phi}(x)$. Thus, $\tilde{\gamma}(x) = -(1 + \tilde{\gamma}'(\xi_x))\tilde{\phi}(x)$, hence

$$\lim_{k\to\infty} \frac{\tilde{\gamma}(b_{n_k})}{\tilde{g}_0(b_{n_k})} = -\lim_{k\to\infty} (1 + \tilde{\gamma}'(\xi_{n_k})) \cdot \frac{\tilde{\phi}(b_{n_k})}{\tilde{g}_0(b_{n_k})} = -A.$$

Thus for each $[h] \in \mathcal{H}^+(L)$, we obtain a well–defined real number

$$\nu[h] = \lim_{k\to\infty} \frac{b_{n_k} - h(b_{n_k})}{b_{n_k} - g_0(b_{n_k})}$$

and $\nu : \mathcal{H}^+(L) \to \mathbb{R}$ is a group homomorphism, nontrivial since $\nu[g_0] = 1$.

By Lemma 7.2, if $h \in \Gamma_L$, then either $h(b_{n_k}) = b_{n_k}$, $\forall$ large values of k, or $\nu[h] \geq D > 0$, or $\nu[h^{-1}] \geq D > 0$. That is, the image of the homomorphism ν must be infinite cyclic.

It is clear that $\nu[h] = 0$, $\forall h \in \Gamma_\# \setminus \{g_0^{\pm 1}, \dots, g_q^{\pm 1}\}$, hence the cohomology class $[\nu \circ \chi]$ is compactly supported in L. □

The proof of Theorem 7.4 is complete.

We need a good holonomy interpretation of the cohomology class $[\nu \circ \chi]$. By Theorem 7.4, there is a smallest positive number $r \in \operatorname{Im}(\nu)$. Normalize ν so that $r = 1$. Then $\Phi = [\nu \circ \chi]$ is a compactly supported, *integral* cohomology class in L. It is clear that Φ takes the value 1 on a suitable loop α in L, hence it is *nondivisible*. By standard theory, Φ is the Poincaré dual of a homology class that is represented by a closed, oriented, connected, nonseparating submanifold $H \subset L$ of codimension 1. We may suppose that our basepoint 0 lies on H. For any loop σ on L, based at 0, the homological intersection number $\sigma \frown H$ is the value of Φ on σ. Let α be a simple loop that meets the handle H once at 0 chosen so that $\alpha \frown H = +1$ and h_α the holonomy around the loop α (a contraction). Let $f = h_\alpha$, g a contraction in $\{g_0, g_1, \dots, g_q\}$, and as usual $b_k = g^k(b)$.

Remark. Since we do not know that $f = h_\alpha$ is an element of $\Gamma_\#$, in order to apply Lemma 7.1 in the proof of the next lemma, we can not take the contraction to be f instead of g.

Lemma 7.12 *If $h \in \Gamma_L$ and $\nu[h] = 0$, then h fixes b_k, for all k sufficiently large.*

Proof. We consider separately the two cases of the proof of Theorem 7.4. In the first case, Lemma 7.5 and the definition of ν in Lemma 7.7 implies that neither h nor h^{-1} is a contraction to 0. In the second case, Lemma 7.2 and the definition of ν in Lemma 7.11 implies that neither h nor h^{-1} is a contraction to 0. Thus Lemma 7.1 implies h fixes b_k, for all k sufficiently large. □

For $0 \leq i \leq q$, there exists a positive integer m_i so that $\nu[g_i] = m_i$.

Lemma 7.13 $g_i(b_k) = f^{m_i}(b_k)$, $0 \leq i \leq q$, *k sufficiently large.*

Proof. Since $\nu[f^{-m_i} \circ g_i] = 0$, it follows that $f^{-m_i} \circ g_i$ fixes b_k, $0 \leq i \leq q$, k sufficiently large, and the lemma follows. □

Lemma 7.14 *Without loss we can assume H and α are contained in the interior of the compact core C_0.*

Proof. Recall our definition of the core of a leaf L from Subsection 4.2. Let $C_k = K_k \cap L$ be a core of L. Choose k sufficiently large that C_k contains both H and α. Renumber the C_j's so that $C_k = C_0$. This gives a new $\Gamma_\#$ containing the previous one, a new set $\{g_0, g_1, \dots, g_q\}$ containing the previous one, and a smaller ε so that all the $\{g_0, g_1, \dots, g_q\}$ are contractions to zero on $[0, \varepsilon)$. Note that $f = h_\alpha \in \{g_0, \dots, g_q\}$. Take $f = g_0$. Let (a, b) be a component of $(0, \varepsilon) \setminus \overline{F}$ and repeat the argument of this section to obtain a compactly supported cohomology class $[\tilde{\nu} \circ \chi]$ and a contraction $\tilde{f}$ so that $\tilde{\nu}[\tilde{f}] = 1$. For $0 \leq i \leq q$, there exists a positive integer $\tilde{m}_i$ so that $\tilde{\nu}[g_i] = \tilde{m}_i$. Therefore $\tilde{\nu}[f] = \tilde{m}_0$ so by Lemma 7.13 $f(b_k) = \tilde{f}^{\tilde{m}_0}(b_k)$, for all k sufficiently large, where $b_k = f^k(b)$. Therefore $\nu[\tilde{f}] = 1/\tilde{m}_0$. Since 1 is the smallest positive number in $\mathrm{Im}(\nu)$, it follows that $\tilde{m}_0 = 1$ or $\tilde{\nu}[f] = 1$. Therefore, since $f^{m_i}(b_k) = g_i(b_k)$, for all k sufficiently large, it follows that $\tilde{\nu}[g_i] = m_i = \nu[g_i]$, $1 \leq i \leq q$. Further, since $\nu[\gamma] = 0 = \tilde{\nu}[\gamma]$ for all $\gamma \in \Gamma_\#$ with neither γ nor γ^{-1} germinally a contraction to 0 and since $\Gamma_\#$ generates Γ_L, it follows that $\tilde{\nu} = \nu$. Therefore $\tilde{H} = H$, and $\tilde{\alpha} = \alpha$. Take $f = h_\alpha$. □

Remark. These are the ε, f, H, and α of Theorem 1.3.

Fix $\Gamma_\#$ and $\{g_0, g_1, \dots, g_q\}$.

Lemma 7.15 *If $\gamma \in \Gamma_\#$ and neither γ nor γ^{-1} is germinally a contraction to zero, then γ fixes b_k, for all $k \geq N$, for a possibly larger value of N.*

Proof. In fact, the last statement of Lemma 7.1, gives this result for all but finitely many elements of $\Gamma_\#$, while the first statement of the Lemma applied to each of these finitely many $\gamma \in \Gamma_\# \setminus \{g_0, g_1, \cdots, g_q\}$ in turn gives the result for the remaining $\gamma \in \Gamma_\#$ such that neither γ nor γ^{-1} is germinally a contraction to zero, for a possibly larger N. □

Remark. This is the N in Theorem 1.3.

Proof. [Proof of Theorem 1.3] Any h_σ as in Theorem 1.3 is a product of $\gamma \in \Gamma_\#$ such that neither γ nor γ^{-1} is germinally a contraction to 0 and the Theorem follows. □

By Section 2, we have also established Theorem 1.1 and Theorem 1.4.

Proposition 7.16 *The handle H, curve α, and contraction $f = h_\alpha$ do not depend on choice of gap (a, b).*

Proof. Starting with a different gap (a^*, b^*) of $(0, \varepsilon) \setminus \overline{F}$, we have the same set of contractions $\{g_0, g_1, \cdots, g_q\} \subset \Gamma_\#$. Take $f = g_0$. Repeat the argument of this section to obtain a compactly supported cohomology class $[\nu^* \circ \chi]$ and a contraction f^* so that $\nu^*[f^*] = 1$. For $0 \leq i \leq q$, there exists a positive integer m_i^* so that $\nu^*[g_i] = m_i^*$. Therefore $\nu^*[f] = m_0^*$ so by Lemma 7.13 $f(b_k) = (f^*)^{m_0^*}(b_k)$, for all k sufficiently large, where $b_k = f^k(b)$. Therefore $\nu[f^*] = 1/m_0^*$. Since 1 is the smallest positive number in $\mathrm{Im}(\nu)$, it follows that $m_0^* = 1$ or $\nu^*[f] = 1$. Therefore, since $f^{m_i}(b_k) = g_i(b_k)$, for all k sufficiently large, it follows that $\nu^*[g_i] = m_i = \nu[g_i]$, $1 \leq i \leq q$. Further, since $\nu[\gamma] = 0 = \nu^*[\gamma]$ for all $\gamma \in \Gamma_\#$ with neither γ nor γ^{-1} germinally a contraction to 0 and since $\Gamma_\#$ generates Γ_L, it follows that $\nu^* = \nu$. Therefore $H^* = H$, $\alpha^* = \alpha$, and $f^* = f$. □

8 Problems

Besides the problems mentioned at the end of Section 1, a number of problems related to exceptional minimal sets have remained very stubborn. As with the problems at the end of Section 1, these are theorems for Markov LMS's [3].

If X is a LMS, an arc transverse to the foliation can be chosen to meet X in a Cantor set C. Fix $x \in C$ and let L be the leaf of $\mathcal{F}$ containing x. The holonomy group of L relative to X is the group $H_x(L, X)$ of germs at x of all holonomy maps (restricted to C) that fix x..

The following question was asked by Dippolito [6, §9].

Question. Let X be an exceptional LMS and $L \subset X$ a leaf. Is $H_x(L, X)$

either trivial or infinite cyclic and generated by the germ of a contraction that is unique in a suitable neighborhood of x in C? Do exactly a countable infinity of leaves in X have $H_x(L, X) \cong \mathbb{Z}$ and are all the semi-proper leaves among these?

Remark. In this paper we have shown that a semiproper leave can not have trivial holonomy but in a subtle sense have not proven that $H_x(L, X) \cong \mathbb{Z}$. For x not on a semi-proper leaf there are no results.

Hector has proposed the following question [11].

Question. If X is an exceptional LMS does X have finitely many semi-proper leaves?

Finally,

Question. If X is an exceptional LMS, does it have Lebesgue measure $|X| = 0$?

9 Examples of Markov minimal sets

In this section we provide a construction of examples of arbitrary Markov minimal sets. One method of doing this, that of "branched staircases", is due to Takamura [12] and Inaba [9]. We present a different method.

9.1 Constructing the plug

The first step in our construction is to construct a plug $(P, \mathcal{F})$ where ∂P consists of two tranverse boundary tori σ and ρ and $\mathcal{F}$ meets σ and ρ in circles. Further $\sigma = \sigma_1 \cup \sigma_2 \cup \cup \sigma_3 \cup \sigma_4$ and $\rho = \rho_1 \cup \rho_2 \cup \rho_3 \cup \rho_4$ in such a way that a part of P is foliated with a product foliation meeting ∂P in $\sigma_1 \cup \rho_1$, another part of P is foliated by a product foliation meeting ∂P in σ_3, and another part of P is foliated by a product foliation meeting ∂P in ρ_3.

Let $N_1 = T \times [0, 2]$ where T is the 2-torus. Foliate $T \times [1, 2]$ with the product foliation and $T \times (0, 1)$ by cylinders spiralling in on the compact leaves $T \times \{0\}$ and $T \times \{1\}$ with logitudes as junctures to obtain a foliation $\mathcal{F}_1$ of N_1.

Thus one can drill out a wormhole X transverse to $\mathcal{F}_1$ from $T \times \{0\}$ to $T \times \{2\}$ and wormholes Y and Z transverse to $\mathcal{F}_1$ with Y and Z in $T \times (0, 1)$ parallel to meridians of $T \times \{0\}$. Spin the foliation $\mathcal{F}_1$ at ∂Y and ∂Z to make ∂Y and ∂Z tangential boundary. Choose the direction of spin so that it is possible to drill out wormholes W and U transverse to $\mathcal{F}_1$ with W from $T \times \{2\}$ to ∂Z and U from ∂Y to $T \times \{0\}$ (see Figure 4). One obtains a foliated manifold $(N_2, \mathcal{F}_2)$. Let H be a punctured torus and foliate $H \times [0, 1]$

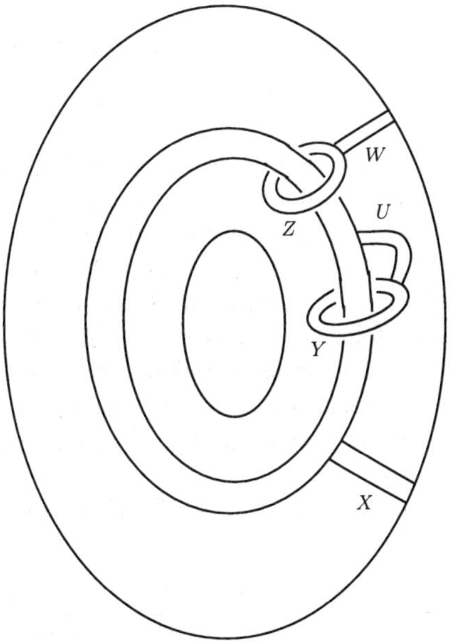

Figure 4. Constructing the plug

with the product foliation. Obtain a foliated manifold $(N_3, \mathcal{F}_3)$ by pasting $H \times \{0\}$ to the once punctured ∂Y and $H \times \{1\}$ to the once punctured ∂Z.

Obtain a foliated manifold $(N_4, \mathcal{F}_4)$ by pasting the (now) twice punctured $T \times \{0\}$ to the (now) twice punctured $T \times \{2\}$ so that one end of W is matched to one end of U and one end of X is matched to the other end of X. The foliated manifold $(N_4, \mathcal{F}_4)$ has boundary consisting of two transverse tori σ and τ foliated by circles. The portion of $(N_4, \mathcal{F}_4)$ coming from the (now) twice punctured $T \times [1, 2]$ is foliated by the product foliation. Denote the portions of σ and τ meeting this part of $(N_4, \mathcal{F}_4)$ by σ_1 and τ_1. Denote the portion of σ meeting $H \times [0, 1]$ by σ_3. Then $\sigma = \sigma_1 \cup \sigma_2 \cup \sigma_3 \cup \sigma_4$ where $\partial W = \sigma_1 \cup \sigma_2$ and $\partial U = \sigma_4$.

Finally obtain the foliated plug $(P, \mathcal{F})$ by taking two copies of the foliated manifold of the previous paragraph $(N_4, \mathcal{F}_4)$ and $(N_4', \mathcal{F}_4')$ and pasting τ to τ' so that τ_1 is pasted to τ_1'. For the sake of notation we rename the second boundary component of P by ρ where $\rho = \rho_1 \cup \rho_2 \cup \rho_3 \cup \rho_4$ with $\rho_1 = \sigma_1'$, $\rho_2 = \sigma_2'$, $\rho_3 = \sigma_3'$, and $\rho_4 = \sigma_4'$. The foliated manifold $(P, \mathcal{F})$ satisfies the properties of the first paragraph of this subsection.

9.2 The construction

Suppose a Markov pseudogroup is given as on page 167 of [3] and use the notation of [3]. Since we can assume $\mathcal{W} = \bigcup_{i=1}^{m}(R(h_i) \cup D(h_i))$ is contained in a compact subset of $\mathbb{R}$, we may assume it is contained in the circle S^1. For $1 \leq i \leq m$, let $A_i = S^1 \setminus D(h_i)$ and $B_i = S^1 \setminus R(h_i)$. Let S be the surface obtained by removing $2m$ discs from the two-sphere S^2. Then ∂S consists of the $2m$ circles $C_1, \ldots, C_m, C'_1, \ldots, C'_m$. Let $(P_1, \mathcal{F}_1), \ldots, (P_m, \mathcal{F}_m)$ be m copies of the foliated plug constructed in Subsection 9.1. Let $S^2 \times S^1$ be foliated by the product foliation. Construct a foliated manifold $(M, \mathcal{F})$ with a Markov minimal set X as in [3] as follows. For $1 \leq i \leq m$, paste the plug $(P_i, \mathcal{F}_i)$ into $S \times S^1$ by pasting σ^i to $C_i \times S^1$ and ρ^i to $C'_i \times S^1$ so that σ^i_1 is pasted to $C_i \times X_i$, σ^i_3 is pasted to $C_i \times A_i$, ρ^i_1 is pasted to $C'_i \times I_i$, and ρ^i_3 is pasted to $C'_i \times B_i$.

Thus, for $1 \leq i \leq m$, h_i maps X_i onto I_i. Further, since σ^i_3 is pasted to $C_i \times A_i$,

$$X_i \supset \bigcup_{\{j|p_{ij}=1\}} I_j, \qquad \text{and} \qquad A_i \supset \bigcup_{\{j|p_{ij}=0\}} I_j$$

and ρ^i_3 is pasted to $C'_i \times B_i$ and

$$B_i \supset \bigcup_{j \neq i} I_j,$$

it follows that pasting in the plug P_i creates no additional holonomy on Z_0 than h_i and h_i^{-1}. Thus $\tau : Z_0 \to Z_0$ generates the holonomy of $\Gamma|Z_0$ and X is a Markov minimal set.

References

1. A. Candel and L. Conlon, Foliations I, Amer. Math. Soc., Providence, RI, 1999.
2. J. Cantwell and L. Conlon, *Poincaré-Bendixson theory for leaves of codimension one*, Trans. Amer. Math. Soc., **265** (1981), 181–209.
3. J. Cantwell and L. Conlon, *Foliations and subshifts*, Tohoku Math. Jour., **40** (1988), 165–187.
4. J. Cantwell and L. Conlon, *Leaves of Markov local minimal sets in foliations of codimension one*, Publicacions Matemàtiques, **33** (1989), 461–484.
5. J. Cantwell and L. Conlon, *Generic Leaves*, Comment. Math. Helv., **73** (1998), 306–336.

6. P. Dippolito, *Codimension one foliations of closed manifolds*, Ann. of Math., **107** (1978), 403–453.
7. E. Ghys, *Topologie des feuilles génériques*, Ann. of Math., **141** (1995), 387–422.
8. G. Hector and U. Hirsch, Introduction to the Geometry of Foliations, Part B, Vieweg and Sohn, Braunschweig, 1983.
9. T. Inaba, *Examples of Exceptional Minimal Sets*, in A Fête of Topology; Papers Dedicated to I. Tamura, ed. Y. Matsumoto and T. Mizutani and S. Morita, Academic Press, 1988, 95–100.
10. R. Sacksteder, *Foliations and pseudogroups*, Amer. J. Math., **87** (1965), 79–102.
11. P. Schweitzer, *Some problems in foliation theory and related areas*, in Lecture Notes in Math., **652**, Springer–Verlag, 1978, 240–252.
12. M. Takamura,, Reeb stability for leaves with uncountable endset (in Japanese), Master Thesis, Hokkaido Univ. 1984.

Received October 24, 2000, revised December 29, 2000.

Proceedings of
FOLIATIONS: GEOMETRY AND DYNAMICS
held in Warsaw, May 29–June 9, 2000
ed. by Paweł WALCZAK *et al.*
World Scientific, Singapore, 2002
pp. 263–273

SOME REMARKS ON PARTIALLY HOLOMORPHIC FOLIATIONS

MARIUSZ FRYDRYCH
Wydział Matematyki Uniwersytetu Łódzkiego,
ul. Banacha 22, 90-238 Łódź, Poland,
e-mail: frydrych@math.uni.lodz.pl

JERZY KALINA
Instytut Matematyki Politechniki Łódzkiej,
Al. Politechniki 11, 93-590 Łódź, Poland,
e-mail: jkalina@math.uni.lodz.pl

The aim of this paper is to introduce the concept of Nijenhuis tensor on the submodule of sections of the normal bundle to the foliation endowed with the complex structure and its relationship to the formal integrability of canonically associated subbundle of complexified tangent bundle of the given foliated manifold.

1 Linear algebraic background

Linear algebraic background introduced in this section will be applied to the the tangent spaces of manifolds in the next parts of this paper.

A complex structure on a real vector space V is a linear endomorphism J such that $J^2 = -1$. The real space V may be considered as a complex vector space by the formula $(a + ib)v = av + bJv, a, b \in \mathbb{R}, v \in V$.

Proposition 1.1 *Let V be a real vector space. There exist canonical correspondences between the following objects:*

(i) *complex structures J on V*

(ii) *idempotents $\tau : V \otimes_{\mathbb{R}} \mathbb{C} \to V \otimes_{\mathbb{R}} \mathbb{C}$ (i.e. $\tau^2 = \tau$) such that $\sigma\tau + \tau\sigma = \sigma$, where $\sigma : V \otimes_{\mathbb{R}} \mathbb{C} \to V \otimes_{\mathbb{R}} \mathbb{C}$ is a natural conjugation in the complexification of V*

(iii) *complex subspaces $W \subset V \otimes_{\mathbb{R}} \mathbb{C}$ such that $V \otimes_{\mathbb{R}} \mathbb{C} = W \oplus \overline{W}$.*

Proof. Assume (i). Let $\tau = \frac{1}{2}(1 - iJ_{\mathbb{C}})$, where $J_{\mathbb{C}} = J \otimes_{\mathbb{R}} 1$ denotes the complexification of the real operator J. Immediately we get that $\tau^2 = \tau$, and

$$\tau\sigma = \frac{1}{2}(1 - iJ_{\mathbb{C}})\sigma = \frac{1}{2}(\sigma - i\sigma J_{\mathbb{C}}) = \frac{1}{2}(\sigma + \sigma i J_{\mathbb{C}}) = \sigma\frac{1}{2}(1 + iJ_{\mathbb{C}}) = \sigma(1 - \tau)$$

which proves (ii). We have used the fact that endomorphism acting on $V \otimes_{\mathbb{R}} \mathbb{C}$ is complexification of an endomorphism on V if and only if it commutes with σ. Assuming (ii) put $W = \operatorname{image} \tau$ Then $\overline{W} = \sigma W$. We check that

$$\begin{aligned} W \cap \overline{W} &= \operatorname{image} \tau \cap \operatorname{image}(\sigma\tau) = \operatorname{image} \tau \cap \operatorname{image}(\sigma - \tau\sigma) \\ &= \operatorname{image} \tau \cap \operatorname{image}(1 - \tau)\sigma = \operatorname{image} \tau \cap \operatorname{image}(1 - \tau) \\ &= \operatorname{image} \tau \cap \ker \tau = 0. \end{aligned}$$

From the assumption it follows the identity $\tau + \sigma\tau\sigma = 1$ which together with the above gives desired decomposition $W \oplus \overline{W} = V \otimes_{\mathbb{R}} \mathbb{C}$ which proves (iii). Assume (ii). Let $\hat{J} = i(2\tau - 1)$. We easily check that

$$\hat{J}^2 = i^2(4\tau^2 - 4\tau + 1) = -(4\tau - 4\tau + 1) = -1.$$

Moreover,

$$\begin{aligned} \sigma\hat{J} &= \sigma i(2\tau - 1) = -i\sigma(2\tau - 1) - i(2(-\tau\sigma + \sigma) - \sigma) \\ &= -i(2(-\tau + 1) - 1)\sigma = i(2\tau - 1)\sigma = \hat{J}\sigma. \end{aligned}$$

From the above we see that $\hat{J} = J \otimes_{\mathbb{R}} 1$ for a unique complex structure $J : V \to V$ which proves (i). Assume (iii). Let τ be a projection onto W along $\overline{W}$. From this we get that $1 - \tau$ is a projection onto $\overline{W}$ along W so $1 - \tau = \sigma\tau\sigma$ which ends the proof. □

Note that the projection τ restricted to the $V \otimes_{\mathbb{R}} 1$ determines a $\mathbb{C}$-linear isomorphism from complex space (V, J) to W. Similarly, the projection $(1 - \tau)|V \otimes_{\mathbb{R}} 1$ is a $\mathbb{C}$-isomorphism from $(V, -J)$ to $\overline{W}$.

2 Bundle structures and Bott connection

Throughout the paper M denotes an n dimensional manifold (real or complex) and $\mathcal{F}$ a foliation of M by submanifolds (real or complex) of dimension p and codimension $q = n - p$. If M is a complex manifold foliated by complex submanifolds, such a foliation will be called *tangentially holomorphic* to emphasis the holomorphic nature of its leaves. All objects in this paper are assumed to be of class C^∞, except where explicitly stated otherwise. In a local computations the little Greek letters range between 1 and p and

Roman letters between 1 and q. Except where explicitly stated otherwise the Einstein summation conventions will be in force.

For a given foliation $\mathcal{F}$ of M, let $T\mathcal{F}$ denote the tangent bundle along the leaves. Then we have the short exact sequence

$$0 \to T\mathcal{F} \to TM \to TM/T\mathcal{F} \to 0. \tag{1}$$

For shortness we will denote the normal bundle $TM/T\mathcal{F}$ by $\nu\mathcal{F}$ and the projection by $TM \xrightarrow{\pi} \nu\mathcal{F}$. The exactness of the sequence 1 is equivalent to the exactness of the sequence of the dual vector bundles

$$0 \to \nu^*\mathcal{F} \to T^*M \to T^*\mathcal{F} \to 0. \tag{2}$$

Note the vector bundle $\nu^*\mathcal{F}$ can be naturally identified with the annihilator of the the tangent bundle of $\mathcal{F}$.

Definition 2.1 We call a bundle mapping $h\colon \nu\mathcal{F} \to TM$ the splitting of the exact sequence 1 if $\pi h = 1$.

The bundle $\mathcal{H} = \text{image}(h)$ which is complementary to the vertical bundle $T\mathcal{F}$ in TM, is called *horizontal*.

Given a splitting of the sequence (1), any vector $Y \in TM$ decomposes into $Y = Y^h + Y^v$ where $Y^h \in \text{image}(h)$ and $Y^v \in T\mathcal{F}$.

Let (U, φ) be a distinguished chart for $\mathcal{F}$ with local coordinates (x^α, y^j). Consider the horizontal parts $Y_j = (\frac{\partial}{\partial y^j})^h$ of vectors $\frac{\partial}{\partial y^j}$. The vectors $Y_1, \ldots, Y_q$ form a basis of the horizontal subspace $\mathcal{H}$. Since $\frac{\partial}{\partial x^\alpha}$ form a local basis of $T\mathcal{F}$, thus $(\frac{\partial}{\partial y^j})^v = \lambda_j^\alpha(x,y)\frac{\partial}{\partial x^\alpha}$ for some smooth functions $\lambda_j^\alpha(x,y)$, therefore we have $Y_j = \frac{\partial}{\partial y^j} - \lambda_j^\alpha(x,y)\frac{\partial}{\partial x^\alpha}$. The local basis $(\frac{\partial}{\partial x^\alpha}, Y_j)$ will be called *$\mathcal{H}$-adapted, associated* to the distinguished chart (U, φ). It is straightforward that (ξ^α, dy^j) is the dual basis when $\xi^\alpha = dx^\alpha + \lambda_j^\alpha(x,y)dy^j$. Now, if we demand, that the horizontal bundle $\mathcal{H}$ is integrable (involutive), we obtain equations

$$d\xi^\alpha \in \text{ideal}\left(\xi^1, \ldots, \xi^p\right)$$

which are equivalent to the system of first order P.D.E. on functions $\lambda_j^\alpha(x,y)$

$$Y_j\lambda_k^\alpha - Y_k\lambda_j^\alpha = 0, \quad j,k = 1,\ldots,q; \ \ \alpha = 1,\ldots,p$$

i.e.

$$\frac{\partial\lambda_k^\alpha(x,y)}{\partial y^j} - \frac{\partial\lambda_j^\alpha(x,y)}{\partial y^k} - \lambda_j^\beta(x,y)\frac{\partial\lambda_k^\alpha(x,y)}{\partial x^\beta} + \lambda_k^\beta(x,y)\frac{\partial\lambda_j^\alpha(x,y)}{\partial x^\beta} = 0,$$

where

$$j,k = 1,\ldots,q; \ \ \alpha = 1,\ldots,p.$$

Definition 2.2 Let $\mathcal{F}$ be a foliation on M. A vector field X on M is called a foliated vector field (an infinitesimal automorphism of $\mathcal{F}$) if the local flow associated to X preserves the foliation.

We have the well known fact

Proposition 2.3 *Let Y be a vector field on the foliated manifold $(M, \mathcal{F})$. The following conditions are equivalent:*

(i) *Y is foliated*

(ii) *$[Y, X] \in \Gamma(T\mathcal{F})$ for all $X \in \Gamma(T\mathcal{F})$*

(iii) *In every distinguished chart with the coordinates (x^α, y^j), Y has the form*

$$f^\alpha(x, y)\frac{\partial}{\partial x^\alpha} + g^j(y)\frac{\partial}{\partial y^j}.$$

Let us recall [4] that a linear connection ∇ in a vector bundle $E \to M$ is a linear homomorphism

$$\nabla : \Gamma(E) \to \Gamma(T^*M \otimes E)$$

such that

$$\nabla fu = \mathrm{d}f \otimes u + f\nabla u, \qquad \text{for} \quad f \in C^\infty(M),\ u \in \Gamma(E).$$

As usual we define $\nabla_X : \Gamma(E) \to \Gamma(E)$ for $X \in \Gamma(TM)$ by $\nabla_X u = (\nabla u)(X)$.

The curvature R of a connection ∇ is the skew-symmetric map

$$\mathrm{R}\colon \Gamma(TM) \times \Gamma(TM) \longrightarrow \Gamma(\mathrm{End}(E))$$

such that

$$\mathrm{R}_{XY} = [\nabla_X, \nabla_Y] - \nabla_{[X,Y]}$$

for any $X, Y \in \Gamma(TM)$.

In the case of the foliated manifold there a is natural partial connection

Definition 2.4 ([1]) For any foliation $\mathcal{F}$ on M there exists a linear partial flat connection

$$\overset{\circ}{\nabla}\colon\ \Gamma(\nu\mathcal{F}) \longrightarrow \Gamma(T^*\mathcal{F} \otimes \nu\mathcal{F})$$

in a normal bundle $\nu\mathcal{F}$ given by

$$\overset{\circ}{\nabla}_X u = \pi[X, Y]$$

for $X \in \Gamma(T\mathcal{F})$, $u \in \Gamma(\nu\mathcal{F})$ and any $Y \in \Gamma(TM)$, such that $\pi(Y) = u$.

From the Jacobi's identity it follows that this partial connection is flat i.e. $\mathrm{R}_{X_1 X_2} = 0$, for $X_1, X_2 \in \Gamma(T\mathcal{F})$.

Let $C^\infty_{\mathcal{F}}$ be the sheaf of germs of functions which are locally constant along the leaves of $\mathcal{F}$.

In the usual way we can define the sheaf $\overset{\circ}{\Gamma}(\nu\mathcal{F})$ of germs of local sections of the normal bundle $\nu\mathcal{F}$ which are covariantly constant with respect to the partial Bott connection. It is evident that $\overset{\circ}{\Gamma}(\nu\mathcal{F})$ forms a sheaf of modules over the sheaf $C^\infty_{\mathcal{F}}$ of rings.

Proposition 2.5 *Every lift in* $\Gamma(TM)$ *of any section* $\overset{\circ}{\Gamma}(\nu\mathcal{F})$ *is a foliated vector field .*

Proof. Take a local section $u \in \overset{\circ}{\Gamma}(\nu\mathcal{F})$ in a neighbourhood of $m \in M$. Let (U, φ) be a distinguished chart around m. Let Y be any lift of u over U. Thus in that chart $Y = a^\alpha(x,y)\frac{\partial}{\partial x^\alpha} + b^j(x,y)\frac{\partial}{\partial y^j}$. Therefore we have,

$$0 = \overset{\circ}{\nabla}_{\frac{\partial}{\partial x^\beta}} u = \pi\left[\frac{\partial}{\partial x^\beta}, Y\right] = \frac{\partial b^j(x,y)}{\partial x^\beta}\pi\left(\frac{\partial}{\partial y^j}\right).$$

Since $\pi(\frac{\partial}{\partial y^1}), \ldots, \pi(\frac{\partial}{\partial y^q})$ form a local basis of $\nu\mathcal{F}|U$, the coefficients $\frac{\partial b^j(x,y)}{\partial x^\beta}$ $= 0$ which proves that the vector field Y is foliated.

We say that the normal bundle $\nu\mathcal{F}$ is *complex* if it is endowed with a bundle endomorphism $J \colon \nu \to \nu$ which is a complex structure over the fibres. Suppose that $\overset{\circ}{\nabla} J = 0$ i.e. $\overset{\circ}{\nabla}_X Ju = J(\overset{\circ}{\nabla}_X u)$ for all $X \in T\mathcal{F}$ and $u \in \Gamma(\nu\mathcal{F})$. Such a connection will be called $\mathbb{C}$-linear. Assuming $\nu\mathcal{F}$ is complex, we introduce the concept of the Nijenhuis tensor $\mathcal{N}$ on the sections of the sheaf $\overset{\circ}{\Gamma}(\nu\mathcal{F})$ as a skew-symmetric $C^\infty_{\mathcal{F}}$ linear mapping

$$\mathcal{N} : \Lambda^2\overset{\circ}{\Gamma}(\nu\mathcal{F}) \longrightarrow \overset{\circ}{\Gamma}(\nu\mathcal{F}).$$

Let u, v be arbitrary germs of sections of $\overset{\circ}{\Gamma}(\nu\mathcal{F})$ at a point. Choose any lifts $\widetilde{u}$, $\widetilde{v}$, $\widetilde{Ju}$, $\widetilde{Jv}$ of u, v, Ju, Jv respectively. Define

$$\mathcal{N}(u,v) = J\pi\left(\left[\widetilde{Ju}, \widetilde{v}\right] + \left[\widetilde{u}, \widetilde{Jv}\right]\right) - \pi\left(\left[\widetilde{Ju}, \widetilde{Jv}\right] - [\widetilde{u}, \widetilde{v}]\right).$$

Since we assumed $\mathbb{C}$-linearity of the partial Bott connection ($\overset{\circ}{\nabla} J = 0$) we see that for every $u \in \overset{\circ}{\Gamma}(\nu\mathcal{F})$ it follows that $Ju \in \overset{\circ}{\Gamma}(\nu\mathcal{F})$. Using the above proposition, it is easy to see that the definition of $\mathcal{N}$ doesn't depend of the choice of the lifts.

Definition 2.6 We say that a complex structure J on the normal bundle $\nu\mathcal{F}$ is integrable if $\overset{\circ}{\nabla} J = 0$ and $\mathcal{N} = 0$.

Let $(\nu\mathcal{F}, J)$ be a complex normal bundle as above. Consider the complexification of exact sequence 1

$$0 \to T\mathcal{F}\otimes_{\mathbb{R}}\mathbb{C} \to TM\otimes_{\mathbb{R}}\mathbb{C} \xrightarrow{\pi\otimes 1} \nu\mathcal{F}\otimes_{\mathbb{R}}\mathbb{C} \to 0.$$

According to Proposition 1.1 in Section 1 we get the following decomposition of the bundle:

$$\nu\mathcal{F}\otimes_{\mathbb{R}}\mathbb{C} = W \oplus \overline{W}.$$

Denote by V the subbundle of $TM\otimes_{\mathbb{R}}\mathbb{C}$ defined by the formula

$$V = \ (\pi \otimes 1)^{-1}(W).$$

It is easy to check that $V = \ker \hat{\pi}$ where

$$\hat{\pi}\colon\ TM\otimes_{\mathbb{R}}\mathbb{C} \longrightarrow \nu\mathcal{F}$$

is given by

$$\hat{\pi}(u \otimes 1 + v \otimes i) = \ \pi(u) \ - \ J(\pi(v)).$$

Note that $\hat{\pi}$ is the composition of $\pi \otimes 1$ with the projection $\nu\mathcal{F}\otimes_{\mathbb{R}}\mathbb{C}$ onto the antiholomorphic subspace $\overline{W}$ in $\nu\mathcal{F}\otimes_{\mathbb{R}}\mathbb{C}$, which is identified with the complex bundle $(\nu\mathcal{F}, -J)$.

For the vector bundle V given by the above construction we have the following

Proposition 2.7 *Let J be a complex structure of $\nu\mathcal{F}$. The subbundle V is formally integrable (i.e. $[\Gamma(V),\ \Gamma(V)] \subset \Gamma(V)$) if and only if J is integrable.*

Proof. For shortness we will denote any vector in complexification $u \otimes 1 + v \otimes i$ by $u + iv$. Suppose that $u + iv \in V$. We get

$$(\pi \otimes 1)(u + iv) = \pi(u) + i\pi(v) \in W = \ker(1 + iJ).$$

This means that $u + iv \in V$ if and only if $\pi(v) = -J(\pi(u))$. Let us now consider to local sections Y_1, Y_2 of the subbundle V. From the above we obtain the following equations

$$\pi(v_k) = -J\left(\pi(u_k)\right), \quad k = 1, 2, \tag{3}$$

where $Y_k = u_k + iv_k$, for $k = 1, 2$. Calculating the complexified Lie bracket of the fields Y_1, Y_2 we get

$$[Y_1, Y_2] = \left([u_1, u_2] - [v_1, v_2]\right) + i\left([u_1, v_2] + [v_1, u_2]\right).$$

Thus $[Y_1, Y_2] \in \Gamma(V)$ means that

$$\pi\left([u_1, v_2] + [v_1, u_2]\right) = -J\left(\pi\left([u_1, u_2] - [v_1, v_2]\right)\right). \tag{4}$$

If $u_1 \in \Gamma(T\mathcal{F})$, we can choose $Y_1 = u_1$ $(v_1 = 0)$ and we get

$$\pi\left([u_1, v_2]\right) = -J\left(\pi\left([u_1, u_2]\right)\right),$$

which means that

$$\overset{\circ}{\nabla}_{u_1}\pi(v_2) = \overset{\circ}{\nabla}_{u_1}(-J\left(\pi\left(u_2\right)\right)) = -J\left(\overset{\circ}{\nabla}_{u_1}\pi\left(u_2\right)\right).$$

This proves that $\overset{\circ}{\nabla} J = 0$.

Let $u, v \in \overset{\circ}{\Gamma}(\nu\mathcal{F})$ and $u_1 = \widetilde{Ju}$, $u_2 = \widetilde{Jv}$, $v_1 = \widetilde{u}$, $v_2 = \widetilde{v}$. Since $\overset{\circ}{\nabla} J = 0$, we see that $u_1 + iv_1, u_2 + iv_2 \in \Gamma(V)$, so substituting them into (4) we get $\mathcal{N}(u, v) = 0$ which proves the necessity.

To prove the sufficiency, let (U, φ) be a distinguished chart for $\mathcal{F}$ with local coordinates (x^α, y^j). The local sections $\pi(\frac{\partial}{\partial y^j}), j = 1, \ldots, q$ form a local basis of $\nu\mathcal{F}$. Let further, $J^l_k(x, y)$, $k, l = 1, \ldots, q$, be the matrix components of J with respect to the above basis. By the assumption, $\overset{\circ}{\nabla} J = 0$ we infer that $\frac{\partial}{\partial x^\alpha} J^l_k(x, y) = 0$, $\alpha = 1, \ldots, p$, which means that $J^l_k(x, y)$, $k, l = 1, \ldots, q$ don't depend on leaf coordinates. It is easy to see that the vector fields $\frac{\partial}{\partial x^\alpha}, \frac{\partial}{\partial y^j} - iJ^k_j(y)\frac{\partial}{\partial y^k}$, $\alpha = 1, \ldots, p$, $j = 1, \ldots, q$, form a set of generators of $\Gamma(V|U)$. To prove the formal integrability of V it is enough to check that space spanned over the last q fields is closed under the Lie bracket, which is equivalent to the vanishing of the Nijenhuis tensor. This ends the proof. □

From that proof we can see that our foliation is given by the local submersions on pieces of complex manifolds in such a manner that the transition diffeomorphisms are biholomorphisms, which means that our foliation is transversely holomorphic.

Conversely, if a foliation is transversely holomorphic then the normal bundle is canonically endowed with the integrable complex structure and, by Proposition 2.7, the bundle V is formally integrable. We summarize the above conclusion as

Corollary 2.8 *The subbundle V is formally integrable if and only if $\mathcal{F}$ is transversely holomorphic.*

Remark. In the dual description we consider the annihilator

$$Q = \overline{V}^\perp \subset T^*M \otimes_{\mathbb{R}} \mathbb{C}$$

of the "antiholomorphic" bundle $\overline{V}$, which satisfies the conditions

$$Q \oplus \overline{Q} = T\mathcal{F}^{\perp} \otimes_{\mathbb{R}} \mathbb{C} \subset T^*M \otimes_{\mathbb{R}} \mathbb{C},$$

$$d\Gamma(Q) \subset \text{ideal}\left(\Gamma\left(Q \oplus \overline{Q}\right)\right).$$

It is straightforward, that formal integrability of V is equivalent to

$$d\Gamma(Q) \subset \text{ideal}(\Gamma(Q)).$$

3 Formal integrability and main results

Using Proposition 2.7 from the previous section we are able to prove the following theorems.

Theorem A *Let M be a complex manifold of dimension m and $\mathcal{F}$ a smooth foliation of M by complex submanifolds of dimension p. Let further $TM \otimes_{\mathbb{R}} \mathbb{C} = U \oplus \overline{U}$ and $T\mathcal{F} \otimes_{\mathbb{R}} \mathbb{C} = F \oplus \overline{F}$, $F \subset U$ be the decompositions in the sense of Proposition 1.1. For the subbundle $V = U \oplus \overline{F}$ the following conditions are equivalent:*

(i) $[\Gamma(V), \Gamma(V)] \subset \Gamma(V)$,

(ii) *$\mathcal{F}$ is holomorphic,*

(iii) *Bott connection is $\mathbb{C}$ linear.*

Proof. Let $\nu\mathcal{F} \otimes_{\mathbb{R}} \mathbb{C} = W \oplus \overline{W}$ be the decomposition of the complexified normal bundle of $(\nu\mathcal{F}, J)$, where J denotes the induced complex structure on the quotient bundle $\nu\mathcal{F}$. Now we prove that

$$V = (\pi \otimes 1)^{-1}(W) = U \oplus \overline{F}$$

so that we can apply Proposition 2.7.

Let $Z = X + iY \in (\pi \otimes 1)^{-1}(W)$, where $X, Y \in TM$, $Z \in TM \otimes_{\mathbb{R}} \mathbb{C}$. Therefore $\pi(X) + i\pi(Y) \in W$ which means that

$$\pi(Y) = -J\pi(X). \quad (*)$$

Decompose Z in the following manner:

$$Z = \left(\frac{X + JY}{2} - iJ\frac{X + JY}{2}\right) + \left(\frac{X - JY}{2} + iJ\frac{X - JY}{2}\right) = Z_1 + Z_2.$$

We observe that $Z_1 \in U$ and $Z_2 \in \overline{U}$ but, by (*), $\pi(Z_2) = 0$ and we conclude that $Z_2 \in \overline{F}$ which implies that $Z \in U \oplus \overline{F}$. Thus we get the inclusion

$$(\pi \otimes 1)^{-1}(W) \subset U \oplus \overline{F}.$$

To prove the opposite inclusion, note that $(\pi \otimes 1)(\overline{F}) = \{0\} \subset W$ and $(\pi \otimes 1)(U) \subset W$ since $(\pi \otimes 1)(X - iJX) = \pi(X) - iJ\pi(X) \in W$, where $X \in TM$. By Proposition 2.7 and the above considerations we get the equivalence between (i) and (iii). The equivalence (i) with (ii) is given by the following lemma. □

Lemma 3.1 *A tangentially holomorphic foliation $\mathcal{F}$ on a complex manifold M (i.e. such that leaves of $\mathcal{F}$ are complex submanifolds of M) is holomorphic if and only if it is transversally holomorphic.*

Proof. If a foliation $\mathcal{F}$ on the complex manifold is holomorphic then it is in particular tangentially and transversally holomorphic. Now suppose that $\mathcal{F}$ is tangentially and transversally holomorphic. Let $m \in M$ be an arbitrary point in M, and

$$(U, \phi), \quad \phi = (w, z), \ w = \left(w^1, \dots, w^p\right), \ z = \left(z^1, \dots, z^q\right),$$

be a complex chart on M around a point m, such that complex submanifolds $\{(w, z) \in U;\ w = \text{const}\}$ are transverse to the given foliation $\mathcal{F}|U$. Because the foliation $\mathcal{F}$ is tangentially holomorphic, we can define complex 1-forms of type (1,0)

$$\zeta^j = dz^j + \lambda\,(w, z)^j_\alpha\, dw^\alpha, \quad j = 1, \dots, q$$

such that

$$Q = \overline{V}^{\perp} = \operatorname{span}\left\{\zeta^1, \dots, \zeta^q\right\}$$

for some complex functions

$$\lambda^j_\alpha \in C^\infty(U, \mathbb{C}), \quad \alpha = 1, \dots, p;\ j = 1, \dots, q$$

(see the Theorem A and Remark). It is evident that

$$\left(dw^1, \dots, dw^p,\ \zeta^1, \dots, \zeta^q,\ d\overline{w}^1, \dots, d\overline{w}^p,\ \overline{\zeta}^1, \dots, \overline{\zeta}^q\right)$$

is a frame of $T^*M \otimes_{\mathbb{R}} \mathbb{C}|U$. Now, the fact that foliation $\mathcal{F}$ is transversely holomorphic implies that

$$d\zeta^j \in \operatorname{ideal}\left(\zeta^1, \dots, \zeta^q\right), \quad j = 1, \dots, q.$$

A simple computation shows that

$$\begin{aligned} d\zeta^j = {} & \frac{\partial \lambda^j_\alpha}{\partial z^l} \zeta^l \wedge dw^\alpha - \left(\lambda^l_\beta \frac{\partial \lambda^j_\alpha}{\partial z^l} - \frac{\partial \lambda^j_\alpha}{\partial w^\beta}\right) dw^\beta \wedge dw^\alpha \\ & + \frac{\partial \lambda^j_\alpha}{\partial \overline{z}^l} \overline{\zeta}^l \wedge dw^\alpha - \left(\overline{\lambda}^l_\beta \frac{\partial \lambda^j_\alpha}{\partial \overline{z}^l} - \frac{\partial \lambda^j_\alpha}{\partial \overline{w}^\beta}\right) d\overline{w}^\beta \wedge dw^\alpha, \end{aligned}$$

so $d\zeta^j \in \text{ideal}\left(\zeta^1, \dots, \zeta^q\right)$ implies

$$\frac{\partial \lambda^j_\alpha}{\partial \overline{z}^l} = 0,$$

and

$$\overline{\lambda}^l_\beta \frac{\partial \lambda^j_\alpha}{\partial \overline{z}^l} - \frac{\partial \lambda^j_\alpha}{\partial \overline{w}^\beta} = 0,$$

which means that functions

$$\lambda\,(z,w)^j_\alpha, \quad j = 1, \dots, q; \quad \alpha = 1, \dots, p,$$

are holomorphic. It follows that the annihilator $\Gamma(Q)$ is generated by holomorphic 1-forms ζ^j, $\ j = 1, \dots, q$. Now the classical complex version of Frobenius theorem ends the proof. □

Finally, we are in a position to prove the following.

Theorem B Let M be a real manifold and V a subbundle of $TM \otimes_{\mathbb{R}} \mathbb{C}$ such that $V + \overline{V}$ and $V \cap \overline{V}$ are formally integrable subbundles. Then there exists on M a flag of foliations $(\mathcal{H}, \mathcal{F})$ and a canonical complex structure on $T\mathcal{H}/T\mathcal{F}$, where $T\mathcal{H} \otimes_{\mathbb{R}} \mathbb{C} = V + \overline{V}$, $T\mathcal{F} \otimes_{\mathbb{R}} \mathbb{C} = V \cap \overline{V}$. Moreover, this complex structure is integrable if and only if V is formally integrable, which means that the foliation $\mathcal{F}$ is transversely holomorphic in $\mathcal{H}$.

Proof. Since $V + \overline{V}$ and $V \cap \overline{V}$ are invariant under complex conjugation, there exist distributions H and F in TM such that $V + \overline{V} = H \otimes_{\mathbb{R}} \mathbb{C}$ and $V \cap \overline{V} = F \otimes_{\mathbb{R}} \mathbb{C}$. Standard algebraic observations give a sequence of natural isomorphisms

$$\frac{H}{F} \otimes_{\mathbb{R}} \mathbb{C} \approx \frac{H \otimes_{\mathbb{R}} \mathbb{C}}{F \otimes_{\mathbb{R}} \mathbb{C}} = \frac{V + \overline{V}}{V \cap \overline{V}} \approx \frac{V}{V \cap \overline{V}} \oplus \overline{\frac{V}{V \cap \overline{V}}}.$$

Proposition 1.1 in Section 1 gives a complex structure J on the quotient bundle H/F. Because of the formal integrability of $V + \overline{V}$ and $V \cap \overline{V}$ the distributions H and F are involutive, so actually determine the flag of foliations $(\mathcal{H}, \mathcal{F})$. Theorem A gives us the last part of the statement. □

We can regard Proposition 2.7 as a geometrical version of the classical Nirenberg theorem ([2], [3]).

References

1. R. Bott, *Lectures on characteristic classes and foliations*, in Lecture Notes in Math. **279**, Springer Verlag, 1972, 1–94.

2. T. Duchamp and M. Kalka, *Invariance of tangentially holomorphic foliations and Monge-Ampere equation*, Mich. Math. J., **35**, (1988), 91–115.
3. L. Nirenberg, *A complex Frobenius theorem*, in Seminars on analytic functions, **I**, Princeton, 1957, 172–189.
4. R.O. Wells, Differential analysis on complex manifolds, Springer Verlag, 1980.

Received December 18, 2000.

Proceedings of
FOLIATIONS: GEOMETRY AND DYNAMICS
held in Warsaw, May 29–June 9, 2000
ed. by Paweł WALCZAK *et al.*
World Scientific, Singapore, 2002
pp. 275–295

FOLIATIONS AND COMPACTLY GENERATED PSEUDOGROUPS

ANDRÉ HAEFLIGER
Section de mathématiques, 2-4 Rue du Lièvre, Genève, Switzerland,
e-mail: andre.haefliger@math.unige.ch

The holonomy pseudogroup of a foliation on a compact manifold satisfies a condition called "compact generation". Most properties enjoyed by the holonomy pseudogroup of a foliation on compact manifolds are also valid for compactly generated pseudogroups. In this talk we illustrate this principle by proving that a compactly generated pseudogroup of holomorphic transformations of a one-dimensional complex manifold has a finite number of closed orbits, unless all the orbits are closed, a property proved for transversely holomorphic foliations of codimension one on compact manifolds by M. Brunella and M. Nicolau [3].

It is well known that the dynamic properties of a foliation $\mathcal{F}$ on a manifold M are encoded in (the equivalence class) of its holonomy pseudogroup. Therefore it is natural to prove let say a theorem A in foliation theory, whose hypothesis and conclusion can be read on the holonomy pseudogroup, by proving first a theorem A' about pseudogroups satisfying suitable conditions reflecting the hypothesis of Theorem A, and then deduce Theorem A from Theorem A' (this is usually easy).

As an example, Epstein, Millet and Tischler [5] proved that for a foliation $\mathcal{F}$ on a manifold M with a countable basis, the subset of M which is the union of leaves with trivial holonomy is a G_δ dense set (Theorem A). The hypothesis that M has a countable basis implies easily that the holonomy pseudogroup $\mathcal{H}$ of $\mathcal{F}$ with respect to a complete transversal T contains an open set T_0 with countable basis meeting all the orbits and such that the restriction of $\mathcal{H}$ to T_0 is countably generated. For such a pseudogroup $\mathcal{H}$ it is immediate that the set of orbits with trivial isotropy groups is a G_δ dense set (Theorem A'). In turn, this implies immediately Theorem A.

Sometimes the hypothesis of Theorem A involve conditions on the fundamental group or on the cohomology of M. This implies conditions on the fundamental group or the cohomology of the holonomy pseudogroup $\mathcal{H}$ of $\mathcal{F}$. The notions of homotopy or cohomology for pseudogroups, or more generally for étale groupoids, have been introduced in Haefliger [11] and [12]; it is imperative that those notions depend only on the equivalence class of the pseudogroup.

The holonomy pseudogroup of a foliation on a *compact* manifold satisfies a certain finiteness condition, called *compact generation*, which should be the hypothesis corresponding to the compactness of M in Theorem A'.

In this talk we first recall the precise definitions of equivalence classes of pseudogroups and of compactly generated pseudogroups.

In the second part we illustrate the above considerations in a particular case. Namely we first state a recent result of Brunella and Nicolau [3] about transversely holomorphic foliations of codimension one on compact manifolds and give the proof of this Theorem A. Then we state a corresponding Theorem A' concerning compactly generated pseudogroups of holomorphic transformations of a one-dimensional complex manifold. The translation of the proof of theorem A uses the notion of Čech cohomology for étale groupoids and we prove a finiteness theorem in 2.5 which is also at the heart of the proof of Theorem A.

In the third section, we consider another étale groupoid attached to a foliation on a manifold M, its fundamental groupoid [1] (whose elements are the homotopy classes of paths in the leaves joining two points of a totally transversal submanifold). We exhibit a presentation for this groupoid and introduce the notion of compact presentation, reflecting conditions imposed on the fundamental groupoid by the compactness of the manifold M.

The author thanks the referee for a constructive remark concerning the growth type of orbits.

1 Compactly generated pseudogroups

1.1 Pseudogroups and their equivalences

A *pseudogroup of transformations* $(\mathcal{H}, T)$ (or simply $\mathcal{H}$) of a topological space T is a collection $\mathcal{H}$ of homeomorphisms from open subsets of T to open subsets of T such that:

1) if $h, h' \in \mathcal{H}$, then the composition[a] hh' and the inverse h^{-1} belong to $\mathcal{H}$,

[a] The composition hh' of the homeomorphism $h : U \to V$ with the homeomorphism $h' : U' \to V'$ is the homeomorphism $h'^{-1}(U \cap V') \to h(U \cap V')$ defined by $x \mapsto h(h'(x))$.

2) the identity map of T belongs to $\mathcal{H}$,

3) if a homeomorphism from an open set of T to an open set of T is locally in $\mathcal{H}$, then it is in $\mathcal{H}$.

If T is a differentiable or a Riemannian manifold, then $\mathcal{H}$ is called a pseudogroup of differentiable or isometric transformations if all its elements are differentiable or are Riemannian isometries.

The restriction of $\mathcal{H}$ to an open subset T_0 of T is the set of elements of $\mathcal{H}$ with source (or domain) and target (or range) in T_0.

The *orbit* $\mathcal{H}.x$ of a point $x \in T$ is the set of the images of x under the elements of $\mathcal{H}$. The space of orbits, with the quotient topology, is noted $\mathcal{H}\backslash T$.

For $x \in T$, the isotropy group $\mathcal{H}_x$ is the group of germs at x of the elements $h \in \mathcal{H}$ such that $h(x) = x$.

Two pseudogroups $(\mathcal{H}_0, T_0)$ and $(\mathcal{H}_1, T_1)$ are *equivalent* if there is a pseudogroup $(\mathcal{H}, \mathcal{T})$ and homeomorphisms f_i from T_i onto open sets T_i' of T, $i = 0, 1$, such that

1) T'_i meets all the orbits of $\mathcal{H}$,

2) f_i induces an isomorphism from $\mathcal{H}_i$ onto the restriction $\mathcal{H}_{|T_i'}$ of $\mathcal{H}$ to T_i'.

If $\mathcal{H}_i$ and $\mathcal{H}$ are pseudogroups of differentiable transformations, then $\mathcal{H}_0$ is differentiably equivalent to $\mathcal{H}_1$ if f_0 and f_1 are diffeomorphisms.

Remarks. Let $(\mathcal{H}_0, T_0)$ and $(\mathcal{H}_1, T_1)$ be two equivalent pseudogroups. Then the spaces $\mathcal{H}_0\backslash T_0$ and $\mathcal{H}_1\backslash T_1$ are homeomorphic.

If T_0 carries an $\mathcal{H}_0$-invariant measure finite on compact subsets, then there is a corresponding $\mathcal{H}_1$-invariant measure on T_1.

Examples.

1) Let T_0 be an open subset of T meeting all the orbits of $\mathcal{H}$. Then the restriction of $\mathcal{H}$ to T_0 is equivalent to $\mathcal{H}$.

2) Let T_0 be the real line $\mathbb{R}$ and T_1 be the circle $\mathbb{R}/\mathbb{Z}$. The pseudogroup of transformations $\mathcal{H}_0$ of $\mathbb{R}$ generated by the translation $x \mapsto x+1$ is equivalent to the pseudogroup of transformations $\mathcal{H}_1$ of $\mathbb{R}/\mathbb{Z}$ generated by the identity map. To see this, consider the pseudogroup $\mathcal{H}$ of transformations of the union T of $T_0 = \mathbb{R}$ and $T_1 = \mathbb{R}/\mathbb{Z}$ generated by the integral translations of $\mathbb{R}$ and the restrictions of the projection $\mathbb{R} \to \mathbb{R}/\mathbb{Z}$ to small open sets of $\mathbb{R}$. The natural inclusions f_i of T_i in T induce isomorphisms from $\mathcal{H}_i$ to $\mathcal{H}_{|T_i}$.

1.2 The holonomy pseudogroup of a foliation

A *foliation* $\mathcal{F}$ of codimension q on a manifold M can be given by a foliated cocycle over an open cover $\mathcal{U} = (U_i)_{i\in I}$, namely submersions p_i from U_i

onto open subsets T_i of R^q, with connected fibers, satisfying the following *Compatibility condition:* for every $x \in U_i \cap U_j$, there is an open neighbourhood $U_{ij}^x \subset U_i \cap U_j$ of x and a homeomorphism h_{ij}^x from $p_j(U_{ij}^x)$ to $p_i(U_{ij}^x)$ such that $p_i = p_j \circ h_{ij}^x$ on U_{ij}^x.

Note that h_{ij}^x is uniquely defined and that for $y \in U_{ij}^x$, we have $h_{ij}^x = h_{ij}^y$ (see the remark below). The *leaves* of $\mathcal{F}$ are the connected components of M endowed with the topology having as a basis of open subsets the intersection of the fibers of the p_i with the open subsets of M.

A foliated cocycle $\{p'_k\}_{k \in K}$ over another open cover $\mathcal{U}' = \{U'_k\}_{k \in K}$ defines the same foliation $\mathcal{F}$ if the above compatibility condition is satisfied for the union of the two foliated cocycles.

Let T be the disjoint union of the open sets T_i; to simplify the notations we identify T_i with the corresponding open set of T. The *holonomy pseudogroup* of $\mathcal{F}$ (with respect to the foliated cocycle $\{p_i\}$) is the pseudogroup of transformations $(\mathcal{H}, T)$ of T generated by the elements h_{ij}^x.

The holonomy pseudogroup of $\mathcal{F}$ associated to another compatible cocycle is equivalent to $(\mathcal{H}, T)$. *Therefore the holonomy pseudogroup of $\mathcal{F}$ is well defined only up to equivalence.* By "the" holonomy pseudogroup of $\mathcal{F}$, we mean any pseudogroup equivalent to the holonomy pseudogroup associated to a foliated cocycle defining $\mathcal{F}$.

Note that if M is a differentiable manifold of class C^r and if the submersions p_i are of class C^r, the elements of the holonomy pseudogroup of $\mathcal{F}$ are differentiable of class C^r.

Remark. It is well known that one can construct a foliated cocycle defining $\mathcal{F}$ such that the fibers of the submersions p_i (called the plaques) are connected, that one plaque in U_i intersects at most one plaque in U_j and that the intersection of any two plaques is connected if it is non empty. In that case there is a unique homeomorphism $h_{ij} : p_j(U_i \cap U_j) \to p_i(U_i \cap U_j)$ such that $p_i = h_{ij} \circ p_j$ on $U_i \cap U_j$. The holonomy pseudogroup of $\mathcal{F}$ is generated by the h_{ij}.

There is a bijective natural correspondence between leaves of $\mathcal{F}$ and orbits of $\mathcal{H}$, holonomy group of a leaf and isotropy subgroup of the corresponding orbit, closed minimal subsets of M saturated by leaves and closed minimal sets of T invariant by $\mathcal{H}$, etc... There is a natural homeomorphism from the space of leaves to the space of orbits $\mathcal{H}\backslash T$.

As the holonomy pseudogroup of a foliation is defined only up to equivalence, we shall respect the following rule.

General Principle. *We shall consider only notions or properties of a pseudogroup $(\mathcal{H}, T)$ which depend only on its equivalence class.*

1.3 Compact generation

Definition. A pseudogroup $(\mathcal{H}, T)$ is *compactly generated* if

a) T is locally compact and contains a relatively compact open subset T_0 meeting all the orbits of $\mathcal{H}$,

b) there is a finite set $S = \{s_1, \dots, s_k\}$ of elements of $\mathcal{H}$ generating $\mathcal{H}_{|T_0}$; moreover each $s_i : V_i \to W_i$ is the restriction of an element $\tilde{s}_i \in \mathcal{H}$ whose domain $\widetilde{V}_i$ contains the closure of V_i.

Remarks. 1) By adding to S a finite number of elements, we can assume that S is closed under inverses, contains the identity of T_0 and that the union $\widetilde{T}$ of the open sets $\tilde{s}_i(\widetilde{V}_i \cap T_0)$ contains $\overline{T_0}$. It follows that the restrictions of the $\tilde{g}_i$ to $\widetilde{T}$ generate $\mathcal{H}_{|\widetilde{T}}$.

2) In the definition above, it is understood that T is Hausdorff. If T is locally compact but not Hausdorff, one can find an equivalent pseudogroup $(\mathcal{H}', T')$, where T' is Hausdorff, for instance the disjoint union of open Hausdorff subsets of T covering T.

According to the above general principle, we have to check that compact generation depends only on the equivalence class of $(\mathcal{H}, T)$. This follows from the following more precise property.

Lemma. *Let $(\mathcal{H}, T)$ be a compactly generated pseudogroup as in the definition above. Let $T'_0 \subset T$ be a relatively compact open subset meeting all the orbits. Then $\mathcal{H}_{|T'_0}$ satisfies the condition b) above.*

Proof. We can assume that the set S of generators of $\mathcal{H}_{|T_0}$ is closed under inverses and contains the identity map of T_0. As T_0 is relatively compact and T'_0 meets each orbit, one can find a finite set K of elements of $\mathcal{H}$ whose domains form an open cover of T_0, whose ranges are contained in T'_0, each element $k : V \to W$ of K being the restriction of an element $\tilde{k} \in \mathcal{H}$ defined on a neighbourhood of $\overline{V}$. Interchanging the role of T_0 and T'_0, one find a finite set K' of elements of $\mathcal{H}$ verifying similar conditions. □

It is easy to see that the finite set S' of homeomorphisms of the form $ksk', (ksk')^{-1}, k_2sk_1^{-1}$, where $s \in S, k, k_1, k_2 \in K, k' \in K'$, generates $\mathcal{H}_{|T'_0}$.

Examples.

1) The holonomy pseudogroup of a foliation on a closed manifold M is compactly generated. To see this, choose a foliated cycle $\{p_i : U_i \to T_i\}_{i \in I}$ defined over a finite cover of M and choose for each $i \in I$ an open set U_i^0 whose closure is contained in U_i , such that the fibers of the restriction of p_i to U_i^0 are connected, and such that the U_i^0 form an open cover of M. Then the disjoint union T_0 of the $p_i(U_i^0)$ is a relatively compact open set

in the disjoint union T of the $p_i(U_i) = T_i$ meeting every orbit of $\mathcal{H}$; the elements h_{ij}^x restricted to T_0 generate $\mathcal{H}_{|T_0}$ and one can extract from this set a finite generating set.

The same argument holds if M is compact and if $\mathcal{F}$ is either transverse or tangent to the boundary of M.

2) The pseudogroup of transformations $\mathcal{H}$ of $T = \mathbb{R}$ generated by $s_1 : x \mapsto x/2, \forall x$, and by $s_2 : x \mapsto -x, \forall x \neq 0$, is compactly generated. Indeed choose $T_0 =]-1, +1[$; then the restrictions of s_1 and s_2 to T_0 generate $\mathcal{H}_{|T_0}$.

3) The pseudogroup of transformations of $T = [-1, +1]$ generated by $x \mapsto -x, \forall x \neq 0$ is not compactly generated.

4) Let Γ be a subgroup of a simply connected Lie group G, and let $\mathcal{H}$ be the pseudogroup of transformations of G generated by the elements of Γ acting by left translations on G. This pseudogroup will be noted $\Gamma \ltimes G$.

If G is nilpotent, then $\Gamma \ltimes G$ is compactly generated if and only if Γ is finitely generated and if Γ is cocompact, i.e. the quotient of G by the closure of Γ is compact. In that case $\Gamma \ltimes G$ is equivalent to the holonomy pseudogroup of a foliation on a closed manifold.

When G is solvable, then $\Gamma \ltimes G$ is compactly generated if Γ satisfies some arithmetic conditions. For instance, let $G = \mathrm{Aff}_1$ be the group of affine transformations of $\mathbb{R}$ preserving the orientation, and let $\mu : \mathrm{Aff}_1 \to \mathbb{R}^*$ be the homomorphism associating to $x \mapsto ax + b$ the number a. Then Gael Meigniez [17] has proved the remarkable theorem that $\Gamma \ltimes G$ is compactly generated if and only if Γ is finitely generated and cocompact, and 1 is an integral combination of elements of $\mu(\Gamma) \cap\,]0, 1[$ and also an integral combination of elements of $\mu(\Gamma) \cap\,]1, \infty[$.

As shown by Meigniez, many of those pseudogroups are holonomy pseudogroups of foliations on closed manifolds, but it is likely that it is not the case for all of them.

In general it is very difficult to decide if $\Gamma \ltimes G$ is compactly generated or not when G is non-compact and semi-simple. For example the case $G = SL(2, \mathbb{R})$ is very mysterious, although many interesting compactly generated examples with Γ dense in G are described in Ghys, Gomez-Mont, Saludes [8] using arithmetic constructions.

5) Cavalier [4] has given a complete classification of equivalence classes of compactly generated pseudogroups of holomorphic transformations of a complex curve, preserving a non trivial meromorphic vector field.

Growth type of orbits. Let $(\mathcal{H}, T)$ be a pseudogroup generated by a finite set S of elements. In general the type of the growth function of the orbit of a point $x \in T$ (associating to a positive integer n the number of images of

x under composition of at most n elements of S or their inverses) is not an invariant of the equivalence class of $\mathcal{H}$ (see Example 2) in 1.1). Nevertheless for compactly generated pseudogroup $(\mathcal{H}, T)$, one can define the growth type of orbits as the type of the growth function of the corresponding orbit of $\mathcal{H}_{|T_0}$ with respect to the finite generating set S satisfying the conditons of the Remark 1) following the definition of compact generation. It is easy to see that this definition does not depend on the choice of such generators and, by the lemma, is invariant under equivalences.

The argument of Plante [19] shows that, for a compactly generated pseudogroup $(\mathcal{H}, T)$, the existence of an orbit with subexponential growth implies the existence of a measure on T invariant by $\mathcal{H}$ and finite on compact subsets.

1.4 Examples of foliations theorems versus pseudogroups

1) The analogue of the *Reeb stability theorem* for compactly generated pseudogroups is the following. *Let $(\mathcal{H}, T)$ be a compactly generated pseudogroup. Let $x \in T$ be such that the orbit $\mathcal{H}.x$ is closed and the isotropy group $\mathcal{H}_x$ is finite. Then there is a neighbourhood U of T and a finite group Γ of homeomorphisms of U such that the restriction of $\mathcal{H}$ to U is generated by Γ. Moreover the map associating to an element of Γ its germ at x is an isomorphism.*

We prove this theorem as an illustration of the use of the condition of compact generation. We first observe that the orbit of x is discrete (this is true for countably generated pseudogroups). We can then choose an open relatively compact neighbourhood T_0 of x meeting all the orbits of $\mathcal{H}$ such that $\overline{T}_0 \cap \mathcal{H}.x = \{x\}$. By the usual argument, we can find an arbitrarily small open neighbourhood U of x and a subgroup Γ of the group of homeomorphisms of U whose elements belong to $\mathcal{H}$ and such that the map associating to an element of Γ its germ at x is an isomorphism. Using the lemma in 1.3, we can choose a finite generating set S of $\mathcal{H}_{T_0}$, stable under inverses, satisfying the condition b) in the definition of compact generation and U small enough such that, for $g \in S$, either $U \cap \overline{\text{domain } g} = \emptyset$ or $U \subseteq \text{domain } g$. In this last case, g has the same germ at x as an element of Γ. We can assume that U is small enough so that those two elements coincide on U.

Let $h \in \mathcal{H}_{|U}$. We claim that h is locally the restriction of an element of Γ. Indeed let $y \in \text{domain } h$. By hypothesis there are elements $g_1, \dots, g_k$ in S such that, on a small neighbourhood of y, we have $h = g_1 \dots g_k$. By the conditions above, each g_i is locally the restriction of an element of Γ.
2) It has been proved by Edwards, Millet and Sullivan [6], generalizing

arguments of Epstein, that for a foliation $\mathcal{F}$ of codimension 2 on a compact manifold M such that all the leaves are compact, then the leaves of $\mathcal{F}$ are the fibers of a generalized Seifert bundle.

The corresponding statement for a compactly generated pseudogroup $(\mathcal{H}, T)$ of transformations of a 2-manifold T should be the following: if all orbits are closed, then $(\mathcal{H}, T)$ is the pseudogroup of change of charts of an orbifold structure on $\mathcal{H}\backslash T$. (I have not checked this statement.)
3) The following theorem is proved in Haefliger [10]. Let $\mathcal{F}$ be a transversely oriented foliation of codimension one on a connected paracompact manifold M.

a) If the rank of the group $\mathrm{Hom}(\pi_1(M), \mathbb{Z})$ is finite, then the set of closed leaves is closed.

b) If M is closed and $\mathcal{F}$ is real analytic, then either there is a finite number of compact leaves, or all the leaves are compact.

The corresponding result for pseudogroups is the following. *Let $(\mathcal{H}, T)$ be a countably generated pseudogroup of orientation preserving transformations of a one -dimensional manifold T such that $\mathcal{H}\backslash T$ is connected.*

a') *If the rank of* $\mathrm{Hom}(\pi_1(\mathcal{H}), \mathbb{Z})$ *is finite (where $\pi_1(\mathcal{H})$ is defined in [11], see also [20] and [2]), then the set of closed orbits is closed.*

b') *If $(\mathcal{H}, T)$ is compactly generated and real analytic, either there is a finite number of closed orbits, or all orbits are closed.*

This statement implies immediately the corresponding statement for foliations because the fundamental group of M surjects onto the fundamental group of $\mathcal{H}$.

2 Pseudogroups of holomorphic transformations of a complex curve

2.1 A theorem of Brunella and Nicolau

The proof of the following theorem of Brunella and Nicolau [3] is an adaptation to the case of transversely holomorphic foliations of the proof of a theorem of Jouanoulou and Ghys on holomorphic foliations of codimension one (with possible singularities) on a compact complex manifold. For the real analytic case, see 3) in 1.4.

Recall that a transversely holomorphic foliation of codimension one on a manifold M is given by a foliated cocycle $\{p_i : U_i \to \mathbb{C}\}$ such that the transition functions h_{ij} are holomorphic.

Theorem A. *Let $\mathcal{F}$ be a transversely holomorphic foliation of codimension one on a compact connected manifold M. Then either there is finite number of compact leaves or all leaves are compact and there is a (non constant) transversely holomorphic meromorphic function constant on the leaves.*

Proof. [3] Consider the following short exact sequence of sheaves

$$0 \to \Omega^1_{\mathcal{F}} \to \Omega^1_{\mathcal{F},\log} \to \mathcal{D}_{\mathcal{F}} \otimes \mathbb{C} \to 0.$$

Here $\Omega^1_{\mathcal{F}}$ (resp. $\Omega^1_{\mathcal{F},\log}$) is the sheaf of germs of transversely holomorphic 1-forms (resp. meromorphic 1-forms with polar singularities of order ≤ 1) locally constant on the leaves. Restricted to U_i, such a form is locally the pull back by p_i of a 1-form $f(z)dz$ where f is holomorphic (resp. meromorphic with poles of order ≤ 1). The third sheaf is the sheaf of germs of functions locally constant on the leaves with support a finite number of plaques. The last map associates to a pole of the germ of a meromorphic 1-form its residue.

Consider the associated long exact cohomology sequence

$$\ldots \to H^0(M, \Omega^1_{\mathcal{F},\log}) \to H^0(M, \mathcal{D}_{\mathcal{F}} \otimes \mathbb{C}) \to H^1(M, \Omega^1_{\mathcal{F}}) \to \ldots$$

The space $H^0(M, \Omega^1_{\mathcal{F},\log})$ of sections of $\Omega^1_{\mathcal{F},\log}$ is the space of transversely meromorphic 1-forms locally constant on the leaves with polar singularities of order at most one. The space $H^0(M, \mathcal{D}_{\mathcal{F}} \otimes \mathbb{C})$ is a vector space with basis the compact leaves of $\mathcal{F}$.

The main point of the proof is the fact that $H^1(M, \Omega^1_{\mathcal{F}})$ is finite dimensional (see Gomez-Mont [9]). Therefore if there are at least $\dim H^1(M, \Omega^1_{\mathcal{F}})$ $+2$ compact leaves, one can find two linearly independent transversely meromorphic 1-forms constant on the leaves. Their quotient is a transversely meromorphic non-constant function which is constant on the leaves. Therefore all the leaves are compact. □

The corresponding theorem for pseudogroups is the following. It is clear that it implies Theorem A.

Theorem A'. *Let $(\mathcal{H}, T)$ be a compactly generated pseudogroup of holomorphic transformations of a one-dimensional complex manifold T such that $\mathcal{H}\backslash T$ is connected. Then either there is a finite number of closed orbits, or all orbits are closed and there is a non-constant $\mathcal{H}$-invariant meromorphic function on T.*

The rest of the section is devoted to the proof of this theorem which is just a translation in terms of pseudogroups of the above proof of Theorem A. This involves the definition of appropriate notions of cohomology with value in $\mathcal{H}$-sheaves on T. It is more natural to define those notions in the framework of étale groupoids, and so we begin by recalling this notion.

2.2 Étale groupoids $(\mathcal{G}, T)$

An *étale groupoid* $(\mathcal{G}, T)$ is a topological category $\mathcal{G}$ with space of objects T, such that all the arrows (morphisms of the category) are invertible and such that the maps $\alpha : \mathcal{G} \to T$ and $\beta : \mathcal{G} \to T$ associating to each arrow in $\mathcal{G}$ its source (initial object) and its target (terminal object) are étale maps, i.e. are locally homeomorphisms (see [12] and [2]). We shall often identify T with the space of units of $\mathcal{G}$.

For instance, a topological space T can be considered as an étale groupoid with space of units T and such that every arrow is a unit. A discrete group Γ can be considered as an étale groupoid with a single unit.

To an étale groupoid $(\mathcal{G}, T)$ is associated a pseudogroup $(\mathcal{H}, T)$ whose elements are precisely the homeomorphisms from open sets U of T to open sets of T obtained by composing a section of α above U with β. Conversely to each pseudogroup $(\mathcal{H}, T)$ is associated an étale groupoid with space of units T, called an étale groupoid of germs; its set of arrows is the set of germs of elements of $\mathcal{H}$ with the usual germ topology, the projection α and β associating to a germ its source and target. The pseudogroup associated to it is again $(\mathcal{H}, T)$. For this reason the notion of pseudogroups and the notion of étale groupoids of germs are equivalent notions, and we shall often use the same notation $(\mathcal{H}, T)$ for a pseudogroup or for its associated étale groupoid of germs.

We now extend to étale groupoids the notion of compact generation.

Definition. An étale groupoid $(\mathcal{G}, T)$ is *compactly generated* if

a) T is Hausdorff locally compact and contains a relatively compact open set T_0 meeting all the $\mathcal{G}$-orbits,

b) there is an open set $S \subset \mathcal{G}$ generating the restriction $\mathcal{G}_{|T_0}$ of $\mathcal{G}$ to T_0 and contained in a finite number of compact subsets[b].

It is easy to check that a pseudogroup $(\mathcal{H}, T)$ is compactly generated if and only if its associated groupoid of germs is compactly generated.

Remark. If $\mathcal{G}$ is Hausdorff (which is in general not the case, even if the associated groupoid of germs is Hausdorff), then S is relatively compact in $\mathcal{G}$.

Localization. Let $\mathcal{U} = \{U_i\}_{i \in I}$ be an open cover of T. Let $T_{\mathcal{U}}$ be the disjoint union of the U_i, i.e. the set of pairs (i, x) with $x \in U_i$. The

[b]We use compact in the Bourbaki sense, i.e. for us a compact space is always assumed to be Hausdorff. In general the union of two compact subsets is compact if and only if this union is Hausdorff.

localization $(\mathcal{G}_{\mathcal{U}}, T_{\mathcal{U}})$ of $(\mathcal{G}, T)$ over $\mathcal{U}$ is the étale groupoid whose elements are the triple (j, g, i) with $g \in \mathcal{G}, \alpha(g) \in U_i, \beta(g) \in U_j$. The projections α and β map (j, g, i) to $(i, \alpha(g))$ and $(j, \beta(g))$ respectively. The composition $(k, g', j)(j, g, i)$ whenever defined is equal to (k, gg', i).

Equivalence. Two étale groupoids $(\mathcal{G}, T)$ and $(\mathcal{G}', T')$ are *equivalent* if there exist open covers $\mathcal{U}$ of T and $\mathcal{U}'$ of T' such that $(\mathcal{G}_{\mathcal{U}}, T_{\mathcal{U}})$ is isomorphic to $(\mathcal{G}'_{\mathcal{U}'}, T'_{\mathcal{U}'})$.

Again it is an easy exercise to check that two pseudogroups are equivalent if and only if their associated étale groupoids are equivalent. Also an étale groupoid $(\mathcal{G}, T)$, with T Hausdorff, equivalent to a compactly generated groupoid, is also compactly generated. The analogue of the lemma in 1.3 is valid for compactly generated étale groupoids.

2.3 $\mathcal{G}$-sheaves

A $\mathcal{G}$-sheaf $\mathcal{A}$ is a sheaf of abelian groups on T with a continuous action of $\mathcal{G}$: for each $g \in \mathcal{G}$ an isomorphism $a \mapsto g.a$ from the stalk of $\mathcal{A}$ over $\alpha(g)$ to the stalk over $\beta(g)$, depending continuously on g and a. The $\mathcal{G}$-sheaves form naturally an abelian category (see Haefliger [12], Kumjian [16]).

As before we are interested only in notions invariant by equivalence. If $(\mathcal{G}', T')$ is equivalent to $(\mathcal{G}, T)$, there is an equivalence between the category of $\mathcal{G}$-sheaves and the category of $\mathcal{G}'$-sheaves. For instance, if $\mathcal{U} = \{U_i\}_{i \in I}$ is an open cover of T, the corresponding $\mathcal{G}_{\mathcal{U}}$-sheaf is the sheaf $\mathcal{A}_{\mathcal{U}}$ whose stalk above (i, x), $x \in U_i$, is $(i, \mathcal{A}_x)$. The action of (j, g, i) on (i, a) where $a \in \mathcal{A}_{\alpha(g)}$ is equal to $(j, g.a)$.

$H^0(\mathcal{G}, \mathcal{A})$ will be the group of $\mathcal{G}$-invariant sections of $\mathcal{A}$.

Example. Let $(\mathcal{G}, T)$ be an étale groupoid of germs of analytic local automorphisms of a complex curve T. We can consider the short exact sequence of $\mathcal{G}$-sheaves

$$0 \to \Omega^1 \to \Omega^1_{\log} \to \mathcal{D} \otimes C \to 0$$

where Ω^1 (resp. $\Omega^1_{\log}$) is the $\mathcal{G}$-sheaf of germs of holomorphic 1-forms (resp. meromorphic 1-forms with poles of order ≤ 1) on T, and $\mathcal{D} \otimes C$ the sheaf of germs of complex valued functions with discrete support. The third homomorphism in the exact sequence associates to a pole of a meromorphic 1-form its residue. The action of $\mathcal{G}$ is the natural action by pull-back.

2.4 Čech Cohomology

For a $\mathcal{G}$-sheaf $\mathcal{A}$, we define the cochain complex [12]

$$0 \to C^0(\mathcal{G},\mathcal{A}) \xrightarrow{\delta^0} C^1(\mathcal{G},\mathcal{A}) \xrightarrow{\delta^1} C^2(\mathcal{G},\mathcal{A}) \to \dots$$

Here $C^0(\mathcal{G},\mathcal{A})$ is the group of sections of $\mathcal{A}$ and $C^k(\mathcal{G},\mathcal{A})$ is the group of continuous functions c associating to a sequence $g_1,\dots,g_k$ of composable elements of $\mathcal{G}$ an element $c(g_1,\dots,g_k) \in \mathcal{A}_{\beta(g_1)}$.

The homomorphism δ^0 associates to a section c of $\mathcal{A}$ the 1-cochain $\delta^0 c(g) = g.c(g) - c(\beta(g))$. The homomorphism $\delta^1 : C^1(\mathcal{G},\mathcal{A}) \to C^2(\mathcal{G},\mathcal{A})$ is defined by

$$\delta^1 c(g_1,g_2)) = g_1.c(g_2) - c(g_1g_2) + c(g_1).$$

$\hat{H}^k(\mathcal{G},\mathcal{A})$ denote the k-th cohomology group of the above cochain complex. It is *not* invariant under equivalence. To remedy this, we pass to a limit. If $\mathcal{U} = \{U_i\}_{i\in I}$ is an open cover of T and $\mathcal{V} = \{V_j\}_{j\in J}$ is a finer open cover, after choosing for each $j \in J$ and element $\tau(j) \in I$ such that $V_j \subseteq U_{\tau(j)}$, we can define a natural map of cochain complexes $C^*(\mathcal{G}_\mathcal{U},\mathcal{A}_\mathcal{U}) \to C^*(\mathcal{G}_\mathcal{V},\mathcal{A}_\mathcal{V})$. As usual, one checks that the homomorphism induced on cohomology is independent of the choice of τ. So the next definition makes sense and is clearly invariant under equivalence.

Definition. The Čech cohomology $\check{H}^k(\mathcal{G},\mathcal{A})$ is defined as

$$\check{H}^k(\mathcal{G},\mathcal{A}) = \lim \hat{H}^k(\mathcal{G}_\mathcal{U},\mathcal{A}_\mathcal{U})$$

where the limit is taken over the open covers $\mathcal{U}$ of T.

In particular $\check{H}^0(\mathcal{G},\mathcal{A}) = \hat{H}^0(\mathcal{G},\mathcal{A})$ is the group $H^0(\mathcal{G},\mathcal{A})$ of invariant sections of $\mathcal{A}$.

Note that, when the étale groupoid is the trivial groupoid equal to its space of units T, then our definition agrees with the usual definition of the Čech cohomology $\check{H}^k(T,\mathcal{A})$. The proof of the following facts is like their proof in the particular case of the classical Čech cohomology.

Facts.

1) A short exact sequence $0 \to \mathcal{A} \to \mathcal{B} \to \mathcal{C} \to 0$ induces an exact sequence

$$\begin{aligned} 0 \to \check{H}^0(\mathcal{G},\mathcal{A}) \to \check{H}^0(\mathcal{G},\mathcal{B}) &\to \check{H}^0(\mathcal{G},\mathcal{C}) \to \check{H}^1(\mathcal{G},\mathcal{A}) \\ &\to \check{H}^1(\mathcal{G},\mathcal{B}) \to \check{H}^1(\mathcal{G},\mathcal{C}) \end{aligned}$$

2) The natural map $\hat{H}^1(\mathcal{G},\mathcal{A}) \to \check{H}^1(\mathcal{G},\mathcal{A})$ is always injective. If $H^1(T,\mathcal{A}) = 0$, it is bijective (Leray).

We prove only the last part of 2). We want to prove that, for every open cover $\mathcal{U} = \{U_i\}_{i\in I}$ of T, the homomorphism $\hat{H}^1(\mathcal{G},\mathcal{A}) \to \hat{H}^1(\mathcal{G}_\mathcal{U},\mathcal{A}_\mathcal{U})$ is surjective.

Let $z \in C^1(\mathcal{G}_\mathcal{U},\mathcal{A}_\mathcal{U})$ be a 1-cocycle; the value of z on (i,g,j) is of the form $(i,a(i,g,j))$, where $a(i,g,j) \in \mathcal{A}_{\beta(g)}$. The cocycle condition means that, for composable elements $(i,g,j),(j,g',k) \in \mathcal{G}_\mathcal{U}$, we have

$$a(i,gg',k) = g.a(j,g',k) + a(i,g,j).$$

We can consider $a(i,1_x,j)$ as a Čech 1-cocycle in $C^1(\mathcal{U},\mathcal{A})$. By hypothesis it is a coboundary (because $H^1(\mathcal{U},\mathcal{A}) \to \check{H}^1(T,\mathcal{A}) = 0$ is injective). Therefore for each $i \in I$, there is a section l_i of $\mathcal{A}$ above U_i such that, for $x \in U_i \cap U_j$, we have $a(i,1_x,j) = l_i(x) - l_j(x)$. The 1-cocycle $Z((i,g,j)) = (i,a(i,g,j)) - l_i(\beta(g)) + l_j(\alpha(g)))$ is cohomologous to z. For $g = 1_x$ we have $Z((i,1_x,j)) = 0$. We claim that $Z((i,g,j))$ depends only on g; this will imply that Z is the image of a 1-cocycle in $C^1(\mathcal{G},\mathcal{A})$. To check the claim, we note that, if $\alpha(g) \in U_j \cap U_{j'}$ and $\beta(g) \in U_i \cap U_{i'}$, the cocycle condition implies

$$Z((i',g,j')) = Z((i',1_{\beta(g)},i)) + Z((i,g,j)) + g.Z((j,1_{\alpha(g)},j')) = Z((i,g,j)).$$

□

Remark. There is a natural homomorphism from $\check{H}^*(\mathcal{G},\mathcal{A})$ to $H^*(\mathcal{G},\mathcal{A})$, the cohomology groups defined in Haefliger [12] (isomorphic to the groups defined in Kumjian [16] and called Grothendieck cohomology groups). Like in the case of usual Čech cohomology for topological spaces, this homomorphism is bijective for $* = 0,1$ and injective for $* = 2$

2.5 A finiteness theorem

The proof of Theorem A' follows the proof of Theorem A. Using the short exact sequence given in 2.3 and the associated cohomology exact sequence (see 2.4), it remains to prove the following finiteness theorem applied to the $\mathcal{G}$-sheaf $\mathcal{A} = \Omega^1$.

Theorem. *Let $(\mathcal{G},T)$ be a holomorphic compactly generated étale groupoid and let $\mathcal{A}$ be the $\mathcal{G}$-sheaf of germs of holomorphic sections of a holomorphic vector bundle over T with a (holomorphic) action of $\mathcal{G}$. Then*

$$\dim \check{\mathrm{H}}^*(\mathcal{G},\mathcal{A}) < \infty \quad \text{for } * = 0,1.$$

Proof. Using the facts mentioned above, the proof is the same as the classical proof of the finiteness of the first cohomology group of a compact complex manifold with coefficient in a locally free sheaf (see for instance paragraph 29 and Appendix B in the book of Foster [7] and the references therein).

After localization, we can assume that T is the disjoint union of balls in $\mathbb{C}^n$. This implies that $H^1(T, \mathcal{A}) = 0$. We can also assume that the vector bundle over T is trivial.

Let T_0 be an open relatively compact subset meeting all the orbits of $\mathcal{G}$, and choose an open set S in $\mathcal{G}$ contained in the union of a finite number of compact subsets and generating $\mathcal{G}_{|T_0}$ (see the definition in 2.2).

On the vector spaces $C^*(\mathcal{G}, \mathcal{A})$ or $C^*(\mathcal{G}_{|T_0}, \mathcal{A}_{|T_0})$, we consider the topology of the convergence on compact sets (they are Fréchet spaces). The subspace of cocycles $Z^*(\mathcal{G}, \mathcal{A})$ is closed.

Note that a 1-cocycle $z \in Z^1(\mathcal{G}_{|T_0}, \mathcal{A}_{|T_0})$ is determined by its restriction to S, because

$$z(g_1 g_2) = g_1.z(g_2) + z(g_1).$$

We have the following commutative diagram where the horizontal arrows are induced by the inclusion:

$$\begin{array}{ccc}
C^0(\mathcal{G}, \mathcal{A}) & \longrightarrow & C^0(\mathcal{G}_{|T_0}, \mathcal{A}_{|T_0}) \\
\downarrow & & \downarrow \\
Z^1(\mathcal{G}, \mathcal{A}) & \xrightarrow{\rho} & Z^1(\mathcal{G}_{|T_0}, \mathcal{A}_{|T_0}) \\
\downarrow & & \downarrow \\
\hat{H}^1(\mathcal{G}, \mathcal{A}) & \longrightarrow & \hat{H}^1(\mathcal{G}_{|T_0}, \mathcal{A}_{|T_0}) \\
\downarrow & & \downarrow \\
0 & & 0
\end{array}$$

The second horizontal arrow ρ is a compact operator due to Montel theorem and the remark above. The last one is an isomorphism by the fact 2) proved in 2.4. Therefore

$$\rho \oplus \delta^0 : C^0(\mathcal{G}_{|T_0}, \mathcal{A}_{|T_0}) \oplus Z^1(\mathcal{G}, \mathcal{A}) \to Z^1(\mathcal{G}_{|T_0}, \mathcal{A}_{|T_0})$$

is surjective. A theorem of Schwartz implies that $\delta^0 = (\rho + \delta^0) - \rho$ has finite codimension, i.e $\check{H}^1(\mathcal{G}, \mathcal{A}) = \hat{H}^1((\mathcal{G}_{|T_0}, \mathcal{A}_{|T_0})$ is finite dimensional. □

Remarks. 1) Let $\mathcal{F}$ be a transversely holomorphic foliation on a manifold M and let $(\mathcal{H}, T)$ be the groupoid of germs of its holonomy pseudogroup. Consider on M the sheaf $\Theta_{\mathcal{F}}$ of germs of transversely holomorphic vector fields on M and let Θ be the $\mathcal{H}$-sheaf of germs of holomorphic vector fields on T. It follows from 3.2 below that the group $\check{H}^1(\mathcal{H}, \Theta)$ is naturally isomorphic to a subgroup of $H^1(M, \Theta_{\mathcal{F}})$.

2) Let $(\mathcal{G}, T)$ be a compactly generated holomorphic étale groupoid; this means that T is a complex manifold and that the associated pseudogroup

(see 2.2) is made up of holomorphic transformations of T. Let Θ be the $\mathcal{G}$-sheaf of germs of holomorphic vector fields on T. One should prove the existence of a germ of a complex space, whose Zariski tangent space is isomorphic to the finite dimensional vector space $H^1(\mathcal{G}, \Theta)$, which is versal for the germs of deformations of $(\mathcal{G}, T)$ in an appropriate sense [1] among holomorphic groupoids.

2.6 Holomorphic dynamic in dimension one

Let $\mathcal{F}$ be a transversely holomorphic foliation on a closed manifold M. E.Ghys, X. Gomez-Mont and J. Saludes [8] have constructed a decomposition of M in dynamically defined components analogous to the decomposition into Fatou and Julia sets for iteration of rational functions. They noticed that this decomposition comes from a decomposition of the holonomy pseudogroup of $\mathcal{F}$ and that their construction works for compactly generated pseudogroups of holomorphic transformations of a complex curve. We briefly describe part of their results in this framework.

Let $(\mathcal{H}, T)$ be a compactly generated pseudogroup of local holomorphic automorphisms of a complex manifold T of dimension one. The complexification $\tau \otimes \mathbb{C}$ of the tangent bundle τ of T splits as the direct sum $\tau^{1,0} \oplus \tau^{0,1}$ of two complex line bundles on which $\mathcal{H}$ acts through the differential. The Beltrami differentials are the sections of the bundle $\tau^{1,0} \oplus \tau^{0,1^*}$. Let $\mathcal{C}(\tau^{1,0})$ be the sheaf of germs of local sections σ of $\tau^{1,0}$ with distributional derivatives locally in L^2 such that $\overline{\partial}\sigma$ is an essentially bounded measurable Beltrami differential.

The $\mathcal{H}$-invariant sections of $\mathcal{C}(\tau^{1,0})$ correspond to $\mathcal{H}$-invariant vector fields σ on T and the hypothesis of compact generation implies that the local flow ϕ_t generated by σ is $\mathcal{H}$-complete. This means that, for any $x \in T$ and $s > 0$, there is a partition $0 = t_0 < t_1 < \ldots < t_k = s$, a sequence $x = x_0, \ldots, x_k$ of points of T and elements $h_0, \ldots, h_{k-1}$ of $\mathcal{H}$ such that, for $i = 0, \ldots, k-1$, the local flow $\phi_{t_{i+1}-t_i}(h_i(x_i))$ is defined and is equal to x_{i+1}. The set of points x of T where the section σ does not vanish is an open $\mathcal{H}$-invariant subset U and $\mathcal{H}$ preserves on U the parallelism $x \mapsto (\sigma(x), \sqrt{-1}\sigma(x))$. The $\mathcal{H}$-completeness of σ implies that the restriction of $\mathcal{H}$ to U is complete in the sense of Salem [20] and one can apply the analogue of Molino theory to the complete pseudogroup $\mathcal{H}_{|U}$ [20].

The *Julia set* of $\mathcal{H}$ is by definition the closed set of points $x \in T$ such that any $\mathcal{H}$-invariant section of $\mathcal{C}(\tau^{1,0})$ vanishes at x. Its complement is the *Fatou set* of $\mathcal{H}$. The Fatou set is an $\mathcal{H}$-invariant open set of T and its connected components are called Fatou components. For a Fatou component U, there are three exclusive cases for the restriction $\mathcal{H}_{|U}$ to U :

1) $\mathcal{H}_{|U}$ has closed orbits; it is equivalent to the pseudogroup generated by the identity map of the space of orbits, a compact Riemann surface minus a finite number of points.

2) $\mathcal{H}_{|U}$ is equivalent to a pseudogroup generated by a subgroup Γ of the group $G = \mathbb{C}$ or Aff_1 acting by complex automorphisms on some strip $\{z \in \mathbb{C} \mid \alpha < \Im(z) < \beta\}$, where $-\infty \leq \alpha < \beta \leq +\infty$ and the closure of Γ is one dimensional.

3) $\mathcal{H}_{|U}$ is equivalent to the pseudogroup generated by a dense subgroup of the Lie group $G = \mathbb{C}$ or Aff_1 acting by left translations on itself (the group Aff_1 has a left invariant complex structure isomorphic to the upper halfspace).

3 The fundamental groupoid and compact presentation

3.1 The fundamental (or monodromy) groupoid of a foliation

Let $\mathcal{F}$ be a foliation of codimension q on a manifold M and let T be a total transversal to $\mathcal{F}$, i.e. a submanifold of M of dimension q transverse to the leaves and cutting every leaf. The fundamental groupoid $(\mathcal{G}, T)$ of $\mathcal{F}$ relative to T is the étale groupoid defined as follows [1]. The set of elements of $\mathcal{G}$ with source a point $x \in T$ and target a point $y \in T$ is non empty if and only if x and y are in a same leaf L; it is the set of homotopy classes in L of continuous paths $c : [0,1] \to L$ joining y to x. Composition and inverse in $\mathcal{G}$ are induced by the corresponding operations for paths. A basis of open sets for the topology on $\mathcal{G}$ is obtained as follows. Let U be an open subset of T and let $c : U \times [0,1] \to M$ be a continuous map such that, for each $u \in U$, the map $c_u : [0,1] \to M$ sending t to $c(u,t)$ is a path in a leaf joining u to a point of T; then the set of homotopy classes in the leaves of the paths c_u is an open set of the basis.

Let $(\mathcal{H}, T)$ be the associated étale groupoid of germs (cf. 2.2). It is equivalent to the holonomy groupoid of $\mathcal{F}$ and is called the holonomy groupoid of $\mathcal{F}$ associated to the transversal T. The kernel of the natural projection from $\mathcal{G}$ to $\mathcal{H}$ is a sheaf of groups on T. Its stalk above $x \in T$ is the fundamental group of the holonomy covering of the leaf through x. (The fundamental groupoid as defined above is the restriction to T of the Lie groupoid called by several people the monodromy groupoid of $\mathcal{F}$.)

The fundamental groupoid plays the leading role in the study of germs of deformations of foliations (cf. Bonatti-Haefliger [1]).

We now describe a *presentation of the fundamental groupoid* $(\mathcal{G}, T)$ of $\mathcal{F}$. We can always choose a foliated cocycle $\{p_i : U_i \to T_i\}_{i \in I}$ for $\mathcal{F}$ satisfying the properties mentionned in the remark in 1.2. Moreover we can assume that the fibers (plaques) of each p_i are simply connected and that there is an embedding $q_i : T_i \to U_i$ such that $p_i q_i$ is the identity of T_i and such that the $q_i(T_i)$ are disjoint. Recall that there is a unique homeomorphism h_{ij} from $p_j(U_i \cap U_j)$ to $p_i(U_i \cap U_j)$ such that $p_i(x) = h_{ij}(p_j(x))$ for each $x \in U_i \cap U_j$. To simplify the notations, we identify T_i to its image $q_i(T_i)$ so that p_i can be considered as a retraction from U_i to T_i and the union T of the T_i as a total transversal to $\mathcal{F}$.

Let $(\mathcal{G}, T)$ be the fundamental groupoid of $\mathcal{F}$ with respect to the transversal T. We now describe a set S of generators $g_{ij}(y) \in \mathcal{G}$ for $(\mathcal{G}, T)$, where $i, j \in I$ and $y \in p_j(U_i \cap U_j)$. The element $g_{ij}(y)$ is the homotopy class of a path in the simply connected subset $p_j^{-1}(y) \cup p_i^{-1}(h_{ji}(y))$ of the leaf through y from $h_{ij}(y)$ to y.

Each element of $\mathcal{G}$ can be expressed as a composition of elements of S. For $x \in U_i \cap U_j \cap U_k$, we have the relation

$$g_{ik}(p_k(x)) = g_{ij}(p_j(x))g_{jk}(p_k(x)).$$

The set R of all such relations is complete, i.e. two words in the elements of S represent the same element of $\mathcal{G}$ if and only one can pass from one to the other by using the relations in R (see Higgins [15]). This follows from the fact that the nerve of the open cover of a leaf L by the plaques contained in it has the same 1-homotopy type as L.

For a general discussion on presentations of étale groupoids, see 3.3.

3.2 Relations with the cohomology of M

As proved in Haefliger [13], the fundamental group of $(\mathcal{G}, T)$ is isomorphic to the fundamental group of M. In this section, we indicate a similar relation between first cohomology groups.

Let $\mathcal{F}$ be a foliation on a manifold M given by a foliated cocycle $p_i : U_i \to T_i$ over an open cover $\mathcal{U} = \{U_i\}_{i \in I}$ of M like in the preceding section and let $(\mathcal{G}, T)$ be the associated fundamental groupoid. Consider M as a trivial étale groupoid $(\mathcal{M}, M)$ with space of units M, where $\mathcal{M}$ is the space of germs of the identity map of M. Let $(\mathcal{M}_{\mathcal{U}}, M_{\mathcal{U}})$ be its localization over $\mathcal{U}$. There is a continuous homomorphism $p : (\mathcal{M}_{\mathcal{U}}, M_{\mathcal{U}}) \to (\mathcal{G}, T)$ mapping $(i, 1_x, j) \in \mathcal{M}_{\mathcal{U}}$, where $x \in U_i \cap U_j$, to $g_{ij}(p_j(x)) \in \mathcal{M}_{\mathcal{U}}$. Let $\mathcal{A}$ be a $\mathcal{G}$-sheaf over T; its pull back $p^*\mathcal{A}$ is a $\mathcal{M}_{\mathcal{U}}$-sheaf over the disjoint union $M_{\mathcal{U}}$ of the U_i. The stalk of $p^*\mathcal{A}$ over $(i, x) \in M_{\mathcal{U}}$ is the group of triples (i, x, a), where $x \in U_i,\ a \in \mathcal{A}_{p_i(x)}$. The action of $(j, 1_x, i) \in \mathcal{M}_{\mathcal{U}}$ on (i, x, a) is given by

$$(j, 1_x, i).(i, x, a) = (i, x, g_{ji}(p_i(x)).a).$$

In fact this sheaf is obtained by localization over $\mathcal{U}$ of a sheaf $\mathcal{A}_{\mathcal{F}}$ on M. The homomorphism p induces a natural homomorphism of chain complexes

$$p^* : C^*(\mathcal{G}, \mathcal{A}) \to C^*(\mathcal{M}, p^*\mathcal{A})$$

mapping a cochain c to the cochain p^*c defined by

$$(p^*c)((i_0, 1_x, i_1), (i_1, 1_x, i_2), \dots) = (i_0, x, c(g_{i_0 i_1}(p_{i_1}(x)), g_{i_1 i_2}(p_{i_2}(x)), \dots)).$$

Note that $\check{H}^*(\mathcal{M}_{\mathcal{U}}, p^*\mathcal{A}) = \check{H}^*(M, \mathcal{A}_{\mathcal{F}})$, the usual Čech cohomology.

Proposition. *The homomorphisms*

$$\check{H}^*(\mathcal{G}, \mathcal{A}) \to \check{H}^*(M, \mathcal{A}_{\mathcal{F}})$$

induced by p are bijective for $* = 0, 1$.

This follows from the following lemma.

Lemma. *In the commutative diagram below the horizontal homomorphisms induced by p*

$$\begin{array}{ccc} C^0(\mathcal{G}, \mathcal{A}) & \longrightarrow & C^0(\mathcal{M}_{\mathcal{U}}, p^*\mathcal{A}) \\ \downarrow & & \downarrow \\ Z^1(\mathcal{G}, \mathcal{A}) & \longrightarrow & Z^1(\mathcal{M}_{\mathcal{U}}, p^*\mathcal{A}) \end{array}$$

are isomorphisms.

Proof. The first horizontal homomorphism is an isomorphism because the restriction of $p^*\mathcal{A}$ to a plaque is a constant sheaf and each plaque is connected.

To check the injectivity of the second homomorphism, let $z \in Z^1(\mathcal{G}, \mathcal{A})$ be such that $p^*(z) = 0$. This implies that

$$0 = p^*(z)((i, 1_x, j)) = (i, z(g_{ij}(p_j(x))),$$

hence $z = 0$, because a 1-cocycle which vanish on the generators is zero.

To prove the surjectivity, we note that a cochain $c \in C^1(\mathcal{M}_{\mathcal{U}}, p^*\mathcal{A})$ associates to $(i, 1_x, j)$ an element of $p^*\mathcal{A}$ of the form $(i, x, a_{ij}(p_j(x)))$, where a_{ij} is a section of $\mathcal{A}$ above $p_j(U_i \cap U_j)$, because the intersection of two plaques is connected. If c is a cocycle, then

$$0 = \delta c((i, 1_x, j), (j, 1_x, k)) = (i, 1_x, j).c((j, 1_x, k)) - c((i, 1_x, k) + c((i, 1_x, j)),$$

i.e

$$a_{ik}(p_i(x)) = g_{ij}(p_j(x)).a_{jk}(p_k(x)) + a_{ij}(p_j(x)).$$

A cocycle $z \in Z^1(\mathcal{G}, \mathcal{A})$ such that $p^*(z) = c$ is defined as follows. For any $g \in \mathcal{G}$, we can find a sequence of indices $i_0, \dots, i_k$ and a sequence of points $x_1, \dots, x_k$ with $x_j \in U_{i_{j-1}} \cap U_{i_j}$ such that

$$g = g_{i_0 i_1}(p_{i_1}(x_1)) \dots g_{i_{k-1} i_k}(p_{i_k}(x_k)).$$

We define

$$z(g) = a_{i_0 i_1}(p_{i_1}(x_1)) + \ldots + g_{i_0 i_1}(p_{i_1}(x_1)) \\ \ldots g_{i_{k-2} i_{k-1}}(p_{i_{k-1}}(x_k)).a_{i_{k-1} i_k}(p_{i_k}(x_k)).$$

This definition is independent of the choice of generators in the expression of g thanks to the relations.

3.3 Compactly presented étale groupoids

We first extend to étale groupoids the construction of groupoids given in Higgins [15] using a presentation by generators and relations.

Given a topological space T, we consider a topological space S, called an alphabet over T, endowed with two étale maps $\alpha : S \to T$ and $\beta : S \to T$. We also consider a topological space S' disjoint from S with a homeomorphism $s \mapsto (s)'$ from S to S' and two étale projections $\alpha, \beta : S' \to T$ defined by $\alpha((s)') = \beta(s)$ and $\beta((s)') = \alpha(s)$. For $(s)' \in S'$ we also define $((s)')' = s$.

Let $W^{(1)}$ be the union of S and S', endowed with the union of the projections α and β to T. For a positive integer n, we consider the topological space $W^{(n)} = W^{(1)} \times_T \ldots \times_T W^{(1)}$ of sequences $(s_1, \ldots, s_n)$ (also noted $s_1 \ldots s_n$) of n elements of $W^{(1)}$ such that $\beta(s_{i+1}) = \alpha(s_i)$. The maps $\alpha : W^{(n)} \to T$ (resp. $\beta : W^{(n)} \to T$) associating to such a sequence the point $\alpha(s_n)$ (resp. $\beta(s_1)$) are étale maps. A point in $W^{(n)}$ is called a word of length n in the alphabet S. We also consider the space $W^{(0)}$ of empty words homeomorphic to T, the point $x \in T$ corresponding to the empty word $\{\ \}_x$ at x, the projections α and β from $W^{(0)}$ mapping the empty word at x to x. The space of words in the alphabet S is the union W_S of the $W^{(n)}$, $n \geq 0$, endowed with the two étale projections α and β to T which are the union of the corresponding projections from the $W^{(n)}$.

We have a continuous étale map $W^{(n)} \times_T W^{(m)} \to W^{(n+m)}$ associating to (w_1, w_2) with $\alpha(w_1) = \beta(w_2)$ the word $w_1 w_2$ obtained by juxtaposition. $W^{(0)}$ acts by the identity. With this partial multiplication W_S is a topological category. If ϕ is an étale map from S to an étale groupoid $(\mathcal{G}, T)$ such that $\alpha\phi = \alpha$ and $\beta\phi = \beta$, it extends to a continuous functor $\Phi : W_S \to \mathcal{G}$ which induces the identity on T. Let $R \subseteq W_S$ be an open subset such that for each $r \in W_S$ we have $\alpha(r) = \beta(r) = x_r$. If for each $r \in R$ we have $\Phi(r) = 1_{x_r}$, we say that the relations R hold in $\mathcal{G}$ under ϕ.

Given a word $w = s_1 \ldots s_n \in W^{(n)}$, where $n \geq 2$ and such that for some $k < n$ we have $s_{k+1} = (s_k)'$, we get a word in $W^{(n-2)}$ by suppressing in w the elements s_k and s_{k+1}. This operation generates in W_S an equivalence relation which is open. The quotient of W_S by this equivalence relation is an étale groupoid $(\mathcal{G}_S, T)$, called the *free étale groupoid* over S. Any étale map ϕ from S to an étale groupoid $(\mathcal{G}, T)$ such that $\alpha\phi = \alpha$ and $\beta\phi = \beta$ extends uniquely to a continuous homomorphism from $(\mathcal{G}_S, T)$ to $(\mathcal{G}, T)$ projecting on the identity of T. If $\Phi(S)$ generates $\mathcal{G}$ and if the kernel of Φ is the normal subgroupoid N generated by R, then $\mathcal{G}$ is isomorphic to the quotient $\mathcal{G}_S/N$ and (S, ϕ, R) is called a presentation of $\mathcal{G}$. Note that the quotient $\mathcal{G}_S/N$ is naturally an étale groupoid because N is open in $\mathcal{G}$; it will also be noted $\langle S; R\rangle$.

We indicate a tentative definition of the notion of compact presentation for an étale groupoid.

Compact presentation. An étale groupoid $(\mathcal{G}, T)$ is said to be compactly presented if:

a) T is Hausdorff locally compact and contains a relatively compact open set T_0 meeting all the $\mathcal{G}$-orbits.

b) There is an open subset $S \subseteq \mathcal{G}$ generating the restriction $\mathcal{G}_{|T_0}$ of $\mathcal{G}$ to T_0 and contained in the union of a finite number of compact subsets. Let $\widetilde{S}$ be an open set containing this union.

c) In the space W_S of words in the alphabet S, identified to an open subset of $W_{\widetilde{S}}$, there is an open set R which is contained in the union of a finite number of compact sets in $W_{\widetilde{S}}$, and such that $(\mathcal{G}, T)$ is isomorphic to $\langle S; R\rangle$.

One checks as in 1.3 that the fundamental groupoid of a foliation on a compact manifold is compactly presented, and that the notion of compact presentation depends only on the equivalence class.

A test for this definition would be to prove the analogue of the global stability theorem of Reeb for compactly presented groupoid of dimension one.

References

1. C. Bonatti and A. Haefliger, *Déformations de feuilletages*, Topology **29** (1990), 205–229.
2. M. Bridson and A. Haefliger, Metric spaces of non-positive curvature, Grundlehren der mathematischen Wissenschaften 319, Springer Verlag, 1999.
3. M. Brunella and M. Nicolau, *Sur les hypersurfaces solutions des équations de Pfaff*, C. R. Acad. Sc. Paris, **329** (1999), 793–795.
4. V. Cavalier, *Pseudogroupes complexes quasi parallélisés de dimension un*, Ann. Inst. Fourier (Grenoble) **44** (1994), 1539–1565.

5. D.B.A. Epstein, K.C. Millet and D. Tischler, *Leaves without holonomy*, J. London Math. Soc. **16** (1977), 548–552.
6. R. Edwards, K. Millett and D. Sullivan, *Foliations with all leaves compact*, Topology **16** (1997), 13–32.
7. O. Forster, Riemannsche Flächen, Heidelberger Taschenbücher Band 184, Springer Verlag, 1977.
8. E. Ghys, X. Gomez-Mont and J. Saludes, *Fatou and Julia components of transversely holomorphic foliations*, preprint ENS-Lyon (2000).
9. X. Gomez-Mont, *Transversal holomorphic structures*, J. Diff. Geom. **15** (1980), 161–185.
10. A. Haefliger, *Variétés feuilletées*, Ann. Scuola Norm. Sup. Pisa **16** (1962), 367–397.
11. A. Haefliger, *Homotopy and Integrability*, in Manifolds, Amsterdam 1970, Lecture Notes in Math. **197**, Springer Verlag, 1971, 133–163.
12. A. Haefliger, *Differentiable Cohomology*, in Differential Topology (Varenna 76), Liguori, Naples, 1979, 19–70.
13. A. Haefliger, *Groupoides d'holonomie et classifiants*, in Structure transverse des feuilletages, Toulouse 1982, Astérisque **116** (1984), 70–97.
14. A. Haefliger, *Pseudogroups of local isometries*, in Differential Geometry, Santiago de Compostela, Pitman Res. Notes in Math. **131**, 1994, 174–197.
15. P.J. Higgins, *Presentations of groupoids and applications to groups*, Proc. Camb. Phil. Soc. **60** (1964), 7–20.
16. A. Kumjian, *On equivariant sheaf cohomology and elementary C^*-bundles*, J. of Operator Algebra **20** (1998), 207–240.
17. G. Meigniez, *Holonomy groups of solvable foliations*, in Integrable systems and foliations, Progress in Mathematics 145, Birkhäuser 1997, 107–146.
18. P. Molino, Riemannian Foliations, Progress in Mathematics 73, Birkkhäuser 1988.
19. J.F. Plante, *Foliations with measure preserving holonomy*, Ann. of Math. **102** (1975), 327–361.
20. E. Salem, *Riemannian foliations and pseudogroups*, in P. Molino, Riemannian foliations, Progress in Mathematics 73, Birkkhäuser 1988, 265–296.

Received November 16, 2000, revised December 28, 2000.

Proceedings of
FOLIATIONS: GEOMETRY AND DYNAMICS
held in Warsaw, May 29–June 9, 2000
ed. by Paweł WALCZAK *et al.*
World Scientific, Singapore, 2002
pp. 297–314

TRACES AND INVARIANTS FOR NON-COMPACT MANIFOLDS

JAMES L. HEITSCH
Mathematics, Statistics, and Computer Science,
University of Illinois at Chicago, 851 S. Morgan Street (M/C 249),
Chicago, Illinois 60607-7045, USA, e-mail: heitsch@math.uic.edu

We define a general trace for complete Riemannian manifolds of bounded geometry. This trace takes values in the Haefliger functions of the manifold and its existence allows the extension of the classical theory of invariants for compact manifolds to complete Riemannian manifolds of bounded geometry.

1 Introduction

The heat equation method for invariants of manifolds is now quite well developed and classical results (index theory, Lefschetz theory, Betti numbers, Novikov-Shubin invariants, torsion, ...) can be extended to new situations (foliations, non-compact manifolds) provided that an appropriate trace function can be defined, and that the asymptotics of the heat operator as $t \to \infty$ (which is in general a global problem) can be handled. The asymptotics of the heat operator as $t \to 0$ present no problem as these are local.

The L^2 Index Theorem of Atiyah [1] is a classical example of this sort of extension. The case considered there is a covering of a compact manifold and an operator lifted to the cover from the compact base. The trace is given by integration over a fundamental domain of the covering. In [18], John Roe provided an extension to non compact manifolds which admit a regular exhaustion. The exhaustion allows one to define a trace by averaging over the manifold, much as we do in the final section of this paper. Alain Connes extended the heat equation method to foliations which admit a transverse invariant measure. See [10] and the references therein. Here the trace is given by integration over the plaques of a cover by foliation charts with the resulting function being integrated against the transverse measure to obtain a real number. In [12] and [14] we showed how to remove the requirement that the foliation admit an invariant transverse measure and obtained a general families index theorem. The trace in this case is given by integrating over the plaques and taking the Haefliger cohomology class of the resulting differential form on the transverse space.

In this paper we define a general trace (the Haefliger trace) for any complete Riemannian manifold M of bounded geometry. This trace takes values in $C_h(M)$ the space of Haefliger functions of M, which is a quotient space of the space of functions on a maximal uniformly separated discrete subset. The basic idea is to abstract the theory for foliations worked out in [13] to the case of a single leaf. The main innovation in this paper is the introduction of the Haefliger trace taking values in the space of Haefliger functions. As an example of the use of this trace, we outline a proof of a Lefschetz theorem for Dirac operators defined on any complete Riemannian manifold of bounded geometry.

We note here that $C_h(M)$ agrees with the uniformly finite coarse homology group $H_0^{uf}(M)$ defined by Block and Weinberger, [7], and that the application of uniformly finite coarse homology to index theory was anticipated in section 9 of [8]. The main idea of this paper is that there is a naturally defined *geometric* trace taking values in $C_h(M)$, and given this, the geometric arguments of [13] and [14] are readily translated to this setting, giving index theorems, Lefschetz theorems, etc. This paper may be seen as alternate approach to invariants for non-compact manifolds which provides immediate access to results not readily obtainable from the Block-Weinberger view: e.g. Lefschetz Theory.

2 Haefliger functions and the trace

In this section, we define the space of Haefliger functions for a complete Riemannian manifold of bounded geometry and a trace which takes values in this space of functions.

2.1 Haefliger functions

Denote by M a complete Riemannian manifold of bounded geometry. This class of manifolds includes all compact manifolds and their covers as well as any leaf (and its covers) of any foliation of a compact manifold. There are also examples of such manifolds which cannot be a leaf of a foliation of any compact manifold, [6], [16].

Let $r > 0$. A *lattice* Γ for M is cover by open balls of radius r whose centres are at least a distance r apart. The bounded geometry of M then implies that there is an integer k so that any element of Γ intersects at most k other elements of Γ non-trivially.

To construct a lattice, let $A \subset M$ be a maximal set of distinct points which are separated by a distance of at least r. Zorn's lemma implies that such a set must exist. Then let Γ be the set of balls of radius r whose centres are the elements of A.

For any lattice Γ set

$$C^0(\Gamma) = \{f : \Gamma \to \boldsymbol{R} \,|\, f \text{ is bounded}\}.$$

The space of Haefliger functions on M is a quotient space of this space. Denote by $d(\cdot,\cdot)$ the distance function on M. A permutation π of Γ is *bounded* if

$$\sup_{U\in\Gamma} d(U, \pi(U)) < \infty.$$

The group G of all bounded permutations acts on $C^0(\Gamma)$ by composition, that is

$$(\pi^* f)(U) = f(\pi^{-1}(U)).$$

Set

$$\mathcal{H}_\Gamma = \text{ linear span of } \{f - \pi^* f \,|\, f \in C^0(\Gamma), \pi \in G\}.$$

The space of *Haefliger functions* $C_h(M)$ for M is the space

$$C_h(M) = C^0(\Gamma)/\overline{\mathcal{H}}_\Gamma,$$

where $\overline{\mathcal{H}}_\Gamma$ is the sup norm closure of $\mathcal{H}_\Gamma$. Denote the class of f in $C_h(M)$ by $[f]$. Then $[f] = 0$ provided that for all $\epsilon > 0$ there is a representative $f_1 \in [f]$ so that $\sup\limits_{U\in\Gamma} |f_1(U)| < \epsilon$.

Note that if $f(U) = -f(\pi(U))$ for all $U \in \widehat{\Gamma} \subseteq \Gamma$, and $\widehat{\Gamma} \cap \pi(\widehat{\Gamma}) = \emptyset$, then we may set $f(U) = 0$ for all $U \in \widehat{\Gamma} \cup \pi(\widehat{\Gamma})$ without changing the Haefliger class of the function f. To see this, define $g(U) = f(U)$ for $U \in \widehat{\Gamma}$ and $g(U) = 0$ otherwise. Then $[g - \pi^* g] = 0$, and $f_1 = f - (g - \pi^* g)$ satisfies $f_1(U) = 0$ for all $U \in \widehat{\Gamma} \cup \pi(\widehat{\Gamma})$, and $f_1(U) = f(U)$ otherwise.

A priori, $C_h(M)$ depends on the choice of lattice Γ. However, this is not the case.

Theorem 1 *$C_h(M)$ does not depend on Γ.*

Proof. Let Γ_1 and Γ_2 be two lattices for M and set $\Gamma = \Gamma_1 \cup \Gamma_2$. Now Γ may not be a lattice as the centres of distinct balls may be arbitrarily close, but there is still a k so that each element of Γ meets at most k other elements of Γ. It addition, $C^0(\Gamma)/\mathcal{H}_\Gamma$ still makes sense. With these observations, it is an easy exercise to show

$$C^0(\Gamma_1)/\mathcal{H}_{\Gamma_1} \simeq C^0(\Gamma)/\mathcal{H}_\Gamma \simeq C^0(\Gamma_2)/\mathcal{H}_{\Gamma_2}.$$

□

Similarly, one can show that $C_h(M)$ does not depend on r.

2.2 The trace

Given a bounded measurable function g on M and any partition of unity $\Psi = \{\psi_U\}$ subordinate to the cover of M given by the elements of the lattice Γ, define $\int g \in C^0(\Gamma)$ to be the function

$$\left(\int g\right)(U) = \int_U \psi_U(x) g(x)\, dx$$

where dx is the volume form on M. It follows easily from the definitions that the Haefliger function $[\int g]$ does not depend on Ψ.

For any bundle E, denote by $C(E)$ the bounded measurable sections of E, by $C^\infty(E)$ the bounded smooth sections, and by $C_0^\infty(E)$ the smooth sections with compact support. Denote by $E \otimes E^*$ the bundle $\pi_1^*(E) \otimes \pi_2^*(E^*)$ over $M \times M$, where $\pi_1, \pi_2 : M \times M \to M$ are the two projections. For any $k \in C(E \otimes E^*)$, $k(x,x)$ is a linear map on E_x and so has a well defined trace $\operatorname{tr}(k(x,x))$ which is a bounded measurable function on M. We define the *Haefliger trace* $\operatorname{Tr}(k)$ of k to be the Haefliger class of the function $\displaystyle\int \operatorname{tr}(k)$, that is

$$\operatorname{Tr}(k) = \left[\int \operatorname{tr}(k)\right].$$

(If M is compact, any $k \in C(E \otimes E^*)$ defines an operator on $C(E)$ by

$$(ks)(x) = \int_M k(x,y)s(y)\, dy,$$

and the Haefliger trace of k is the usual trace of k as in this case, $C_h(M) = \boldsymbol{R}$)

Denote by $C_s^\infty(E \otimes E^*)$ the set of $k \in C^\infty(E \otimes E^*)$ so that there is $s > 0$ with $k(x,y) = 0$ for all (x,y) with $d(x,y) \geq s$. In addition we require that k and all its derivatives be bounded on $M \times M$. Any $k \in C_s^\infty(E \otimes E^*)$

defines a bounded smoothing operator on $L^2(E)$ with finite propagation speed and $\mathrm{tr}(k(x,x))$ is a bounded smooth function on M.

An element $k \in C^\infty(E \otimes E^*)$ is *basic* if there are $\widehat{\Gamma} \subset \Gamma$ and $\pi \in G$ so that for all $U_1 \neq U_2 \in \widehat{\Gamma}$, $U_1 \cap U_2 = \emptyset$ and $\pi(U_1) \cap \pi(U_2) = \emptyset$ and

$$\sup(k) \subset \cup_{\widehat{\Gamma}} U \times \pi(U).$$

Note that if k is basic, then $\sup(k) \cap (U \times \pi(U))$ is a compact subset of the open set $U \times \pi(U)$.

Proposition 2 *Any $k \in C_s^\infty(E \otimes E^*)$ may be written as a finite sum $k = k_1 + \cdots + k_q$ where each k_i is basic.*

Proof. Choose $s > 0$ so that $k(x,y) = 0$ for all x and y with $d(x,y) > s$. Because of the bounded geometry of M and the fact that the centres of the elements of Γ are at least a distance r apart, Γ can be partitioned into a finite number of sets, $\widehat{\Gamma}_1, \dots \widehat{\Gamma}_n$, so that for all distinct $U_1, U_2 \in \widehat{\Gamma}_\ell$, $\ell = 1, \dots n$, $d(U_1, U_2) > 4r + 2s$. This of course implies that $U_1 \cap U_2 = \emptyset$. If $(U_1 \times U_3) \cap \sup(k) \neq \emptyset$, then $d(U_1, U_3) \leq s$ so it further guarantees that if $(U_1 \times U_3) \cap \sup(k) \neq \emptyset$ and $(U_2 \times U_4) \cap \sup(k) \neq \emptyset$ then $U_3 \cap U_4 = \emptyset$. This condition guarantees that the elements $\pi_{i,j} \in G$ chosen below actually exist.

Again by the bounded geometry of M and the fact that the centres of the elements of Γ are at least a distance r apart, there is an integer p so that for each $U \in \Gamma$ there are at most p elements of Γ whose product with U intersects $\sup(k)$ non-trivially. For a given U, denote these sets by $\widehat{U}_1, \dots, \widehat{U}_p$. (If there are fewer than p such sets, we complete the collection to a set of p elements by adding the empty set the required number of times). For $i = 1, \dots, n$, $j = 1, \dots, p$, choose $\pi_{i,j} \in G$ so that $\pi_{i,j}(U) = \widehat{U}_j$ for all $U \in \widehat{\Gamma}_i$. Let $\{\psi_U\}$ be a partition of unity subordinate to the cover given by Γ and set

$$k_{i,j}(x,y) = \sum_{U \in \widehat{\Gamma}_i} \psi_U(x) \psi_{\pi_{i,j}(U)}(y) k(x,y).$$

Then each $k_{i,j}$ is basic and $k = \sum_{i,j} k_{i,j}$. □

Let $k_1(x,y) \in C(E \otimes E^*)$, and $k_2(x,y) \in C_s^\infty(E \otimes E^*)$, and define $k_1 \circ k_2 \in C(E \otimes E^*)$ by

$$k_1 \circ k_2(x,y) = \int_M k_1(x,z) k_2(z,y)\, dz,$$

and similarly for $k_2 \circ k_1$. We have immediately

Proposition 3 *$k_1 \circ k_2$ and $k_2 \circ k_1$ are elements of $C(E \otimes E^*)$.*

We now come to the central result of the paper, namely the fact that Tr satisfies the following trace property. It is this property which allows us to extend classical results from compact manifolds to complete manifolds of bounded geometry.

Theorem 4 *Suppose* $k_1(x,y) \in C(E \otimes E^*)$, *and* $k_2(x,y) \in C_s^\infty(E \otimes E^*)$. *Then*

$$Tr(k_1 \circ k_2) = Tr(k_2 \circ k_1).$$

Proof. By Proposition 2 we may assume that k_2 is basic with corresponding $\pi, \widehat{\Gamma}$. Choose a partition of unity $\{\psi_U\}$ subordinate to Γ so that for all $U \in \widehat{\Gamma}$, $\psi_U \times \psi_{\pi(U)} \mid \sup(k_2) \cap (U \times \pi(U)) = 1$. Then the Haefliger function $\mathrm{Tr}(k_1 \circ k_2)$ is represented by the function $f \in C^0(\Gamma)$ whose value at $U \in \Gamma$ is

$$\begin{aligned} f(U) &= \int_U \psi_U(x) \mathrm{tr}\Big(\int_M k_1(x,y) k_2(y,x) dy \Big)\, dx \\ &= \int_U \int_M \mathrm{tr}\,(k_1(x,y)k_2(y,x))\, dy\, dx \end{aligned}$$

which, because k_2 is basic, equals

$$\begin{aligned} \int_U \int_{\pi(U)} \mathrm{tr}\,(k_1(x,y)k_2(y,x))\, dy\, dx &= \int_{\pi(U)} \int_U \mathrm{tr}\,(k_2(y,x)k_1(x,y))\, dx\, dy \\ &= \int_{\pi(U)} \int_M \mathrm{tr}\,(k_2(y,x)k_1(x,y))\, dx\, dy \end{aligned}$$

again because k_2 is basic, and this equals

$$\int_{\pi(U)} \psi_{\pi(U)}(y) \int_M \mathrm{tr}\,(k_2(y,x)k_1(x,y))\, dx\, dy.$$

This last integral is the value at $\pi(U)$ of the function $g \in C^0(\Gamma)$ which represents $\mathrm{Tr}(k_2 \circ k_1)$, i.e. it is $(\pi^{-1})^*(g)(U)$. But g and $(\pi^{-1})^*(g)$ represent the same Haefliger function. Thus $\mathrm{Tr}(k_1 \circ k_2) = \mathrm{Tr}(k_2 \circ k_1)$. □

3 Dirac complexes and Lefschetz functions

The Haefliger trace adapts extremely well to any situation where the heat equation method is used. As an example of this, we will outline a proof of a Lefschetz theorem for complete Riemannian manifolds of bounded geometry. This of course includes various index theorems as special cases.

Let M be a complete Riemannian manifold of bounded geometry. Fix a lattice Γ for M. We assume that each bundle over M comes equipped with a metric, and if the bundle is a complex bundle, that the metric is Hermitian. Unless otherwise stated, all objects considered here (e.g. metrics, connections, bundle maps, ...) are assumed to be bounded in the sense that there are global bounds for the local expressions of the object and all its derivatives is any local orthonormal framing of the relevant bundle.

3.1 Dirac complexes

A Dirac complex (E, d) on M consists of the following

1. $E = (E_0, E_1, \dots, E_\ell)$, a family of smooth finite dimensional complex vector bundles over M.
2. $d = (d_0, d_1, \dots, d_{\ell-1})$, a family of differential operators where
$$d_i : C_0^\infty(E_i) \to C_0^\infty(E_{i+1}),$$
and $d_{i+1}\, d_i = 0$.
3. Set $E = \oplus_i E_i$ and denote the adjoint of d_i by d_i^*. We assume that E is a Clifford bundle over the Clifford algebra of T^*M, and that there is an Hermitian connection ∇ on E, compatible with Clifford multiplication, so that the operator
$$D = \oplus(d_i + d_{i-1}^*) : C_0^\infty(E) \to C_0^\infty(E)$$
is given by the composition
$$C_0^\infty(E) \xrightarrow{\nabla} C_0^\infty(T^*M \otimes E) \xrightarrow{m} C_0^\infty(E)$$
where m is Clifford multiplication.

All the classical complexes (de Rham, Signature, Dolbeault, and Spin) give rise to Dirac complexes on M provided M supports the necessary geometric structures for these complexes to be defined.

Denote the space of L^2 sections of E_i by $L^2(E_i)$. The operators d_i and d_{i-1}^* extend to densely defined unbounded operators
$$d_i : L^2(E_i) \to L^2(E_{i+1}) \quad \text{and} \quad d_{i-1}^* : L^2(E_i) \to L^2(E_{i-1}).$$
As usual with possibly non-compact manifolds, we define
$$H^i(M, d) = \ker(d_i)/\overline{(\operatorname{im} d_{i-1} \cap \operatorname{dom} d_i)}.$$

It is necessary to mod out by $\overline{\operatorname{im} d_{i-1} \cap \operatorname{dom} d_i}$, the closure of the image of d_{i-1} in the domain of d_i, as $\operatorname{im} d_{i-1} \cap \operatorname{dom} d_i$ is in general not a closed subspace.

Define

$$\Delta_i = d_{i-1}\, d^*_{i-1} + d^*_i\, d_i : C^\infty_0(E_i) \to C^\infty_0(E_i).$$

The operator $D = \oplus(d_i \oplus d^*_{i-1})$ is a Dirac operator and the fact that M is complete implies that D is essentially self adjoint [9]. Thus $D^2 = \oplus\Delta_i$ is also essentially self adjoint, and we can apply the spectral mapping theorem to each of the Δ_i.

Hodge Theory extends to complete manifolds and we have the following facts:

- $L^2(E_i) = \ker \Delta_i \oplus \overline{\operatorname{image} d_{i-1}} \oplus \overline{\operatorname{image} d^*_i}$.
- The natural map $\ker \Delta_i \to H^i(E, d)$ is an isomorphism.

3.2 Geometric endomorphisms

An endomorphism $T = (T_0, \dots, T_\ell)$ of a Dirac complex (E, d) is a collection of complex linear maps

$$T_i : C^\infty(E_i) \to C^\infty(E_i)$$

so that $d_i\, T_i = T_{i+1}\, d_i$. The most interesting of these are the so-called geometric endomorphisms [2] defined as follows. Let $f : M \to M$ be a smooth diffeomorphism of bounded distortion. For $i = 0, \dots, \ell$, let $A_i : f^*(E_i) \to E_i$ be a smooth bounded bundle map. For $s \in C^\infty(E_i)$ define

$$(T_i s)(x) = A_{i,x}(s(f(x))).$$

We assume that the A_i are chosen so that $T = (T_0, \dots, T_\ell)$ defines an endomorphism of (E, d).

Example. Let (E, d) be the de Rham complex of M and $f : M \to M$ a diffeomorphism of bounded distortion. For A_i take the ith exterior power of the dual df^* of the differential df. Then for each i, T_i is just the map f^* on the differential i forms on M.

As $d_i\, T_i = T_{i+1}\, d_i$, T_i induces a well defined map $T^\#_i : H^i(E, d) \to H^i(E, d)$. We would like our Lefschetz theorem to relate $\sum_i (-1^i)\mathrm{tr}\; T^\#_i$ to indices defined on the fixed point sets of f. In general, $H^i(E, d)$ is not finite dimensional, so the classical trace does not make sense. To get our theorem, we must recast $T^\#_i$ in terms of an operator on $L^2(E_i)$ in order to apply the more general Haefliger trace.

Denote by $P_i : L^2(E_i) \to \ker(\Delta_i)$ the projection, and define

$$T_i^* = P_i T_i P_i : L^2(E_i) \to \ker(\Delta_i) \subset L^2(E_i).$$

We also denote by T_i^* the map $P_i T_i P_i \circ j : \ker(\Delta_i) \to \ker(\Delta_i)$ where $j : \ker(\Delta_i) \to L^2(E_i)$ is the injection. It follows from results of [13] that the Haefliger trace $\mathrm{Tr}\,(T_i^*)$ exists and is finite.

Definition 5 The Lefschetz function $L(T)$ of the geometric endomorphism T is the Haefliger function for M given by

$$L(T) = \sum_{i=0}^{\ell} (-1)^i \mathrm{Tr}\,(T_i^*).$$

That this is a reasonable definition is a result of the fact that the diagram

$$\begin{array}{ccc} \ker \Delta_i & \xrightarrow{\quad T_i^* \quad} & \ker \Delta_i \\ \downarrow & & \downarrow \\ H^i(E,d) & \xrightarrow{\quad T_i^{\#} \quad} & H^i(E,d) \end{array}$$

commutes.

4 Fixed point indices and the Lefschetz theorem

In this section, we describe the restrictions we place on the fixed point set N of f and state our Lefschetz theorem.

We assume that N is a disjoint union of closed submanifolds $N = \cup N_i$ of M. As they are disjoint and closed, each N_i admits a positive function $\epsilon_i(x)$ so that

$$\mathcal{N}_\epsilon(N_i) = \{x \in M \,|\, d(x, N_i) < \epsilon_i(x)\}$$

is an embedded normal disc bundle in M and we may assume that these disc bundles are pairwise disjoint.

We further assume that f is non-degenerate on N. This means that f_* restricted to the normal bundle of N in M does not have 1 as an eigen value at any point in N, i.e. f has no transverse invariant directions. Note that the identity map I_M of M satisfies this condition.

Suppose that a is a smooth volume form on N, and denote its restriction to N_i by a_i. We define the Haefliger function $\int_N a$ as follows. Let $\{\psi_U\}$ be a partition of unity subordinate to the cover by the elements of the lattice

Γ. Then $\int_N a = [g]$ where g is the function on Γ

$$g(U) = \sum_i \int_{N_i \cap U} \psi_U \, a_i.$$

The fact that only a finite number of $N_i \cap U \neq \emptyset$ follows from the facts that $\overline{U}$ is compact, $f|_N = I_N$ and that the N_i are disjoint and closed.

Theorem 6 (Lefschetz Theorem) *Let $M, (E,d), f, A$ and T be as above. Associated to N is a smooth volume form a which depends only on f, A, the symbols of the Δ_i, the metrics and their derivatives to a finite order only on N, so that in $C_h(M)$*

$$L(T) = \int_N a.$$

For the classical complexes and $T = f^*$ acting on forms, a is given by the classical local integrand of the Atiyah-Singer G Index Theorem [4]. So, if M is a compact manifold, special cases of this theorem recover the Atiyah-Singer Index Theorem, [5], the Atiyah-Singer G Index Theorem [4], and the Atiyah-Bott Lefschetz Theorem for geometric endomorphisms, [2].

If (E,d) is an arbitrary Dirac complex and $N_i = \{x\}$ is a single point, then

$$a_i = \frac{\sum_{i=0}^{\ell} (-1^i) \operatorname{tr} A_{i,x}}{|\det(I_x - df_x)|}.$$

For the classical complexes with $T = f^*$ and $N_i = \{x\}$, the a_i may be further identified. For this see [2] and [3].

If $f = I_M$, then $N = M$ and we define $I(D)$, the index of D by $I(D) = L(I_M)$. As an example of the theorems available in this case, we state only the following.

Theorem 7 *Let M be an even dimensional complete Riemannian Spin manifold of bounded geometry. Let $\not{D}_E^+$ be the generalized Atiyah-Singer operator, [15], associated to the complex vector bundle E. Then in $C_h(M)$*

$$I(\not{D}_E^+) = \left[\int ch(E) \widehat{A}(M) \right]$$

where $ch(E)$ is the Chern class of E, and $\widehat{A}(M)$ is the $\widehat{A}$ genus of M.

There are similar theorems for the other classical operators.

5 An example

In this section, we give an example of a non-compact complete Riemannian manifold M which is a covering of Σ_4, the surface of genus 4, and a diffeomorphism $f : M \to M$ of bounded distortion which has two fixed points in the interior of each fundamental domain of the covering $M \to \Sigma_4$. This example is a single generic leaf of the foliation defined in Section 4 of [13], to which the reader is referred for the proofs of the claims made here.

Let $\Pi \subset SL_2\boldsymbol{R}$ be the subgroup generated by the elements $\alpha_j = \theta^{-j}\alpha\,\theta^j$, $j = 0, \ldots, 7$ where θ is rotation by $\pi/16$ and

$$\alpha = \begin{bmatrix} d & 0 \\ 0 & d^{-1} \end{bmatrix},$$

where d is chosen so that $\Pi\backslash SL_2\boldsymbol{R}/SO_2 = \Sigma_4$. Let $\Pi_1 \subset \Pi$ be generated by $\{\alpha_i^{\pm 1}\alpha_j^{\pm 1}\,|\,i, j = 0, 3, 4, 7\} \cup \{\alpha_i^{\pm 1}\alpha_j^{\pm 1}\,|\,i, j = 1, 2, 5, 6\}$. Then $M = \Pi_1\backslash SL_2\boldsymbol{R}/SO_2$ and $\Pi/\Pi_1 \simeq \boldsymbol{Z}_2 * \boldsymbol{Z}_2 \simeq D_\infty$ so M is non compact. Note that D_∞ is solvable so also amenable.

Let f be the diffeomorphism of M determined by ρ, rotation of the Poincaré disc by $\pi/2$. Take for the fundamental domains of Σ_4 a regular 16-gon centred at zero in the Poincaré disc and all its translates under Π. This determines a set of fundamental domains in M. A careful reading of Section 4 of [13] shows that f fixes each fundamental domain of M, (i.e. for any fundamental domain D in the Poincaré disc, there is an element $\gamma \in \Pi_1$ so that $\rho(D) = \gamma(D)$). f acts on each fundamental domain of M by rotation by $\pi/2$. In each such fundamental domain, the only fixed points are the centre and the 16 vertices. But the vertices are identified in groups of 8 by Π_1 so in each of these fundamental domains there are exactly three fixed points and at each fixed point, df acts by rotation by $\pi/2$. By slightly altering the fundamental domains we have chosen (namely adding a small neighbourhood of every other vertex and deleting a corresponding neighbourhood of the other vertices), we obtain new fundamental domains so that each has exactly two fixed points contained in its interior.

M has all four classical complexes defined on it, and the local indices at the fixed points for $T = f^*$ are all non trivial. For the de Rham complex they are all -2, for the Signature complex $-2i$, for the Dolbeault complex with $k = 0$, they are $1/(1-i)$, and for $k = 1$, they are $i/(1-i)$, and for the Spin complex they are $\pm i/\sqrt{2}$ depending on which lifting of df one uses.

6 Proof of Lefschetz theorem

As noted above, each Δ_i is an essentially self adjoint operator on $L^2(E_i)$. In addition, it is non-negative, so any bounded Borel function g on $[0,\infty)$ applied to Δ_i yields a well defined bounded operator $g(\Delta_i)$ on $L^2(E_i)$ whose Schwartz kernel we denote by $k_g(x,y)$. Denote by $\mathcal{S}(\boldsymbol{R}^+)$ the Schwartz class functions on $\boldsymbol{R}$ restricted to $[0,\infty)$. If g is a Schwartz class function on $\boldsymbol{R}$, we denote its Fourier transform by $\widehat{g}$.

We now recall three theorems from [13]. The proofs for the case considered here are essentially the same as those given in [13] and so are omitted.

Theorem 8 *If $g \in \mathcal{S}(\boldsymbol{R}^+)$, then $k_g(x,y) \in C^\infty(E \otimes E^*)$, (so $Tr(g(\Delta_i))$ exists and is finite), and the map $g \to k_g$ is continuous. If $\widehat{g} \in C_0^\infty(\boldsymbol{R})$, the compactly supported functions, then $k_g \in C_s^\infty(E \otimes E^*)$.*

Theorem 9 *Suppose that $g \in \mathcal{S}(\boldsymbol{R}^+)$, B is a differential operator on E_i (whose coefficients, with respect to orthonormal bases, and all their derivatives are uniformly bounded), and $T = (T_0, \dots, T_\ell)$ is a geometric endomorphism of (E,d). Then $Tr(BT_ig(\Delta_i))$ and $Tr(T_iBg(\Delta_i))$ exist and are finite.*

Corollary 10 *For $i = 0, \dots, \ell$ and $t > 0$, $Tr(T_ie^{-t\Delta_i})$ exists and is finite.*

Theorem 11 $Tr(T_i^*) = \lim_{t\to\infty} Tr(T_ie^{-t\Delta_i})$.

Note that the convergence in Theorem 11 is pointwise and not uniform, so when applying a linear functional to $C_h(M)$ as in Section 7 below, one needs to check that the functional commutes with taking limits. This does not happen in general. For an explicit counter example see [19], and for a systematic discussion, [11].

As the proof of the following theorem is where the trace property of Tr is most central, we will outline that proof here. For details see [13], Section 5.

Theorem 12 *For all $t > 0$,*

$$\sum_{i=0}^{\ell}(-1)^i\, Tr(T_ie^{-t\Delta_i})$$

is independent of t.

Proof. We use Quillen's formalism of the supertrace [17]. Let $E = \oplus E_i$ and $E^+ = \oplus E_{2i}$ and $E^- = \oplus E^{2i+1}$. An operator A on E which preserves this splitting is called even and one that reverses it is called odd. The super trace applied to an even operator A is defined to be

$$\mathrm{Tr}_s(A) = \mathrm{Tr}(A\,|_{E^+}) - \mathrm{Tr}(A\,|_{E^-}).$$

Note that if A and B are both even operators, then $\mathrm{Tr}_s(AB) = \mathrm{Tr}_s(BA)$, while if both are odd, $\mathrm{Tr}_s(AB) = -\mathrm{Tr}_s(BA)$.

Set

$$T = \oplus_{i=0}^{\ell} T_i, \quad d = \oplus_{i=0}^{\ell-1} d_i, \quad d^* = \oplus_{i=0}^{\ell-1} d_i^*, \quad \Delta = \oplus_{i=0}^{\ell} \Delta_i.$$

T and Δ are even while d and d^* are odd and these operators satisfy

$$\Delta = dd^* + d^*d, \quad d\Delta = \Delta d, \quad d^*\Delta = \Delta d^*, \text{ and } Td = dT.$$

It is not difficult to show that if $\varphi(x)$ is an analytic function whose restriction to $\boldsymbol{R}^+$ is in $\mathcal{S}(\boldsymbol{R}^+)$, then

$$d\varphi(\Delta) = \varphi(\Delta)d \quad \text{and} \quad d^*\varphi(\Delta) = \varphi(\Delta)d^*.$$

For $s > t > 0$ set

$$\varphi(x) = e^{-tx} - e^{-sx} \quad \text{and} \quad \psi(x) = \varphi(x)/x$$

and note that both restricted to $\boldsymbol{R}^+$ are in $\mathcal{S}(\boldsymbol{R}^+)$. As

$$\sum_{i=0}^{\ell} (-1)^i \mathrm{Tr}\,(T_i^* e^{-t\Delta_i}) = \mathrm{Tr}_s(Te^{-t\Delta}),$$

we must only show that $\mathrm{Tr}_s(T\varphi(\Delta)) = 0$.

By Theorem 9, we have that both $\mathrm{Tr}_s(Tdd^*\psi(\Delta))$ and $\mathrm{Tr}_s(Td^*d\psi(\Delta))$ exist and are finite, so we have

$$\mathrm{Tr}_s(Tdd^*\psi(\Delta)) + \mathrm{Tr}_s(Td^*d\psi(\Delta)) = \mathrm{Tr}_s(Tdd^*\psi(\Delta) + Td^*d\psi(\Delta)).$$

But from the definitions, this last equals

$$\mathrm{Tr}_s(T\Delta\psi(\Delta)) = \mathrm{Tr}_s(T\varphi(\Delta)),$$

which is what we want to show is zero. As $\mathrm{Tr}_s(Tdd^*\psi(\Delta)) = \mathrm{Tr}_s(dTd^*\psi(\Delta))$, this will follow if we show that

$$\mathrm{Tr}_s(dTd^*\psi(\Delta)) = -\mathrm{Tr}_s(Td^*d\psi(\Delta)).$$

Theorem 8 implies that if $\psi_n \in \mathcal{S}(\boldsymbol{R}^+)$ converges to ψ, then

$$\lim_{n\to\infty} \mathrm{Tr}(dTd^*\psi_n(\Delta)) = \mathrm{Tr}(dTd^*\psi(\Delta))$$

and

$$\lim_{n\to\infty} \mathrm{Tr}(Td^*d\psi_n(\Delta)) = \mathrm{Tr}(Td^*d\psi(\Delta)).$$

Set $\psi_1(x) = (1/2)\psi(x/2)$ and $\psi_2(x) = e^{-tx/2} + e^{-sx/2}$. Then ψ_1, ψ_2 restricted to $\boldsymbol{R}^+$ are in $\mathcal{S}(\boldsymbol{R}^+)$, and $\psi = \psi_1\psi_2$. Choose sequences $\psi_{i,n} \in \mathcal{S}(\boldsymbol{R}^+)$ converging to ψ_i with $\widehat{\psi}_{i,n} \in C_0^\infty(\boldsymbol{R})$. Then

$$\begin{aligned}\mathrm{Tr}_s(dTd^*\psi(\Delta)) &= \lim_{n\to\infty} Tr_s(dTd^*(\psi_{1,n}\psi_{2,n})(\Delta)) \\ &= \lim_{n\to\infty} Tr_s(dTd^*\psi_{1,n}(\Delta)\psi_{2,n}(\Delta)) \\ &= \lim_{n\to\infty} Tr_s(\psi_{2,n}(\Delta)dTd^*\psi_{1,n}(\Delta))\end{aligned}$$

by Theorem 4 (so it is here that we use the trace property of Tr). But

$$\lim_{n\to\infty} Tr_s(\psi_{2,n}(\Delta)dTd^*\psi_{1,n}(\Delta)) = -\lim_{n\to\infty} Tr_s(Td^*\psi_{1,n}(\Delta)\psi_{2,n}(\Delta)d),$$

since both $\psi_{2,n}(\Delta)d$ and $Td^*\psi_{1,n}(\Delta)$ are odd operators. This last term equals

$$\begin{aligned}-\lim_{n\to\infty} Tr_s(Td^*d\psi_{1,n}(\Delta)\psi_{2,n}(\Delta)) &= -\lim_{n\to\infty} Tr_s(Td^*d(\psi_{1,n}\psi_{2,n})(\Delta)) \\ &= -\mathrm{Tr}_s(Td^*d\psi(\Delta))\end{aligned}$$

which completes the proof of Theorem 12. □

Corollary 13 $L(T) = \lim_{t\to 0}\sum_{i=0}^{\ell}(-1)^i Tr(T_i^* e^{-t\Delta_i})$

To finish the proof of our Lefschetz theorem, we have

Theorem 14 *Associated to N is a smooth volume form a which depends only on f, A, the symbols of the Δ_i, the metrics and their derivatives to a finite order only on N so that in $C_h(M)$*

$$\lim_{t\to 0}\sum_{i=0}^{\ell}(-1)^i Tr(T_i^* e^{-t\Delta_i}) = \int_N a.$$

The proof of this theorem is a highly technical local computation and is identical to that given in [13] and so is omitted.

7 Functionals on $C_h(M)$

The space $C_h(M)$ is somewhat difficult to understand. In the special case that M is compact, $C_h(M) = \boldsymbol{R}$. One of the main results of [7], Theorem 3.1, is that $C_h(M) \neq 0$ if and only if M satisfies Condition (15) below. However, even in the simple case of $M = \boldsymbol{R}$, it is difficult to give a good description of this space. To obtain usable invariants, we need to exhibit interesting linear maps from $C_h(M)$ to more tractable spaces, especially, maps to $\boldsymbol{R}$ or $\boldsymbol{C}$. Given the result of Block and Weinberger quoted above, it should not be surprising that $C_h(M)$ is particularly susceptible to averaging operations, one of which we give below. For more in this vein, see [18] and [1].

Suppose that M is a complete Riemannian manifold of bounded geometry. Given $x_0 \in M$ and $s \geq 0$, denote by $B(x_0, s)$ the ball in M of radius s and centre x_0. Assume that for some $x_0 \in M$ and any fixed t,

$$\lim_{s \to \infty} \frac{\mathrm{vol}(B(x_0, s))}{\mathrm{vol}(B(x_0, s+t))} = 1. \tag{15}$$

Let Γ be a lattice in M and denote the centre of $U \in \Gamma$ by x_U. Set

$$\Gamma_s = \{U \in \Gamma \mid d(x_0, x_U) < s\}.$$

For any $[f] \in C_h(M)$, define $A : C_h(M) \to \boldsymbol{R}$ as follows.

$$A([f]) = \lim_{s \to \infty} \frac{\sum_{\Gamma_s} f(U)}{\mathrm{vol}(B(x_0, s))}.$$

Proposition 16 *$A : C_h(M) \to \boldsymbol{R}$ is well defined.*

Proof. We first show that

$$F(s) = \frac{\sum_{\Gamma_s} f(U)}{\mathrm{vol}(B(x_0, s))}$$

converges as $s \to \infty$. Since the centres of the Us are uniformly separated and M has bounded geometry, there is a constant C so that for any difference of two concentric balls, $B = B(x_0, s+t) - B(x_0, t)$, the number of elements of Γ with centres in B is bounded by $C\,\mathrm{vol}(B)$. If $|f|$ is bounded by C_f, it follows immediately that

$$\left| \sum_{x_U \in B} f(U) \right| \leq C_f\, C\, \mathrm{vol}(B).$$

In particular, this holds if $t = 0$, so $F(s)$ is a bounded function and we need only show that for bounded $t > 0$,

$$\lim_{s \to \infty} F(s+t) - F(s) = 0.$$

Note that t may vary with s, but it is a bounded function of s. Given any set $B \subset M$, denote by $f(B)$ the sum

$$f(B) = \sum_{x_U \in B} f(U)$$

and write $B(s)$ for $B(x_0,s)$ and $vB(s)$ for $\mathrm{vol}(B(x_0,s))$. Then

$$\begin{aligned}
| F(s+t) - F(s)| &= \left|\frac{f(B(s+t))}{vB(s+t)} - \frac{f(B(s))}{vB(s)}\right| \\
&\leq \left|\frac{f(B(s))}{vB(s+t)} - \frac{f(B(s))}{vB(s)}\right| + \left|\frac{f(B(s+t)-B(s))}{vB(s+t)}\right| \\
&\leq C_f C\, vB(s)\left|\frac{1}{vB(s+t)} - \frac{1}{vB(s)}\right| + C_f C\left|\frac{vB(s+t)-vB(s)}{vB(s+t)}\right| \\
&= 2C_f C\frac{vB(s+t)-vB(s)}{vB(s+t)} = 2C_f C(1-\frac{vB(s)}{vB(s+t)})
\end{aligned}$$

which by our assumption goes to zero as $s \to \infty$.

To show that $A([f])$ is independent of the choice of f, we need only show that for any $f \in C^0(\Gamma)$ and $\pi \in G$,

$$\lim_{s\to\infty} \frac{\sum_{\Gamma_s}[f(U) - \pi^* f(U)]}{vB(s)} = 0.$$

Let C_f be a bound for f (and so also for $\pi^* f$), and let t be a bound on the distance that π moves elements of Γ, i.e. for all $U \in \Gamma$, $d(x_U, x_{\pi(U)}) < t$. Then

$$\left|\frac{\sum_{\Gamma_s}[f(U) - \pi^* f(U)]}{vB(s)}\right| = \left|\frac{f(B(s)) - \pi^* f(B(s))}{vB(s)}\right|.$$

Since π moves elements of Γ at most a distance t, each element in $f(B(s-t))$ occurs in $\pi^* f(B(s))$. Thus $f(B(s))$ and $\pi^* f(B(s))$ differ on at most the number of points in $B(s) - B(s-t)$. Thus

$$\left|\frac{f(B(s)) - \pi^* f(B(s))}{vB(s)}\right| \leq \quad 2C_f C\frac{[vB(s) - vB(s-t)]}{vB(s)}.$$

As above, this last goes to zero as $s \to \infty$. □

Recall that the manifold M of our example above supports the four classical complexes. It is not difficult to see that M also satisfies Condition 15. For the classical complexes, the operators Δ_i commute with the action of the covering group of M over Σ_4. The diffeomorphism f also commutes with this action, so we are in the situation considered by Atiyah in [1]. It follows easily that for the operators in Theorem 11, the linear functional A is the Atiyah trace (given by integration over a single fundamental domain, which is a compact set) divided by the volume of the fundamental domain. As the Schwartz kernel of the $T_i e^{-t\Delta_i}$ is uniformly bounded and converges pointwise to the Schwartz kernel of T_i^* and A is integration over a compact set, A does commute with the limit in Theorem 11. Thus we have the following.

Theorem 17 *Let M and f be as in the example above and denote by V the volume of a fundamental domain for Σ_4 in the Poincaré disc. Then*

- *For the de Rham complex,* $A(L(f^*)) = \dfrac{-4}{V}$
- *for the Signature complex* $A(L(f^*)) = \dfrac{-4i}{V}$
- *for the Dolbeault complex with* $k = 0$, $A(L(f^*)) = \dfrac{2}{(1-i)V}$, *and for* $k = 1$, $A(L(f^*)) = \dfrac{2i}{(1-i)V}$
- *for the Spin complex* $A(L(f^*)) = \dfrac{\pm i\sqrt{2}}{V}$ *depending on which lifting of* df *one uses.*

References

1. M.F. Atiyah, *Elliptic operators, discrete groups and von Neumann algebras*, Astérisque, **32/33** (1976), 43–72.
2. M.F. Atiyah and R. Bott, *A Lefschetz fixed point formula for elliptic complexes I*, Ann. of Math., **86** (1967), 374–407.
3. M.F. Atiyah and R. Bott, *A Lefschetz fixed point formula for elliptic complexes II*, Ann. of Math., **88** (1968), 451–491.
4. M.F. Atiyah and G.B. Segal, *The index of elliptic operators: II*, Ann. of Math., **87** (1968), 531–545.
5. M.F. Atiyah and I.M. Singer, *The index of elliptic operators: I*, Ann. of Math., **87** (1968), 484–530.
6. O. Attie and S. Hurder, *Manifolds which cannot be leaves of foliations*, Topology, **35** (1996), 335–353.
7. J. Block and S. Weinberger, *Aperiodic tilings, positive scalar curvature, and amenability*, Journal AMS, **5** (1992), 907–918.
8. J. Block and S. Weinberger, Large scale homology theories and geometry, in Geometric Topology, ed. W.H. Kazez, Amer. Math. Soc., Providence, R.I., 1997.
9. P. Chernoff, *Essential self adjointness of powers of generators of hyperbolic equations*, J. Func. Anal., **12** (1973), 401–404.
10. A. Connes, Geometrie Non-Commutative, InterEditions, Paris, 1990.
11. D. Guido and T. Isola, *Noncommutative Riemann integration and Novikov-Shubin invariants for open manifolds*, J. Func. Anal., **176** (2000), 115–152.

12. J.L. Heitsch, *Bismut Superconnections and the Chern Character for Dirac Operators on Foliated Manifolds*, K-Theory, **9** (1995), 507–528.
13. J.L. Heitsch and C. Lazarov, *A Lefschetz theorem for foliated manifolds*, Topology, **29** (1990), 127–162.
14. J.L. Heitsch and C. Lazarov, *A general families index theorem*, K-Theory, **18** (1999), 181–202.
15. H.B. Lawson and M.-L. Michelsohn, Spin Geometry Princeton University Press, Princeton, N.J., 1989.
16. A. Phillips and D. Sullivan, *Geometry of leaves*, Topology, **20** (1981), 209–218.
17. D. Quillen, *Superconnections and the Chern Character*, Topology, **24** (1985), 89–95.
18. J. Roe, *An index theorem on open manifolds I*, J. Diff. Geo., **27** (1988), 87–113.
19. J. Roe, *An index theorem on open manifolds II*, J. Diff. Geo., **27** (1988), 115–136.

Received July 31, 2000, revised November 17, 2000 and February 14, 2001.

Proceedings of
FOLIATIONS: GEOMETRY AND DYNAMICS
held in Warsaw, May 29–June 9, 2000
ed. by Paweł WALCZAK *et al.*
World Scientific, Singapore, 2002
pp. 315–332

HILBERT MODULES OF FOLIATED MANIFOLDS WITH BOUNDARY

MICHEL HILSUM
Institut de Mathématiques, C.N.R.S., 175,
rue du Chevaleret, 75 013 Paris,
e-mail: hilsum@math.jussieu.fr

We define in this article the notion of boundary Hilbert module over a C*-algebra B and show that the class in $K_*(B)$ defined by such a module is null. This generalizes the bordism invariance of the index of elliptic operators. Some consequences on foliated manifolds are derived, such as obstructions on the existence of a riemannian metric with longitudinal positive scalar curvature

1 Introduction

The bordism invariance of the index of the Dirac operator on smooth spin manifolds has been established by M.F. Atiyah and I.M. Singer, and play an important role in the first proof of the index theorem for elliptic operators [17]. It has been generalized in various geometric situations: for smooth families of Dirac operators by W. Shih [20], for coverings of a manifold with principal countable discrete groups by J. Rosenberg [19]; for correspondences between smooth manifolds, using bivariant K-theory, by A. Connes and G. Skandalis [4]; alternative analytical proofs where given of [20] by R. Melrose and P. Piazza [14] and by L. Nicolaescu [16], and of [19] by E. Leichtnam and P. Piazza [12]. In each of these situations, the corresponding index is a class in the K-theory group of a C*-algebra.

Here we wish to settle an general result in the context of Hilbert modules over C*-algebras. We introduce a notion of *Hilbert module with boundary* over some C*-algebra B. We also define *partitioned Hilbert module* over B. These objects behave well with respect to processes of pasting and cutting, similarly as cutting a partitioned manifold along the separating manifold gives two manifolds with boundary.

The main result of this article states that the index class in $K_*(B)$ defined by the boundary of such an Hilbert module is null (Theorem 4.3). There are in fact two cases, the even-dimensional case, in which the boundary module is $\mathbf{Z}/\mathbf{Z}_2$-graded, and the index has value in $K_0(B)$, and the odd-dimensional, with no grading, where $K_1(B)$ is concerned. We have treated here only the odd case.

For proving this, we develop an idea due to N. Higson which relates this problem to *Callias* or *Dirac-Shrödinger* type operator on partitioned manifolds [6], [18]. These operators have been first studied by C. Callias, and have been since the subject of several works by N. Anghel, J. Bruning, H. Moscovici, U. Bunke and M. Lesch (*cf ref.* in monograph [13]).

Given an even partitioned module, Theorem 4.2 states the equality between the index class in $K_1(B)$ of the Callias type operator and the index class of the operator on the partitioning module.

These notions are quite general and it appears that index bordism invariance is a rather analytical property.

We specialize then these results to Hilbert modules arising from foliated manifolds.

Let (W, F) be a spin foliated closed manifold of class $C^{0,\infty}$. We use the longitudinal Dirac operator A acting on some Hilbert module $\mathcal{E}$ over the C*-algebra of the foliation $C^*(W, F)$, defined by A. Connes [3] (cf. minicourse in these proceedings by H. Moriyoshi [15]).

We obtain then a bordism invariance property of the longitudinal index class $[\mathcal{E}, A] \in K_*(C^*(W, F))$ (Theorem 5.2).

Let (W_∞, g) be a smooth Riemannian manifold, and $\kappa : W_\infty \to \mathbf{R}$ the scalar curvature of g, which is a smooth function. One important problem in Riemannian geometry is to determine topological obstructions to the existence on W_∞ of a Riemannian structure with positive scalar curvature (cf. [11] for a survey). Recall A. Lichnerowicz's Theorem stating that if W_∞ is closed and spin with $\kappa \geq 0$, then

$$\langle \hat{A}(W_\infty), [W_\infty] \rangle = 0, \tag{1.1}$$

where $\hat{A}(W)$ is the $\hat{A}$*-genus* of the tangent bundle of the manifold.

Then Theorems 4.2 and 4.3 gives as corollaries obstructions for a manifold (W, F) to be equivalent by a transverse bordism to a foliated manifold with strictly positive scalar curvature (Proposition 6.2), which straightens an earlier result of J. Rosenberg [19]. Similar conditions occur when (W, F) separates a foliation with strictly positive scalar curvature and Proposition 6.3 generalizes a theorem of J. Roe [18]. When the foliation is of class $C^{\infty,\infty}$, using the fundamental class of A. Connes sect. III.7 [3], we derive, in Section 7, cohomological obstructions, among them A. Lichnerowicz's (1.1).

2 Preliminaries

Let B a C*-algebra and $\mathcal{E}$ a Banach right B-module ; then $\mathcal{E}$ is said to be a Hilbert module over B if it comes with a sesquilinear map $\langle .,. \rangle : \mathcal{E} \times \mathcal{E} \to B$ such that for $\xi, \eta \in \mathcal{E}$, $a \in B$, one has $\langle \xi, \eta \rangle^* = \langle \eta, \xi \rangle$, $\langle \xi, \xi \rangle \geq 0$,$\langle \xi, a\eta \rangle = \langle \xi, \eta \rangle a$, and the map $\xi \to \| \langle \xi, \xi \rangle \|^{\frac{1}{2}}$ is the norm on $\mathcal{E}$. As usual we denote by $\mathcal{L}(\mathcal{E})$ the C*-algebra of bounded adjointable operators of the B-module $\mathcal{E}$, and by $\mathcal{K}(\mathcal{E})$ the closed ideal of compact operators of $\mathcal{E}$, generated by the rank one operators of the form $\zeta \to \xi \langle \eta, \zeta \rangle$, for $\xi, \eta \in \mathcal{E}$. Let D a densely defined closed operator on $\mathcal{E}$, commuting with the action of B. The adjoint of D is the operator with domain the set of $\xi \in \mathcal{E}$ such that there exists $\eta \in \mathcal{E}$ satisfying for every $\zeta \in \mathcal{E}$ the equality $\langle D\zeta, \xi \rangle = \langle \zeta, \eta \rangle$ and then $D^*\xi = \eta$. The operator is said to be symmetric if $\langle D\xi, \eta \rangle = \langle \xi, D\eta \rangle$ for $\xi, \eta \in \operatorname{dom} D$, and selfadjoint if $D = D^*$. For a densely defined operator D of B-module, we define $\mathcal{L}(D)$ the subalgebra of $\mathcal{L}(\mathcal{E})$ of operators $a \in \mathcal{L}(\mathcal{E})$ such that $a(\operatorname{dom} D) \subset \operatorname{dom} D$ and such that the graded commutator $[D, a]$ extends to an element of $\mathcal{L}(\mathcal{E})$.

Lemma 2.1 *Let D be a closed densely defined B-operator on $\mathcal{E}$ such that D^* is densely defined. For any $a \in \mathcal{L}(D)$, one has $a^* \in \mathcal{L}(D^*)$, and $[a, D]^*$ is the closure of $[D^*, a^*]$.*

Proof. For $\xi \in \operatorname{dom} D$, $\eta \in \operatorname{dom} D^*$, the equality $\langle D\xi, a^*\eta \rangle = \langle [a, D]\xi, \eta \rangle + \langle a\xi, D^*\eta \rangle$ shows that $a^*\eta \in \operatorname{dom} D^*$ and $(D^*a^* - a^*D^*)\eta = [a, D]^*\eta$. □

Given a closed linear operator D on $\mathcal{E}$, we shall denote by $\mathcal{W}(D)$ the Hilbert B-module $\operatorname{dom} D$ equipped with the B-product $\langle \xi, \eta \rangle + \langle D\xi, D\eta \rangle$. Recall that a *core* or an *essential domain* for a closed operator D is a dense subspace of $\mathcal{W}(D)$.

An operator D is said to be *regular* or *affiliated* if $1 + D^*D$ and $1 + DD^*$ are surjective operators of $\mathcal{E}$, and in that case the inverse operators $(1 + D^*D)^{-1}$ and $(1 + DD^*)^{-1}$ belongs to $\mathcal{L}(\mathcal{E})$ [10]. If $D = D^*$, then these conditions are equivalent to the surjectivity of $\operatorname{im}(i + T)$. The following is elementary

Lemma 2.2 *An closed densely defined B-operator D on $\mathcal{E}$ is regular if and only if the natural continuous injection $\mathcal{W}(D) \to \mathcal{E}$ is adjointable*

We precise now some terms which will be used in the sequel.

i) An *unbounded module over B* is the data of a Hilbert module $\mathcal{E}$ over B with an unbounded closed densely defined regular operator T commuting with B. We shall call $(\mathcal{E}, T, \tau)$ an even unbounded module if τ is a unitary involution such that $\tau T + T\tau = 0$.

ii) A *symmetric module over B* is an unbounded module $(\mathcal{E}, T)$ with $T \subset T^*$.

iii) A *closed module over* B is an unbounded module $(\mathcal{E}, T)$ with $T = T^*$ and $(i+T)^{-1} \in \mathcal{K}(\mathcal{E})$.

Recall that if $(\mathcal{E}, T)$ is a closed module, we can associate an element $[\mathcal{E}, T] \in K_1(B)$ as follows [21]: as the operator $F = T(1+T^2)^{-\frac{1}{2}}$ satisfies $F^* - F \in \mathcal{K}(\mathcal{E})$ and $F^2 - 1 \in \mathcal{K}(\mathcal{E})$ and its image in the quotient C*-algebra $\mathcal{L}(\mathcal{E})/\mathcal{K}(\mathcal{E})$ is a projection, it gives a class in $K_0(\mathcal{L}(\mathcal{E})/\mathcal{K}(\mathcal{E}))$. Then $[\mathcal{E}, T]$ is the image of this class by the index map $K_0(\mathcal{L}(\mathcal{E})/\mathcal{K}(\mathcal{E})) \mapsto K_1(\mathcal{K}(\mathcal{E}))$, followed by the natural inclusion $K_1(\mathcal{K}(\mathcal{E})) \to K_1(B)$.

Analogously, to a closed even module $(\mathcal{E}, T, \tau)$ corresponds a class in $K_0(B)$.

We end this section with a lemma which will be useful later. Let $\mathcal{E}_1$, and $\mathcal{E}_2$ two Hilbert B-modules, T_1 a selfadjoint regular operator on $\mathcal{E}_1$ and T_2 a selfadjoint operator on $\mathcal{E}_2$ possibly singular. Let $f \in \mathcal{L}(\mathcal{E}_1, \mathcal{E}_2)$ such that $f(\operatorname{dom} T_1) \subset \operatorname{dom} T_2$, and such that the operator $T_2 f - f T_1$ defined on $\operatorname{dom} T_1$ extends to a bounded adjointable map which we shall denote by a. By the lemma above, the adjoint of a is the closure of the densely defined $f^* T_2 - T_1 f^*$. The map $u_f(\xi) = f\xi$ for $\xi \in \operatorname{dom} T_1$ defines a continuous map from $\mathcal{W}(T_1)$ to $\mathcal{W}(T_2)$, commuting with B and of norm less than $\sqrt{\|a\|^2 + \|f\|^2}$.

Lemma 2.3 *The map* u_f *is adjointable and we have, for* $\eta \in \operatorname{dom} T_2$

$$u_f^* \eta = f^* \eta + (i+T_1)^{-1} a^* \eta + (1+T_1^2)^{-1} a^* (i+T_2)\eta.$$

Proof. Let denote by w_f the operator on the right hand side above. For $\xi \in \operatorname{dom} T_1$ and $\eta \in \operatorname{dom} T_2$, we have $w_f \eta \in \operatorname{dom} T_1$. Put $\zeta = (i+T_2)\eta$ and $\theta = (i+T_1)\xi$, we have then

$$\begin{aligned}
&\langle \xi, w_f \eta \rangle + \langle T_1 \xi, T_1 w_f \eta \rangle \\
&= \langle \xi, (i+T_1)^{-1}(f^* + (T_1 - i)^{-1} a^*)(T_2 + i)\eta \rangle \\
&+ \langle T_1 \xi, T_1 (i+T_1)^{-1}(f^* + (T_1 - i)^{-1} a^*)(T_2 + i)\eta \rangle \\
&= \langle (T_1 + i)^{-1}\theta, (T_1 + i)^{-1}(f^* + (T_1 - i)^{-1} a^*)\zeta \rangle \\
&+ \langle (T_1 + i)^{-1} T_1 \theta, (T_1 + i)^{-1} T_1 (f^* + (T_1 - i)^{-1} a^*)\zeta \rangle \\
&= \langle \theta, (f^* + (T_1 - i)^{-1} a^*)\zeta \rangle \\
&= \langle (f + a(T_1 + i)^{-1})\theta, \zeta \rangle \\
&= \langle (T_2 + i) f (T_1 + i)^{-1}\theta, \zeta \rangle \ = \ \langle (T_2 + i) f\xi, (T_2 + i)\eta \rangle \\
&= \langle u_f \xi, \eta \rangle + \langle T_2 u_f \xi, T_2 \eta \rangle.
\end{aligned}$$

□

Let us finally recall the definition of the *external tensor product* of two regular operators T_i acting on $\mathcal{E}_i$, for $i = 1, 2$, this is an operator S acting on the external tensor product of Hilbert modules $\mathcal{E}_1 \otimes \mathcal{E}_2$ with domain defined as follows: let $\mathcal{W}(T_1) \otimes \mathcal{W}(T_2)$ be the tensor product ohf Hilbert module, and $I_j : \mathcal{W}(T_1) \to \mathcal{E} - j$ the canonical injection. Then $I = I_1 \otimes I_2 : \mathcal{W}(T_1) \otimes \mathcal{W}(T_2) \to \mathcal{E}_1 \otimes \mathcal{E}_2$ is adjointable [8, 10]. Then let $\operatorname{dom} S = \operatorname{im} I$ and for $\xi_j \in \operatorname{dom} T_j$, one set $S(\xi \otimes \xi_2) = T_1\xi_1 \otimes \xi_2 + \xi_1 \otimes T_2\xi_2$. It is essentially defined on the algebraic tensor product $\mathcal{D}_1 \otimes \mathcal{D}_2$ for any essential domains $\mathcal{D}_i$ of T_i, and moreover, if $(1 + T_i^* T_i)^{-1} \in \mathcal{K}(\mathcal{E}_i)$, for $i = 1, 2$, then $(1 + S^*S)^{-1} \in \mathcal{K}(\mathcal{E}_1 \otimes \mathcal{E}_2)$.

3 Partitioned and boundary modules

We define now an operator useful for the sequel. Let $(\mathcal{E}, T)$ be an unbounded module over B, with $T = T^*$. Let τ_i, $i = 0, 1$, be the unitary on $\mathbf{C}^2$, given by

$$\tau_i = \begin{pmatrix} 0 & (-1)^i \\ 1 & 0 \end{pmatrix}.$$

Let $\Psi_\infty(T)$ acting on the tensor product of Hilbert modules $\mathcal{E}_b \otimes L^2(\mathbf{R}) \otimes \mathbf{C}^2$ be defined by the formula

$$\Psi_\infty(T) = T \otimes 1 \otimes \tau_0 + 1 \otimes \partial_1 \otimes \tau_1,$$

where $\partial_1 f(u) = \frac{df}{du}$. Then $\Psi_\infty(T)$ is essentially selfadjoint on the *algebraic* tensor productf $\operatorname{dom} T \otimes C_c^\infty(\mathbf{R}) \otimes \mathbf{C}^2$, and for ξ_1, ξ_2 in that subspace, one has the more familiar form

$$\Psi_\infty(T) = \begin{pmatrix} 0 & -\frac{\partial}{\partial u} + T \\ \frac{\partial}{\partial u} + T & 0 \end{pmatrix} \begin{pmatrix} \xi_1 \\ \xi_2 \end{pmatrix} = \begin{pmatrix} -\frac{\partial}{\partial u}\xi_2 + T\xi_2 \\ \frac{\partial}{\partial u}\xi_1 + T\xi_1 \end{pmatrix}. \tag{3.1}$$

We denote by $\Psi(T)$ acting on $\mathcal{E} \otimes L^2(\mathbf{R}) \otimes \mathbf{C}^2$ be the closure of the restriction $\Psi_\infty(T)$ on the algebraic tensor product $\operatorname{dom} T \otimes C_c^\infty(]0,1[) \otimes \mathbf{C}^2$. Given $\varphi \in C([0,1])$, we denote by

$$m(\varphi) = 1 \otimes \varphi \otimes 1$$

the operator in $\mathcal{L}(\mathcal{E} \otimes L^2([0,1])) \otimes \mathbf{C}^2)$ obtained by multiplication by φ on the second term of the tensor product.

Lemma 3.1 *The operators $\Psi(T)$ and $\Psi_\infty(T)$ are regular. For $\theta \in C^\infty([0,1])$ one has $m(\theta) \in \mathcal{L}(\Psi(T)^*)$ and moreover if $\theta(0) = \theta(1) = 0$, then $m(\theta) \operatorname{dom} \Psi(T)^* \subset \operatorname{dom} \Psi(T)$. If $(i+T)^{-1} \in \mathcal{K}(\mathcal{E})$, then $(1+\Psi(T)^*\Psi(T))^{-1}$ is a compact operator.*

Proof. As $\Psi(T)$ is an external tensor product, the first and last assertions follow. As $m(\theta) \in \mathcal{L}(\Psi(T))$, Lemma 2.1 gives the third assertion. It suffices then to prove that $\theta\xi \in \operatorname{dom} \Psi_\infty(T)$ for $\operatorname{support} \theta \subset]0,1[$. Let

$\xi \in \operatorname{dom}\Psi(T)^*$ and $\varphi \in C_c^\infty(]0,1[)$ such that $\varphi\theta = \theta$ and $\varphi'\theta = 0$. Then from Lemma 2.1, for $\eta \in \operatorname{dom}\Psi_\infty(T)$ we get the following equality which shows that $\theta\xi \in \operatorname{dom}\Psi_\infty(T)$

$$\langle \Psi_\infty(T)\eta, \theta\xi\rangle = \langle \Psi_\infty(T)\varphi\eta, \theta\xi\rangle$$

$$= \langle \Psi(T)\varphi\eta, \theta\xi\rangle = \langle \varphi\eta, \Psi(T)^*\theta\xi\rangle = \langle \eta, \Psi(T)^*\theta\xi\rangle.$$

□

We introduce now the notion of partitioned module. Let

- $(\mathcal{E}, D, \tau)$ be an even unbounded module with $D = D^*$.
- $(\mathcal{E}_b, T)$ a closed module over B.

Suppose that there exist projections q^-, p and q^+ in $\mathcal{L}(\mathcal{E})$ such that $1 = q^- + p + q^+$ and

$$\operatorname{im} p \simeq \mathcal{E}_b \otimes L^2([0,1]) \otimes \mathbf{C}^2.$$

Assume that the involution τ commutes with $q^\pm$ and restricts on $\operatorname{im} p$ to $1 \otimes 1 \otimes \tau_0\tau_1$. For any $\psi \in C([0,1])$, let $a(\psi) \in \mathcal{L}(\mathcal{E})$ be the operator given by the formula

$$a(\psi) = \psi(0)q^- + m(\psi) + \psi(1)q^+.$$

Definition 3.2 With these hypothesis, the unbounded module $(\mathcal{E}, D, \tau)$ is partitioned by $(\mathcal{E}_b, T)$ if the following properties hold

a. For every $\varphi \in C_c^\infty([0,1])$, we have $a(\varphi)\operatorname{dom} D \subset \operatorname{dom} D$

b. If $\varphi \in C_c^\infty(]0,1[)$, and if $\xi \in \operatorname{dom} D$ or $\xi \in \operatorname{dom}\Psi(T)$, then

$$Da(\varphi)\xi = \Psi(T)\varphi\xi.$$

c. If $\varphi, \psi \in C^\infty([0,1])$ such that $\operatorname{support}\varphi \cap \operatorname{support}\psi = \emptyset$, then

$$a(\varphi)Da(\psi) = 0.$$

From this definition, we see that the operator D is well known on $\operatorname{im} p$ and condition *b)* gives a kind of locality property of this operator. There is no interplay of the action of D on $\operatorname{im} q^-$, $\operatorname{im} p$, and $\operatorname{im} q^+$.

Example 3.3 The simplest construction is given by $\mathcal{E} = \mathcal{E}_b \otimes L^2(\mathbf{R}) \otimes \mathbf{C}^2$ with $D = \Psi_\infty(T)$, with obvious formulae for p and $q^\pm$.

Lemma 3.4 *For any $\varphi \in C^\infty([0,1])$, we have $a(\varphi) \in \mathcal{L}(D)$. If $\varphi(0) = \varphi(1) = 0$, then $a(\varphi)(D+i)^{-1} \in \mathcal{K}(\mathcal{E})$.*

Proof. Suppose first support$(\varphi) \subset]0,1[$, and let $\varphi \in C_c^\infty(]0,1[)$ such that $\psi\varphi = \varphi$. Then by properties b, c of the definition, one has $[D, a(\varphi)] = [D, a(\varphi)]\psi = [\Psi(T), a(\varphi)]\psi$. Thus $[D, a(\varphi)] = (1 \otimes \varphi' \otimes \tau_1)p$ is bounded.

Suppose then $\varphi(0) = \varphi(1) = 0$. Then there exists a sequence φ_k in $C_c^\infty(]0,1[)$ such that φ_k' converges uniformly to φ'. For $\xi \in \operatorname{dom} D$, one has $\lim D\varphi_k\xi = \lim([D,\varphi_k] + \varphi_k D)\xi = \lim(1 \otimes \varphi_k' \otimes \tau_1)p\xi + \lim \varphi_k D\xi = (1 \otimes \varphi' \otimes \tau_1)p\xi + \varphi D\xi$. As D is closed, $D\varphi\xi = \lim D\varphi_k\xi$, and thus $[D,\varphi] = (1 \otimes \varphi' \otimes \tau_1)p$ is bounded.

In general every φ can be written $\varphi = \varphi_1 + \varphi_2$ where support $\varphi_1 \subset [0, \frac{1}{2}[$ and $\varphi_2(0) = \varphi_2(1)$. Suppose then that support $\varphi \subset [0, \frac{1}{2}[$ and let $\tilde{\varphi}(t) = \varphi(t)$ on $[0, \frac{1}{2}[$ and $\tilde{\varphi}(t) = \varphi(1-t)$ on $]\frac{1}{2}, 0]$. Then one has $a(\tilde{\varphi}) \in \mathcal{L}(D)$ and thus $a(\varphi) \in \mathcal{L}(D)$ by property c.

To prove that $a(\varphi)(D+i)^{-1} \in \mathcal{K}(\mathcal{E})$, it suffices by continuity to take $\varphi \in C_c^\infty(]0,1[)$. The closure $\mathcal{W}_0$ on $\mathcal{W}(D)$ of $\mathcal{E}_b \otimes C_c^\infty(]0,1[) \otimes \mathbf{C}^2$ is isomorphic $\mathcal{W}(\Psi(T))$, and we have a factorization of $a(\varphi)(i+D)^{-1}$ by

$$\mathcal{E} \xrightarrow{(i+D)^{-1}} \mathcal{W}(D) \xrightarrow{a(\varphi)} \mathcal{W}_0 \longrightarrow \mathcal{E}\cdot$$

By Lemma 2.3, the map $\xi \to a(\varphi)\xi$ from $\mathcal{W}(D)$ to $\mathcal{W}_0$ is adjointable, and the injection $\mathcal{W}_0 \to \mathcal{E}$ is compact. □

We come now to the main notion of this article. Let

- $(\mathcal{E}, S, \tau)$ be an even symmetric module over B.
- $(\mathcal{E}_b, T)$ a closed module over B.

Suppose that there exist projections q, p in $\mathcal{L}(\mathcal{E})$ such that $1 = q + p$ and

$$\operatorname{im} p \simeq \mathcal{E}_b \otimes L^2([0,1]) \otimes \mathbf{C}^2.$$

Assume that the involution τ commutes with q and restricts on $\operatorname{im} p$ to $1 \otimes 1 \otimes \tau_0\tau_1$. For any $\psi \in C([0,1])$, let $b(\psi) \in \mathcal{L}(\mathcal{E})$ be the operator given by the formula

$$b(\psi) = m(\psi) + \psi(1)q.$$

Definition 3.5 With these notations, the closed module $(\mathcal{E}_b, T)$ is the boundary of $(\mathcal{E}, S, \tau)$ if the following properties hold

a. For every $\varphi \in C_c^\infty([0,1])$, with $\varphi(0) = 0$, one has $b(\varphi)(\operatorname{dom} S^*) \subset \operatorname{dom} S$, and $S^* b(\varphi) = S b(\varphi)$.

b. If $\varphi \in C_c^\infty(]0,1[)$, and if $\xi \in \operatorname{dom} S$ or $\xi \in \operatorname{dom} \Psi(T)$, then

$$S\varphi\xi = \Psi(T)\varphi\xi.$$

c. For every $\psi \in C^\infty([0,1])$ with $\operatorname{support}\psi \cap \operatorname{support}\varphi = \emptyset$, one has $b(\varphi)S^*b(\psi) = 0$.

Example 3.6 We can take $\mathcal{E} = \mathcal{E}_b \otimes L^2([0,+\infty[) \otimes \mathbf{C}^2$ with S the closure on $C_c(]0,+\infty[)$ of the matrix given by the formula (3.1), in which case the boundary is given by $(\mathcal{E}_b, T)$.

Remark 3.7 Suppose $(\mathcal{E}, S, \tau)$ and $(\mathcal{E}_b, T)$ satisfies the preceding conditions, but with $b(\psi)$ replaced by $\tilde{b}(\psi) = m(\psi) + \psi(1)q$ and the condition a replaced by the analogous condition $\tilde{a}$ with $\varphi(1) = 0$. The one says that $(\mathcal{E}_b, T)$ is a right boundary. The two definition are related as follows: we see that under the unitary U of $\mathcal{E}_b \otimes L^2(\mathbf{R}, \mathbf{C})$ implemented by the map

$$U : \begin{pmatrix} \xi_1(u) \\ \xi_2(u) \end{pmatrix} \to \begin{pmatrix} \xi_1(1-u) \\ -\xi_2(1-u) \end{pmatrix},$$

from $L^2([0,1]) \otimes \mathbf{C}^2$ to $L^2([0,1]) \otimes \mathbf{C}^2$, the operator $\Psi(-T)$ becomes

$$U^*\Psi(-T)U = \begin{pmatrix} 0 & -\frac{\partial}{\partial u} + T \\ \frac{\partial}{\partial u} + T & 0 \end{pmatrix}.$$

We see then that if $(\mathcal{E}_b, -T)$ is the boundary of $(\mathcal{E}, S, \tau)$ in the sens of the definition 3.8, then replacing $b(\psi)$ by $\tilde{b}(\psi) = (q + pU)b(\psi)(q + pU)$, the closed module $(\mathcal{E}_b, T)$ becomes the right boundary of $(\mathcal{E}, S, \tau)$ (and inversely).

Lemma 3.8 *Let $(\mathcal{E}, S, \tau)$ be an unbounded module with boundary as in the definition above. Then for any $\varphi \in C^\infty([0,1])$, one has $b(\varphi) \in \mathcal{L}(S)$ and $b(\varphi) \in \mathcal{L}(S^*)$.*

Proof. The proof of $b(\varphi) \in \mathcal{L}(S)$ is the same as in Lemma 3.4. The last assertion follows then from Lemma 2.1. □

We relate now the two notions of partition and boundary. Let $(\tilde{\mathcal{E}}, D, \tau)$ be an unbounded module partitioned by $(\mathcal{E}_b, T)$ with projections q^-, p, q^+ and satisfying to the definition 3.2. We can form the Hilbert module $\mathcal{E} = \operatorname{im} p + q^+$ and set $b(\varphi) = (p + q^+)a(\varphi)$ for $\varphi \in C([0,1])$. Let S the operator obtained by taking the closure of the restriction of D to the subspace linearly spanned by $\xi \in \operatorname{dom} D$ such that $a(\varphi)\xi = 0$ for some $\varphi \in C_c(]0,1])$.

Proposition 3.9 *The triple $(\mathcal{E}, S, \tau)$ is a symmetric module which admits $(\mathcal{E}_b, T)$ as boundary. Moreover, if $q^-(1 + D^2)^{-1} \in \mathcal{K}(\tilde{(\mathcal{E})})$, then one has $(1 + S^*S)^{-1} \in \mathcal{K}(\mathcal{E})$.*

Proof. For the first assertion, the only point to check is the regularity of S. Let $\theta \in C_c^\infty([0,1[)$ such that $\theta \equiv 1$ in a neighbourhood of 0. Then for

$\xi \in \operatorname{dom} D$, $\eta \in \operatorname{dom} \Psi(T)$, one has $a(1-\theta)\xi \in \operatorname{dom} S$, $b(\theta)\eta \in \operatorname{dom} S$ and

$$S(a(1-\theta)\xi + b(\theta)\eta) = Da(1-\theta)\xi + \Psi(T)b(\theta)\eta.$$

From Lemma 2.3, we see that that the map $(\xi, \eta) \to (a(1-\theta)\xi, b(\theta)\eta)$ from $\mathcal{W}(D) \oplus \mathcal{W}(\Psi(T))$ to $\mathcal{W}(S)$ is a surjective and adjointable and the formally same map to $\mathcal{E}$ is adjointable too. By a classical result of Mishchenko, there exists a right inverse to this map, and thus the canonical injection $\mathcal{W}(S) \to \mathcal{E}$ which factors through a product of adjointable maps, is in $\mathcal{L}(\mathcal{W}(S), \mathcal{E})$.

For the second assertion, let $\mathcal{W} \subset \mathcal{W}(D)$ be the closed subspace generated by the elements $\xi \in \operatorname{dom} D$ such that $q^+\xi = 0$. The hypothesis implies that the natural injection $\mathcal{W} \to \tilde{\mathcal{E}}$ is compact but as $\mathcal{W} = \mathcal{W}(S)$, this is equivalent to suppose that the natural injection $\mathcal{W} \to \mathcal{E}$ is compact. □

Inversely, let $(\mathcal{E}, S, \tau)$ a symmetric module with boundary $(\mathcal{E}_b, T)$ and projectors p, q satisfying to the definition 3.5. Let $(\mathcal{E}_1, S_1)$ another symmetric module with *right* boundary $(\mathcal{E}_b, -T)$, p_1, q_1 the projectors, as in the remark above. We can form the Hilbert module $\tilde{\mathcal{E}} = \operatorname{im} q_1 \oplus \mathcal{E}_b \otimes L^2([0,1], \mathbf{C}^2) \oplus \operatorname{im} q$. Let $\theta \in C^\infty([0,1])$ such that $\theta \equiv 1$ in a neighbourhood of 1, and $\theta \equiv 0$ near 0, and set D the operator with $\operatorname{dom} D = b_1(\theta) \operatorname{dom} S_1 + b(1-\theta) \operatorname{dom} S$ and for $\xi \in \operatorname{dom} S$ and $\zeta \in \operatorname{dom} S_1$

$$D(b_1(\theta)\xi + b(1-\theta)\zeta) = (1 \otimes U)S_1 b_1(\theta)\xi + Sb(1-\theta)\zeta. \tag{3.2}$$

In order to ensure that D is well defined, we have to show that $b_1(\theta)\xi + b(1-\theta)\zeta = 0$ implies $S_1 b_1(\theta)\xi + Sb(1-\theta)\zeta = 0$. But there exists $\psi \in C_c^\infty(]0,1[)$ such that $\psi \equiv 1$ on support$(\theta(1-\theta)$ and thus $S_1 b_1(\theta)\xi + Sb(1-\theta)\zeta = \Psi(T)\psi(b_1(\theta)\xi + b(1-\theta)\zeta) = 0$. Analogously, Lemma 3.1 shows that the operator D is symmetric and does not depend of the choice of θ.

Proposition 3.10 *The triple $(\tilde{\mathcal{E}}, D, \tilde{\tau})$ is an unbounded module with $D = D^*$ and is partitioned by $(\mathcal{E}_b, T)$.*

Proof. Let us show that D is selfadjoint, and let $\zeta \in \operatorname{dom} D^*$. We have identically $\langle D\xi, \zeta\rangle = \langle \xi, D^*\zeta\rangle$ for $\xi = b_1(\theta)\xi + b(1-\theta)\eta$, and then by letting $\eta = 0$, Lemma 3.8 implies $b_1(\theta)\zeta \in \operatorname{dom} S_1^*$, and thus $b_1(\theta)\zeta \in \operatorname{dom} S_1$. Analogously, one has $b(\theta)\zeta \in \operatorname{dom} S$, which shows that $\operatorname{dom} D = \operatorname{dom} D^*$.

It follows then from Lemma 2.1 that the map $(\xi, \eta) \to (b_1(\theta)\xi, b(1-\theta)\eta)$ from $\mathcal{W}(S_1) \oplus \mathcal{W}(S)$ to $\mathcal{W}(D)$ is a surjective and adjointable map. There exists a right inverse to this map, and thus the canonical injection $\mathcal{W}(D) \to \tilde{\mathcal{E}}$ is adjointable. □

4 Callias type operators

In this section we introduce Callias type operator for partitioned Hilbert modules (cf. mem. [13]). Usually, such operators are defined for odd dimensional manifolds. The present definition concerns the even dimensional case, and agrees with U. Bunke's work [2].

Let now an even unbounded module $(\mathcal{E}, D, \tau)$ over B, with D selfadjoint. For $a \in \mathcal{L}(\mathcal{E})$, the operator $D + \tau a$ is a bounded perturbation of a regular operator, and is also regular, and let

$$F(a) = (D^2+1)^{-\frac{1}{2}}(D+\tau a).$$

As $(D+\tau)^2 = D^2+1$, one has $(D+\tau)^{-1} \in \mathcal{L}(\mathcal{E})$ and $F(1) = (D^2+1)^{-\frac{1}{2}}(D+\tau)$ is a unitary. The equality $F(a) = F(1) + \tau(D^2+1)^{-\frac{1}{2}}(a-1)$, shows then $F(a) \in \mathcal{L}(\mathcal{E})$.

Lemma 4.1 *Let $a = a^* \in \mathcal{L}(D)$, satisfying $[D,a](D+i)^{-1} \in \mathcal{K}(\mathcal{E})$ and $(a^2-1)(D+i)^{-1} \in \mathcal{K}(\mathcal{E})$. Then the following relations hold $F(a) - F(a)^* \in \mathcal{K}(\mathcal{E})$ and $F(a)^2 - 1 \in \mathcal{K}(\mathcal{E})$.*

Proof. We have $D(D^2+1)^{-\frac{1}{2}} = (D^2+1)^{-\frac{1}{2}}D$ and $(D^2+1)^{-\frac{1}{2}}\tau = \tau(D^2+1)^{-\frac{1}{2}}$. Thus to prove the first relation, it remains to show that $[a,(D^2+1)^{-\frac{1}{2}}] \in \mathcal{K}(\mathcal{E})$. It suffices to prove that $[a,(D^2+1)^{-1}] \in \mathcal{K}(\mathcal{E})$ and as $(i+D)^{-1}(-i+D)^{-1} = (1+D^2)^{-1}$, we are reduced to show that $[a,(D\pm i)^{-1}] \in \mathcal{K}(\mathcal{E})$. For $\xi \in \mathcal{E}$, one has $(i+D)^{-1}\xi \in \operatorname{dom} D$ and $[a,(i+D)^{-1}]\xi = (i+D)^{-1}[D,a](i+D)^{-1}\xi$, and by hypothesis this last operator is compact.

For the second relation, we have that $F(a)^2 = F(a)F(a)^*$ modulo $\mathcal{K}(\mathcal{E})$ and thus the following operator is compact

$$\begin{aligned} F(a)F(a)^* - 1 &= (D^2+1)^{-\frac{1}{2}}(D^2+a^2+\tau[D,a])(D^2+1)^{-\frac{1}{2}} - 1 \\ &= \quad (D^2+1)^{-\frac{1}{2}}(\tau[D,a]+a^2-1)(D^2+1)^{-\frac{1}{2}}. \end{aligned}$$

□

This lemma shows that under these conditions the couple $(\mathcal{E}, F(a))$ determines a class in $K_1(B)$ [21]. Recall that to the unbounded module $(\mathcal{E}_b, T)$ with $T = T^*$ and $(i+T)^{-1} \in \mathcal{K}(\mathcal{E})$ is also associated a class in $K_1(B)$ as follows: the operator $F = T(1+T^2)^{-1}$ satisfies $F - F^* \in \mathcal{K}(\mathcal{E}_b)$ and $F^2 - 1 \in \mathcal{K}(\mathcal{E}_b)$, and the class of $(\mathcal{E}_b, T)$ is by definition the one of the couple $(\mathcal{E}_b, F)$ in $K_1(B)$. For any $\psi \in C^\infty([0,1])$ such that $\psi(1) = -1$ and $\psi(0) = 1$, $\psi'(0) = \psi'(1) = 0$, we set $F_\psi = F(a(\psi))$. From the Lemma 4.1, the couple $(\mathcal{E}, F_\psi)$ determines also a class in $K_1(B)$.

Theorem 4.2 *Let $(\mathcal{E}_b, T)$ be an unbounded module over B, which partitions the unbounded module $(\mathcal{E}, D, \tau)$. For any ψ as above, the classes in $K_1(B)$ of $(\mathcal{E}_b, T)$ and of $(\mathcal{E}, F_\psi)$ are equal.*

Proof. We prove first the assertion in the particular case where $\mathcal{E} = \mathcal{E}_b \otimes L^2(\mathbf{R}, \mathbf{C}^2)$ and $D = \Psi_\infty(T)$. We know that $K_i(B)$ is canonically isomorphic with the bivariant group $KK^i(\mathbf{C}, B)$ of G. Kasparov. Let $\alpha \in KK^1(\mathbf{C}, C_0(\mathbf{R}))$ and $\beta \in KK^1(C_0(\mathbf{R}), \mathbf{C})$ be the dual Dirac and Dirac elements as introduced in [9]. They are mutually inverse, in the sense that $\alpha \otimes_{C_0(\mathbf{R})} \beta = Id$ and $\beta \otimes_{\mathbf{C}} \alpha = Id$. Let $z = [\mathcal{E}_b, T]$, and form the external cup-product $\beta \otimes_{\mathbf{C}} z \in KK^0(C_0(\mathbf{R}), B)$.

Then, by using the characterization of cup products by connections [3, IV.A], one finds that the class of F_ψ in $KK^1(\mathbf{C}, B)$ is the cup product of α with $\beta \otimes_{\mathbf{C}} z$. By associativity of the cup-product, one finds

$$[\mathcal{E}, F_\psi] = \alpha \otimes_{C_0(\mathbf{R})} (\beta \otimes_{\mathbf{C}} z) = (\alpha \otimes_{C_0(\mathbf{R})} \beta) \otimes_{\mathbf{C}} z = z.$$

For the general case, one can suppose that $\mathcal{E} = \mathcal{E}_b \otimes L^2(\mathbf{R}) \otimes \mathbf{C}^2$. Indeed,let $\mathcal{E}_1 = \mathcal{E} \oplus \mathcal{E}_0$ where $\mathcal{E}_0 = B \otimes L^2(\mathbf{R}) \otimes \mathbf{C}^2$, $\tau_1 = \tau \oplus \tau_0$ and $D_1 = D \oplus 1 \otimes L$ where $L = \frac{\partial}{\partial t} \otimes \tau_1$. By the first step the module $(\mathcal{E}_0, 1 \otimes L)$ gives the null class in $K_1(B)$, and the classes defined by D and D_1 in $K_1(B)$ are then equal. By the stabilisation theorem of G. Kasparov ([21], Theorem 6.2), the modules $\ker(\tau_1 \pm 1)$ are stabilized, and isomorphic to $\mathcal{E}_b \otimes L^2(\mathbf{R})$. Let now for $t \in]0,1]$ the Hilbert module given by

$$\mathcal{E}_t = \operatorname{im} q^- \oplus (\mathcal{E}_b \otimes L^2([-t^{-1}, t^{-1}], \mathbf{C}^2)) \oplus \operatorname{im} q^+,$$

and $\mathcal{E}_0 = \mathcal{E}_b \otimes L^2(\mathbf{R}, C^2)$. By using the Propositions 3.10 and 3.9, one gets an evident selfadjoint operator D_t on $\mathcal{E}_t$. Let $D_0 = \Psi_\infty(T)$. Then with obvious notations the family $F_t = (D_t + a_t(\psi))(D^2+1)^{-\frac{1}{2}}$ gives a homotopy between F_0 and F_1 in $KK^1(\mathbf{C}, B)$, and thus $[F_1] = [F_0] = [\mathcal{E}_b, T]$. □

Theorem 4.3 *If the closed module $(\mathcal{E}_b, T)$ is the boundary of a symmetric module $(\mathcal{E}, S, \tau)$ such that $(1 + S^*S)^{-1} \in \mathcal{K}(\mathcal{E})$, then one has $[\mathcal{E}_b, T] = 0$ in the group $K_1(B)$.*

Proof. Let $\hat{\mathcal{E}}$ be the partitioned Hilbert module obtained by adding $\mathcal{E}_b \otimes L^2(] - \infty, 0]) \otimes \mathbf{C}^2$ and let D the corresponding operator. For any $\psi \in C^\infty[0,1]$ with $\psi(0) = 1$, $\psi(1) = -1$ and $\psi'(0) = \psi'(1) = 0$, we have $a(\psi) - 1 \in \mathcal{K}(D)$, and thus $F(a(\psi)) - F(1) \in \mathcal{K}(\hat{\mathcal{E}})$, where $F(a) = (D^2+1)^{-\frac{1}{2}}(D + \tau a)$ is defined as before. As $F(1)$ is invertible, by Theorem 4.2, we have $[\mathcal{E}_b, T] = [\hat{\mathcal{E}}, F(1)] = 0$ in $K_1(B)$. □

5 Bordism of foliations

Let (W, F) be a closed (i.e. compact boundaryless) foliated manifold of class $C^{0,\infty}$. We suppose for simplicity that $\dim F$ is odd.

We need the longitudinal index class of the foliation (W, F) [3], which is the subject of the minicourse by H. Moriyoshi in these proceedings [15]. We recall briefly its definition.

Let G be the holonomy groupoid of the foliation (W, F) with source and target maps $r, s : G \to V$, and $C^*(G)$ the reduced C*-algebra of G, also denoted $C^*(W, F)$. We assume that the tangent bundle to the foliation F is spin and let S be a spinor bundle, and A_l the longitudinal Dirac operator acting $C^{0,\infty}(W, S)$.

Then by taking the pull back $s^*(S)$ of S on G, one can form a Hilbert module $\mathcal{E}$ over B, and the pull back of A_l on G is a closable selfadjoint regular operator on $\mathcal{E}$. Letting A its closure, then the couple $(\mathcal{E}, A)$ is a closed module.

The class of $(\mathcal{E}, A)$ in $K_1(C^*(W, F))$ is the longitudinal index class of the spin structure on F.

Let (V, E) be a connected compact foliated manifold such that W is a union of connected component of the boundary V, and that the foliation E is transverse to W and intersects W along F.

By [5], Lemma 3.5, one has a tubular neighbourhood U of W with a $C^{0,\infty}$-diffeomorphism $\theta : U \to W \times [0, 1]$, such that $E \simeq F \oplus (U \times \mathbf{R})$ on U w.r.t. θ. We may choose a riemannian structure g along the leaves on V satisfying

$$g = dt^2 \oplus h,$$

where h is a riemannian structure along the leaves of F, and we fix a spinor bundle S_E for E compatible with S, in the sens that the restriction of S_E to U is equal to the product of S by the standard spin bundle of $T[0, 1] = [0, 1] \times \mathbf{R}$. Now let D_l be the longitudinal Dirac operator acting on $C^{0,\infty}(V, S_E)$. Then one has an isomorphism $C^{0,\infty}(U, S_E) \simeq C^{0,\infty}(W, S) \otimes C_c^\infty([0, 1]) \otimes \mathbf{C}^2$, and one has [13]

$$D_l = \begin{pmatrix} 0 & -\frac{\partial}{\partial u} + A_l \\ \frac{\partial}{\partial u} + A_l & 0 \end{pmatrix}. \tag{5.1}$$

Let H the holonomy groupoid of the foliation (V, E) and let $H_W^W = \{\gamma; r(\gamma), s(\gamma) \in W\} \subset H$. Then G is a closed and open subgroupoid of H_W^W, in general non trivial, and H_W^W is a closed subgroupoid of H.

The restriction of the source map $s : H \to V$ to H_W^W is a covering of the foliation F. The pullback by this map of S and of A_l gives rise to a closed module $(\mathcal{E}_b, \tilde{A})$ over $C^*(H_W^W)$.

As G is open and closed in H_W^W, one has a morphism $C^*(W,F) \to C^*(H_W^W)$, and a map $\Phi : K_*(C^*(W,F)) \to K_*(C^*(H_W^W))$, and by construction, one has

$$\Phi([\mathcal{E}, A]) = [\mathcal{E}_b, \tilde{A}]. \tag{5.2}$$

Let $H_V^W = \{\gamma; s(\gamma) \in V, r(\gamma) \in W\} \subset H$. Then similarly, there is a Hilbert $C^*(H_W^W)$-module $\mathcal{J}$ associated to $s^*(S_E)$, and the longitudinal Dirac operator on E determines an operator D on $\mathcal{J}$. Let $\mathcal{E}_+ \subset \mathcal{J}$ the submodule generated by section of $s^*(S_E)$ whith support in the complementary of $r^*(U)$.

Recall that a submanifold W of V transverse to E is said to be *faithful* if it intersects all the leaves of E.

Lemma 5.1 *Using $\theta : U \to [0,1] \times W$, one has an isomorphism $\mathcal{J} = (\mathcal{E}_b \otimes L^2([0,1], \mathbf{C}^2)) \oplus \mathcal{E}^+$. With respect to this isomorphism, the closed module $(\mathcal{E}_b, \tilde{A})$ is the boundary of $(\mathcal{J}, D, \tau)$.*

Moreover, if W is a faithful transverse to the foliation on V, then $(1 + D^*D)^{-1}$ is a compact operator of $\mathcal{J}$.

Proof. The restriction groupoid H_V^W gives rise to a $C^*(V,E) - C^*(H_W^W)$ Hilbert bimodule. Let D_V be the longitudinal Dirac operator on $C^*(V,E)$. Then $D = D_V \otimes_{C^*(V,E)} Id$.

On the complete manifold $\hat{V} = (]-\infty, 0] \times W) \cup V$ one has a foliation $\hat{E}$ of class $C^{0,\infty}$ which extends trivially E. With evident notations, by [Proposition 5.2] [7], the longitudinal operator $\hat{D}_V$ is regular on $\hat{E}$. Thus, by Proposition 3.9, D_V is regular and $(1 + D_V^* D_V)^{-1}$ is compact. This implies that D is regular.

If W is faithful, there the bimodule gives a Morita equivalence between $C^*(V,E)$ and $C^*(H_W^W)$. Then as $(1 + D_V^* D_V)^{-1}$ is compact, the operator $(1 + D^*D)^{-1}$ is compact. □

This last lemma and Theorem 4.3 give now immediately

Theorem 5.2 *Suppose that $\partial V = W$ and that W is a faithful tranverse to (V,E). Then one has $\Phi([\mathcal{E}, A]) = 0$ in $K_*(C^*(H_W^W))$.*

Remark 5.3 Last theorem is not true if the hypothesis W faithful is dropped. Here is a counter-example communicated to us by the referee whom we thanks gratefully.

Let T^2 be the two-torus and $F \subset T(T^2)$ a Kronecker foliation of irrationnal slope. Then the Thom isomorphism in K-theory for $\mathbf{R}$-actions [3] shows that the longitudinal index class from the spin structure is a generator in $K_1(C^*(T^2, F)) = \mathbf{Z}^2$.

Let $V_1 = T^2 \times [\frac{1}{2}, 1]$ and let $W = T^2 \times \{1\}$ and $W_1 = T^2 \times \{\frac{1}{2}\}$. There exists a two dimensional foliation on V_1 transverse to W and inducing F, and for which W_1 is a leaf: this foliation can be obtained by extending the product foliation near W and tabulating leaves by rotating along W_1 (leaves wind around W_1). Let E the foliation on the solid torus $V = D^2 \times S^1$ obtained by extending this foliation by a Reeb component on the complementary of $V_1 - W_1$.

Then W is not faithful, $C^*(H_W^W)$ is Morita equivalent to $C^*(T^2, F)$, and $\Phi([\mathcal{E}, A]) \neq 0$. Therefore, the conclusion of the theorem is not satisfied.

Now suppose that $\partial V = W = W_1 \cup W_2$ where $(W_1, F_1), (W_2, F_2)$ be are closed manifold with spin foliation of class $C^{0,\infty}$, and that there exist a spin transverse foliated bordism (V, E) between them. Let B_j be the reduced C*-algebra of (W_j, F_j) and $\tilde{B}_j$ the reduced C*-algebra of the groupoid $G_{W_j}^{W_j}$, and $\Phi_j : K_*(B_j) \to K_*(\tilde{B}_j)$. Again , we assume that W is faithful; the under these conditions

Corollary 5.4 *If* $\Phi_1([\mathcal{E}_1, A_1]) \neq 0$, *then* $[\mathcal{E}_2, A_2] \neq 0$.

Proof. There is a Hilbert module $\mathcal{E}_j$ over $C^*(H_W^W)$ built from $H_{W_j}^W = s^{-1}(W_j) \cap r^{-1}(W)$, with a representation $C^*(B_j) \to \mathcal{L}(\mathcal{E}_j)$. This bimodule gives a class in the bivariant K-theory group $[\mathcal{E}_j] \in KK(\tilde{B}_j, C^*(H_W^W))$, such that the associated homomorphism $K_1(\tilde{B}_j) \to K_1(C^*(H_W^W))$ is injective. Keeping the above notations, one has $C^*(W, F) = B_1 \oplus B_2$ and $\Phi = [\mathcal{E}_1] \circ \Phi_1 - [\mathcal{E}_2] \circ \Phi_2$. By the last proposition and using injectivity, one gets the conclusion. □

Remark 5.5 As mentioned in the introduction, there are analytical proofs of bordism invariance in various geometric situations [4, 1, 14, 16, 12]. It relies essentially on the exact sequence in bivariant K-theory for B a C*-algebra of G. Kasparov [9]

$$\to KK^0(C_0(V - W), B) \to KK^1(C(W), B) \to KK^1(C(V), B) \to,$$

where $\partial V = W$ and B is a separable C*-algebra.

It is less elementary as it uses pseudodifferential calculus. However, one can guess that its works for $B = C^*(V, E)$ and this would give in that situation an alternative proof of Theorem 5.2.

6 Foliation with longitudinal positive scalar curvature

Let (W, F) be a foliated manifold of class $C^{0,\infty}$ and g a $C^{0,\infty}$-riemannian structure along the leaves. The longitudinal scalar curvature is the smooth function $\kappa_l : V \to \mathbf{R}$ the value of which at $x \in V$ is equal to the scalar curvature of the restriction of g to leaf which contains x.

One may asks under which condition κ_l should be positive. J. Rosenberg [19] and A. Connes [3] obtained the following

Proposition 6.1 *Let (W, g) be a smooth spin foliated manifold and g a longitudinal riemannian metric with strictly positive scalar curvature. Then the longitudinal index class of the spin structure in $K_*(C^*(W, F))$ is nul.*

Using the result of the last section , one then obtains

Corollary 6.2 *Let $(W_1, F_1), (W_2, F_2)$ be closed manifold with spin foliations of class $C^{0,\infty}$, and let (V, F) a transverse faithful bordism between (W_1, F_1) and (W_2, F_2). If $\Phi_1([\mathcal{E}_1, A_1]) \neq 0$, then (W_2, F_2) cannot carry a riemannian structure with strictly positive longitudinal scalar curvature.*

Now we focus on the following situation: (V, E) is foliated manifold of class $C^{0,\infty}$ (V is not assumed to be compact), and (W, F) a codimension one closed foliated submanifold, faithfully transverse to E. We choose a tubular neighboorhood $U \simeq [-1, 1] \times W$ and the foliation E and F are assumed to have compatible spin structure. We define as above $(\mathcal{E}, A)$ (resp. $(\mathcal{E}_b, \tilde{A})$, $(\mathcal{J}, D, \tau)$) the longitudinal index classes of $C^*(W, F)$ (resp. $C^*(G_W^W)$).

Suppose now that $V$is *separated* or *partitioned* by W, which means that $V - W = V_+ \cup V_-$ where V_-, V_+ are open and $V_- \cap V_+ = \emptyset$.

Then as in Lemma 5.1, the unbounded module $(\mathcal{J}, D, \tau)$ is partitioned by $(\tilde{\mathcal{E}}, \tilde{A})$.

In this situation, we obtain a generalization of a previous results of M. Gromov-B. Lawson [11] and of J. Roe [18].

Proposition 6.3 *Let (W, F) be a closed spin foliated manifold of class $C^{0,\infty}$ which separates faithfully (V, E) as in conditions above. If there exists $c > 0$ such that the longitudinal scalar curvature satisfies $\kappa > c$ on V, then $\Phi([\mathcal{E}, A]) = 0$ in $K_*(C^*(H_W^W))$.*

Proof. The Bochner-Weitzenböck formula tells us that

$$D^2 = \tilde{\nabla}^* \tilde{\nabla} + \tilde{\kappa},$$

where $\tilde{\kappa} = \kappa \circ r$ and $\tilde{\nabla}$ are the pullbacks of the Clifford connection and of the scalar curvature on $\mathcal{J}$.

Let ψ be a smooth function on V which modulo $C_c(V)$ is equal to 1 on V_- and to -1 on V_- and such that $\|d\psi\| \leq \frac{\kappa}{2}$. Let then $F_\psi = (1 + D^2)^{\frac{1}{2}}(D + \tau\tilde{\psi})$ is invertible, with $\tilde{\psi}$ is the pullback of ψ. One $\|F_\psi F_\psi^* - 1\| = \|(1 + D_V^2)^{\frac{1}{2}}(\tilde{\psi}^2 - 1 + [D_V, \tilde{\psi}])(1 + D_V^2)^{\frac{1}{2}}\| < 1$, which shows that F_ψ is invertible. By Theorem 4.2, it follows that $[\tilde{\mathcal{E}}, \tilde{A}] = 0$ in $K_1(C^*(H_W^W))$. □

7 Example: the smooth case

We illustrate this by several geometrical examples. There are spin foliated manifolds for which we know that $[\mathcal{E}, A] \neq 0$, for example all foliations which comes from a free action of a simply connected solvable Lie group II.C [3]. Here are another examples.

Suppose that (W, F) is a foliation of classe $C^{\infty,\infty}$. Let $\mathcal{R} \subset H^*(W, \mathbf{R})$ be the subring generated by Pontryagin classes of the quotient vector bundle TW/F, Chern classes of holonomy invariant complex vector bundles on W, and the image of WO_q in $H^*(W, \mathbf{R})$ by the Weil homomorphism ($q = \dim W - \dim F$). Then A. Connes has proved that every $\omega \in \mathcal{R}$ determines a homomorphism $\omega_* : K_*(B) \to \mathbf{R}$ such that III.7 [3]

$$\omega_*([\mathcal{E}, A]) = \langle \hat{A}(F) \wedge \omega, [W] \rangle, \tag{7.1}$$

where $\hat{A}(F)$ is the *$\hat{A}$-genus* of the tangent bundle of the manifold, a universal polynomial in the Pontryagyn classes of F.

Proposition 7.1 *Let (W, F) be a foliated manifold of classe $C^{\infty,\infty}$ which is a faithful tranverse boundary component of a compact foliated manifold of class $C^{0,\infty}$. If there exists $\omega \in \mathcal{R}$ such that $\langle \hat{A}(F) \wedge \omega, [W] \rangle \neq 0$, then $\Phi([\mathcal{E}, A]) \neq 0$ in $K_1(C^*(H_W^W))$.*

Proof. Form the construction of the fundamental class [3], Section III.7, as H_W^W is itself a groupoid of local tranverse diffeomorphisms of (W, F), each $\omega \in \mathcal{R}$ defines similarly a map $\tilde{\omega}_* : K_*(C^*(H_W^W)) \to \mathbf{C}$ such $\tilde{\omega}_*([\mathcal{E}_b, \tilde{A}]) = \langle \hat{A}(F) \wedge \omega, [W] \rangle$. The assertion follows now from relation (5.2). □

Thus if (W, F) satisfies conditions of Theorem 5.2, or of Proposition 6.3, the longitudinal index class implies the annulation of all the caracteristic numbers in the r.h.s. of Equation (7.1).

There always exists $\omega \in H^*(W, \mathbf{R})$ such that $\hat{A}(F) \wedge \omega = \hat{A}(W)$; in particular, if $\Phi([\mathcal{E}, A]) = 0$, then one recovers Lichnerowicz's condition (1.1).

Remark 7.2 In the case where W is the boundary of V, the construction works if we only suppose that the foliations are oriented, by using the signature operator alongs the leaves instead of the Dirac operator. The condition in the last proposition becomes $\langle L(W) \wedge \omega, [W] \rangle \neq 0$, where $L(W)$ is the L-genus of the bundle F.

Furthermore, if (V, E) is assumed to be a class $C^{\infty,\infty}$ then Stokes' Theorem gives directly $\langle \hat{A}(F) \wedge \omega, [W] \rangle = 0$ for every $\omega \in \mathcal{R}$.

References

1. P. Baum and R.G. Douglas, *Relative K homology and C^*-algebras*, K-Theory, **5** (1991), no. 1, 1–46.
2. U. Bunke, *A K-theoretic relative index theorem and Callias-type Dirac operators*, Math. Ann., **303** (1995), 241–279.
3. A. Connes, Non-Commutative Geometry, Academic Press, New-York, 1994.
4. A. Connes and G. Skandalis, *The longitudinal index theorem*, Tohôku J. Math., **317** (1994), 521–526.
5. C. Godbillon, *Feuilletages: études géometriques*, Progress in Mathematics, no. 98, Birkhaüser Verlag, Leipzig, 1991.
6. N. Higson, *A note on the cobordism invariance of the index*, Topology, **30** (1991), 3.
7. M. Hilsum, *Fonctorialité en K-théorie bivariante pour les variétés lipschitziennes*, K-theory, **3** (1989), 401–440.
8. K. Jensen and K. Thomsen, Elements of KK-theory, Birkhaüser, Boston, 1991.
9. G.G. Kasparov, *The operator K-functor and extensions of C^*-algebras*, Math. U.S.S.R. Izv., **16** (1981), 513–572.
10. E.C. Lance, Hilbert C*-modules, L. N. Ser., no. 210, London Mathematical Society, Cambridge University Press, 1994.
11. B. Lawson and M.-L. Michelsohn, Spin Geometry, London Mathematical Society. L. N. Ser., no. 210, Cambridge University Press, Saint Louis, 1994.
12. E. Leichtnam and P. Piazza, *Spectral sections and higher Atiyah-Patodi-Singer index theory on Galois coverings*, G.A.F.A., **8** (1996), 3–26.
13. M. Lesch, Operators of Fuchs type, conical singularities, and asymptotic methods, Teubner-Texte zür Mathematik, no. 136, B.G. Teubner, Stuttgart, 1997.
14. R.B. Melrose and P. Piazza, *Families of Dirac operators, boundaries and the b-calculus*, J. Diff. Geom., **46** (1997), 99–179.
15. H. Moriyoshi, *Operators algebras and the index theorem for foliated manifolds*, Foliations: Geometry and Dynamics (Warsaw), in these proceedings, World Scientific, 2001, 127–155.
16. L. Nicolaescu, *Generalized symplectic geometries and the index of families of elliptic problems.*, Mem. Amer. Math. Soc., **609** (1997).
17. R.S. Palais, Seminar on the Atiyah-Singer index theorem, Annals of Math Studies, no. 21, Princeton University Press, Princeton, 1965.
18. J. Roe, *Partinioning noncompact manifolds and the dual Toeplitz prob-*

lem, Operator algebras and applications, no."?" London Math. Soc. Lecture Notes Ser., 135, Cambridge University Press, 1988.
19. J. Rosenberg, *C*-algebras, positive scalar curvature, and the Novikov conjecture.*, Publ. Math., Inst. Hautes Etud. Sci., **58** (1983), 409–424.
20. W. Shih, *Fiber cobordism and the index of a family of elliptic differential operators*, Bull. Amer. Math. Soc., **72** (1966), 984–991.
21. N.E. Wegge-Olsen, K-theory and C*-algebras, Oxford University Press, 1993.

Received November 3, 2000, revised July 3, 2001.

Proceedings of
FOLIATIONS: GEOMETRY AND DYNAMICS
held in Warsaw, May 29–June 9, 2000
ed. by Paweł WALCZAK *et al.*
World Scientific, Singapore, 2002
pp. 333–349

NON-EUCLIDEAN AFFINE LAMINATIONS

VADIM A. KAIMANOVICH
CNRS UMR-6625, IRMAR,
Université Rennes-1, France,
e-mail: kaimanov@univ-rennes1.fr

The purpose of the present paper is to discuss examples of affine Riemann surface laminations which do not admit a leafwise Euclidean structure. The first example of such a lamination was constructed by Ghys [5]. Our discussion is based on the geometric methods developed by Lyubich, Minsky and the author [13], [12], which rely on the observation that any affine surface A gives rise in a natural way to a hyperbolic 3-manifold $\mathfrak{H}A$ with a distinguished point at infinity. In particular, we give a new interpretation and a generalization of the example of Ghys.

1 Affine and hyperbolic laminations

In this Section we recall the basic facts on the relationship between affine and hyperbolic laminations. Although our exposition is self-contained, more details on this relationship can be found in [12].

1.A Affine and Euclidean surfaces

By endowing a Riemann surface S with an atlas of coordinate charts with transition maps from a given pseudo-group $\mathcal{C}$ (contained in the pseudo-group of all holomorphic maps) one can define finer geometric structures on S.

Definition 1.1 We shall say that S is

(i) an *affine Riemann surface*, if $\mathcal{C}$ is the group of all complex affine maps $z \mapsto az + b,\ a, b \in \mathbb{C}, a \neq 0$;

(ii) a *Euclidean surface*, if $\mathcal{C}$ is the group of all maps $z \mapsto az + b,\ a, b \in \mathbb{C}, |a| = 1$ (so that the transitions are Euclidean motions).

If S is an affine surface, then its tangent and cotangent bundles (and hence all tensor bundles) are endowed with a natural flat connection. Being *parallel* with respect to this connection means to have constant coefficients in any affine coordinate chart (the reader is referred to [3] and [8] for general notions from the theory of affine manifolds). So, one can talk about parallel vector fields, forms, Riemannian metrics, etc. on S. In these terms a Euclidean surface is just an affine surface endowed with a parallel conformal metric.

An affine Riemann surface structure is the same as a *complex affine structure*, or, in "real terms", a *projective Euclidean* ($\equiv$ *similarity*) structure. In particular, an *affine plane* is $\mathbb{R}^2$ endowed with the class of all multiples of a given Euclidean structure.

Any complete affine surface is a quotient of the affine plane by a freely acting discrete group of Euclidean motions. Therefore, for any such surface the affine structure can be refined to a Euclidean one, i.e. there exists a parallel conformal metric. However, this is no longer true if we pass from a single surface to a collection of surfaces assembled into a lamination.

1.B Hyperbolic space

Before describing a relationship between affine surfaces and 3-dimensional hyperbolic geometry, let us first recall the basic notions concerning the hyperbolic space.

The *sphere at infinity* $\partial\mathbf{H}^3$ is the boundary of the visibility compactification of the 3-dimensional hyperbolic space $\mathbf{H}^3$. The space $\mathbf{H}^3$ with a distinguished boundary point $q \in \partial\mathbf{H}^3$ is called *pointed at infinity*.

By $\mathcal{P}_q = \partial\mathbf{H}^3 \setminus \{q\}$ we denote the *punctured visibility sphere*. The hyperbolic space $\mathbf{H}^3$ is fibred over $\mathcal{P}_q$ by means of the projection $\mathfrak{p}_q$ which assigns to any $h \in \mathbf{H}^3$ the uniquely determined point $\mathfrak{p}_q(h) \in \mathcal{P}_q$ such that h lies on the geodesic joining q and $\mathfrak{p}_q(h)$.

Definition 1.2 ([9]) Let $(\mathbf{H}^3, q)$ be a pointed at infinity hyperbolic space. The choice of the point $q \in \partial\mathbf{H}^3$ determines the *Busemann cocycle* β_q on $\mathbf{H}^3 \times \mathbf{H}^3$ by the formula

$$\beta_q(h_1, h_2) = \lim_h \left[d(h_1, h) - d(h_2, h)\right] ,$$

where $h \in \mathbf{H}^3$ converges to q in the visibility topology.

Definition 1.3 The *horosphere* centred at q and passing through a point $h \in \mathbf{H}^3$ is the level set of the Busemann cocycle:

$$\mathrm{Hor}_q(h) = \left\{h' \in \mathbf{H}^3 : \beta_q(h, h') = 0\right\} .$$

By $\mathrm{Hor}(\mathbf{H}^3, q)$ we denote the *space of horospheres centred at* q, and by

$$\mathrm{Hor}(\mathbf{H}^3) = \bigcup_{q \in \partial \mathbf{H}^3} \mathrm{Hor}(\mathbf{H}^3, q)$$

the *space of all horospheres in* $\mathbf{H}^3$. The center of a horosphere $\Upsilon \in \mathrm{Hor}(\mathbf{H}^3)$ is denoted $\Upsilon_\infty \in \partial \mathbf{H}^3$.

Below we shall usually omit the subscript q when the point at infinity q is fixed.

These notions are best illustrated by looking at the *upper half-space model*

$$\mathbf{H}^3 \cong \mathbb{C} \times \mathbb{R}_+ = \{(z,t) : z \in \mathbb{C},\, t > 0\}, \qquad d\rho^2 = \frac{|dz|^2 + dt^2}{t^2}, \tag{1.4}$$

where $|dz|^2$ is the standard Euclidean metric on $\mathbb{C}$.

The geodesics in this model are either Euclidean half-circles orthogonal to the boundary plane or vertical lines, and the visibility sphere is the union of the distinguished point at infinity $q = \infty$ and the boundary plane

$$\mathcal{P} = \{(z,t) : z \in \mathbb{C},\, t = 0\} \cong \partial \mathbf{H}^3 \setminus \{q\}.$$

The Busemann cocycle with respect to the point $q = \infty$ is

$$\beta(h_1, h_2) = \log \frac{t(h_2)}{t(h_1)}. \tag{1.5}$$

Thus, the horosphere $\mathrm{Hor}(h)$ is the horizontal coordinate plane passing through h, and the map $\mathfrak{p}$ is the coordinate projection $h \mapsto z(h)$ (see Figure 1).

1.C Hyperbolization of affine surfaces

Proposition 1.6 *The punctured visibility sphere* $\mathcal{P} = \partial \mathbf{H}^3 \setminus \{q\}$ *of a pointed at infinity hyperbolic space* $(\mathbf{H}^3, q)$ *is endowed with a natural structure of an affine plane.*

Proof. The projection $\mathfrak{p}$ allows one to identify any horosphere $\mathrm{Hor}(h)$ with the punctured visibility sphere $\mathcal{P}$. Denote by ε_h the Riemannian metric on $\mathcal{P}$ obtained by restricting the hyperbolic metric onto $\mathrm{Hor}(h)$ and then projecting it onto $\mathcal{P}$. By formula (1.4)

$$d\varepsilon_h^2 = \frac{|dz|^2}{t^2}, \tag{1.7}$$

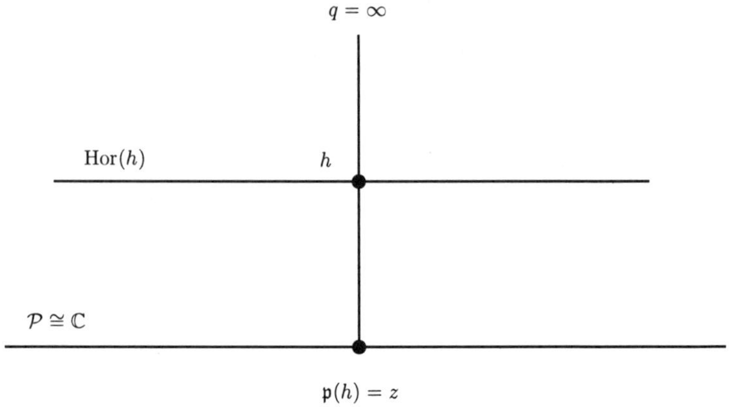

Figure 1.

so that $(\mathcal{P}, \varepsilon_h)$ is a Euclidean plane. Further, by (1.5) the structures ε_h for different h are all proportional, and

$$\frac{\varepsilon_{h_1}}{\varepsilon_{h_2}} = \exp\left[\beta(h_1, h_2)\right] . \tag{1.8}$$

□

Conversely, let us show that any affine surface gives rise to a pointed at infinity hyperbolic 3-manifold.

Definition 1.9 Let S be a Riemann surface. The elements of the *scaling bundle*

$$\mathfrak{p} : \mathfrak{H}S \to S$$

over S are conformal circles in the tangent spaces T_zS, $z \in S$. Any circle $h \in \mathfrak{H}S$ can be considered as the unit circle of the associated conformal Euclidean metric ε_h on $T_{\mathfrak{p}h}S$. Below we shall often identify h and ε_h, and consider $\mathfrak{H}S$ as the *bundle of conformal Euclidean metrics on tangent spaces* T_zS, $z \in S$.

We can use formula (1.8) to define $\beta(h_1, h_2)$ for any two points $h_1, h_2 \in \mathfrak{H}S$ from the same fibre. Therefore, any fibre of $\mathfrak{p}$ is endowed with the metric

$$d(h_1, h_2) = \left|\beta(h_1, h_2)\right| , \qquad \mathfrak{p}h_1 = \mathfrak{p}h_2 . \tag{1.10}$$

Proposition 1.11 *If A is an affine plane, then $\mathfrak{H}A$ is given a natural structure of a pointed at infinity hyperbolic space.*

Proof. We shall realize A as the boundary plane of the hyperbolic space in the upper half-space model. The fibres of the bundle $\mathfrak{p} : \mathfrak{H}A \to A$ are endowed with the metric (1.10), and the points of $\mathfrak{H}A$ are themselves metrics on tangent spaces $T_z A$. In order to combine them and produce a metric on $\mathfrak{H}A$ we shall need the affine connection over A.

Since A is simply connected, the affine connection (being flat) gives natural bijections between all fibres of $\mathfrak{p}$ which preserve the distance (1.10). These bijections are obtained by the parallel transport of metrics ε_h, $h \in \mathfrak{H}A$ and provide us with a product structure on $\mathfrak{H}A$. In other words, any metric ε_h extends from the tangent space $T_{\mathfrak{p}h}A$ to a Euclidean metric (also denoted ε_h) on A (note that because of this formula (1.8) for the Busemann cocycle now makes sense for all $h_1, h_2 \in \mathfrak{H}A$, not just for those from the same fibre).

Now taking the product of the metric ε_h and the fibre metric (1.10) gives precisely the Riemannian metric (1.4). □

We shall call the constructed correspondence $\mathfrak{H}$ between affine planes and pointed at infinity hyperbolic spaces the *hyperbolization functor*.

Remark **1.12** In the same way one can easily see that the hyperbolization functor $\mathfrak{H}$ is bijective between the category of complete affine surfaces and the category of complete hyperbolic 3-manifolds obtained by factorizing a pointed at infinity hyperbolic space $(\mathbf{H}^3, q)$ by a freely acting discrete horospheric group (i.e. the one which preserves the horospheres centred at q).

1.D Affine and Euclidean laminations

Let now $\mathcal{L}$ be a *Riemann surface lamination*, i.e. the leaves of $\mathcal{L}$ are Riemann surfaces, and there is an atlas of local charts ("flow boxes") such that the transition maps are conformal and transversely continuous. Further restricting the class of leafwise transition maps one gets *affine* and *Euclidean* laminations (cf. Definition 1.1). Their leaves are affine and Euclidean surfaces, respectively. If the leaves of an affine lamination are isomorphic to the standard affine plane $\mathbb{C}$, we also call it a $\mathbb{C}$-*lamination*.

Now one can ask whether, given an affine lamination $\mathcal{A}$, its structure can be refined to that of a Euclidean lamination. In other words, whether one can choose a Euclidean ($\equiv$ parallel) metric on every leaf of $\mathcal{A}$ (in [5] this is called *uniformization*).

As we have already mentioned, a single leaf of an affine lamination is always uniformizable. However, the problem is that the leafwise Euclidean metrics have to be consistent with the lamination structure, i.e. have to be "well-behaved" when one passes from one leaf to another one. In the lamination setup (with no transverse smooth structure) one can deal either with the continuous or with the Borel category.

Definition 1.13 An affine lamination is *uniformizable* in the continuous (resp. Borel) category if it admits a leafwise parallel metric which is transversely continuous (resp. Borel).

Below for simplicity we shall only deal with the affine $\mathbb{C}$-laminations, although all the considerations carry over to general affine laminations as well (cf. Remark 1.12 and Remark 1.17 below).

Let $\mathcal{A}$ be a $\mathbb{C}$-lamination. Then the application of the hyperbolization functor $\mathfrak{H}$ to the leaves L of $\mathcal{A}$ gives rise to the lamination $\mathcal{H} = \mathfrak{H}\mathcal{A}$ (the *hyperbolization* of $\mathcal{A}$) whose leaves are pointed at infinity hyperbolic 3-spaces $\mathfrak{H}L$. The lamination $\mathcal{H}$ is a 1-dimensional fibre bundle $\mathfrak{p} : \mathcal{H} \to \mathcal{A}$ over $\mathcal{A}$. The leafwise Busemann cocycles piece together a global *Busemann cocycle* β on $\mathcal{H}$,

The elements of $\mathcal{H}$ can be considered as Euclidean conformal metrics on the tangent spaces to points of $\mathcal{A}$ (cf. Definition 1.9). Therefore, *sections* $\sigma : \mathcal{A} \to \mathcal{H}$ of the fibre bundle $\mathfrak{p}$ are in one-to-one correspondence with leafwise conformal Riemannian metrics on $\mathcal{A}$. Sections corresponding to parallel ($\equiv$ Euclidean) leafwise metrics are precisely those which consist of leafwise horospheres, i.e. those for which $\beta(\sigma(z_1), \sigma(z_2)) = 0$ for any z_1, z_2 from the same leaf in $\mathcal{A}$.

Theorem 1.14 *A $\mathbb{C}$-lamination $\mathcal{A}$ is uniformizable in the continuous (resp. Borel) category if and only if the Busemann cocycle on the lamination $\mathcal{H}$ is cohomologically trivial in the transversely continuous (resp. Borel) cohomology, i.e. if and only if there exists a transversely continuous (resp. Borel) function $f : \mathcal{H} \to \mathbb{R}$ such that*

$$\beta(h_1, h_2) = f(h_2) - f(h_1) \tag{1.15}$$

for any two points h_1, h_2 from the same leaf in $\mathcal{H}$.

Proof. If $\sigma : \mathcal{A} \to \mathcal{H}$ is the section corresponding to a parallel leafwise metric on $\mathcal{A}$, then

$$\begin{aligned}\beta(h_1, h_2) &= \beta\big(h_1, \sigma \circ \mathfrak{p}(h_1)\big) + \beta\big(\sigma \circ \mathfrak{p}(h_1), \sigma \circ \mathfrak{p}(h_2)\big) + \beta\big(\sigma \circ \mathfrak{p}(h_2), h_2\big) \\ &= f(h_2) - f(h_1)\end{aligned}$$

with $f(h) = \beta\big(\sigma \circ \mathfrak{p}(h), h\big)$. Therefore, the graph of the section σ coincides with the level set $f^{-1}(0)$.

Conversely, if β satisfies (1.15), then the level set $f^{-1}(0)$ is the graph of a uniquely defined parallel section σ.

Clearly, the function f and the section σ are continuous (resp. Borel) simultaneously. □

Corollary 1.16 *Let an affine lamination $\mathcal{A}$ be a quotient of a $\mathbb{C}$-lamination $\widetilde{\mathcal{A}}$ with respect to a discrete group G of automorphisms. Then $\mathcal{A}$ is uniformizable in the continuous (resp. Borel) category if and only if the Busemann cocycle on $\widetilde{\mathcal{A}}$ is cohomologically trivial by means of a certain transversely continuous (resp. Borel) G-invariant function on $\mathfrak{H}\widetilde{\mathcal{A}}$.*

Remark **1.17** In order to make Theorem 1.14 valid for general affine laminations one has to pass to the leafwise de Rham cohomology by replacing the Busemann cocycle with the appropriate differential 1-form (the differential of the Busemann cocycle with respect to the second argument). Then the above Corollary would follow from Theorem 1.14 directly applied to the quotient lamination $\mathcal{A}$.

2 Foliations and laminations associated with the hyperbolic space

2.A Tautological foliations

The simplest building blocks of an affine lamination are the standard affine planes. As we have seen, such a plane arises as the punctured visibility sphere $\mathcal{P}_q \cong \partial\mathbf{H}^3 \setminus \{q\}$ of a pointed at infinity hyperbolic space $(\mathbf{H}^3, q)$. Conversely, the hyperbolization functor $\mathfrak{H}$ allows one to recover the space $(\mathbf{H}^3, q)$ from $\mathcal{P}_q$. Varying the boundary points $q \in \partial\mathbf{H}^3$ we obtain a family of affine planes $\mathcal{P}_q$, $q \in \partial\mathbf{H}^3$.

Definition 2.1 The *tautological $\mathbb{C}$-foliation* $\mathcal{A}_0$ is the foliation of the locally compact total space

$$\partial^2\mathbf{H}^3 = \partial\mathbf{H}^3 \times \partial\mathbf{H}^3 \setminus \mathrm{diag} = \bigcup_{q\in\partial\mathbf{H}^3} \mathcal{P}_q \times \{q\} \tag{2.2}$$

with the leaves $\mathcal{P}_q$. Its hyperbolization $\mathcal{H}_0 = \mathfrak{H}\mathcal{A}_0$ is called the *tautological pointed at infinity hyperbolic foliation*. The total space of $\mathcal{H}_0$ is

$$\mathbf{H}^3 \times \partial\mathbf{H}^3 = \bigcup_{q\in\partial\mathbf{H}^3} \mathbf{H}^3 \times \{q\}$$

and the leaves are pointed at infinity hyperbolic spaces $(\mathbf{H}^3, q)$. The Busemann cocycle on $\mathcal{H}_0$ is

$$\beta\big((h_1, q), (h_2, q)\big) = \beta_q(h_1, h_2)\ .$$

2.B Parametrizations of the unit tangent bundle

Denote the *unit tangent bundle* of the hyperbolic space by $U\mathbf{H}^3$ with the canonical projection

$$p : U\mathbf{H}^3 \to \mathbf{H}^3 ,$$

and let $\gamma = \{\gamma^\tau\}_{\tau\in\mathbb{R}}$ be the *geodesic flow* on $U\mathbf{H}^3$. The endpoints of the geodesic determined by a tangent vector $v \in U\mathbf{H}^3$ are denoted $\gamma^{\pm\infty}(v) \in \partial\mathbf{H}^3$. By

$$\mathrm{Hor}(v) = \mathrm{Hor}_{\gamma^\infty(v)}\big(p(v)\big) = \{h \in \mathbf{H}^3 : \beta_{\gamma^\infty(v)}\big(p(v)v, h\big) = 0\}$$

we denote the horosphere centred at the point $\gamma^\infty(v)$ and passing through the point $p(v)$. Clearly,

$$\mathrm{Hor}(v_1) = \mathrm{Hor}(v_2) \iff \mathrm{Hor}(\gamma^\tau v_1) = \mathrm{Hor}(\gamma^\tau v_2)\ \forall\, \tau \in \mathbb{R} ,$$

so that the formula

$$\gamma^\tau\, \mathrm{Hor}(v) = \mathrm{Hor}(\gamma^\tau v) , \qquad \tau \in \mathbb{R} \tag{2.3}$$

determines an action of the geodesic flow on the space $\mathrm{Hor}(\mathbf{H}^3)$. There are

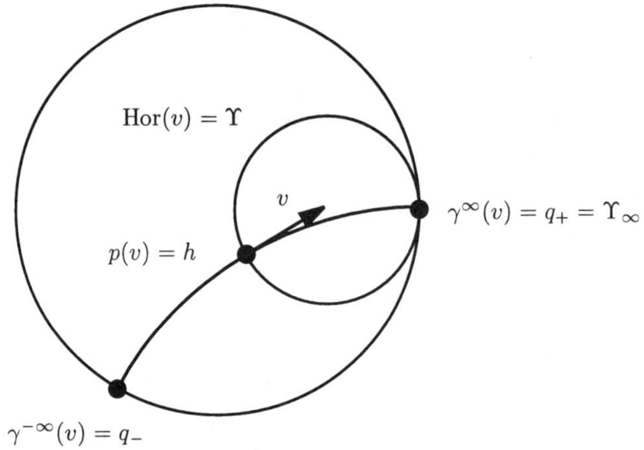

Figure 2.

two natural parametrizations of the space $U\mathbf{H}^3$ (see Figure 2).

Proposition 2.4 *The map*

$$v \mapsto \big(p(v), \gamma^{\infty}(v)\big) \tag{2.5}$$

from $U\mathbf{H}^3$ to the space $\mathbf{H}^3 \times \partial\mathbf{H}^3$ is a diffeomorphism. For any $(h, q_+) \in \mathbf{H}^3 \times \partial\mathbf{H}^3$ the associated vector $v \in U\mathbf{H}^3$ is the directing vector of the geodesic ray issued from the point h in the direction q_+.

Proposition 2.6 *The map*

$$v \mapsto \big(\gamma^{-\infty}(v), \mathrm{Hor}(v)\big) \tag{2.7}$$

from $U\mathbf{H}^3$ to the space

$$\partial\mathbf{H}^3 \times \mathrm{Hor}(\mathbf{H}^3) \setminus \{(q, \Upsilon) : q = \Upsilon_\infty\} = \bigcup_{\Upsilon \in \mathrm{Hor}(\mathbf{H}^3)} \mathcal{P}_{\Upsilon_\infty} \times \{\Upsilon\} \tag{2.8}$$

is a diffeomorphism. For any $q_- \in \partial\mathbf{H}^3, \Upsilon \in \mathrm{Hor}(\mathbf{H}^3)$ with $q_- \neq \Upsilon_\infty$ the associated vector $v \in U\mathbf{H}^3$ is the tangent vector to the geodesic joining q_- with Υ_∞ at the point of its intersection with the horosphere Υ.

2.C Stable and strongly stable foliations

Recall the definitions of two natural foliations associated with the geodesic flow on $U\mathbf{H}^3$:

Definition 2.9 Two vectors $v_1, v_2 \in U\mathbf{H}^3$ belong to the same leaf of the *stable foliation* $\mathcal{W}^s$ of the geodesic flow if

$$\limsup_{t\to+\infty} \mathrm{dist}(\gamma^t v_1, \gamma^t v_2) < \infty\,, \tag{2.10}$$

and to the same leaf of the *strongly stable foliation* $\mathcal{W}^{ss}$ if

$$\lim_{t\to+\infty} \mathrm{dist}(\gamma^t v_1, \gamma^t v_2) = 0\,, \tag{2.11}$$

where dist denotes the metric on $U\mathbf{H}^3$.

Condition (2.10) means that $\gamma^\infty(v_1) = \gamma^\infty(v_2)$. Therefore,

Proposition 2.12 *The identification* (2.5) *establishes an isomorphism between the foliations $\mathcal{H}_0$ and $\mathcal{W}^s$.*

Condition (2.11) means that $\mathrm{Hor}(v_1) = \mathrm{Hor}(v_2)$ (this is why $\mathcal{W}^{ss}$ is also often called *horosphere foliation*). Therefore,

Proposition 2.13 *Under the identification* (2.7) *the foliation $\mathcal{W}^{ss}$ is isomorphic to the foliation of the space* (2.8) *with the leaves $\mathcal{P}_{\Upsilon_\infty} \times \{\Upsilon\}$.*

Corollary 2.14 *The foliation $\mathcal{W}^{ss}$ is a Euclidean foliation.*

Proof. Indeed, the leaves of $\mathcal{W}^{ss}$ can be identified with horospheres in $\mathbf{H}^3$, so that they are endowed with the natural Euclidean structures induced by the Riemannian metric on $\mathbf{H}^3$. □

In terms of the coordinates (q_-, Υ) (2.7) the geodesic flow on $U\mathbf{H}^3$ takes the form

$$\gamma^\tau(q_-, \Upsilon) = (q_-, \gamma^\tau \Upsilon) , \tag{2.15}$$

so that it acts by affine laminar automorphisms of $\mathcal{W}^{ss}$ mapping (not isometrically!) concentric horospheres one onto the other. The foliation $\mathcal{A}_0$ is the result of factorization of $\mathcal{W}^{ss}$ by the geodesic flow (so that the space of $\mathcal{A}_0$ has one dimension less than the space of $\mathcal{W}^{ss}$). This is the reason why the leaves of $\mathcal{A}_0$ carry just a *natural* affine structure (the "scale" having been lost as a result of the action of the geodesic flow). On the hand, $\mathcal{A}_0$ is, of course, uniformizable.

The total space of the hyperbolization $\mathfrak{H}\mathcal{W}^{ss}$ is

$$\bigcup_{\Upsilon \in \mathrm{Hor}(\mathbf{H}^3)} \mathfrak{H}\mathcal{P}_{\Upsilon_\infty} \times \{\Upsilon\} = \mathbf{H}^3 \times \mathrm{Hor}(\mathbf{H}^3) .$$

The Busemann cocycle on $\mathfrak{H}\mathcal{W}^{ss}$ is

$$\beta\big((h_1, \Upsilon), (h_2, \Upsilon)\big) = \beta_{\Upsilon_\infty}(h_1, h_2) . \tag{2.16}$$

The action of the geodesic flow on $U\mathbf{H}^3$ (2.15) induces its action on $\mathfrak{H}\mathcal{W}^{ss}$

$$\gamma^\tau(h, \Upsilon) = \big(h, \gamma^\tau \Upsilon\big) \tag{2.17}$$

by isometries between leaves.

2.D Laminations associated with Kleinian groups

The roles of two factors $\partial\mathbf{H}^3$ in the definition (2.2) of the total space $\partial^2\mathbf{H}^3$ of the foliation $\mathcal{A}_0$ are quite different: the first one (the "leafwise direction") is indispensable if we want to have a $\mathbb{C}$-lamination, whereas nothing prevents us from replacing the second one (the "transverse direction") with an arbitrary subset of $\partial\mathbf{H}^3$. Therefore, for any subset $X \subset \partial\mathbf{H}^3$ the space

$$\mathcal{A}_X = \bigcup_{q \in X} \mathcal{P}_q \times \{q\}$$

is endowed with a lamination structure (this is not a foliation unless X is a submanifold of $\partial\mathbf{H}^3$).

Definition 2.18 ([12]) Let G be a Kleinian group. The lamination $\mathcal{A}_G = \mathcal{A}_{\Lambda(G)}$, where $\Lambda(G) \subset \partial\mathbf{H}^3$ is the limit set of G, is called the *affine lamination associated with the group G*. The corresponding *hyperbolic lamination*

$\mathcal{H}_G = \mathfrak{H}\mathcal{A}_G$ *associated with the group* G is the product lamination of the total space $\mathbf{H}^3 \times \Lambda(G)$.

Since the limit set is G-invariant, the group G acts on $\mathcal{A}_G$ by laminar affine maps and on $\mathcal{H}_G$ by laminar isometries. Moreover, the action of G on $\mathbf{H}^3$ (and, therefore, on $U\mathbf{H}^3 \cong \mathbf{H}^3 \times \partial\mathbf{H}^3$) is discontinuous, so that it is also discontinuous on $\mathcal{H}_G \cong \mathbf{H}^3 \times \Lambda(G)$. Denote by $\mathcal{M}_G$ the corresponding quotient hyperbolic lamination.

Remark **2.19** Although the quotient hyperbolic lamination $\mathcal{M}_G$ always makes sense, the quotient of the affine lamination $\mathcal{A}_G$ by the action of the group G is well-defined ($\equiv$ the action of G on $\mathcal{A}_G$ is discontinuous) only if G is an elementary group.

2.E Non-triviality of the Busemann cocycle

If the group G is non-elementary, then the Busemann cocycle of the lamination $\mathcal{M}_G$ is non-trivial for an obvious reason: it is non-trivial for the leaves in $\mathcal{M}_G$ corresponding to fixed points of the hyperbolic elements in G. Denote by $\Lambda_0 = \Lambda_0(G) \subset \Lambda$ the set of all such fixed points. It is not hard to see that even if we discard the set Λ_0 (i.e. remove from $\mathcal{M}_G$ all leaves with non-trivial Busemann cocycle), then the Busemann cocycle of the lamination

$$\mathcal{M}'_G = \mathcal{H}'_G/G \,, \qquad \mathcal{H}'_G = \mathbf{H}^3 \times \big(\Lambda(G) \setminus \Lambda_0(G)\big)$$

is non-trivial in the continuous cohomology. We shall prove a stronger result:

Theorem 2.20 *For any non-elementary Kleinian group* G *the Busemann cocycle of the lamination* $\mathcal{M}'_G$ *is non-trivial in the Borel cohomology.*

Proof. Fix for convenience a reference point $o \in \mathbf{H}^3$ (its choice is irrelevant for what follows) and take a probability measure μ on the group G such that the first moment $\sum_{g\in G} d(o, go)\,\mu(g)$ is finite and $\mu(g) > 0$ for all $g \in G$. Denote by μ^∞ the product measure on the space G^∞ of sequences $\boldsymbol{g} = (g_1, g_2, \dots)$. Every $\boldsymbol{g} \in G^\infty$ gives rise to the (random) sequence $h_n = h_n(\boldsymbol{g}) = g_1 g_2 \dots g_n o \in \mathbf{H}^3$ which μ^∞-a.e. has the following properties (see [11]):

(i) There exists a limit

$$h_\infty = h_\infty(\boldsymbol{g}) = \lim_{n\to\infty} h_n \in \partial\mathbf{H}^3 \,,$$

and the image ν of the measure μ^∞ under the map $\boldsymbol{g} \mapsto h_\infty(\boldsymbol{g})$ is purely non-atomic.

(ii) If U denotes the Bernoulli shift $(g_1, g_2, \dots) \mapsto (g_2, g_3, \dots)$ in the space (G^∞, μ^∞), then

$$h_\infty(U\boldsymbol{g}) = g_1^{-1} h_\infty(\boldsymbol{g}) .$$

(iii) There exists a number $l = l(\mu) > 0$ (the same for a.e. $\boldsymbol{g} \in G^\infty$) such that

$$\frac{1}{n} d(o, h_n) \underset{n\to\infty}{\longrightarrow} l .$$

(iv) The distance between h_n and the geodesic ray joining the points $o \in \mathbf{H}^3$ and $h_\infty \in \partial \mathbf{H}^3$ is $o(n)$.

Combination of (iii) and (iv) implies that μ^∞-a.e.

$$\frac{1}{n} \beta_{h_\infty}(h_0, h_n) \to l . \tag{2.21}$$

Assuming that the Busemann cocycle on $\mathcal{M}'_G$ is trivial, i.e. there exists a G-invariant Borel function f on $\mathcal{H}'_G$ such that

$$\beta_q(h_1, h_2) = f(h_2, q) - f(h_1, q) . \tag{2.22}$$

put

$$F(\boldsymbol{g}) = f(o, h_\infty) .$$

As it follows from (i) above, $\nu(\Lambda_0) = 0$, so that the function F is μ^∞-a.e. well-defined. Then

$$\begin{aligned} \beta_{h_\infty}(o, h_n) &= f(h_n, h_\infty) - f(o, h_\infty) = f(g_1 g_2 \dots g_n o, h_\infty) - f(o, h_\infty) \\ &= f(o, g_n^{-1} \dots g_2^{-1} g_1^{-1} h_\infty) - f(o, h_\infty) = F(U^n \boldsymbol{g}) - F(\boldsymbol{g}) . \end{aligned}$$

Since U preserves the measure μ^∞, (2.21) would be then impossible by the Poincaré recurrence theorem, which gives the sought for contradiction. □

Corollary 2.23 *There is no Borel G-equivariant map assigning to every point $q \in \Lambda(G) \setminus \Lambda_0(G)$ a horosphere centred at q.*

3 An example of a non-Euclidean affine foliation

The first example of an affine foliation $\mathcal{A}$ which is not Euclidean was given by Ghys [5] (also see [6]) on the base of a construction of non-standard deformations of Fuchsian groups due to Goldman [7] and Ghys [4] (note that a completely different example is given in [12]). Here we shall recast the example of Ghys by making more transparent its connection with the foliations associated with the geodesic flow on $\mathbf{H}^3$. The dimension of our example is lower as instead of the group $SL(2,\mathbb{C})$ of isometries of $\mathbf{H}^3$ considered by Ghys in [5] we deal directly with the space $U\mathbf{H}^3$ (on the other hand, our example is real, whereas the foliation of Ghys is holomorphic). The argumentation is also different: it is based on the recent results [2], [10] on the ergodicity of the horocycle foliation on abelian covers of hyperbolic manifolds rather than on the more restrictive ergodicity of the geodesic flow used by Ghys (which allows us to drop the compactness assumption, see Remark 3.13).

For an arbitrary Kleinian group G the Busemann cocycle on the hyperbolic foliation $\mathcal{M}_G$ is cohomologically non-trivial (Theorem 2.20); however, $\mathcal{M}_G$ does not correspond to any affine foliation (see Remark 2.19). On the other hand, recall that the tautological affine lamination $\mathcal{A}_0$ can be considered as a quotient of the strongly stable foliation $\mathcal{W}^{ss}$ of the geodesic flow with respect to the action of the flow (see formula (2.15) and the ensuing discussion). This factorization preserves the leafwise affine structure, but destroys the Euclidean structure. The example exploits the same idea, but, in order to have a discontinuous action, we replace the "whole" geodesic flow with a one-dimensional representation ($\equiv$ character) of the group G.

3.A Twisted action

Take a compact hyperbolic manifold H and put $G = \pi_1(H)$. The actions of G and of the geodesic flow γ on $U\mathbf{H}^3$ commute (note that only the first of these actions is isometric!). We shall now define a new "twisted" action of G on $U\mathbf{H}^3$ by combining the original action of G with the geodesic flow. *From now on we shall assume that the first Betti number of G is positive, i.e. the group* $\mathrm{Hom}(G,\mathbb{R}) \cong H^1(H,\mathbb{R})$ *of additive real-valued characters of G is non-trivial.*

Remark **3.1** Although $\mathrm{Hom}(G,\mathbb{R})$ is trivial for certain cocompact Kleinian groups, it is plausible (according to [15, p. 98]) that if G is a lattice (in particular, if G is cocompact), then it always has the so-called *Millson property*: there exists a finite index subgroup $G' \subset G$ such that $\mathrm{Hom}(G',\mathbb{R})$ is non-trivial. This property has been proved for several classes of lattices.

Definition 3.2 The *twisted action* T_χ of the group G on $U\mathbf{H}^3$ determined by a character $\chi \in \mathrm{Hom}(G,\mathbb{R})$ is

$$T_\chi^g v = g \circ \gamma^{\chi(g)}(v) = \gamma^{\chi(g)} \circ g(v) \, , \tag{3.3}$$

where in the right-hand side $v \mapsto gv$ is the standard action of G on $U\mathbf{H}^3$ (see Figure 3). We shall also use the same notation T_χ for the action of the group G on the space of horospheres $\mathrm{Hor}(\mathbf{H}^3)$ defined by the formula

$$T^g_\chi \Upsilon = \gamma^{\chi(g)} \circ g(\Upsilon) = g \circ \gamma^{\chi(g)}(\Upsilon) ,$$

where γ now stands for the action (2.3) of the geodesic flow on $\mathrm{Hor}(\mathbf{H}^3)$. Then in the coordinates (q, Υ) (2.7) on $U\mathbf{H}^3$

$$T^g_\chi(q, \Upsilon) = (gq, T^g_\chi \Upsilon) .$$

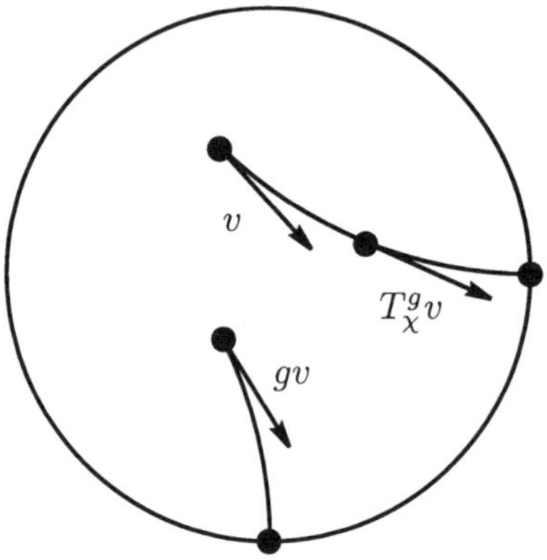

Figure 3.

3.B Admissible characters

Definition 3.4 A character $\chi \in \mathrm{Hom}(G, \mathbb{R})$ is *admissible* if the action T_χ is free, proper and totally discontinuous.

One can easily see that all characters sufficiently close to the identity in $\mathrm{Hom}(G, \mathbb{R})$ are admissible. The following complete description due to Salein [14] was pointed out to us by Ghys. Note that although Salein deals with compact hyperbolic surfaces only, his proof based on a criterion of Benoist [1] almost *verbatim* carries over to compact hyperbolic manifolds of higher dimensions as well.

Definition 3.5 The *stable norm* on $\mathrm{Hom}(G, \mathbb{R})$ is

$$\|\chi\| = \sup_{g \in G} \frac{\chi(g)}{l(g)} ,$$

where

$$l(g) = \min\{d(h, gh) : h \in \mathbf{H}^3\}\ , \qquad g \in G\ ,$$

denotes the length of the closed geodesic on H associated with the conjugacy class of g in the group G.

Proposition 3.6 *The action T_χ is admissible iff $\|\chi\| < 1$.*

Remark **3.7** As a motivation for this result note that if $\chi(g) = -l(g)$ for a certain element $g \in G$, then $T^g_\chi v = v$ for any vector $v \in U\mathbf{H}^3$ tangent to the axis of g.

3.C Absence of Euclidean structures

The action T_χ preserves the foliation $\mathcal{W}^{ss}$ and its affine (but not Euclidean, unless $\chi = 0$!) structure.

Definition 3.8 For an admissible character $\chi \in \mathrm{Hom}(G, \mathbb{R})$ denote by $\mathcal{B}_\chi$ the affine foliation of the quotient manifold $U\mathbf{H}^3/T_\chi$ obtained by factorizing the strongly stable foliation $\mathcal{W}^{ss}$ by the action T_χ.

For $\chi = 0$ the foliation $\mathcal{B}_0$ is just the strongly stable foliation of the geodesic flow on $U\mathbf{H}^3/G$, so that it is endowed with a natural leafwise Euclidean structure.

Theorem 3.9 *If $\chi \neq 0$, then the affine foliation $\mathcal{B}_\chi$ does not admit any Borel Euclidean structure.*

Proof. Since $\mathcal{B}_\chi$ is defined as the quotient of the strongly stable foliation $\mathcal{W}^{ss}$ by the action T_χ (3.3), by Corollary 1.16 we have to prove that the Busemann cocycle (2.16) on the hyperbolization $\mathfrak{H}\mathcal{W}^{ss}$ is not cohomological to zero by means of a certain Borel T_χ-invariant function:

$$\begin{aligned} \beta\big((h_1, \Upsilon), (h_2, \Upsilon)\big) &= f(h_2, \Upsilon) - f(h_1, \Upsilon)\ , \\ f\big(T^g_\chi(h, \Upsilon)\big) &= f(h, \Upsilon) \qquad \forall g \in G,\ (h, \Upsilon) \in \mathfrak{H}\mathcal{W}^{ss}\ . \end{aligned} \tag{3.10}$$

We shall prove it under the only assumption that $\chi \neq 0$ (i.e. without requiring that the character χ be necessarily admissible).

Indeed, if (3.10) were satisfied, then, as it follows from formula (2.16) for the Busemann cocycle on $\mathfrak{H}\mathcal{W}^{ss}$, the function f would be expressed as

$$f(h, \Upsilon) = \varphi(\Upsilon) + \beta_{\Upsilon_\infty}(\Upsilon, h)\ , \tag{3.11}$$

where φ is a function on the space of horospheres $\mathrm{Hor}(\mathbf{H}^3)$, and $\beta_{\Upsilon_\infty}(\Upsilon, h)$ denotes the common value $\beta_{\Upsilon_\infty}(h', h)$, $h' \in \Upsilon$.

On the other hand, as it follows from (2.17), the action T_χ on the hyperbolization $\mathfrak{H}W^{ss}$ has the form

$$T^g_\chi(h, \Upsilon) = (gh, T^g_\chi \Upsilon)\ .$$

Therefore, using (3.11) and the fact that $(\gamma^\tau \Upsilon)_\infty = \Upsilon_\infty$ for any $\tau \in \mathbb{R}$ and $\Upsilon \in \mathrm{Hor}(\mathbf{H}^3)$, we obtain

$$\begin{aligned} f\big(T_\chi^g(h,\Upsilon)\big) = f(gh, T_\chi^g\Upsilon) &= \varphi(T_\chi^g\Upsilon) + \beta_{g\Upsilon_\infty}(T_\chi^g\Upsilon, gh) \\ &= \varphi(T_\chi^g\Upsilon) + \beta_{\Upsilon_\infty}(\gamma^{\chi(g)}\Upsilon, h) \\ &= \varphi(T_\chi^g\Upsilon) + \beta_{\Upsilon_\infty}(\gamma^{\chi(g)}\Upsilon, \Upsilon) + \beta_{\Upsilon_\infty}(\Upsilon, h) \\ &= \varphi(T_\chi^g\Upsilon) - \chi(g) + \beta_{\Upsilon_\infty}(\Upsilon, h) \,. \end{aligned}$$

The latter formula compared with (3.11) implies that f is T_χ-invariant iff the function φ on $\mathrm{Hor}(\mathbf{H}^3)$ satisfies the relation

$$\varphi(T_\chi^g\Upsilon) - \chi(g) = \varphi(\Upsilon) \qquad \forall g \in G,\, \Upsilon \in \mathrm{Hor}(\mathbf{H}^3) \,. \tag{3.12}$$

In particular, the function φ has to be invariant with respect to the standard action of the kernel $G_0 = \ker\chi$ of the homomorphism χ. On the other hand, the horosphere foliation on the abelian cover $\mathbf{H}^3/G_0$ of the compact manifold $H = \mathbf{H}^3/G$ is ergodic with respect to the smooth measure class, which is equivalent to ergodicity (again with respect to the smooth measure class) of the action of G_0 on the space $\mathrm{Hor}(\mathbf{H}^3)$, see [2], [10]. Therefore, the function φ must be a.e. constant, which is impossible for $\chi \neq 0$ in view of formula (3.12). □

Remark **3.13** The only property of the manifold H used in the proof of Theorem 3.9 is ergodicity of the horosphere foliation on its homology cover. It was proved in [10] that if the geodesic flow on a hyperbolic manifold $H = \mathbf{H}^3/G$ is ergodic with respect to the Liouville measure ($\equiv$ the group G is of divergent type), then the horosphere foliation on the homology cover of H is also ergodic. Therefore, Theorem 3.9 is valid not just for compact manifolds, but also for all hyperbolic manifolds with ergodic geodesic flow.

References

1. Y. Benoist, *Actions propres sur les espaces homogènes réductifs*, Ann. of Math., **144** (1996), 315–347.
2. M. Babillot and F. Ledrappier, *Geodesic paths and horocycle flow on abelian covers*, 1–32, Proceedings of the International Colloquium on Lie groups and Ergodic Theory, Bombay, 1996, eds. S.G. Dani, Narosa Publ. House, Bombay, 1998.
3. Y. Carrière, *Un survol de la théorie des variétés affines*, Séminaire de théorie spectrale et géométrie Chambéry–Grenoble 1987–1988, 9–22.
4. E. Ghys, *Flots d'Anosov dont les feuilletages stables sont différentiables*, Ann. Sci. École Norm. Sup., **20** (1987), 251–270.

5. E. Ghys, *Sur l'uniformisation des laminations paraboliques*, 73–91, in Intagrable systems and foliations, Progress in Math., **145**. Birkhaüser, Boston, 1997.
6. E. Ghys, *Laminations par surfaces de Riemann*, Panoramas et Synthèses, **8** (1999), 49–95.
7. W. Goldman, *Non-standard Lorentz space forms*, J. Diff. Geom., **21** (1985), 301–308.
8. W. Goldman, *Projective geometry on manifolds*, Preprint, 1988.
9. V.A. Kaimanovich, *Invariant measures of the geodesic flow and measures at infinity on negatively curved manifolds*, Ann. Inst. H. Poincaré, Phys. Théor., **53** (1990), 361–393.
10. V.A. Kaimanovich, *Ergodic properties of the horocycle flow and classification of Fuchsian groups*, J. Dynam. Control Sys., **6** (2000), 21–56.
11. V.A. Kaimanovich, *The Poisson formula for groups with hyperbolic properties*, Ann. of Math., **152** (2000), 659–692.
12. V.A. Kaimanovich and M. Lyubich, *Conformal and harmonic measures on laminations associated with rational maps*, Stony Brook, preprint 2001-5.
13. M. Lyubich and Y. Minsky, *Laminations in holomorphic dynamics*, J. Diff. Geom., **47** (1997), 17–94.
14. F. Salein, *Variétés anti-de Sitter de dimension 3 possédant un champ de Killing non trivial*, C. R. Acad. Sci. Paris, Sér. I **324** (1997), 525–530.
15. E.B. Vinberg, V.V. Gorbatsevich and O.V. Shvartsman, *Discrete subgroups of Lie groups*, 1–124, in Lie Groups and Lie Algebras II (A.L. Onishchik, E.B. Vinberg eds.), Springer, Berlin, 2000.

Received November 7, 2000, revised September 20, 2001.

Proceedings of
FOLIATIONS: GEOMETRY AND DYNAMICS
held in Warsaw, May 29–June 9, 2000
ed. by Paweł WALCZAK *et al.*
World Scientific, Singapore, 2002
pp. 351–354

TRANSVERSE LUSTERNIK-SCHNIRELMANN CATEGORY AND NON-PROPER LEAVES

REMI LANGEVIN

Laboratoire de Topologie, Departement de Mathematiques,
Université de Bourgogne, B.P. 47 870, 21078 Dijon, France,
e-mail: langevin@u-bourgogne.fr

PAWEŁ G. WALCZAK

Wydział Matematyki, Uniwersytet Łódzki,
ul. Banacha 22, 90-238 Łódź, Poland,
e-mail: pawelwal@math.uni.lodz.pl

The transverse Lusternik-Schnirelmann category for foliated manifolds (as well as its saturated version) has been defined in [2]. The article [1] contains the definitions, motivation and the survey of results concerning these notion obtained until now. In this note, we present some observations made recently by the authors and related to the presence of non-proper leaves in foliated manifolds. We would like to mention that our results are valid in any codimension on arbitrary, *a priori* non-compact, manifolds.

To begin with let us accept the following definition: A leaf L on a foliated manifold $(M, \mathcal{F})$ has *pathwise non-trivial holonomy* when there exists a loop $\gamma : [0,1] \to L$ on L such that the holonomy map $h_\gamma : V \to T$, V being a neighbourhood of the origin $\gamma(0)$ in a transversal T, admits no non-trivial paths $\delta : [0,\epsilon) \to T$ consisting of points fixed by h_γ. More precisely, if δ is such a path, then there exists a sequence (s_n) of positive reals such that $s_n \to 0$ as $n \to \infty$ and $h_\gamma(\delta(s_n)) \neq \delta(s_n)$ for any n.

Proposition 1 *If a leaf L has pathwise non-trivial holonomy, U is an open neighbourhood of L in M and $H : U \times [0,1] \to M$ is a foliated homotopy for which $H_0 = \mathrm{id}_U$ and $H_1(U) \subset L'$, L' being a leaf, then $L' = L$ and $H_t(L) \subset L$ for any $t \in [0,1]$.*

Proof. Let $A \subset [0,1]$ be the set of all t such that $H_s(L) \subset L$ for any $s \leq t$. Clearly, A is non-empty and closed. Assume that $t_0 = \max A < 1$.

Take a loop $\gamma_0 : [0,1] \to L$ with pathwise non-trivial holonomy. Let $x_0 = \gamma(0)$, $x_t = H_t(x_0)$ and $\gamma_t = H_t \circ \gamma_0$ for $t \in [0,1]$. Since γ_{t_0} is a loop on L one can consider the holonomy map along it. This holonomy map is conjugated to that one along γ_0, so holonomy along γ_{t_0} is also pathwise non-trivial. The curve $[t_0, t_0 + \epsilon) \ni t \to x_t$ is non-trivial in the transverse direction, therefore there exists a sequence (s_n) of reals such that $s_n > t_0$, $s_n \to t_0$ as $n \to \infty$ and

$$h_{\gamma_{t_0}}(x_{s_n}) \neq x_{s_n} \tag{1}$$

for any n. Using a foliated chart around x_{t_0} one can see that (1) implies that the leaf distances $d_{\mathcal{F}}(y_n, x_{t_n})$, where $y_n = h_{\gamma_{t_0}}(x_{x_n})$ satisfy the condition

$$d_{\mathcal{F}}(y_n, x_{s_n}) \geq \Delta, n \in \mathbb{N}, \tag{2}$$

where $\Delta > 0$ is a constant which depends on the choice of the foliated chart only. (See Figure 1, where $t = s_n$ for a large n.)

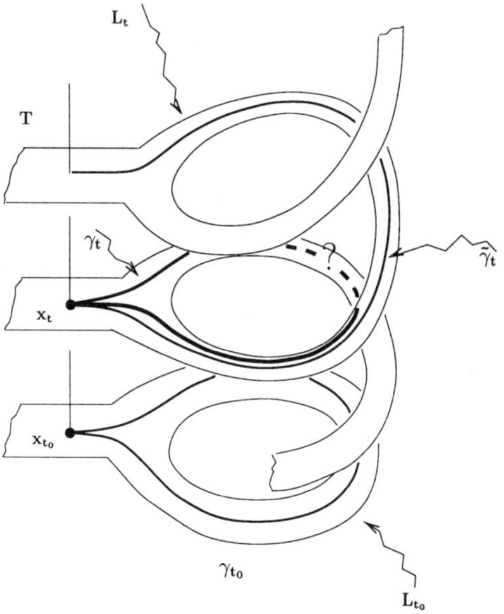

Figure 1.

On the other hand, the leaf curves γ_t and $\tilde{\gamma}_t$, $\tilde{\gamma}_t(s) = h_{\gamma_{t_0}|[0,s]}(x_t)$, have the same origin x_t, stay in the same leaf and – for t sufficiently close to t_0 – project to curves in L. From the uniform continuity of $H|\gamma_{t_0}([0,1]) \times [0,1]$ it follows that these projections are uniformly close in L, therefore, γ_t and $\tilde{\gamma}_t$ are uniformly close in L_t, the leaf through x_t, for t close to t_0. Consequently,

$$d_{\mathcal{F}}(y_n, x_{s_n}) \to 0 \text{ as } n \to \infty. \tag{3}$$

Clearly, (3) contradicts (2). □

Proposition 2 *If L is non-proper and has pathwise non-trivial holonomy, then L admits no open neighbourhoods which can be deformed into a leaf by a foliated homotopy.*

Proof. Assume that U is an open neighbourhood of L and $H : U \times [0,1] \to M$ is a foliated homotopy with $H_0 = \mathrm{id}_U$ and $H_1(L) \subset L'$, L' being a leaf. By Proposition 1, $L' = L$ and $H_t(L) \subset L$ for any t.

Fix $x_0 \in L$ and set $\gamma(t) = H_t(x)$ for $t \in [0,1]$. γ is path in L, so one can choose a chain $V_0, V_1, \ldots, V_k$ of foliated charts, members of a good foliated atlas on M, along γ. We can assume that $V_i \subset \overline{V_i} \subset U$. We can also choose another good foliated chart W_0 around x_0 and make it small enough to satisfy the condition

$$(\forall t \in [0,1])(\exists i \in \{0,1,\ldots,k\})\ \overline{W_t} \subset V_i, \tag{4}$$

where $W_t = H_t(W_0)$. (If we like, we can add that $\overline{W_1} \subset V_k$.)

Since W_0 is connected, so is W_1, and therefore W_1 is contained in a single plaque of V_k. On the other hand, since L is non-proper, $L \cap W_0$ consists of infinitely (but countable) many plaques. Therefore, there exist $j \in \{0, 1 \ldots, k\}$ and arbitrarily close real numbers t_1 and t_2 such that $t_1 < t_2$, $W_{t_1} \cup W_{t_2} \subset V_j$, $W_{t_1} \cap L$ intersects at least two plaques while W_{t_2} is contained in a single plaque of V_j (see Figure 2), a contradiction. □

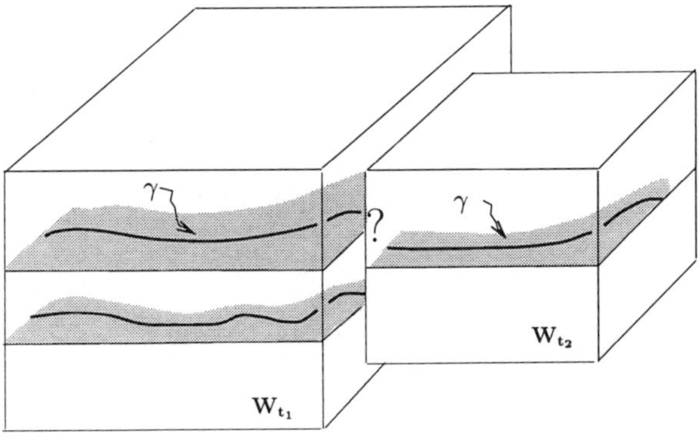

Figure 2.

The above observation can be expressed in terms of transverse Lusternik-Schnirelmann categories as follows:

Corollary 1 *If a foliated manifold $(M, \mathcal{F})$ contains a non-proper leaf with pathwise non-trivial holonomy, then the saturated transverse Lusternik-Schnirelmann category of $\mathcal{F}$ is infinite.* □

In the case of foliations of codimension-one we obtain also the following.

Corollary 2 *If a codimension-one* C^2*-foliation* $\mathcal{F}$ *of a compact manifold* M *contains an exceptional minimal set* E*, then the saturated transverse Lusternik-Schnirelmann category of* $\mathcal{F}$ *is infinite.*

In fact, by Sacksteder Theorem, E contains a resilient leaf with linearly contracting holonomy, such a leaf is non-proper and has pathwise non-trivial holonomy.

Comparing the above observations with those of [2] (and other papers listed in [1]) one can expect that any basic function on a foliated manifold $(M, \mathcal{F})$ has infinite number of critical leaves provided that $\mathcal{F}$ has a non-proper leaf with pathwise non-trivial holonomy.

Acknowledgments

While preparing this article the second author was supported by the Université de Bourgogne and by the Polish KBN grant P03A 033 18.

References

1. H. Colman, *LS-categories for foliated manifolds*, in Foliations: Geometry and Dynamics, eds. P. Walczak et. al., World Sci. Publ., 2001, 17–28.
2. H. Colman, E. Macias-Virgós, *The transverse Lusternik-Schnirelmann category of a foliated manifold*, Topology, **40** (2000), 419–430.

Received June 18, 2001.

Proceedings of
FOLIATIONS: GEOMETRY AND DYNAMICS
held in Warsaw, May 29–June 9, 2000
ed. by Paweł WALCZAK *et al.*
World Scientific, Singapore, 2002
pp. 355–370

STRUCTURALLY STABLE DIFFEOMORPHISMS HAVE NO CODIMENSION ONE PLYKIN ATTRACTORS ON 3-MANIFOLDS

VLADISLAV MEDVEDEV
Department of Applied Mathematics, Nizhny Novgorod State Technical University
24 Minina Str., Nizhny Novgorod 603600, Russia,
e-mail: medvedev@unn.ac.ru

EVGENY ZHUZHOMA
Department of Applied Mathematics, Nizhny Novgorod State Technical University
24 Minina Str., Nizhny Novgorod 603600, Russia,
e-mail: zhuzhoma@focus.nnov.ru

Let f be a structurally stable diffeomorphism of a closed 3-manifold M^3. We show that the spectral decomposition of f has no non-orientable codimension one expanding attractors. This means that a structurally stable diffeomorphism of a closed 3-manifold has no codimension one Plykin attractors.

Introduction

The most investigated structurally stable systems are codimension one Anosov diffeomorphisms and Morse-Smale systems (see reviews [1], [15], and [25]). For example, any codimension one Anosov diffeomorphism is conjugate to a hyperbolic torus automorphism ([4], [14]). The natural generalization of Anosov diffeomorphisms are diffeomorphisms satisfying the axiom A of Smale [25], so called A-diffeomorphisms. A structurally stable A-diffeomorphism of the n-torus T^n ($n \geq 2$) with an orientable codimension one expanding attractor Ω can be obtained by Smale's surgery ([25], p.788-789) a codimension one Anosov diffeomorphism. They are called DA-*diffeomorphisms*, where the abbreviation DA means *Derived-from-Anosov*.

Note that due to the orientability of Ω, the accessible boundary of any component of $T^n - \Omega$ from within $T^n - \Omega$ consists of pairs of codimension one unstable manifolds, so-called *2-bunches* (see definitions in Section 1).

A first example of a codimension one non-orientable expanding attractor was constructed by Plykin [19] on a two-sphere S^2. A feature of Plykin's structurally stable diffeomorphism $f : S^2 \to S^2$ is the existence of so-called *1-bunches*. To be precise, if Ω is a non-orientable codimension one (i.e., one-dimensional) expanding attractor of f, then there exists a component of $S^2 - \Omega$ with an accessible boundary consisting of a unique one-dimensional unstable manifold. In [21], Plykin constructed a codimension one non-orientable attractors with 1-bunches on open n-manifolds, $n \geq 3$ (in the same paper, he classified, up to conjugacy, diffeomorphisms restricted on some neighbourhoods of attractors). Bearing in mind these examples, we call that a codimension one expanding attractor Ω is a *codimension one Plykin attractor* whenever Ω has 1-bunches.

Let us give the sketch of Plykin's construction. Denote by $J : T^n \to T^n$ the involution $\vec{x} \to -\vec{x}$ (mod 1) which has 2^n fixed points $v_1, \ldots, v_{2^n}$. Take the DA-diffeomorphism $g : T^n \to T^n$ with the codimension one orientable expanding attractor Ω_g such that g commutes with J and has the fixed points $v_1, \ldots, v_{2^n}$. Due to Theorem 2.3 [21] (see the careful construction in [11], [23] for $n = 2$ and [5] for $n = 3$ as well), such g exists. Denote by $q : T^n \to T^n/J$ the natural projection which is a double branched covering with the branch points $v_1, \ldots, v_{2^n}$. It is not hard to see that the quotient space T^2/J is a 2-sphere. Since $g(-\vec{x}) = -g(\vec{x})$, we obtain an induced diffeomorphism $f : S^2 \to S^2$ with the one-dimensional Plykin attractor $q(\Omega_g)$. As to $n \geq 3$, the branch points must be removed to get a codimension one Plykin attractor on a smooth manifold. Thus, $M = q(T^n - \cup_{i=1}^{2^n} v_i)$ is an open smooth n-manifold, $n \geq 3$, which admits an induced diffeomorphism with the codimension one Plykin attractor $q(\Omega_g)$.

As to closed n-manifolds, $n \geq 3$, up to now, there are no examples of diffeomorphisms with codimension one Plykin attractors. Nevertheless, such diffeomorphisms were investigated. In [12], one proved that if a closed n-manifold M^n, $n \geq 3$, admits a codimension one expanding attractor (orientable or non-orientable), then M^n has a nontrivial fundamental group. In particular, there no diffeomorphisms of n-sphere S^n, $n \geq 3$, with codimension one Plykin attractors. The more precise result was obtained in [21], where one proved that $\pi_1(M^n)$ contains a subgroup isomorphic to the integer lattice $\mathbb{Z}^n$. In [17], Plante proved that an orientable codimension one expanding attractor defines a nontrivial element of homology group. In particular, a closed manifold M^n has a nontrivial homology group $H_1(M^n)$.

In this paper we prove that a structurally stable diffeomorphism of a closed 3-manifold has no non-orientable codimension one expanding attractors.

Theorem 3.1 *Let $f : M^3 \to M^3$ be a structurally stable diffeomorphism of a closed 3-manifold M^3. Then the spectral decomposition of f does not contain non-orientable codimension one expanding attractors.*

This theorem and the following lemma

Lemma 2.2 *A codimension one expanding attractor Ω of an A-diffeomorphism $f : M^3 \to M^3$ is nonorientable iff Ω is a Plykin attractor.*

imply that a structurally stable diffeomorphism of a closed 3-manifold has no codimension one Plykin attractors.

1 Main definitions

We begin by recalling several definitions. Further details may be found in the books [11], [16], [23], and reviews [1], [25].

A-diffeomorphisms.

Throughout the paper M^n is a closed n-dimensional manifold endowed with some Riemannian metric d. Recall that a diffeomorphism $f : M^n \to M^n$ is said to be an *A-diffeomorphism* if the nonwandering set $NW(f)$ of f is hyperbolic and periodic points are dense in $NW(f)$. The *stable manifold* $W^s(x)$ of a point $x \in NW(f)$ is defined to be the set of points $\{y \in M^n$ such that $d(f^j x, f^j y) \to 0$ as $j \to +\infty\}$. A stable manifold is an injectively immersed Euclidean space for each $x \in NW(f)$. The *unstable manifold* $W^u(x)$ of x is the stable manifold of x for the diffeomorphism f^{-1}. By definition, let $W^s_\varepsilon(x) \subset W^s(x)$ (resp. $W^u_\varepsilon(x) \subset W^u(x)$) be an ε-neighbourhood of x in the intrinsic topology of the manifold $W^s(x)$ (resp. $W^u(x)$), where $\varepsilon > 0$.

Basic sets.

Let f be an A-diffeomorphism of a n-dimensional manifold M^n. It was shown by Smale [25] that $NW(f)$ is a finite union of pairwise disjoint f-invariant closed sets $\Omega_1, \dots, \Omega_k$ such that every restriction $f|\Omega_i$ is topologically transitive. These Ω_i are called the *basic sets* of f. A basic set Ω is called an *expanding attractor* if there exists a closed neighbourhood U of Ω such that $f(U) \subset int\ U$, $\cap_{j\geq 0} f^j(U) = \Omega$, and the topological dimension $\dim\Omega$ of Ω is equal to the dimension $\dim(E^u_\Omega)$ of the unstable splitting E^u_Ω. If $\dim\Omega = n-1$, then Ω is called an *expanding attractor of codimension one.* It is well known (see e.g., [18], [26]) that a codimension one expanding attractor consists of the $(n-1)$-dimensional unstable manifolds $W^u(x)$,

$x \in \Omega$ and is locally homeomorphic to the product of $(n-1)$-dimensional Euclidean space and a Cantor set.

For $\dim E^s_\Omega = 1$, we shall denote by $(x, y)^s$ (resp. $[x, y]^s$) an open (resp. closed) arc of $W^s(z)$, $z \in \Omega$, with endpoints x, $y \in W^s(z)$.

By definition, let $W^s_\varepsilon(x) \subset W^s(x)$ (resp. $W^u_\varepsilon(x) \subset W^u(x)$) be the ε-neighbourhood of x in the intrinsic topology of the manifold $W^s(x)$ (resp. $W^u(x)$), where $\varepsilon > 0$. Following [6], a basic set Ω is called *orientable* if for any $\alpha > 0$, $\beta > 0$ the index of the intersection $W^s_\alpha(x) \cap W^u_\beta(x)$ does not depend on a point of this intersection (it is either $+1$ or -1).

Structurally stable diffeomorphisms.

Let $Diff^1(M^n)$ be the space of C^1 diffeomorphisms of M^n endowed with the uniform C^1 topology. Two diffeomorphisms f, $g \in Diff^1(M^n)$ are called *conjugate* if there is a homeomorphism $\varphi : M^n \to M^n$ such that $\varphi \circ f = g \circ \varphi$. A diffeomorphism f is said to be *structurally stable* if there exists a neighbourhood U of f in $Diff^1(M^n)$ such that any diffeomorphism $g \in U$ is conjugate to f.

According to [22], [13], a diffeomorphism f is structurally stable if and only if it is an A-diffeomorphism and for any points x, $y \in NW(f)$ the manifolds $W^s(x)$, $W^u(y)$ have transversal intersections (if non-empty). The last condition is called the *strong transversality condition.*

Boundary periodic points.

Let Ω be an expanding attractor of codimension one. A point $p \in \Omega$ is called a *boundary point* if at least one component of $W^s(p) - p$ does not intersect Ω. Boundary points exist (see e.g., [6], [14], [19]) and satisfy to the following conditions:

- There are finitely many boundary points.
- Any boundary point $p \in \Omega$ is periodic.
- There is a unique component of $W^s(p) - p$ denoted by $W^s_\emptyset(p)$ which does not intersect Ω.
- Given a point $x \in W^u(p) - p$, there is a unique arc $(x, y)^s$ denoted by $(x, y)^s_\emptyset$ such that $(x, y)^s \cap \Omega = \emptyset$, $y \in \Omega$.

Bunches of codimension one expanding attractors.

Let $G \subset M$ be an open set with a boundary ∂G. The subset $\delta(G) \subset \partial G$ is called an *accessible boundary* from G if for every point $x \in \delta(G)$ there exists an open arc a which is in G and x is an endpoint of a.

Let Ω be a codimension one expanding attractor of f. Due to [6], [20], the accessible boundary $\delta(M - \Omega)$ is a finite union of unstable codimension one manifolds, each one is called a *boundary unstable manifold* through a boundary periodic point of Ω. The boundary unstable manifolds of Ω splits into a finite number of so called bunches in the following way. The family of pairwise disjoin unstable manifolds $W^u(p_1), \ldots, W^u(p_k)$ is said to be a *k-bunch* if there are points $x_i \in W^u(p_i)$ and arcs $(x_i, y_i)^s_\emptyset$, $y_i \in W^u(p_{i+1})$, $1 \le i \le k$, where $W^u(p_{k+1}) = W^u(p_1)$ (thus, $y_k \in W^u(p_1)$) and the number k can't be increased (i.e., there are no $(k+1)$-bunches containing the given one), see Figure 1. The points $p_1, \ldots, p_k$ are called *associated.*

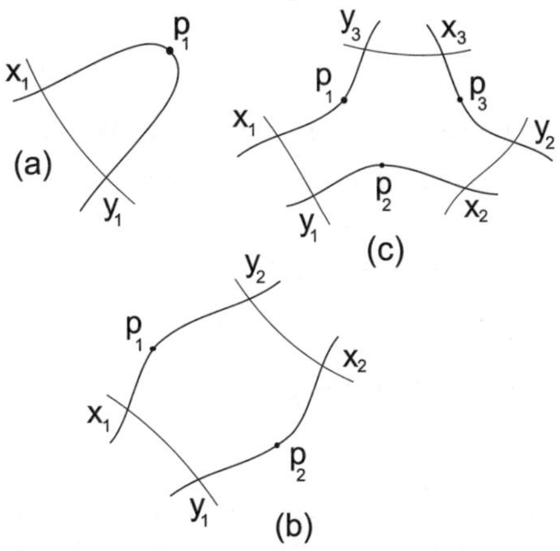

Figure 1. (a) 1-bunch, (b) 2-bunch, and (c) 3-bunch.

Plykin [21] proved that if $n = \dim M \ge 3$, then any bunch of Ω is either a 1-bunch or 2-bunch. It is not difficult to see that if Ω is orientable, then every bunch of Ω is a 2-bunch (see the beginning of the proof of Lemma 2.2). So the boundary periodic points of orientable Ω split into disjoint pairs of associated points.

A codimension one expanding attractor Ω is called a *codimension one Plykin attractor* if Ω has 1-bunches.

2 Preliminaries

In the next lemmas we keep the following notation. Let Ω be an expanding attractor of codimension one and let B be a bunch of Ω. According to [21], B is either a 1-bunch or 2-bunch. In the last case, B consists of two unstable manifolds $W^u(p)$ and $W^u(q)$, where p, q are boundary associated periodic points. If B is a 1-bunch, we will assume that $W^u(p) = W^u(q)$ and $p = q$. The following lemma holds.

Lemma 2.1 *There is a homeomorphism*

$$\varphi : (W^u(p) - p) \cup (W^u(q) - q) \to (W^u(p) - p) \cup (W^u(q) - q)$$

with the following properties

- *Given any point $x \in W^u(p) - p$,*
$$\varphi(x) \in (W^u(q) - q) \cap W^s(x)$$
and moreover, $(x, \varphi(x))^s = (x, \varphi(x))^s_\emptyset$ i.e., the arc $[x, \varphi(x)]^s$ intersects Ω at the ends points x, $\varphi(x)$ only.
- *If m is the period of p, then m is the period of q as well and moreover,*
$$\varphi \circ f^{nm}|_{(W^u(p)-p)\cup(W^u(q)-q)} = f^{nm} \circ \varphi|_{(W^u(p)-p)\cup(W^u(q)-q)}, \qquad (1)$$
where $n \in \mathbb{Z}$. In particular, the maps f^m, φ commute on $(W^u(p) - p) \cup (W^u(q) - q)$.
- *φ extends to the homeomorphism (denoted by the same letter)*
$$\varphi : W^u(p) \cup W^u(q) \to W^u(p) \cup W^u(q),$$
by letting $\varphi(p) = q$ and $\varphi(q) = p$.
- *$\varphi^2 = id$.*

Proof. This lemma is a consequence of Lemmas 2.1, 2.2 [6], and Theorem 2.1 [21]. For the reader's convenience we give the sketch of the proof. Given any point $x \in W^u(p)$ there is a unique point $y \in W^u(q)$ such that $(x, y)^s = (x, y)^s_\emptyset$, and vise versa (see Section 1). Let the map

$$\varphi : (W^u(p) - p) \cup (W^u(q) - q) \to (W^u(p) - p) \cup (W^u(q) - q)$$

be given by $\varphi(x) = y$ whenever $(x, y)^s = (x, y)^s_\emptyset$. By this definition, $\varphi^2 = id$. Thus,

$$\varphi(W^u(p) - p) = (W^u(q) - q) \text{ and } \varphi(W^u(q) - q) = (W^u(p) - p).$$

According to the theorem on the continuous dependence of stable manifolds on initial conditions [25], φ is a homeomorphism. Due to the invariance of Ω, the point q has the same period m as p. Since $W^s(\Omega)$ is invariant under f, we have that

$$f^m \circ \varphi|_{(W^u(p)-p)\cup(W^u(q)-q)} = \varphi \circ f^m|_{(W^u(p)-p)\cup(W^u(q)-q)}.$$

It implies the relation (1).

Because of the restriction $f^m|_{W^u(p)-p}$ has the only hyperbolic fixed point p, there is a closed disk $D_p \subset W^u(p)$ bounded by the closed curve $S^1_p = \partial D_p$ homeomorphic to the circle such that $p \in int(D_p)$ and $D_p \subset$

$int(f^m(D_p))$. Since φ is a homeomorphism, $S_2^1 = \varphi(S_1^1)$ is a circle. According to Shoenflies's Theorem, S_2^1 bounds in $W^u(q)$ the disk denoted by D_q. It follows from the relation (1) that $q \in int(D_q)$.

The restriction $f^m|_{W^u(q)-q}$ has the only fixed point q, which is hyperbolic and repelling. Hence,

$$\bigcap_{j\leq 0} f^{jm}(D_p) = p, \quad \bigcap_{j\leq 0} f^{jm}(D_q) = q.$$

Due to (1),

$$\varphi \circ f^{jm}(D_p - p) = f^{jm} \circ \varphi(D_p - p) = f^{jm}(D_q - q).$$

Similarly, $\varphi \circ f^{jm}(D_q - q) = f^{jm}(D_p - p)$. Hence, φ extends to the homeomorphism $W^u(p) \cup W^u(q) \to W^u(p) \cup W^u(q)$ if we put $\varphi(p) = q$ and $\varphi(q) = p$. □

The following theorem is the crucial step to prove the main theorem.

Theorem 2.1 *Let $f : M^3 \to M^3$ be an A-diffeomorphism of a closed 3-manifold M^3. If f has a codimension one Plykin attractor Ω, then M^3 is non-orientable.*

Proof. Because of Ω is a Plykin attractor, Ω has 1-bunches. Let $W^u(m_0) \subset \Omega$ be the unstable manifold forming a 1-bunch of Ω, where m_0 is a boundary periodic point. Without loss of generality we can assume that m_0 is a fixed point of f.

According to Lemma 2.1, there is the homeomorphism $\varphi : W^u(m_0) \to W^u(m_0)$ such that:

1. For $x \in W^u(m_0) - m_0$, $\varphi(x) \in (W^u(m_0) - m_0) \cap W^s(x)$.
2. For $x \in W^u(m_0) - m_0$,
$$(x, \varphi(x))^s = (x, \varphi(x))^s_{\emptyset} \subset W^s(x).$$
3. $\varphi \circ f|_{W^u(m_0)} = f \circ \varphi|_{W^u(m_0)}$.
4. $\varphi^2 = id$ and m_0 is the only fixed point of φ.

For the sake of simplicity, we divide our proof into several steps. The end of the proof of each step will be denoted by $\diamond$.

Take a point $x_0 \in W^u(m_0)$ such that $x_0 \neq m_0$. Let $D_\alpha \subset W^u(m_0) - m_0$ be a family of closed disks (the index α ranges over the interval $[0;1]$) with the following properties:

- $x_0 \in D_0$, $\varphi(x_0) \notin D_0$.
- $D_0 \cap \varphi(D_0) = \emptyset$.

- $D_{\alpha_1} \subset D_{\alpha_2}$ whenever $\alpha_1 < \alpha_2$.
- The boundary ∂D_α depends continuously on α.
- $\varphi(x_0) \in D_1$.

Such a family of the disks D_α exists because the unstable manifold $W^u(m_0)$ is homeomorphic to the Euclidean plane $\mathbb{R}^2$ and $x_0 \neq \varphi(x_0)$.

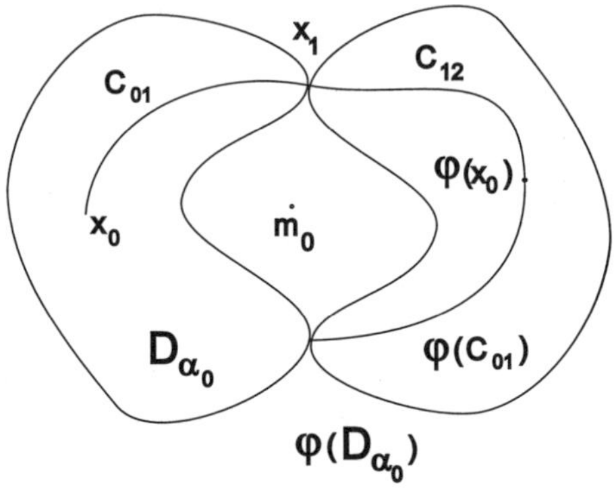

Figure 2. The curve $C = C_{01} \cup C_{12}$.

Step 2.1 *There exists the index* $\alpha_0 \in (0;1)$ *such that*

$$\emptyset \neq D_{\alpha_0} \cap \varphi(D_{\alpha_0}) = \partial D_{\alpha_0} \cap \partial(\varphi(D_{\alpha_0})).$$

Proof of Step 2.1. If $\alpha_1 < \alpha_2$, then $\varphi(D_{\alpha_1}) \subset \varphi(D_{\alpha_2})$ because φ is a homeomorphism. Since $D_0 \cap \varphi(D_0) = \emptyset$, we have that $D_\alpha \cap \varphi(D_\alpha) = \emptyset$ for α is sufficiently small. On the other hand, $D_{\alpha_1} \cap \varphi(D_{\alpha_1}) \neq \emptyset$ because $\varphi(x_0) \in D_1$. Since the boundary ∂D_α depends continuously on α and φ is a homeomorphism, the boundary $\partial\varphi(D_\alpha) = \varphi(\partial D_\alpha)$ depends continuously on α as well. Hence there exists $\alpha_0 \in (0;1)$ such that the disks D_{α_0}, $\varphi(D_{\alpha_0})$ have a non-empty intersection at points of their boundaries only (Figure 2). ◇

Step 2.2 *There exists a curve* $C \subset W^u(m_0) - m_0$ *with no self-intersections such that* $C \cap \varphi(C) = \{x_0, \varphi(x_0)\}$, *where* x_0, $\varphi(x_0)$ *are the endpoints of* C.

Proof of Step 2.2. According to Step 2.1, there is a point $x_1 \in \partial D_{\alpha_0} \cap \varphi(\partial D_{\alpha_0})$ for some $\alpha_0 \in (0,1)$. Since $\varphi^2 = id$ and m_0 is the only fixed point of φ, we have that $\varphi(x_1) \in \partial D_{\alpha_0} \cap \varphi(\partial D_{\alpha_0})$ and $\varphi(x_1) \neq x_1$.

Note that D_{α_0} is homeomorphic to a disk. Hence there is a path $C_{01} \subset D_{\alpha_0}$ with no self-intersections and such that C_{01} connects the points x_0, x_1. Due to step 2.1, the interior of D_{α_0} and the image of the interior under

φ have no intersections. Therefore, $\varphi(C_{01}) \subset \varphi(D_{\alpha_0})$ is a path with no self-intersections. Moreover, $\varphi(C_{01})$ connects the points $\varphi(x_0)$, $\varphi(x_1)$ and $C_{01} \cap \varphi(C_{01}) = x_1$. Obviously, the curve $\varphi(C_{01})$ does not divide the disk $\varphi(D_{\alpha_0})$. Hence there is a path C_{12} with no self-intersections connecting the points $\varphi(x_0)$, x_1. Moreover, we can assume that $\varphi(C_{01}) \cap C_{12} = \varphi(x_0)$. By definition, put $C = C_{01} \cap C_{12}$. Since $\varphi^2 = id$, it follows that C has no self-intersections and $C \cap \varphi(C) = \{x_0, \varphi(x_0)\}$, where x_0, $\varphi(x_0)$ are the endpoints of C ⋄

Step 2.3 *The set $M_c = \cup_{x \in C}(x,y)^s_\emptyset$ is homeomorphic to an open Möbius band. Moreover, any arc $(x,y)^s_\emptyset$ is transversal (in topological sense) to the middle line of the Möbius band.*

Proof of Step 2.3. Consider the open rectangle $P = \cup_{x \in C - x_0}(x,y)^s_\emptyset$. Then the set M_c is obtained by adding to P the arc $(x_0, \varphi(x_0))^s_\emptyset$. By the relation $\varphi^2 = id$, the addition above one can represent as gluing two opposite sides in the closed rectangle $clos(P)$ by an orientation reversing homeomorphism. Hence, M_c is homeomorphic to an open Möbius band. By construction, any arc $(x,y)^s_\emptyset$ is transversal (in topological sense) to the middle line of the Möbius band. ⋄

By Step 2.2, the union $C \cup \varphi(C) = c_0$ is a simple closed curve on the unstable manifold $W^u(m_0)$. Therefore, c_0 bounds on $W^u(m_0)$ an open disk denoted by D_c.

Step 2.4 $m_0 \in D_c$.

Proof of Step 2.4. The exterior of c_0 on $W^u(m_0)$ is homeomorphic to an open annulus. By construction, $\varphi(c_0) = c_0$. Since φ is a homeomorphism, it follows that $\varphi(D_c) = D_c$ and $\varphi(c_0 \cup D_c) = c_0 \cup D_c$. Hence, φ has a fixed point on $c_0 \cup D_c$. By lemma 2.1, m_0 is the unique fixed point of φ. Therefore, $m_0 \in D_c$. ⋄

Step 2.5 *There exists an open subset $C^3 \subset M^3$ that is homeomorphic to the product $M_b \times (0,1)$, where M_b is an open Möbius band.*

Proof of Step 2.5. We keep the notation of the steps above. Take an open simple arc $l_0 \subset D_c$ such that $l_0 \cap \varphi(l_0) = \emptyset$ and one of endpoints of l_0 is x_0. Because of $\varphi(x_0) \neq x_0$ and φ is a homeomorphism, such an arc exists. Since the intersection $\varphi(C) \cap C$ consists of only two points x_0 and $\varphi(x_0)$, it follows that there is the open strip $P_1 \subset D_c$ bounded by the curve C and the arcs l_0, $\varphi(l_0)$ such that $\varphi(P_1) \cap P_1 = \emptyset$. By Lemma 2.1, $m_0 \notin \varphi(P_1) \cup P_1$. Therefore $\varphi(P_1)$ is the open strip bounded by the curve $\varphi(C)$ and arcs l_0, $\varphi(l_0)$. As a consequence, the union $P = \varphi(l_0) \cup l_0 \cup \varphi(P_1) \cup P_1$ is an open strip in D_c. Moreover, one of the component of the boundary ∂P is the boundary $\partial D_c = \varphi(C) \cup C$ of the disk D_c (Figure 3).

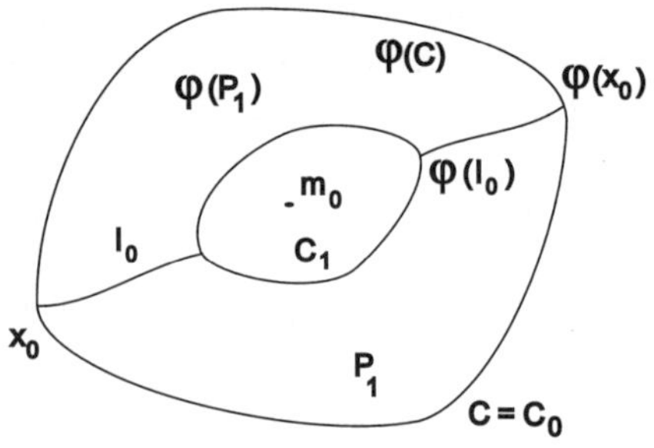

Figure 3. The disk D_c and the strip $P = P_1 \cup \varphi(P_1) \cup l_0 \cup \varphi(l_0)$.

Let C_1 be the side of the rectangle $P_1 \subset D_c$ which is opposite to the side $C \stackrel{def}{=} C_0$. Note that the other pair of the opposite sides l_0 and $\varphi(l_0)$ of P_1 is invariant under φ (i.e., one side maps onto other by φ). Therefore there exists a splitting of the closed rectangle $\varphi(l_0) \cup l_0 \cup P_1 \cup C_0 \cup C_1$ into pairwise disjoint arcs C_α (the index α ranges over the interval $[0,1]$) such that the endpoints of each arc maps onto other by φ. Moreover, without loss of generality we can assume that the arcs C_α form a topological foliation on the closed rectangle $\varphi(l_0) \cup l_0 \cup P_1 \cup C_0 \cup C_1$. It follows from the equality $\varphi(P_1) \cap P_1 = \emptyset$ that the union $C_\alpha \varphi(C_\alpha)$ is a closed simple curve for any $\alpha \in [0,1]$. As a consequence, the family $C_\alpha \varphi(C_\alpha)$, $\alpha \in [0,1]$, forms a topological foliation on the strip P with leaves $C_\alpha \varphi(C_\alpha)$ are closed simple curves and each one is invariant under φ.

Similarly to the proof of Step 2.3, one can prove that each closed curve $C_\alpha \varphi(C_\alpha)$ is a boundary of the set $M_\alpha = \cup_{x \in C_\alpha \cup \varphi(C_\alpha)} (x, \varphi(x))^s_\emptyset$, which is homeomorphic to an open Möbius band M_b. Since stable manifolds are pairwise disjoint, we see that the sets M_α are pairwise disjoint as well. The theorem on the continuous dependence of invariant manifolds on initial conditions [25] implies that the sets M_α depend continuously on α. Since the strip P is homeomorphic to the product $S^1 \times (0,1)$ (here S^1 is a circle), it follows that the open subset

$$C^3 = \cup_{x \in P} (x, \varphi(x))^s_\emptyset = \cup_{\alpha \in (0,1)} M_\alpha$$

is homeomorphic to the product $M_b \times (0,1)$. $\diamond$

It is well known that a Möbius band is a non-orientable surface. Hence, C^3 is a non-orientable 3-manifold. As a consequence, M^3 is a non-orientable 3-manifold as well [9]. This concludes the proof. □

Corollary 2.1 *Suppose the condition of Theorem 2.1 holds and R is a component of $M^3 - \Omega$ such that the accessible boundary $\delta(R)$ of R contains* *a 1-bunch of* Ω*; then the fundamental group* $\pi_1(R)$ *is nontrivial.*

Proof. By the proof of Theorem 2.1, there is an open subset $C^3 \subset R \cap (W^s(\Omega) - \Omega)$ homeomorphic to the product $M_b \times (0,1)$, where M_b is an open Möbius band. Therefore, R is a non-orientable open manifold. It follows the nontriviality of $\pi_1(R)$. □

Recall that a codimension one foliation $\mathcal{F}$ on M^3 is called a *Reeb foliation* if every leaf of $\mathcal{F}$ is homeomorphic to a plane $\mathbb{R}^2$ (in the intrinsic topology) [24].

The following lemma says that the nonorientability of a codimension one expanding attractor Ω is equivalent to Ω being a Plykin attractor.

Lemma 2.2 *A codimension one expanding attractor* Ω *of an A-diffeomorphism* $f : M^3 \to M^3$ *is nonorientable iff* Ω *is a Plykin attractor.*

Proof. Let Ω is a codimension one Plykin attractor. Then Ω has a 1-bunch $W^u(m_0)$, where m_0 is a boundary periodic point. Hence there exists an arc

$$(x,y)^s_\emptyset = (x,y)^s \subset W^s(x), \quad (x,y)^s \cap \Omega = \emptyset, \quad x, y \in W^u(m_0) - m_0.$$

Take a disk $D \subset W^u(m_0)$ such that $x, y \in D$. Since $D \subset \Omega$, D has a neighbourhood U with a local product structure such that the component of $U \cap W^u(m_0)$ containing D divides U into two open domains, say U_1 and U_2, each one homeomorphic to a 3-ball.

Note that the unstable manifold $W^u(m_0)$ can be endowed with a normal orientation because $W^u(m_0)$ is homeomorphic to Euclidean space $\mathbb{R}^2$. Therefore, is we suppose that the arc $[x,y]^s_\emptyset$ intersects $W^u(m_0)$ at x, y with the same index of intersection, then $(x,y)^s_\emptyset$ has to intersect both U_1 and U_2. Since $(x,y)^s \cap \Omega = \emptyset$ and U has a local product structure, $U_i \cap \Omega = \emptyset$, $i = 1, 2$. This contradicts the property of Ω being locally homeomorphic to the product of a Cantor set and a ball [18], [26]. As a consequence, arc $[x,y]^s_\emptyset$ intersects $W^u(m_0)$ at x, y with opposite indexes of intersection, and thus Ω is nonorientable.

Let us assume now that Ω is a nonorientable codimension one expanding attractor. We have to prove that Ω is a Plykin one. Suppose not. Then all bunches of Ω are 2-bunches. Take a 2-bunch $B \subset \Omega$ and denote by K the component of $M^3 - \Omega$ such that the accessible boundary of K contains B, $\subset \delta K$. Since f is structurally stable, it follows that $K \cup B$ is homeomorphic to the product $\mathbb{R}^2 \times [0;1]$ (in the intrinsic topology) [8]. This fact and Plykin's Theorem on the embedding of Ω into a codimension one foliation (see the Theorem on invariant foliations [21], p.93) imply that Ω can be embedded into a Reeb foliation denoted by $\mathcal{F}$. Moreover, since the unstable manifolds of Ω form a C^1 lamination [10] (see also [23], Ch. 12), we can assume that $\mathcal{F}$ is a C^1 foliation as well.

In [24], Rosenberg proved that if a closed three-manifold M^3 admits a C^2 Reeb foliation, then M^3 is homeomorphic to the three dimensional torus T^3. The main technic in the paper [24] is a moving of an embedded 2-disk $D^2 \subset M^3$ in a general position under the C^2 Reeb foliation $\mathcal{F}$ so that $\mathcal{F}$ induces on D^2 a foliation with Morse's singularities of the center and saddle types. Franks (see the proof of Lemma 5.1 [4]) showed that Rosenberg's technic works for C^1 foliation as well. So, $M^3 = T^3$.

Let us show that $\mathcal{F}$ is transversally orientable. Let $\pi : \mathbb{R}^3 \to T^3 \cong \mathbb{R}^3/\mathbb{Z}^3$ be a universal cover, where $\mathbb{Z}^3$ is the group of integer shifts of $\mathbb{R}^3$ which is isomorphic to the fundamental group $\pi_1(T^3)$. Let $\overline{\mathcal{F}}$ be the covering foliation for $\mathcal{F}$ under π. Obviously, $\overline{\mathcal{F}}$ is a Reeb foliation. Since $\mathbb{R}^3$ is simply connected, $\overline{\mathcal{F}}$ is transversally orientable. Due to [2], one can assume that $\overline{\mathcal{F}}$ is defined by a Pfaff 1-form of the kind

$$dz = P(x, y, z)dx + Q(x, y, z)dy,$$

where (x, y, z) are Cartesian coordinates. It follows that $\mathcal{F}$ is transversally oriented because the group $\mathbb{Z}^3$ of covering maps consists of Euclidean translations which preserve orientation of $\mathbb{R}^3$. Hence the lamination formed by the unstable manifolds of points of Ω is transversally oriented as well. This implies that Ω is an orientable attractor. The contradiction concludes the proof. □

3 Proof of the main theorem

Theorem 3.1 *Let $f : M^3 \to M^3$ be a structurally stable diffeomorphism of a closed 3-manifold M^3. Then the spectral decomposition of f does not contain non-orientable codimension one expanding attractors.*

Proof. Assume the converse. Then the spectral decomposition of f contains a codimension one Plykin attractor, say Ω. By definition of Plykin attractors, there is an open component $R \subset M^3 - \Omega$ such that the accessible boundary of R contains the 1-bunch $W^u(m_0) \subset \Omega$, where m_0 is a boundary periodic point. Without loss of generality we can assume that m_0 is a fixed point, passing if necessary to some degree of f.

According to Theorem 2.1, M^3 is non-orientable. Let $\tilde{M}^3$ be an orientable manifold such that $p : \tilde{M}^3 \to M^3$ is an (nonbranched) double covering for M^3. Then there exists a diffeomorphism $\tilde{f} : \tilde{M}^3 \to \tilde{M}^3$ which cover f i.e., $f \circ p = p \circ \tilde{f}$. Let us prove that $\tilde{f}$ is an A-diffeomorphism.

It is obvious that if $x \in \tilde{M}^3$ is a nonwandering point of $\tilde{f}$, then $p(x)$ is a nonwandering point of f because p is a local homeomorphism. Therefore,

$NW(\tilde{f}) \subset p^{-1}(NW(f))$, where $NW(g)$ means the nonwandering set of the map g. Because f is an A-diffeomorphism, the periodic points of f are dense in $NW(f)$. Since p is a double covering, we see that a preimage of any periodic orbit of f is either one periodic orbit or two periodic orbits of $\tilde{f}$. As a consequence, the periodic orbits of $\tilde{f}$ are dense in $p^{-1}(NW(f))$. Hence, $p^{-1}(NW(f)) \subset NW(\tilde{f})$. This proves the equality $NW(\tilde{f}) = p^{-1}(NW(f))$. The hyperbolicity of $NW(f)$ implies the hyperbolicity of $NW(\tilde{f})$ because p is a local diffeomorphism. Thus, $\tilde{f}$ is an A-diffeomorphism.

Because of f is a structurally stable diffeomorphism, f satisfies to the strong transversality condition [13]. Since p is a local diffeomorphism, we see that $\tilde{f}$ satisfies to the strong transversality condition as well. By Robinson's theorem [22], $\tilde{f}$ is a structurally stable diffeomorphism.

Let us show that the preimage $p^{-1}(\Omega)$ contains a codimension one expanding attractor. By definition of a basic set, there is an orbit $O(x_0)$ of some point $x_0 \in \Omega$ which is dense in Ω. The preimage $p^{-1}(x_0)$ consists of two points $\tilde{x}_1$, $\tilde{x}_2$. Clearly, the union $O(\tilde{x}_1) \cup O(\tilde{x}_2)$ is dense in $p^{-1}(\Omega)$, where $O(\tilde{x}_i)$ is the orbit of the point $\tilde{x}_i$ $(i = 1, 2)$. Thus,

$$clos\ O(\tilde{x}_1) \cup clos\ O(\tilde{x}_2) = clos\ (O(\tilde{x}_1) \cup O(\tilde{x}_2)) = p^{-1}(\Omega),$$

where $clos\ (N)$ means a topological closure of N. Obviously, each of the set $clos\ O(\tilde{x}_1)$, $clos\ O(\tilde{x}_2)$ is closed, invariant, and contains an everywhere dense orbit. As a consequence, $clos\ O(\tilde{x}_1)$ and $clos\ O(\tilde{x}_2)$ are basic sets of $\tilde{f}$. Since a spectral decomposition is unique [25], it follows that either

$$p^{-1}(\Omega) = clos\ O(\tilde{x}_1) = clos\ O(\tilde{x}_2)$$

is one basic set or $clos\ O(\tilde{x}_1)$, $clos\ O(\tilde{x}_2)$ are different basic sets. In both cases, we denote $clos\ O(\tilde{x}_1)$ by $\tilde{\Omega}$.

Because Ω is a codimension one expanding attractor, Ω is locally homeomorphic to the product of a Cantor set and the plane $\mathbb{R}^2$. Hence, $\tilde{\Omega}$ is locally homeomorphic to the product of a Cantor set and the plane $\mathbb{R}^2$ as well. It follows that the topological dimension of $\tilde{\Omega}$ is equal to the dimension of the unstable splitting restricted on $\tilde{\Omega}$. Since p is a local diffeomorphism, $\dim E^u_{\tilde{x}} = 2$, $\tilde{x} \in \tilde{\Omega}$. As a consequence, $\tilde{\Omega}$ is an expanding attractor of codimension one.

Recall that the manifold $\tilde{M}^3$ is orientable. Due to Theorem 2.1, the expanding attractor $\tilde{\Omega}$ has no 1-bunches. Plykin [21] proved that for the manifold dimensions $\dim M^n \geq 3$ any codimension one expanding attractor can contain 1-bunches and 2-bunches. Hence, $\tilde{\Omega}$ contains only 2-bunches.

Let $\tilde{m}_1 \in \tilde{\Omega}$ be a lift under p of the point m_0. Because p is a double covering, $\tilde{m}_1$ is a boundary periodic point of the codimension one attractor

$\tilde{\Omega}$. Since every bunch of $\tilde{\Omega}$ is a 2-bunch, it follows that there is the boundary periodic point $\tilde{m}_2 \in \tilde{\Omega}$, $\tilde{m}_2 \neq \tilde{m}_1$, associated with $\tilde{m}_1$. Let $W^s_\emptyset(\tilde{m}_i)$ $(1 = 1, 2)$ be the component of $W^s(\tilde{m}_i) - m_i$ such that $\tilde{\Omega} \cap W^s_\emptyset(\tilde{m}_i) = \emptyset$. In [8], it was shown that $W^s_\emptyset(\tilde{m}_1)$ and $W^s_\emptyset(\tilde{m}_2)$ belong to the unstable manifolds $W^u(\tilde{\alpha}_1)$ and $W^u(\tilde{\alpha}_{k+1})$ respectively of the repelling periodic points $\tilde{\alpha}_1$, $\tilde{\alpha}_{k+1}$ (note that the proof of this result uses the fact that $\tilde{M}^3$ is closed), Figure 4. Moreover, there are repelling periodic points $\tilde{\alpha}_1, \ldots, \tilde{\alpha}_{k+1}$ and saddle periodic points $P_1, \ldots, P_k$ of index 2 such that the following conditions hold:
1. The set

$$a_{12} = \{\tilde{m}_1\} \cup W^s_\emptyset(\tilde{m}_1) \cup \tilde{\alpha}_1 \cup W^s(P_1) \cup P_1 \cup \tilde{\alpha}_2 \cup \ldots \cup P_k \cup \tilde{\alpha}_{k+1} \cup W^s_\emptyset(\tilde{m}_2) \cup \{\tilde{m}_2\}$$

is homeomorphic to an arc with no self-intersections whose endpoints are $\tilde{m}_1$ and $\tilde{m}_2$. 2. $a_{12} - (\tilde{m}_1 \cup \tilde{m}_2) \subset \tilde{M}^3 - \tilde{\Omega}$. 3. The repelling periodic points $\tilde{\alpha}_i$ alternate with saddle periodic points P_i on the arc a_{12}.

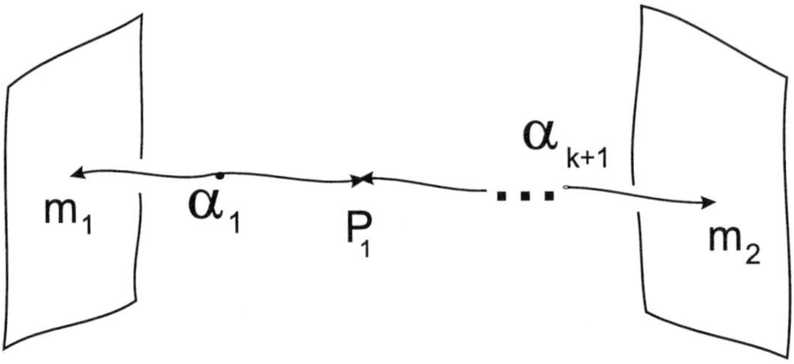

Figure 4. The arc a_{12}.

Let us prove that $p(\tilde{m}_2) = m_0$. According to Lemma 2.1, if $\tilde{x}_n \to \tilde{m}_1$ as $n \to \infty$, then $\tilde{y}_n \to \tilde{m}_2$, where $\tilde{x}_n \in W^u(\tilde{m}_1)$, $\tilde{y}_n \in W^s(\tilde{m}_2)$, and $(\tilde{x}_n, \tilde{y}_n)^s = (\tilde{x}_n, \tilde{y}_n)^s_\emptyset$. Since $(\tilde{x}_n, \tilde{y}_n)^s \cup \tilde{\Omega} = \emptyset$ and $p(\tilde{\Omega}) = \Omega$, it follows that

$$p((\tilde{x}_n, \tilde{y}_n)^s_\emptyset) = (p(\tilde{x}_n), p(\tilde{y}_n))^s_\emptyset.$$

The inclusion $p(\tilde{x}_n) \in W^u(m_0)$ implies $p(\tilde{y}_n) \in W^u(m_0)$ because of $W^u(m_0)$ is a 1-bunch. Since p is a continuous map, $p(\tilde{x}_n) \to p(\tilde{m}_1) = m_0$ and $p(\tilde{y}_n) \to p(\tilde{m}_2)$. From the inclusion $p(\tilde{y}_n) \in W^u(m_0)$ and Lemma 2.1 it follows that $p(\tilde{y}_n) \to m_0$. Hence, $p(\tilde{m}_2) = m_0$.

Because of the relation $p \circ f = \tilde{f} \circ p$, p maps the invariant manifolds of $\tilde{f}$ into the invariant manifolds of f. Then

$$p(W^s_\emptyset(\tilde{m}_1)) = p(W^s_\emptyset(\tilde{m}_2)), p(\tilde{\alpha}_1) = p(\tilde{\alpha}_{k+1}),$$

$$p(W^s(P_1)) = p(W^s(P_k)), p(P_1) = p(P_k), p(\tilde{\alpha}_2) = p(\tilde{\alpha}_k), \ldots .$$

Since the repelling periodic points $\tilde{\alpha}_i$ alternate with saddle periodic points P_i on the arc a_{12}, we see that the number of periodic points on a_{12} is odd because the endpoints both are of saddle type. As a consequence, there is either a periodic point $\tilde{\alpha}_i$ with $p(W^s(P_{i-1})) = p(W^s(P_i))$ or a periodic point P_i with $p(W_1^s(P_i)) = p(W_2^s(P_i))$, where $W_1^s(P_i)$, $W_2^s(P_i)$ are different components of $W^s(P_i) - P_i$. In both cases, there is a point ($\tilde{\alpha}_i$ or P_i) at which p is not a local homeomorphism. This contradiction concludes the proof. □

Acknowledgments

The research was partially supported by the INTAS grant 97-1843 and RFBR grant 99-01-00230. The authors are grateful to D. Anosov, S. Aranson, V. Grines and M. Malkin for useful discussions. We thank a referee for useful remarks.

References

1. D. Anosov. and V. Solodov, *Hyperbolic sets*, in Sbornik of ser. "Modern Problems of Math." ed. D. Anosov, Dynam. Systems-9, **66** (1991), 12–99.
2. S. Aranson and E. Zhuzhoma, *On topological equivalence of codimension one Reeb foliations on a three dimensional torus*, in Methods of Qual. Theor. of Diff. Eq., Gorky, (1978), 41–44 (in Russian).
3. R. Bowen, *Periodic points and measures for axiom A diffeomorphisms*, Trans. Amer. Math. Soc., **154** (1971), 337–397.
4. J. Franks, *Anosov diffeomorphisms*, in Global Analysis, Proc. Symp. in Pure Math., AMS **14** (1970), 61–94.
5. J. Franks and C. Robinson, *A quasi-Anosov diffeomorphism that is not Anosov*, Trans. Amer. Math. Soc., **223** (1976), 267–278.
6. V. Grines, *On topological conjugacy of diffeomorphisms of two a dimensional manifold onto one-dimensional orientable basic sets I*, Trans. Moscow Math. Soc., **32** (1975), 31–56.
7. V. Grines, *On topological conjugacy of diffeomorphisms of two a dimensional manifold onto one-dimensional orientable basic sets II*, Trans. Moscow Math. Soc., **34** (1977), 237–245.
8. V. Grines and E. Zhuzhoma, *Structurally stable diffeomorphisms with codimension one basic sets*, preprint, Universite de Bourgogne, Laboratoire de Topologie, Dijon, no 223 (2000).

9. J. Hempel, 3-manifolds, Princeton, Annals Math. Studies, **86** (1976).
10. M. Hirch and C. Pugh, *Stable manifolds and hyperbolic sets*, in Global Analysis, Proc. Symp., Amer. Math. Soc., **14** (1970), 133–163.
11. A. Katok and B. Hasselblatt, Introduction to the Modern Theory of Dynamical Systems, Encyclopedia of Math. and its Appl., Cambridge Univ. Press.
12. H. Kollmer, *On hyperbolic attractors of codimension one*, in Geometry and Topology, Lect. Notes of Math., **597** (1976), 330–334.
13. R. Mañé, *A proof of* C^1 *stability conjecture*, Publ. Math. IHES, **66** (1988), 161–210.
14. S. E. Newhouse, *On codimension one Anosov diffeomorphisms*, Amer. J. Math., **92** no 3 (1970), 761–770.
15. I. Nikolaev and E. Zhuzhoma, Flows on 2-dimensional manifolds, Lect. Notes in Math.,**1705**, Springer Verlag, 1999.
16. Z. Nitecki, Differentiable dynamics, MIT Press, Cambridge, 1971.
17. J. Plante, *The homology class of an expanded invariant manifolds*, Lect. Notes Math., **468** (1975), 251–256.
18. R. V. Plykin, *On the topology of basic sets of Smale diffeomorphisms*, Math. Sbornik, **84** (1971), 301–312 (in Russian).
19. R. V. Plykin, *Sources and sinks of A-diffeomorphisms of surfaces*, Mat. Sbornik, **94** (1974), 223–253.
20. R. V. Plykin, *On hyperbolic attractors of diffeomorphisms*, Usp. Math. Nauk, **35** 3 (1980), 94–104 (in Russian).
21. R. V. Plykin, *On the geometry of hyperbolic attractors of smooth cascades*, Usp. Math. Nauk, **39** (1984), 75–113 (in Russian).
22. C. Robinson, *Structural stability of* C^1 *diffeomorphisms*, J. Diff. Eq., **22** no 1 (1976), 28–73.
23. C. Robinson C, Dynamical Systems: stability, symbolic dynamics, and chaos, Studies in Adv. Math., Sec. edition, CRC Press.
24. H. Rosenberg H, *Foliations by planes*, Topology, **7** (1968), 131–138.
25. S. Smale, *Differentiable dynamical systems*, Bull. Amer. Math. Soc., **73** 1 (1967), 741–817.
26. R. Williams, *Expanding attractors*, Publ. Math. I.H.E.S., **43** (1974), 169–203.

Received June 7, 2000, revised January 5, 2001

Proceedings of
FOLIATIONS: GEOMETRY AND DYNAMICS
held in Warsaw, May 29–June 9, 2000
ed. by Paweł WALCZAK *et al.*
World Scientific, Singapore, 2002
pp. 371–386

ON EXACT POISSON MANIFOLDS OF DIMENSION 3

TADAYOSHI MIZUTANI
Department of Mathematics, Faculty of Science,
Saitama University, 255 Shimo-Ookubo Urawa, Saitama 338-8570, Japan,
e-mail: tmiztani@rimath.saitama-u.ac.jp

A Poisson manifold is called "exact" if its Poisson bi-vector field represents a trivial element in the Poisson cohomology. We investigate some topological properties of the foliation which is associated with an exact Poisson manifold and construct explicit examples of exact Poisson structures on closed 3-manifolds which are different from homogeneous ones.

1 Introduction

A Poisson manifold is a pair (M, Π) of a C^∞-manifold and a 2-vector field on it, which satisfies

$$[\Pi, \Pi] = 0,$$

where $[\cdot\, , \,\cdot]$ denotes the Schouten bracket.

We call the condition $[\Pi, \Pi] = 0$ for a 2-vector field Π the *Poisson condition* and Π the *Poisson bi-vector field.* The Poisson bracket $\{f, g\}$ of $f, g \in C^\infty(M)$ is then defined by

$$\{f, g\} = \Pi(df, dg).$$

It satisfies the following well-known properties.

(1) $(f, g) \mapsto \{f, g\}$ $f, g \in C^\infty(M)$ gives a Lie algebra structure (over $\boldsymbol{R}$) on $C^\infty(M)$, that is, the pairing $\{f, g\}$ is skew-symmetric bilinear and satisfies the Jacobi identity $\{f, \{g, h\}\} + \{g, \{h, f\}\} + \{h, \{f, g\}\} = 0$,

(2) $\{f, gh\} = \{f, g\}h + g\{f, h\}$ holds for $f, g, h \in C^\infty(M)$.

For any 2-vector field Π on M, we define a homomorphism of bundles

$$I = I_\Pi : T^*M \to TM$$

which is given by

$$I_x(\alpha_x) = \Pi_x(\alpha_x, \cdot) = i_{\alpha_x}\Pi, \qquad \alpha_x \in T_x^*M,$$

at each point $x \in M$. Here, we used the notation of interior product to express a contraction of tensors.

The rank of the linear map I_x is called the *rank of* Π *at* x and it is denoted by rank Π_x. If the rank Π_x is constant on the whole manifold, (M, Π) is called *regular*. In this paper, we are mainly concerned with regular Poisson manifolds.

One of the geometric aspects of a Poisson manifold (M, Π) is the fact that the distribution (plane field) given by

$$\mathrm{Image} I_x \subset T_xM$$

is integrable ([9]), and hence it defines a smooth foliation when (M, Π) is regular. We denote this foliation (=integrable distribution) by $\mathcal{F} = \mathcal{F}_\Pi$. It is called the *characteristic foliation* of (M, Π), and its leaves are called *symplectic leaves*, since Π restricted to each leaf naturally defines a symplectic structure on it.

It is natural to ask to what extent $\mathcal{F}_\Pi$ reflects the properties of Π.

In Section 2, we review basic facts on the Schouten bracket, especially its relationship with the generalized divergence. We also give a condition for a plane field defined by a regular 2-vector field to be integrable. It is expressed in terms of the Schouten bracket and the generalized divergence of Π. In Sections 3, 4, 5, we consider exact Poisson manifolds. A Poisson manifold (M, Π) is called an *exact Poisson manifold*, if there exists a vector field Z such that $[Z, \Pi] = -\Pi$ ([1], [9]). From the view point of Poisson cohomology, this means that a Poisson 2-cocycle defined by Π is a coboundary of a 1-cochain. We ask ourselves if this condition imposes any topological conditions on the characteristic foliation and try to understand which codimension one foliation of a closed 3-manifold has such an exact Poisson structure. In Section 3, we will give homogeneous examples of such structure and characterize them in terms of the *homothetic vector field* Z. In Section 4, we investigate the properties of the homothetic vector field Z and prove it is tangent to the foliation under a mild condition. In Section 5, we will carefully construct an example of an exact Poisson structure which is different from homogeneous ones. Its characteristic foliation is the one known as a Hirsch foliation and it enables us to get examples with exceptional leaves.

In this paper, we tried to use vector field description where possible rather than differential forms since the Poisson structure itself is of vector field nature. All the manifolds are assumed to be C^∞. Λ^pTM denotes the p-th exterior space bundle of the tangent bundle of M and $\Gamma(\Lambda^p TM)$ denotes the set of smooth section of it, that is, the space of p-vector fields.

2 Generalized divergence and the Schouten bracket

Let M be a smooth manifold and P a p-vector field, that is, $P \in \Gamma(\Lambda^p(TM))$ $(p \geq 0)$. Let

$$c : \Gamma((T^*M)) \otimes \Gamma(\Lambda^p(TM)) \to \Gamma(\Lambda^{(p-1)}(TM))$$

denote the contraction.

Definition 1 Let ∇ be a connection (covariant differentiation) on M and P a p-vector filed. Then the $(p-1)$-vector field $\mathrm{Div}_\nabla P$ given by

$$\mathrm{Div}_\nabla P = c(\nabla P)$$

is called a generalized divergence of P associated with the connection ∇.

It is shown that if ∇ is the Levi-Civita connection of a Riemannian metric and $P = X$ is a vector field, $\mathrm{Div}_\nabla X$ coincides with the usual divergence $\mathrm{div} X$ with respect to the Riemannian volume Ω, i.e. $L_X \Omega = (\mathrm{div} X)\Omega$ (L_X is the Lie derivation).

Although the generalized divergence of a p-vector field depends on the choice of the connection ∇, we often omit ∇ and write $\mathrm{Div} P$ for $\mathrm{Div}_\nabla P$.

It is not always true that $\mathrm{Div}^2 = \mathrm{Div} \circ \mathrm{Div} = 0$. It is proved, however, if one chooses a connection which preserves a volume form that $\mathrm{Div}^2 = 0$ holds. In fact, if ∇ preserves a volume form Ω, one can see the following relation of Div and d(= exterior differential) holds.

$$d(\Omega(P)) = (-1)^p \Omega(\mathrm{Div} P), \quad (p = \deg P).$$

One of the definition of the Schouten bracket $[P, Q]$ is the following ([6]).

Definition 2 Let Div be a generalized divergence associated with a torsion free connection of M. Let P, Q be p-vector field and q-vector field on M, respectively. The $(p+q-1)$-vector field $[P, Q]$ defined by

$$[P, Q] = \mathrm{Div}(P \wedge Q) - (\mathrm{Div} P \wedge Q + (-1)^p P \wedge \mathrm{Div} Q)$$

is called the Schouten bracket of P and Q.

It is proved that $[P, Q]$ is well-defined, namely, it is independent of the choice of the *torsion free* connection involved.

The following is a list of some basic properties of the Schouten bracket ([9]). Here, f, g are smooth functions and P, Q, R are a p-vector field, a q-vector field and an r-vector field, respectively. Also, we use the interior product notation for the contraction.

1. For $f, g \in C^\infty(M)$, $[f, g] = 0$,
2. $[f, Q] = i_{df}Q$, more generally, $[fP, Q] = (-1)^p P \wedge (i_{df}Q) + f[P, Q]$,
3. $[P, Q] = (-1)^{pq}[Q, P]$,
4. Let $P = X$ be a vector field, then $[X, Q] = L_X Q$ (the Lie derivative),
5. $[P, Q \wedge R] = [P, Q] \wedge R + (-1)^{(p-1)q} Q \wedge [P, R]$,
6. $\mathrm{Div}[P, Q] = -[\mathrm{Div}P, Q] - (-1)^p [P, \mathrm{Div}Q]$, when $\mathrm{Div}^2 = 0$,
7. $[P, [Q, R]] = (-1)^{p-1}[[P, Q], R] + (-1)^{(p-1)(q-1)}[Q, [P, R]]$ (generalized Jacobi identity).

2.1 Integrability of a plane field

Here, we briefly discuss the integrability of the plane field which is given as the image of a 2-vector field.

Let Π be a 2-vector field on M. Assume the rank of Π is equal to $2l$ $(0 \leq 2l \leq \dim M)$ everywhere on M. Recall that Π defines a distribution $\mathcal{F}_\Pi$ which gives the following subspace of $T_x M$;

$$\mathcal{F}_{\Pi,x} = \{\Pi_x(\alpha_x, \cdot) \in T_x M | \alpha_x \in T^* M\}.$$

We prove the following

Theorem 1 *The distribution $\mathcal{F}_\Pi$ defined by a regular 2-vector field Π whose rank is $2l$, is integrable if and only if $[\Pi, \Pi^l] = [\Pi, \Pi \wedge \cdots \wedge \Pi] = 0$ holds.*

We first prove following formula.

Lemma 2 *If* rank $\Pi = 2l$, *then*

$$[\Pi, \Pi^l] = -2\mathrm{Div}\Pi \wedge \Pi^l,$$

where $\mathrm{Div}\Pi$ *is defined by choosing any torsion free connection on TM.*

Proof. Since $\Pi \wedge \Pi^l = 0$, we have

$$[\Pi, \Pi^l] = -\mathrm{Div}\Pi \wedge \Pi^l - \Pi \wedge \mathrm{Div}\Pi^l. \tag{1}$$

Plugging the following

$$\mathrm{Div}\Pi^l = \mathrm{Div}\Pi \wedge \Pi^{l-1} + \Pi \wedge \mathrm{Div}\Pi^{l-1} + [\Pi, \Pi^{l-1}] \tag{2}$$

into the above (1), we have

$$[\Pi, \Pi^l] = -2\mathrm{Div}\Pi \wedge \Pi^l - \Pi^2 \wedge \mathrm{Div}\Pi^{l-1} - \Pi \wedge [\Pi, \Pi^{l-1}]. \tag{3}$$

Again plugging (2) for $l - 1$ into (3), we obtain

$$-3\mathrm{Div}\Pi \wedge \Pi^l - \Pi^3 \wedge \mathrm{Div}\Pi^{l-2} - \Pi^2 \wedge [\Pi, \Pi^{l-2}] - \Pi \wedge [\Pi, \Pi^{l-1}]. \tag{4}$$

Repeating this we have

$$[\Pi, \Pi^l] = -(l+1)\mathrm{Div}\Pi \wedge \Pi^l - \sum_{i=1}^{l-1} \Pi^i \wedge [\Pi, \Pi^{l-i}] \qquad (5)$$

Using $[\Pi, \Pi^k] = k\,[\Pi, \Pi] \wedge \Pi^{k-1}$ for $k \geq 1$, we get

$$\frac{l(l+1)}{2}[\Pi, \Pi] \wedge \Pi^{l-1} = -(l+1)\,\mathrm{Div}\Pi \wedge \Pi^l.$$

From this, we obtain

$$[\Pi, \Pi^l] = l\,[\Pi, \Pi] \wedge \Pi^{l-1} = -2\mathrm{Div}\Pi \wedge \Pi^l.$$

□

Proof of Theorem 1. Let $\mathcal{F}_\Pi$ be of codimension q and assume it is defined by a local equation of 1-forms

$$\alpha_1 = \cdots = \alpha_q = 0.$$

Then we have

$$\Pi(\alpha_j, \cdot) = 0, \quad j = 1, \ldots, q.$$

Taking the covariant derivative, we have

$$(\nabla\Pi)(\alpha_j, \cdot) + \Pi(\nabla\alpha_j, \cdot) = 0.$$

By contraction, we obtain

$$(\mathrm{Div}\Pi)(\alpha_j) + \Pi(d\alpha_j) = 0. \qquad (6)$$

Thus, if $\{\alpha_1 \ldots \alpha_q\}$ satisfies the Frobenius integrability condition, $\Pi(d\alpha_j) = 0$ and we have

$$\mathrm{Div}\Pi(\alpha_j) = 0.$$

This shows that $\mathrm{Div}\Pi$ is a vector field tangent to $\mathcal{F}_\Pi$ and $\mathrm{Div}\Pi \wedge \Pi^l = 0$, since $\mathrm{rank}\Pi = 2l$.

Conversely, if $\mathrm{Div}\Pi \wedge \Pi^l = 0$, from $0 = \langle \alpha_j, \mathrm{Div}\Pi \wedge \Pi^l \rangle = (\mathrm{Div}\Pi)(\alpha_j) \wedge \Pi^l$ we can see $(\mathrm{Div}\Pi)(\alpha_j) = 0$ and hence by (6), we get $\Pi(d\alpha_j) = 0$ for each j. This means that each $d\alpha_j$ should be of the form

$$\sum_{k=1}^{q} \alpha_k \wedge \beta_{kj}$$

for some 1-forms $\{\beta_{kj}\}$. This shows $\{\alpha_1, \ldots, \alpha_q\}$ satisfies the Frobenius integrability condition. □

We get the following well-known fact

Corollary 3 *Let (M, Π) be a regular Poisson manifold. Then the characteristic distribution is integrable.*

Proof. Let rank $\Pi = l$. Since $[\Pi, \Pi] = 0$, $[\Pi, \Pi^l] = 2l[\Pi, \Pi] \wedge \Pi^{l-1} = 0$. Thus, the result follows. □

Remark. In a similar way, if $\dim M = 2l + 1$ and $\mathrm{rank}\Pi = 2l$, we can show that the characteristic distribution is a contact plane field if and only if $[\Pi, \Pi^l]$ is nowhere zero.

3 Exact Poisson manifolds of special kind

If a Poisson manifold (M, Π) has a vector field Z which satisfies $L_Z\Pi = [Z, \Pi] = -\Pi$, it is called *exact.* In this case Z is called a *homothetic vector vector field* of (M, Π) ([1]). From the view point of the Poisson cohomology, an exact Poisson manifold is a Poisson manifold whose Poisson bi-vector field Π represents 0 in $H^2_{LP}(M)$. Recall that the Poisson cohomology is a cohomology whose p-th cochain group is $\Gamma(\Lambda^p TM)$ and the coboundary operator $\sigma : \Gamma(\Lambda^p TM) \to \Gamma(\Lambda^{p+1}TM)$ is given by $\sigma(P) = -[\Pi, P]$ ([9]).

It is not difficult to give examples of exact Poisson manifolds which are non-compact. The following examples are standard.

Example 1 (Cotangent bundles) Let $(T^*M, d\lambda)$ be the canonical symplectic structure of the cotangent bundle of M, where λ is the Liouville form. Let Π be the 2-vector field on T^*M such that $\Pi = (d\lambda)^{-1}$. Here we regard $d\lambda$ as an isomorphism $T(T^*M) \to T^*(T^*M)$. Note that this means $I_\Pi(d\lambda) = -\Pi$. Using a general formula

$$(d\omega)(X, \cdot\cdot) = L_X\omega - d(\omega(X, \cdot))$$

for a 2-form ω and a vector field X, we see $L_{\Pi(df)}d\lambda = 0$ and Π satisfies the Poisson condition $[\Pi(df, \cdot), \Pi] = 0$.

Let $Z = \Pi(\lambda, \cdot)$. Then by the above, we have $L_Z d\lambda = d(d\lambda(\Pi(\lambda))) = d\lambda$. From this

$$0 = L_Z(I_\Pi \circ d\lambda) = L_Z(I_\Pi) \circ d\lambda + I_\Pi \circ L_Z d\lambda = L_Z(I_\Pi) \circ d\lambda + I_\Pi \circ d\lambda.$$

This means

$$[\Pi(\lambda, \cdot), \Pi] = -\Pi.$$

Thus, $(T^*M, \Pi, \Pi(\lambda, \cdot))$ is an exact Poisson manifold.

Example 2 (Lie Poisson structure) Let $(\mathfrak{g}^*, \Pi)$ be the Lie Poisson structure on the dual space of a Lie algebra $\mathfrak{g}$. The Poisson bi-vector field Π is defined as follows. We have an identification $T^*\mathfrak{g}^* \approx \mathfrak{g}^* \times \mathfrak{g}^{**} = \mathfrak{g}^* \times \mathfrak{g}$ by translations. If $T^*\mathfrak{g}^* \ni p, q$ are represented as $p = (\alpha, x), q = (\alpha, y)$, under this identification, then Π is given by

$$\Pi_\alpha(p,q) = \alpha([x,y]).$$

Let Z be the radial vector field on $\mathfrak{g}^*$. Namely, $Z_\alpha(\alpha \in \mathfrak{g}^*)$ is a tangent vector which corresponds to the curve $e^t \cdot \alpha$ in $\mathfrak{g}^*$. We regard p, q as the constant 1-forms. In other words, we consider p as a 1-form on $\mathfrak{g}^*$, given by $\alpha \mapsto (\alpha, x)$, where $x \in \mathfrak{g} = \mathfrak{g}^{**}$ is independent of α.

Then we have

$$L_Z(\Pi(p,q)) = L_Z(\alpha([x,y])) = \frac{d}{dt}\,|_{t=0}(e^t\alpha([x,y])) = \alpha([x,y]) = \Pi(p,q).$$

On the other hand,

$$\begin{aligned} L_Z(\Pi(p,q)) &= (L_Z\Pi)(p,q) + \Pi(L_Zp,q) + \Pi(p,L_Zq) \\ &= (L_Z\Pi)(p,q). \end{aligned}$$

since p and q are constant fields.

From these we have

$$(L_Z\Pi)(p,q) = \Pi(p,q).$$

This shows $(\mathfrak{g}^*, \Pi, -Z)$ is an exact Poisson structure.

Of course, the homothetic vector field in the above is not unique. In fact if Z, Z' are both homothetic vector field for Π, then the Lie derivative $L_{Z-Z'}\Pi$ vanishes. Thus the set of homothetic vector field of a Poisson structure forms an affine subspace of the vector space of all the vector fields, whose associated vector space is the space of vector fields which preserve the Poisson bi-vector field. Recall that any Hamiltonian vector field $I(df)$, $f \in C^\infty(M)$ preserves Π.

Now, we are interested in the following problem:

Problem. On a closed manifold what kind of codimension one foliations do appear as underlying foliations of exact Poisson manifolds ?

We will consider this problem in the case where M is a closed 3-dimensional manifold.

First, we note that every orientable foliation of dimension 2 is an underlying foliation of a Poisson structure. In fact, let $(M, \mathcal{F})$ be a foliation whose leaves are 2-dimensional and let $\Pi \in \Gamma(\Lambda^2\mathcal{F})$ be a non-zero cross section. Then Π is naturally considered as a 2-vector field on M.

It is easily checked that the image of I_Π coincides with $\mathcal{F}$. The Poisson condition on Π is satisfied since in this dimension, it is equivalent to the integrability of $\mathcal{F}$ (see Section 2).

3.1 Exact Poisson structure of special kind

In this subsection, we give two examples of exact Poisson manifolds which we call 'special'. The underlying manifolds are closed and quotients of 3-dimensional Lie groups.

Example 3 Let X_1, X_2, X_3 be the right invariant vector field of $G = \mathrm{SL}(2, \boldsymbol{R})$ corresponding to $\frac{1}{2}\begin{pmatrix} -1 & 0 \\ 0 & 1 \end{pmatrix}, \begin{pmatrix} 0 & 1 \\ 0 & 0 \end{pmatrix}, \begin{pmatrix} 0 & 0 \\ 1 & 0 \end{pmatrix}$ respectively. They satisfy the following bracket relations;

$$[X_1, X_2] = -X_2, \quad [X_1, X_3] = X_3, \quad [X_2, X_3] = -2X_1.$$

The 2-vector field $\Pi = X_1 \wedge X_2$ satisfies the Poisson condition $[\Pi, \Pi] = 0$ hence defines a Poisson structure. If we choose a uniform discrete subgroup Γ of $G = \mathrm{SL}(2, \boldsymbol{R})$, we obtain an induced Poisson structure on $M = G/\Gamma = \mathrm{SL}(2, \boldsymbol{R})/\Gamma$ which is a closed manifold. The underlying foliation $\mathcal{F}_\Pi$ is known as an Anosov foliation spanned by X_1 and X_2. It is also known that each leaf of this foliation is dense in M.

Let $Z = X_1 + aX_2$, (a is a constant) then

$$\begin{aligned} L_Z\Pi = [Z, \Pi] &= [X_1, X_1 \wedge X_2] \\ &= X_1 \wedge [X_1, X_2] = -X_1 \wedge X_2 = -\Pi. \end{aligned}$$

Thus (M, Π, Z) is a closed exact Poisson manifold.

Similarly, we have the following second example.

Example 4 Let G be a simply connected 3-dimensional solvable Lie group whose Lie algebra is generated by X_1, X_2, X_3 with the relations

$$[X_1, X_2] = -X_2, \quad [X_1, X_3] = X_3, \quad [X_2, X_3] = 0.$$

Like as in the case of Example 3, let $\Pi = X_1 \wedge X_2, Z = X_1 + aX_2$ be the right invariant fields on G. By the same computation, we see that Π defines an exact Poisson structure and Z is a homothetic vector field. Also, if we choose a uniform discrete subgroup Γ, we obtain an exact Poisson structure on a closed 3-dimensional manifold. In this case, $M = G/\Gamma$ is known to be diffeomorphic to a T^2-bundle over S^1 and the foliation is a suspension of a dense linear foliation of T^2, hence the leaves of $\mathcal{F}_\Pi$ are all dense again.

Note that, in both cases of the above examples, the symplectic leaves of the characteristic foliations are generated by the vector fields X_1, X_2 with the relation $[X_1, X_2] = -X_2$, which generate the Lie algebra of 2-dimensional affine group GA. From this, we can see that the leaves are the orbits of a locally free actions of GA.

We call the Poisson manifold which is obtained in the above examples a *special* exact Poisson manifold. It has a homogeneous structure which is induced from an invariant Poisson structure on a 3-dimensional Lie group.

One of the property of the homothetic vector field of our special Poisson manifold is the fact that its divergence (with respect to the canonical volume form) vanishes everywhere. Indeed, let Ω be the volume form on M, such that $\Omega(X_1 \wedge X_2 \wedge X_3) \equiv 1$. Then by an easy computation we can see $L_Z\Omega = 0$ which is, by definition, equal to $(\mathrm{div}Z)\Omega$ and $\mathrm{div}Z = 0$.

In the following , we will show that this property characterizes the special exact Poisson manifolds.

Theorem 4 *Let (M, Π, Z) be a regular exact Poisson manifold, where M is a closed 3-dimensional. Suppose that the homothetic vector field Z is divergence free with respect to some volume form Ω on M. Then (M, Π, Z) is diffeomorphic to a special exact Poisson manifold.*

Proof. Choose a Riemannian metric on M whose associated volume form is equal to Ω. We will use the generalized divergence with respect to the Riemannian connection of this metric. In the next section, we investigate some topological properties of the foliation associated with a codimension one exact Poisson structure. Especially, we prove the homothetic vector field on a closed manifold is necessarily tangent to leaves everywhere (see Theorem 9). If we admit this fact, we have $Z \wedge \Pi = 0$, $\mathrm{Div}Z = \mathrm{div}Z = 0$, hence,

$$-\Pi = [Z, \Pi] = \mathrm{Div}(Z \wedge \Pi) - \mathrm{Div}Z \wedge \Pi + Z \wedge \mathrm{Div}\Pi = Z \wedge \mathrm{Div}\Pi.$$

Since Π is nowhere vanishing, this shows that Z and $\mathrm{Div}\Pi$ are two vector fields tangent to the leaves of $\mathcal{F}_\Pi$, and are linearly independent at each point of M. Taking Div of both sides of $[Z, \Pi] = -\Pi$ (see Section 2), we have

$$[Z, \mathrm{Div}\Pi] = -\mathrm{Div}\Pi.$$

This shows that on M there exists an locally free action of the 2-dimensional affine group GA. Since $\mathrm{div}Z = 0$ and $\mathrm{div}(\mathrm{Div}\Pi) = \mathrm{Div}^2\Pi = 0$, we have $L_Z\Omega = L_{\mathrm{Div}\Pi}\Omega = 0$. This means that the action of GA preserves the volume Ω. Now a theorem of Ghys ([2]) concerning the rigidity of the action of GA on 3-manifolds says that this action is smoothly conjugate to one of the standard ones. That is, it is equivalent to a natural action of GA which is the action on one of the quotient manifolds G/Γ in the examples of this section. This means that there is a diffeomorphism $\varphi : M \to G/\Gamma$ sending Z to X_1 and $\mathrm{Div}\Pi$ to X_2. □

4 Exact Poisson structures on closed manifolds

As before, we consider regular Poisson structures and find some topological properties of exact Poisson manifolds. The main result is Theorem 9 in which we show that the homothetic vector field is always tangent to leaves provided the manifold is closed and the associated foliation is of codimension one.

We start with the following

Lemma 5 *Let (M, Π, Z) be a regular exact Poisson manifold. Then the homothetic vector field Z preserves the foliation $\mathcal{F}_\Pi$.*

Proof. Let the foliation $\mathcal{F}_\Pi$ be defined locally by Pfaffian forms

$$\alpha_1, \ldots, \alpha_q$$

which span $\mathrm{Ker} I_\Pi$. The relation $\Pi(\alpha_i, \cdot) = 0$ leads to the equation

$$(L_Z\Pi)(\alpha_i) + \Pi(L_Z\alpha_i, \cdot) = 0.$$

Since $(L_Z\Pi)(\alpha_i) = -\Pi(\alpha_i, \cdot) = 0$, we have

$$\Pi(L_Z\alpha_i, \cdot) = 0, \quad (i = 1, \ldots, q).$$

Thus each $L_Z\alpha_i$ is a functional linear combination of $\alpha_1, \ldots, \alpha_q$. Now let X be a local vector field which is tangent to the leaves. Then we have

$$\alpha_i(L_Z X) = L_Z(\alpha_i(X)) - (L_Z\alpha_i)(X) = 0, \quad (i = 1, \ldots, q).$$

Thus $L_Z X$ is also tangent to the leaves. This means Z preserves the foliation $\mathcal{F}_\Pi$. □

By the above lemma, the subset of M, where Z is transverse to $\mathcal{F}_\Pi$, is an open *saturated subset* (the subset which is a union of leaves) of M.

Lemma 6 *Let (M, Π, Z) be a codimension one exact Poisson manifold. That is, M is an exact Poisson manifold such that $\mathfrak{F}_\Pi$ is a codimension one foliation. Let U be an open saturated subset of M, where Z is transverse to the foliation $\mathfrak{F}_\Pi$. Then the foliation $\mathcal{F}_\Pi|_U$ restricted on U is defined by a closed 1-form.*

Proof. Take a 1-form α which satisfies $I_\Pi(\alpha) = 0$ and $\alpha(Z) \equiv 1$ on U. It is easy to see that

$$d\alpha = L_Z\alpha \wedge \alpha.$$

By Lemma 5, $L_Z\alpha$ is a functional multiple of α, hence we have $d\alpha = 0$ on U. □

Lemma 7 *Let (M, Π, Z) be a regular exact Poisson manifold. If L is a leaf of $\mathcal{F}_\Pi$ such that Z is tangent to L, then L is a non-compact leaf.*

Proof. Take a leafwise symplectic 2-form ω on M such that $I_\Pi(\omega) = \Pi$. If rank$\Pi = 2k$, the $2k$-form

$$\omega^k = \omega \wedge \cdots \wedge \omega \quad (k\text{-times})$$

restricts to a volume form on each leaf. Since the pairing $\langle \omega^k, \Pi^k \rangle$ is a non-zero constant and $L_Z\Pi^k = -k\Pi^k$, $\langle L_Z\omega^k, \Pi^k\rangle$ is also a non-zero constant. Thus the restriction $(L_Z\omega^k)|_L$ of $L_Z\omega^k$ to L is a non-zero multiple of $\omega^k|_L$. Since Z is tangent to L, this shows $\mathrm{div}_{(\omega^k|L)}Z = C$ (the divergence with respect to the volume $\omega^k|_L$). for some non-zero constant C. This is impossible when L is compact. □

Lemma 8 *Let (M, Π, Z) be a codimension one exact Poisson manifold, where M is a closed manifold. Then the subset of M, which is the union of leaves where the homothetic vector field Z is tangent to each leaf is a non-empty closed saturated set.*

Proof. Closedness of the set is clear. If it is empty, Z is transverse to $\mathcal{F}_\Pi$ everywhere on M and $Z \wedge \Pi^k$ is nowhere zero on M ($2k$ is the rank of Π). Let Ω be a volume form on M dual to $Z \wedge \Pi^k$ (i.e. Ω satisfies $\Omega(Z \wedge \Pi^k) \equiv 1$). Then it is easily seen that $\mathrm{div}_\Omega Z \equiv k$ which is impossible on a closed manifold M. □

Let (M, Π, Z) be an exact Poisson manifold of a closed manifold, which is of codimension one. We put rank $\Pi = 2k$. As we have seen in Lemma 7, the vector field Z is not tangent to a compact leaf. Assume that Z is transverse to a compact leaf L. Since the 1-parameter subgroup ϕ_t generated by Z preserves the foliation $\mathcal{F}_\Pi$, the union $\cup_{t\in\boldsymbol{R}}\phi_t(L)$ consists of compact leaves which are diffeomorphic to L. If $\cup_{t\in\boldsymbol{R}}\phi_t(L)$ is not whole M, there exists a leaf which is the limit leaf of a subset of $\cup_{t\in\boldsymbol{R}}\phi_t(L)$, which itself should be compact and Z must be transverse to it. This implies that it has to be contained in $\cup_{t\in\boldsymbol{R}}\phi_t(L)$ which should be whole M (we assume M is connected). This contradicts the above Lemma 8

Thus, we can conclude that (M, Π, Z) has no compact leaves as long as M is closed. Therefore, for example, there is no exact Poisson structures on S^3 since every codimension one foliation of S^3 has a compact leaf diffeomorphic to T^2 ([7]). Also, we saw Z is not everywhere transverse to $\mathcal{F}_\Pi$. Moreover in the case of special exact Poisson manifold, the homothetic vector field is everywhere tangent to leaves on the whole manifold. Hence it is natural to ask the following question:

Question. Are there any examples of (M, Π, Z) on which Z is tangent to the leaves of $\mathcal{F}_\Pi$ on one part and transverse to them on the other part?

The following theorem shows there is no such example on a closed manifold provided (M, Π, Z) is of codimension one.

Theorem 9 *Let (M, Π, Z) be an exact Poisson structure of a closed manifold, which we assume regular and of codimension one. Then the homothetic vector field Z is tangent to the foliation $\mathcal{F}_\Pi$ everywhere on M.*

Proof. To the contrary, we assume that there exists an open subset of M, where Z is transverse to $\mathcal{F}_\Pi$. Let $M = U \cup U'$ be the partition into two part; on U, Z is transverse to $\mathcal{F}_\Pi$ and on U', Z is tangent to $\mathcal{F}_\Pi$. By Lemma 5, both U and U' are saturated subsets. By Lemma 8, U' is a non-empty closed set. Since $\mathcal{F}_\Pi$ has no compact leaves, the closed saturated subset $U' \neq M$ has to contain an exceptional minimal set. Let E denote it. By a theorem of Sacksteder([8]), if a codimension one foliation is of class C^2, the exceptional minimal set contains a leaf which has a contracting holonomy. Take such a leaf L contained in E. Choose a point $x \in L$ and a transverse small arc $I \approx (-1, 1)$ through x, where 0 corresponds to x. The contracting holonomy gives a germ of a map $\psi : (-\epsilon, \epsilon) \to (-1, 1)$ at 0 ($\epsilon > 0$ is small). The intersection $I \cap E$ is a Cantor set and $I \cap U$ is a union of open intervals. Let $\{\varphi_t\}$ be the 1-parameter subgroup of Z. Since φ_t maps a leaf into a leaf, we can choose and fix a small t_0 so that φ_{t_0} induces a local diffeomorphism $\tilde{\varphi}$ of I at x. (consider I as the set of plaques near x).

Clearly, $\tilde{\varphi}$ fixes $I \cap E$ and preserves the open intervals of $I \cap U$. It turns out that the germs of $\tilde{\varphi}$ and ψ at x commute each other. To see this, we take a loop ℓ in L starting and ending at x which gives the holonomy ψ. Then we choose a chain of distinguished neighbourhoods

$$\{V_0, V_1, \ldots, V_{k-1}, V_k = V_0, \quad V_i \cap V_{i+1} \neq \emptyset (i = 0, \ldots, k-1)\}$$

which covers ℓ. Each V_i is assumed to be diffeomorphic to $D^{n-1} \times I_i, I_i = (-1, 1)$ and $D^{n-1} \times \{t\}, (t \in I_i)$ is a plaque in V_i and $D^{n-1} \times \{0\}$ is the plaque where $\ell \cap V_i$ lies ($\dim M = n$).

The diffeomorphism φ_{t_0} for small t_0 induces a germ of diffeomorphism of each I_i at 0, which we denote by ϕ_i. On the other hand, we have a local holonomy translation along ℓ which gives a germ $h_{i+1,i} : (I_i, 0) \to (I_{i+1}, 0)$. Since the diffeomorphism φ_{t_0} preserves the foliation, we have $h_{i+1,i} \circ \phi_i = \phi_{i+1} \circ h_{i+1,i}$ for each $i = 0, 1 \ldots k-1$. From this relation we have $\psi \circ \tilde{\varphi} = \tilde{\varphi} \circ \psi$ because $\psi = h_{k,k-1} \circ \cdots \circ h_{1,0}$ and $\tilde{\varphi} = \phi_0 = \phi_k$.

Since $\tilde{\varphi}$ has fixed points accumulating to x, by a lemma of Kopell ([5]), ψ can not be of class C^2, contradicting our assumption that the foliation is C^∞ . This proves the theorem. □

In the above theorem, we have in fact proved the following

Theorem 10 *Let $(M, \mathcal{F})$ be a codimension one smooth foliation of a closed manifold without compact leaves. Let Z be a vector field on M whose 1 - parameter group of diffeomorphisms preserves $\mathcal{F}$. If Z is tangent to $\mathcal{F}$ on* *a non-empty set then Z is everywhere tangent to $\mathcal{F}$.*

5 A construction of exact Poisson manifolds

In this section, we will give an explicit example of an exact Poisson structure which is different from previous ones. The manifold we will construct is a closed, 3-dimensional and the underlying foliation of the Poisson structure is a so-called *Hirsch foliation* ([4]).

We begin with describing such a type of codimension one foliations.

Let Σ_0 be an orientable 2-dimensional compact manifold whose boundary is a circle (for example 2-disk D^2). Make a product $S^1 \times \Sigma_0$ and choose an embedding $j : S^1 \to S^1 \times \Sigma_0$ whose image intersects each $\{t\} \times \Sigma_0$, $(t \in S^1)$ exactly at 2-points. Thus the composition

$$p \circ j : S^1 \to S^1,$$

where $p : S^1 \times \Sigma_0 \to S^1$ is the projection to the first factor, is a double covering. We choose j so that this double covering is the natural one and Image j is in the interior of $S^1 \times \Sigma_0$. Delete a small open tubular neighbourhood of Image j from $S^1 \times \Sigma_0$. Let N denote the resulting manifold. It will be helpful to note that N is also obtained as a mapping torus of a diffeomorphism of a 3-times punctured surface Σ_1 and is a fibre bundle over S^1. There is a codimension one foliation on N defined by the fibres of this bundle. Let $\partial_{in} N$ denote the *'interior boundary'* of N. We fix a trivialization of the bundle $\partial_{in} N \to S^1$ as the boundary of the tubular neighbourhood of Image j. Similarly, let $\partial_{ex} N$ denote the *'exterior boundary'* and we will fix a trivialization $\partial_{ex} N \to S^1$ as the boundary of $S^1 \times \Sigma_0$.

Let ϕ be an element of DiffS^1. Then we have a diffeomorphism

$$\tilde{\phi} = \phi \times \mathrm{id} : S^1 \times S^1 \to S^1 \times S^1$$

which gives a diffeomorphism

$$f : \partial_{in} N \to \partial_{ex} N.$$

Identifying the boundary tori of N through f, we obtain a closed 3-manifold M. It has a smooth codimension one foliation $\mathcal{F}_\phi$ which is induced from that of N. The foliation $\mathcal{F}_\phi$ obtained in this way, is called a *Hirsch foliation.*

It is easy to see that the leaves of $\mathcal{F}_\phi$ are all non-compact. If $\phi = id_{S^1}$, for example, all the leaves of $\mathcal{F}_\phi$ are dense in M. One can also choose ϕ so that $\mathcal{F}_\phi$ has exceptional leaves ([4]).

Now we are going to construct an exact Poisson structure on M. In the next lemma, we use the following notations. Let U_0 (resp. U_1) be an 'exterior'(resp. 'interior') collar neighbourhood of the unit circle in the Euclidean plane. (r, θ) is the polar coordinate on $\boldsymbol{R}^2 - \{0\}$. Let Σ_1 be the 3-times punctured surface at the beginning of this section and $\partial\Sigma_1 = C_0 \cup C_1 \cup C_2$ denote the union of circles where C_0 is the fibre of the exterior boundary of N while C_1 and C_2 are those of the interior boundary of N.

Lemma 11 *On Σ_1, we have a 1-form η which satisfies the following*

(1) $d\eta$ is a volume form of Σ_1,

(2) On the neighbourhood of C_0, η is diffeomorphic to $(1/2)r^2 d\theta|_{U_0}$ and on some neighbourhood of C_1, C_2, η is diffeomorphic to $(1/2)r^2 d\theta|_{U_1}$.

Proof. We choose a volume form Ω on Σ_1 which is described as follows. First, around the boundary C_0, we take a collar neighbourhood which is diffeomorphic to U_0, around C_1 and C_2, we take collar neighbourhoods diffeomorphic to U_1. Then, we introduce the Euclidean volume form on these collar neighbourhoods by the above identification. We extend these forms to a volume form Ω on the whole Σ_1 in such a way that they are unchanged in smaller neighbourhoods of the boundary; i.e. $\Omega = rdr \wedge \theta$ near the boundary. This Ω is an orientation we consider on Σ_1. If it is needed, we multiply Ω by a suitable positive function and may assume

$$\int_{\Sigma_1} \Omega = \pi.$$

Again Ω should be left unchanged in a small neighbourhood of the boundary. Let η' be any 1-form on Σ_1 which is equivalent to $(1/2)r^2 d\theta|_{U_0}$ near C_0 and to $(1/2)r^2 d\theta|_{U_1}$ near C_1 and C_2. Then we have

$$\begin{aligned}
\int_{\Sigma_1} d\eta' &= \int_{\partial\Sigma_1} \eta' \\
&= \int_{C_0} \eta' + \int_{C_1} \eta' + \int_{C_2} \eta' \\
&= -(1/2)\int_{S^1} d\theta + \int_{S^1} d\theta = \pi.
\end{aligned}$$

(Recall that the orientation of C_i is determined by taking the interior product $i_X\Omega$ by an outward normal X.)

Take the difference $\Omega - d\eta'$ is a closed 2-form whose support is contained in the interior of Σ_1. By the above it represents zero in $H^2_{\text{compact}}(\text{Int}\Sigma_1)$. Namely, there exists a 1-form η'' whose support is in $\text{Int}\Sigma_1$ which satisfies $\Omega - d\eta' = d\eta''$.

Put

$$\eta = \eta' + \eta''.$$

Then η satisfies the required conditions (1) and (2). □

Now, we construct an exact Poisson structure on M.

Let Π_0 be the 2-vector field on Σ_1 such that $\langle \Omega, \Pi_0 \rangle = 1$ and Z_0 the vector field such that

$$i_{Z_0}\Omega = \eta.$$

Since $L_{Z_0}\Omega = di_{Z_0}\Omega = d\eta = \Omega$, we have

$$\begin{aligned} 0 = L_{Z_0}\langle \Omega, \Pi_0 \rangle &= \langle L_{Z_0}\Omega, \Pi_0 \rangle + \langle \Omega, L_{Z_0}\Pi_0 \rangle \\ &= \langle di_{Z_0}\Omega, \Pi_0 \rangle + \langle \Omega, L_{Z_0}\Pi_0 \rangle = 1 + \langle \Omega, L_{Z_0}\Pi_0 \rangle. \end{aligned}$$

This shows that

$$L_{Z_0}\Pi_0 = -\Pi_0.$$

Now it is not difficult to get a 2-vector field Π and homothetic vector field Z on M. To see this we note that N is obtained from

$$[0,1] \times \Sigma_1$$

by pasting $\{0\} \times \Sigma_1$ and $\{1\} \times \Sigma_1$ by a diffeomorphism $k : \Sigma_1 \to \Sigma_1$, which is an involution interchanging C_1 and C_2.

Taking $\frac{1}{2}(\eta + k^*\eta)$ instead of η if necessary, we can assume everything is k-invariant. Consider the obvious liftings of Π_0 and Z_0 onto the $[0,1] \times \Sigma_1$. Then the fields we are considering on the top and the bottom of the product manifold fit together under the diffeomorphism k. Thus N has a well-defined 2-vector filed and a vector field corresponding to Π_0 and Z_0. Finally, pasting the boundary of by a diffeomorphism

$$f : \partial_{in} N \to \partial_{ex} N,$$

we obtain a 2-vector field Π and the homothetic vector field Z on $(M, \mathcal{F})$. By our construction, (Π, Z) clearly satisfies the relation $L_Z\Pi = -\Pi$.

This finishes our construction of an exact Poisson structure on M.

By choosing different diffeomorphisms for f which are essentially diffeomorphisms of S^1, we get various underlying foliation. For example, when f is 'identity', we get a foliation with dense leaves. It is also possible to choose f so that the foliation contains exceptional leaves ([4]).

Remark. It seems interesting to ask if a similar construction is possible in higher dimensions. That is, 'is it possible to construct an exact Poisson manifold starting from a higher dimensional symplectic manifold in stead of Σ_0, in a similar way as in the preceding construction?'

Of course the following simple procedure is possible. Let (M_1, Π_1, Z_1), (M_2, Π_2, Z_2) be two exact Poisson manifolds. Let us denote the liftings of Π_1, Π_2, Z_1 and Z_2 to the product manifold by the same letters. Then we obtain an exact Poisson manifold $(M_1 \times M_2, \Pi_1 + \Pi_2, Z_1 + Z_2)$.

References

1. P. Dazord, A. Lichnerowicz and Ch.-M. Marle, *Structure locale des variétés de Jacobi*, J. Math. Pures et Appl., **70** (1991), 101–152.
2. E. Ghys, *Actions localement libres du groupe affine*, Invent. Math., **82** (1985), 479–526.
3. G. Hector, E. Macias and M. Saralegi, *Lemme de Moser feuilleté et classification des variétés de Poisson régulières*, Publicacions Matematiques, **33** (1989), 423–430.
4. M. Hirsch, *A stable analytic foliation with only exceptional minimal sets*, in Dynamical Systems, Warwick, 1974, Lecture Notes in Math., **468**, Springer Verlag, 1975, 9–10.
5. N. Kopell, *Commuting diffeomorphisms*, Proc. Pure Math. XIV, AMS, (1970), 165–184.
6. J.L. Koszul, *Crochet de Schouten-Nijenhuis et cohomologie*, Astérisque, hors serie (1985), 257–271.
7. S.P. Novikov, *The topology of foliations*, Trans. Moscow Math. Soc., **14** (1965), 268–304.
8. R. Sacksteder, *Limit sets of foliations*, Amer. J. Math., **87** (1965), 79–102.
9. I. Vaisman, Lectures on the Geometry of Poisson Manifolds, Progress in Mathematics, **118**, Birkhäuser, 1994.

Received November 14, 2000.

Proceedings of
FOLIATIONS: GEOMETRY AND DYNAMICS
held in Warsaw, May 29–June 9, 2000
ed. by Paweł WALCZAK *et al.*
World Scientific, Singapore, 2002
pp. 387–401

FOLIATION CONES CORRESPONDING TO SOME PRETZEL LINKS

YASUHARU NAKAE
Graduate School of Mathematical Sciences, University of Tokyo,
Komaba Meguro, Tokyo 153, Japan,
e-mail: nakae@ms.u-tokyo.ac.jp

The concept of "foliation cones" was introduced by J. Cantwell and L. Conlon. They showed that for a taut, transversely oriented foliation on a compact, connected, oriented sutured 3-manifold there are finitely many, closed convex polyhedral cones on the space of 1-dimensional cohomology classes. This extends a theorem of Thurston for fibred 3-manifolds to depth one foliations. We classify the foliation cones for the sutured manifold obtained from the complement of a 3-component pretzel link with an even number of positive twists by cutting apart along its Seifert surface.

1 Introduction

In [10] Thurston introduced a norm x on the space of 2-dimensional homology classes of a 3-manifold. He showed that a unit ball B_x with respect to this norm is a polyhedron whose vertices are lattice points. For a 3-manifold M which fibres over S^1, he showed that the ray determined by the homology class of the fibre passes through the interior of a top-dimensional face of ∂B_x. Therefore these fibrations $\mathcal{F}$ correspond, up to isotopy, to certain "fibred" rays $[\mathcal{F}] \subset H^1(M, \mathbb{R})$.

The concept of "foliation cones" was introduced by J.Cantwell and L.Conlon in [2]. In [3], they showed that isotopy classes of taut, transversely oriented foliations $\mathcal{F}$ on a compact, connected, oriented sutured 3-manifold (M, γ) with holonomy only on several leaves contained in ∂M correspond to rays in $H^1(M)$. Furthermore in [2] they showed that there are finitely many, closed convex polyhedral cones in $H^1(M)$ having disjoint interiors such that the rays corresponding to such foliations are exactly those lying in the interior of these cones. This extends the above result of

Thurston for fibred 3-manifolds to depth one foliations.

In this paper, we classify the foliation cones for the sutured manifold $M_S(\kappa)$, obtained from the complement $E(\kappa)$ of a 3-component pretzel link κ with an even number of positive twists, by cutting apart along its Seifert surface S.

Theorem 1.1 (Main theorem) *Let $\kappa = (2l+2, 2m+2, 2n+2), (l, m, n = 1, 2, \dots)$ be the three-component $(2l+2, 2m+2, 2n+2)$-type pretzel link with positive twists, and S be the standard Seifert surface for κ. Let $M_S(\kappa)$ be the handlebody of genus 2 obtained from the complement of κ by cutting along the Seifert surface, and $\mathbf{e}_1$ and $\mathbf{e}_2$ be the basis of $H_2(M_S(\kappa), \partial M_S(\kappa))$ represented by the disks compressing simple closed curves in $\partial M_S(\kappa)$ in $M_S(\kappa)$. Let $\mathbf{e}_0$ be an element of $H_2(M_S(\kappa), \partial M_S(\kappa))$ such that $\mathbf{e}_0 + \mathbf{e}_1 + \mathbf{e}_2 = 0$. Then the foliation cones corresponding to foliations of $M_S(\kappa)$ in $H_2(M_S(\kappa), \partial M_S(\kappa))$ are spanned by*

$$\langle \mathbf{e}_1, -\mathbf{e}_0 \rangle, \langle \mathbf{e}_2, -\mathbf{e}_0 \rangle, \langle \mathbf{e}_2, -\mathbf{e}_1 \rangle, \langle -\mathbf{e}_1, \mathbf{e}_0 \rangle, \langle -\mathbf{e}_2, \mathbf{e}_0 \rangle, \langle \mathbf{e}_1, -\mathbf{e}_2 \rangle.$$

In Thurston's theorem for fibred 3-manifold, the cones are related to the ball of the Thurston norm and thus symmetric. But by Cantwell-Conlon's paper [2] there are examples where the foliation cones are not symmetric, e.g. $\kappa = (2, 2, 2)$ and $(2, 2, 4)$. It will be shown that the foliation cones for $(2, 4, 4)$ are not symmetric.

In the proof of this theorem we shall consider a one-dimensional foliation $\mathcal{L}$ transverse to $\mathcal{F}$, look at the dynamics induced on a non-compact leaf of $\mathcal{F}$. Then we shall consider the Markov partitions for this dynamics on a non-compact leaf of depth one foliation and construct a core of the non-compact leaf where the dynamics is visualized. We find that the figure of this core is essentially the same for such κ therefore consider a typical figure of the core. This dynamics is deeply related to structures of the depth one foliation which is constructed by a procedure of Gabai's theorem, as well as to a lamination on the non-compact leaf of this foliation constructed by using Handel-Miller's extension (cf. [4]) of Thurston's theory [1] on the diffeomorphism of surfaces to non-compact case.

The author would like to express his gratitude to Prof. Ken'ichi Ohshika for his helpful advices and encouragement. He also thank the referee for many helpful comments, in particular for pointing out the symmetry of the cases.

2 Computing examples

In this section, we shall compute several examples to show the strategy of the proof of our main theorem. For the definition and the construction of foliation cones, the procedure of calculation for a foliation cone and some notations, the readers are referred to the paper of Cantwell-Conlon [2]. We work under the hypotheses of [2]. If a sutured manifold M is completely disk decomposable then by a theorem of Gabai [7] we obtain a depth one, taut foliation $\mathcal{F}$ and its structure is well understood. In particular for a pretzel link κ with $n+1$ components, the manifold $M_S(\kappa)$ obtained by cutting off a complement space $E(\kappa)$ of κ along a Seifert surface S becomes a handle body with genus n and $M_S(\kappa)$ is completely disk decomposable. We consider its taut depth one foliation $\mathcal{F}$, a non-compact leaf L of $\mathcal{F}$ and the element h of an isotopy class for the first return map of the transverse flow with respect to L. Let $\{R_1, \cdots, R_n\}$ be a Markov partition for $\mathcal{Z} \cap L$ and h, here $\mathcal{Z}$ is a set of orbits of the transverse flow which do not intersect $\partial M_S(\kappa)$.

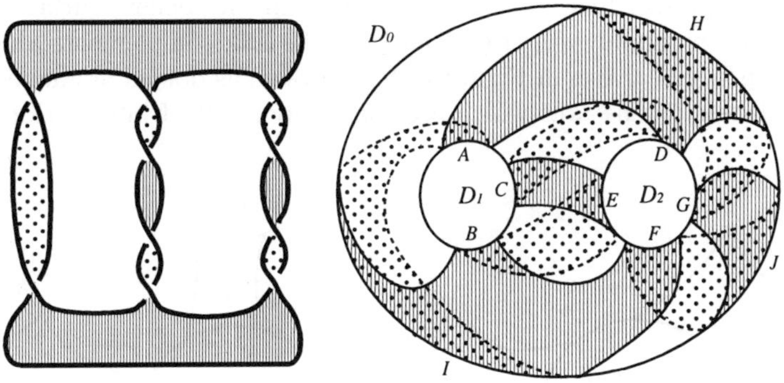

Figure 1. Seifert surface of $(2, 4, 4)$ and its sutured manifold

We define the matrix $A = (a_{ij})$ such that

$$a_{ij} = \begin{cases} 1 & h(R_i) \cap R_j \neq \emptyset , \\ 0 & h(R_i) \cap R_j = \emptyset , \end{cases}$$

and then

$$\Sigma_A = \{(\ldots, i_k, i_{k+1}, \ldots) | a_{i_k i_{k+1}} = 1 \quad \forall k \in \mathbb{Z}\}$$

is the allowable set defined in section 4 of [2]. If $\iota \in \Sigma_A$ is a periodic element of Σ_A then the homology class Γ_ι of the corresponding closed orbit of the transverse flow is defined. Let $\{\Gamma_\iota\}$ be the classes corresponding to minimal loops, then a cone is defined by the inequalities $\{\Gamma_\iota \geq 0\}$. When we perform the disk decomposition of $M_S(\kappa)$ by disks $D_1, \cdots, D_n \subset M$ we define the loops α_i which intersects to these disks transversely such that $\alpha_i \cdot D_j = \delta_{ij}$, $1 \leq i, j \leq n$, with respect to the homological intersection "$\cdot$". We take the homology classes $\mathbf{e}_i = [D_i] \in H_2(M_S(\kappa), \partial M_S(\kappa))$ for $1 \leq i \leq n$ as a basis of $H_2(M_S(\kappa), \partial M_S(\kappa))$, and let D_0 be a disk in S^3 such that for $\mathbf{e}_0 + \mathbf{e}_1 + \cdots + \mathbf{e}_n = 0$ where $\mathbf{e}_0 = [D_0]$. In the figures of examples, the embedded handle body in S^3 is viewed from inside. We set notations that S_+ is the component of $\partial_\tau M$ oriented outwardly and S_- is that oriented inwardly. For the orientation of decomposition disks $D_1, \cdots, D_n$, if the orientation of the boundary of a disk D_i is counterclockwise in the figure then its orientation is defined as positive and expressed by $+D_i$, and if it is clockwise then negative and by $-D_i$. We fix a choice of Seifert surfaces of κ as in Figure 1.

Example 2.1 We compute the foliation cones corresponding to the 3-component pretzel link $\kappa = (2, 4, 4)$. This link, its Seifert surface S and the sutured manifold $M_S(\kappa)$ cutting from the link complements along the Seifert surface is drawn in Figure 1. S_+ is drawn as dotted or shaded part. We set labels on the arcs which are the components of intersection of S_+

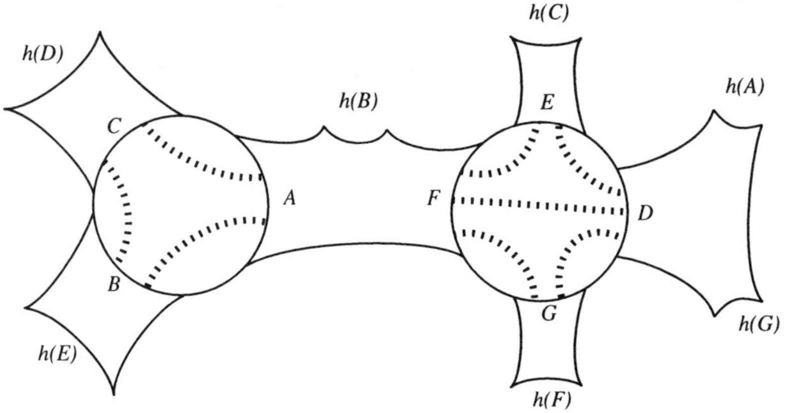

Figure 2. Core of $(2, 4, 4)$

and decomposing disks as in Figure 1.

First we look at the case when we perform a disk decomposition by using disks $\{+D_1, -D_2\}$. Then its core K becomes as in Figure 2. We see by these figures how the laminations $\Gamma_\pm$ behaves. We express the Markov partition $\{R_1, R_2, \cdots, R_8\}$ in a form

$$[\text{ rectangle : labels of stable boundaries on disk }],$$

as follows.

$$\begin{array}{llll} R_1 : \text{AC} & R_2 : \text{BC} & R_3 : \text{AB} & R_4 : \text{DE} \\ R_5 : \text{EF} & R_6 : \text{DF} & R_7 : \text{FG} & R_8 : \text{DG} \end{array}$$

Then the matrix $A = (a_{ij})$ is obtained as follows

$$A = \begin{pmatrix} 0\,0\,0\,1\,0\,0\,0\,0 \\ 0\,0\,0\,0\,1\,0\,0\,0 \\ 0\,0\,0\,0\,0\,1\,0\,0 \\ 0\,1\,0\,0\,0\,0\,0\,0 \\ 0\,0\,1\,0\,0\,0\,1\,0 \\ 1\,0\,0\,0\,0\,0\,1\,0 \\ 0\,0\,0\,0\,0\,0\,0\,1 \\ 1\,0\,0\,0\,0\,1\,0\,0 \end{pmatrix}.$$

Thus the minimal loops defined by this symbolic dynamics are

$$\Gamma_{(\dots 1425361\dots)} = 3(\alpha_1 - \alpha_2),$$
$$\Gamma_{(\dots 678\dots)} = -\alpha_2$$

and then the cone to which $[\mathcal{F}]$ belongs is defined by the inequalities

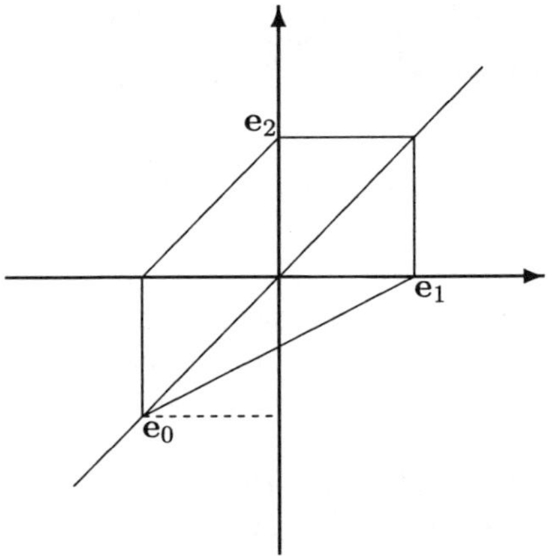

Figure 3. Foliation cones of $(2, 4, 4)$

$$\{\alpha_2 \leq \alpha_1\}, \{\alpha_2 \leq 0\}.$$

The corresponding bases of the cone in $H_2(M, \partial M)$ is the segment $[\mathbf{e}_1, \mathbf{e}_0]$ as in Figure 3.

By symmetry one only has to consider the additional cases $\{-D_1, +D_2\}$ and $\{+D_0, -D_1\}$, then obtains foliation cones corresponding to the pretzel link $\kappa = (2, 4, 4)$ as in Figure 3.

Example 2.2 Secondly we compute the foliation cones corresponding to the pretzel link $\kappa = (6, 6, 6)$. This link κ, its Seifert surfaces and the sutured manifold $M_S(\kappa)$ obtained from the link complements by cutting along the Seifert surfaces are drawn in Figure 4.

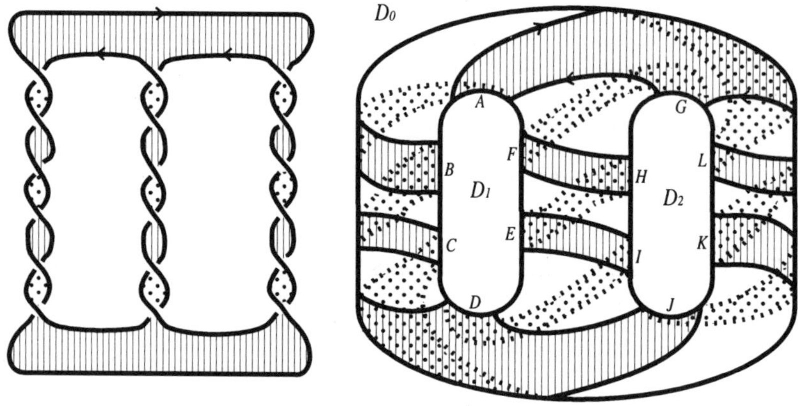

Figure 4. Seifert surface of (6, 6, 6) and its sutured manifold

We set labels on the arcs which are the components of intersection of S_+ and decomposing disks as in Figure 4. Now we perform a disk decomposition for $\{-D_1, +D_2\}$. Then the rectangles of the Markov partition are the following

$$\begin{array}{lll} R_1 : AB & R_2 : BC & R_3 : CD \\ R_4 : AD & R_5 : AE & R_6 : AF \\ R_7 : LG & R_8 : LH & R_9 : LI \\ R_{10} : LJ & R_{11} : LK & R_{12} : KJ \end{array}$$

This Markov partition and the pseudo-Anosov automorphism defines the minimal loops

$$\Gamma_{(\dots R_1 R_2 R_3 R_4 \dots)} = 4(-\alpha_1)$$
$$\Gamma_{(\dots R_{10} R_{11} R_{12} \dots)} = 3\alpha_2.$$

Then the cones to which $\mathcal{F}$ belongs is defined by the inequalities

$$\{\alpha_1 \leq 0\}, \{\alpha_2 \geq 0\},$$

and a base for this cone in $H_2(M, \partial M)$ is the segment $[-\mathbf{e}_1, \mathbf{e}_2]$. By symmetry, without calculating other decompositions we obtain the foliation cones corresponding to the pretzel link $\kappa = (6, 6, 6)$ as in Figure 5.

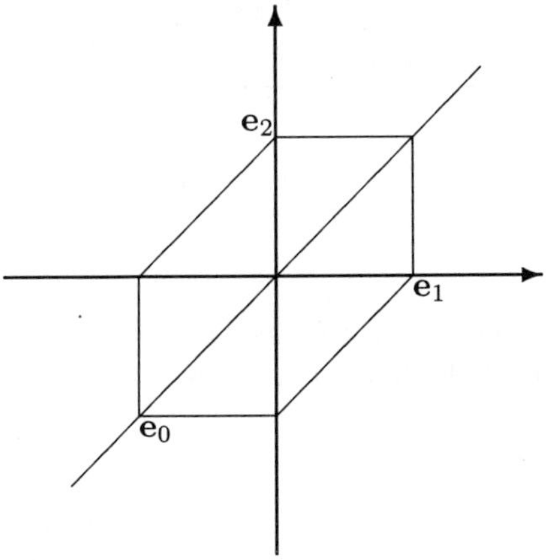

Figure 5. Foliation cones of $(6, 6, 6)$

Now we show the following theorem.

Theorem 2.3 *For* $n \geq 2$ *let* $\kappa = (2, 2, \cdots, 2)$ *be the* $n + 1$ *component pretzel link, let* $\mathbf{e}_i = [D_i]$ *be the homology class of the disk* D_i *in the disk decomposition. For the basis* $\{\mathbf{e}_i\}$ *of* $H_2(M_S(\kappa), \partial M_S(\kappa))$ *the foliation cones corresponding to the pretzel link* κ *are spanned by*

$$\langle \mathbf{e}_1, \mathbf{e}_2, \dots, \mathbf{e}_n \rangle,$$
$$\langle \mathbf{e}_1, \mathbf{e}_2, \dots, \mathbf{e}_{i-1}, \mathbf{e}_0, \mathbf{e}_{i+1}, \dots, \mathbf{e}_n \rangle \quad (i = 1, 2, \cdots, n).$$

Proof. The decomposition disk D_i intersects the suture γ at four points, then we set labels of intersections between ∂D_i and S_+ such that $D_i \cap S_+ = x_i \cup y_i$ for $(1 \leq i \leq n)$ as in Figure 6.

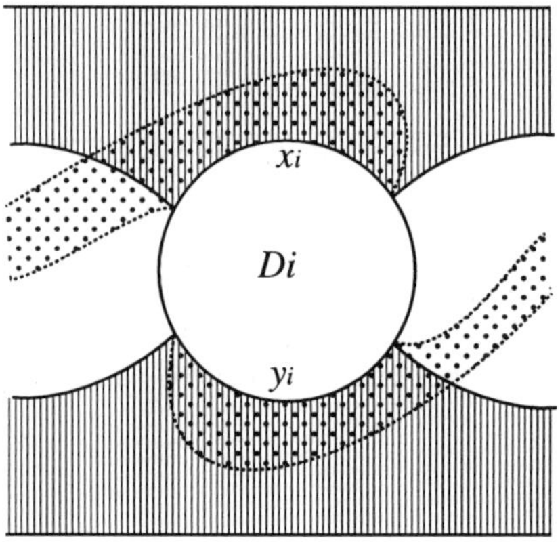

Figure 6. Labels

Then the local situations of the decompositions by $+D_i$ or $-D_i$ are as in Figure 7. The proof the Theorem 2.3 is divided into the following two cases.

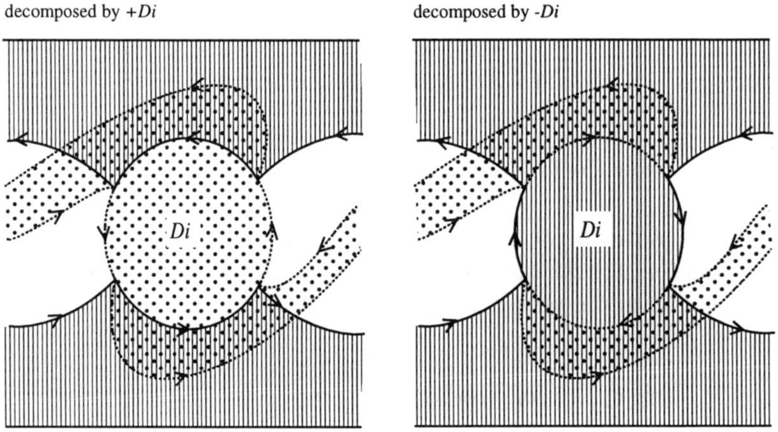

Figure 7. Local situation for decomposition $+D_i$ and $-D_i$

Case 1. The core K corresponding to the decomposition by only the positive disks $\{+D_1, +D_2, \cdots, +D_n\}$ is as in Figure 8.

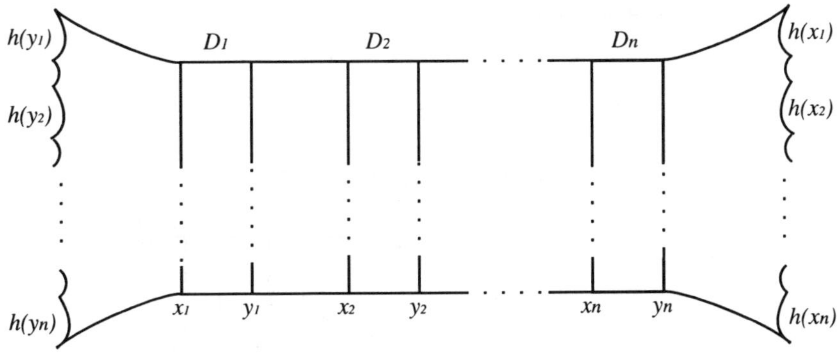

Figure 8. Core K of Case 1

The matrix $A = (a_{ij})$ corresponding to this case is $a_{ij} = 1$ for $1 \leq i, j \leq n$, and then the minimal loops are

$$\Gamma_{(\cdots kk \cdots)} = \alpha_k, \quad k = 1, 2, \cdots, n.$$

Thus the cone corresponding to this decomposition is defined by the inequality $\{\alpha_k \geq 0\}$ for $k = 1, 2, \cdots, n$, and this cone is spanned by $\langle \mathbf{e}_1, \mathbf{e}_2, \cdots, \mathbf{e}_n \rangle$.

Case 2. For some i, we decompose by using the negative disk $-D_i$ and all other decomposition is by positive disks $+D_j$, i.e.

$$\{+D_1, +D_2, \cdots, +D_{i-1}, -D_i, +D_{i+1}, \cdots, +D_n\}.$$

The core K corresponding to this decomposition is as in the Figure 9 and the corresponding matrix $A = (a_{kl})$ is given by

$$a_{kl} = \begin{cases} 1 & k = i \text{ or } l = i \ , \\ 0 & \text{otherwise} \ . \end{cases}$$

Then the minimal loops are

$$\Gamma_{(\cdots ikik \cdots)} = \alpha_k - \alpha_i, (k = 1, \cdots, i-1, i+1, \cdots, n),$$
$$\Gamma_{(\cdots ii \cdots)} = -\alpha_i$$

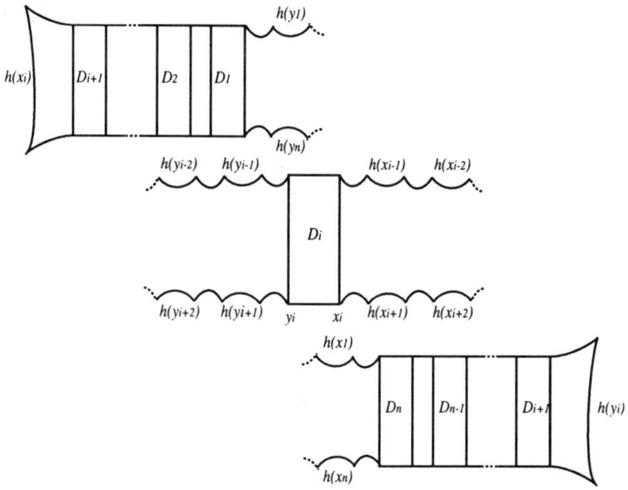

Figure 9. Core K in the Case 2.

and the foliation cone corresponding to this decomposition is defined by the inequalities

$$\{\alpha_k \geq \alpha_i\}, (k = 1, \cdots, i-1, i+1, \cdots, n),$$
$$\{\alpha_i \leq 0\}.$$

In $H_2(M, \partial M)$ this core corresponds to the cone spanned by

$$\langle \mathbf{e}_1, \mathbf{e}_2, \dots, \mathbf{e}_{i-1}, \mathbf{e}_0, \mathbf{e}_{i+1}, \dots, \mathbf{e}_n \rangle.$$

The set of cones obtained from Case 1 and Case 2 for $i = 1, 2, \cdots, n$ fill up the space of the homology classes $H_2(M_S(\kappa), \partial M_S(\kappa))$, thus we obtain Theorem 2.3. □

3 Proof of main theorem

Now we prove the main theorem of this paper.

Proof of main theorem. We take the Seifert surface for the pretzel link $\kappa = (2l+2, 2m+2, 2n+2)$ as in Figure 10.

Then by theorems of Murasugi [9] and Gabai [5] this Seifert surface has minimal genus, and by a theorem of Gabai [8] this link is not fibred. For this Seifert surface the sutured manifold $M_S(\kappa)$ obtained by cutting

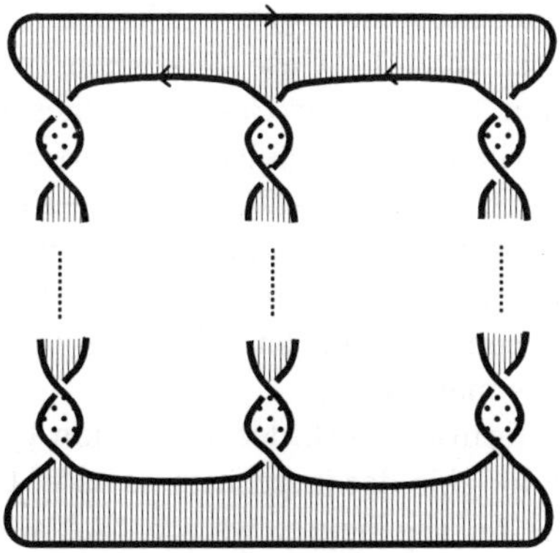

Figure 10. Seifert surface of the pretzel link

apart from the link complement along this S is as in Figure 11. As in the previous section, in Figure 11 we take the labels on the components of the intersection of the decomposing disks D_i and the copy S_+ of the Seifert surface S.

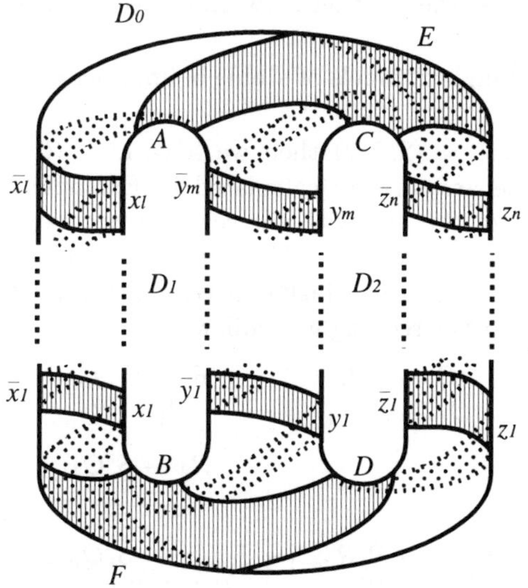

Figure 11. Sutured manifold $M_S(\kappa)$

We divide the proof into three cases. Case 1 is the case when all of l, m and n are greater than 1. Case 2 is when two of them are 1 and the rest is greater than 1. The last Case 3 is when one of them is 1 and the others are greater than 1. In each case we perform six disk decompositions $\{+D_1, -D_2\}$, $\{-D_1, +D_2\}$, $\{+D_0, -D_1\}$, $\{-D_0, +D_1\}$, $\{+D_0, -D_2\}$, $\{-D_0, +D_2\}$, and we prove that the foliation cone for a disk decomposition depends only on the choice of decomposition disks but does not depend on (l, m, n). For each decomposition, let $\mathcal{F}$ be the taut depth one foliation constructed by Gabai's method [7].

Case 1. First we perform the disk decomposition by $\{+D_1, -D_2\}$. Then the figure of the core for this decomposition is as in Figure 12.

Using these labels on the figure, we see that we can take a Markov partition as follows:

$$\begin{array}{cccc}
P : D\bar{z}_1 & Q : DC & R : x_1\bar{y}_m & S : x_1x_2 \\
P_1 : \bar{z}_1\bar{z}_2 & Q_1 : Dy_1 & R_1 : x_1B & S_1 : x_2x_3 \\
\vdots & \vdots & R_2 : x_1\bar{y}_1 & \vdots \\
\vdots & \vdots & \vdots & S_{l-1} : x_lA \\
P_n : \bar{z}_nC & Q_m : Dy_m & R_m : x_1\bar{y}_{m-1} & S_l : x_1A,
\end{array}$$

where we express the relation between an rectangle and labels as

$$[\text{ rectangle } : \text{ labels of stable boundaries on disk }].$$

If one sets $(l, m, n) = (2, 2, 2)$ then we have $\kappa = (6, 6, 6)$, and its core and foliation cone are already calculated in Example 2.2 This foliation cone corresponds to the one with the segment $[\mathbf{e}_1, -\mathbf{e}_2]$ as base. For the rectangles X, Y and Z, let $X \to Y, Z$ denote the relation $h(X) \cap Y \neq \phi$ and $h(X) \cap Z \neq \phi$. For this Markov partition, the relations between the rectangle and the other rectangle as follows:

$$\begin{array}{cccc}
P \to P_1 & Q \to P, R & R \to S, Q_m & S \to S_1 \\
P_1 \to P_2 & Q_1 \to P, R_1 & R_1 \to S & S_1 \to S_2 \\
\vdots & \vdots & R_2 \to S, Q_1 & \vdots \\
\vdots & \vdots & \vdots & S_{l-2} \to S_{l-1} \\
P_{n-1} \to P_n & Q_{m-1} \to P, R_{m-1} & R_{m-1} \to S, Q_{m-2} & S_{l-1} \to S_l, Q \\
P_n \to Q, R & Q_m \to P, R_m & R_m \to S, Q_{m-1} & S_l \to S, Q.
\end{array}$$

Since the size of the matrix A corresponding to the symbolic dynamics (Σ_A, σ_A) is big, we express the diagram of relations as in Figure 13.

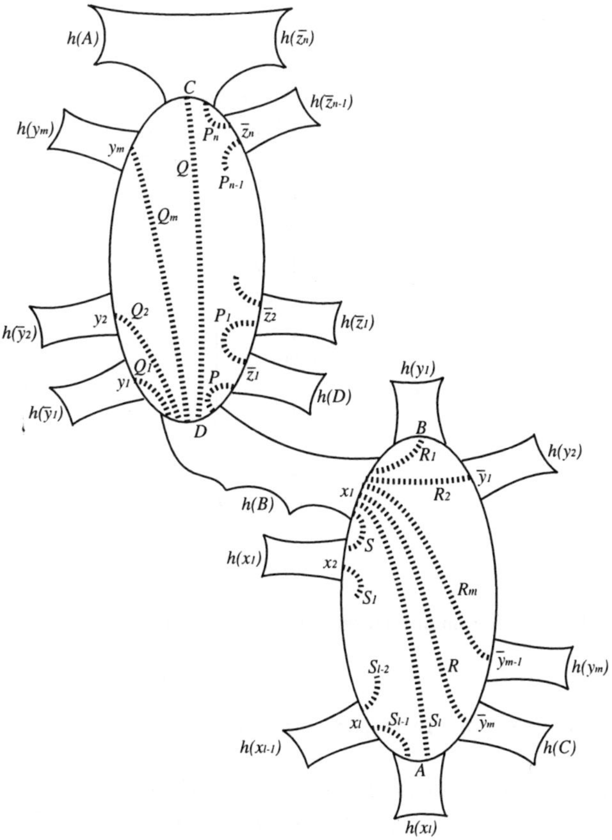

Figure 12. Core K

Then this symbolic dynamics defines the minimal loops

$$\Gamma_{(\dots PP_1P_2\dots P_{n-1}P_nQ\dots)} = (n+2)(-\alpha_2)$$
$$\Gamma_{(\dots PP_1\dots P_nRQ_m\dots)} = \alpha_1 + (n+2)(-\alpha_2)$$
$$\Gamma_{(\dots QRSS_1\dots S_{l-1}\dots)} = (l+1)\alpha_1 + (-\alpha_2)$$
$$\Gamma_{(\dots SS_1\dots S_l\dots)} = (l+1)\alpha_1.$$

Then the cone to which $[\mathcal{F}]$ belongs is defined by the inequalities

$$\{\alpha_2 \leq 0\}, \{\alpha_1 \geq 0\},$$

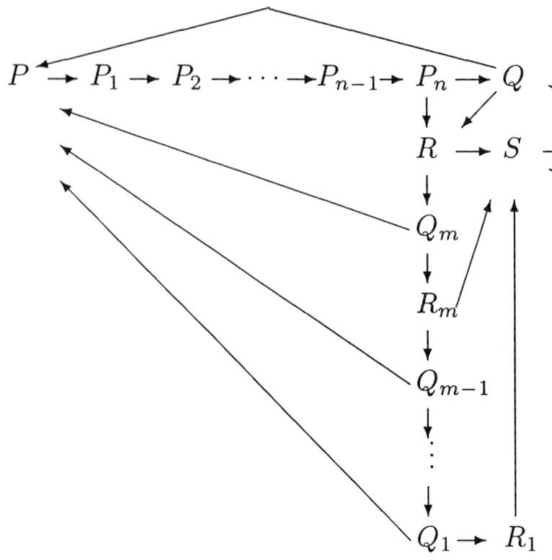

Figure 13. The diagram of relation of the symbolic dynamics in the Case 1-a

and the bases of the cone in $H_2(M, \partial M)$ corresponding to it is the segment $[\mathbf{e}_1, -\mathbf{e}_2]$.

It is not necessary to calculate cones for the other decompositions because there are orientation preserving homeomorphism of $\mathbb{R}^3$ giving arbitrary permutations of the components of κ and these same homeomorphisms give arbitrary permutations of basis $\mathbf{e}_1$, $\mathbf{e}_2$ and $\mathbf{e}_0$, hence calculations for another decompositions is essentially the same by symmetry.

The proofs in the two special cases, Case 2 and Case 3 are minor modifications of the proof in Case 1. The details are left to the reader. □

References

1. S.A. Bleiler and A.J. Casson, Automorphisms of surfaces after Nielsen and Thurston, Cambridge Univ. Press, Cambridge, 1988.
2. J. Cantwell and L. Conlon, J. Cantwell and L. Conlon, *Foliation Cones*, in Proceedings of the Kirbyfest, 35–86, Geometry and Topology Monographs, 1999.
3. J. Cantwell and L. Conlon, *Isotopies of depth one foliations*, in Geo-

metric Study of Foliations, World Sci. Publ., Singapore 1994, 153–173.
4. S. Fenley, *Endperiodic surface homeomorphisms and 3-manifolds*, Math. Z., **224** (1997), 1–24.
5. D. Gabai, *Genera of the Alternating Links*, Duke Math J., **53** (1986), no. 3, 677–681.
6. D. Gabai, *Foliations and the topology of 3-manifolds*, J. Diff. Geom., **18** (1983), 445–503.
7. D. Gabai, *Foliations and genera of links*, Topology, **23** (1984), 381–394.
8. D. Gabai, *Detecting fibred links in S^3*, Comment. Math. Helv., **61** (1986), 519–555.
9. K. Murasugi, *On the Genus of the Alternating Knot I, II*, Math. Soc. of Japan, **10** (1958), 94–105, 235–248.
10. W. Thurston, *A norm for the homology of three manifolds*, Mem. Amer. Math. Soc., **59** (1986), 56–88.

Received October 27, 2000, revised March 15, 2001.

Proceedings of
FOLIATIONS: GEOMETRY AND DYNAMICS
held in Warsaw, May 29–June 9, 2000
ed. by Paweł WALCZAK *et al.*
World Scientific, Singapore, 2002
pp. 403–419

REGULAR PROJECTIVELY ANOSOV FLOWS WITHOUT COMPACT LEAVES

TAKEO NODA

*Graduate School of Mathematical Sciences, University of Tokyo,
Komaba Meguro, Tokyo 153, Japan,
e-mail: noda@ms513yel.ms.u-tokyo.ac.jp*

TAKASHI TSUBOI

*Graduate School of Mathematical Sciences, University of Tokyo,
Komaba Meguro, Tokyo 153, Japan,
e-mail tsuboi@ms.u-tokyo.ac.jp*

This paper concerns projectively Anosov flows φ_t with smooth stable and unstable foliations $\mathcal{F}^s$ and $\mathcal{F}^u$ on a 3 dimensional manifold M. We assume that the manifold M is either a torus bundle over the circle or the unit tangent bundle over the closed surface of genus greater than 1. These manifolds are known to be the only manifolds (up to finite cover) which admits an Anosov flow with smooth stable and unstable foliations. We show that if the foliations $\mathcal{F}^s$ and $\mathcal{F}^u$ do not have compact leaves, then the flow is isotopic to an Anosov flow. For projectively Anosov flows with smooth stable and unstable foliations on the torus bundles over the circle, since the case where the $\mathcal{F}^s$ or $\mathcal{F}^u$ has a compact leaf is already settled by the first author, we obtain the complete classification of them.

1 Introduction and the statement of the result

A non singular flow φ_t on a closed 3-dimensional manifold M is a projectively Anosov flow if there exist a continuous Riemannian metric on M, a continuous splitting $\hat{E}^u \oplus \hat{E}^s$ of $TM/T\varphi$ invariant under the action of φ, and a positive real number C such that the following inequality holds for $t \geq 0$, $v^u \in \hat{E}^u$ and $v^s \in \hat{E}^s$:

$$\frac{\|T\varphi_t(v^u)\|}{\|T\varphi_t(v^s)\|} \geq e^{Ct}\frac{\|v^u\|}{\|v^s\|}.$$

This definition was given in [2], where Eliashberg and Thurston called it a conformally Anosov flow. The same flow was investigated by Mitsumatsu [13] (see also [14]) and was called a projectively Anosov flow, and we adopt the latter terminology because the action of the flow on the projectivized normal bundle of the flow looks essential.

The invariant line bundles $\hat{E}^u$ and $\hat{E}^s$ give rise to the invariant plane fields E^u and E^s over M. As is remarked in [2], E^u and E^s are continuous and integrable, but frequently they are not uniquely integrable. It is, however, interesting to investigate the case where E^u and E^s are smooth, and then E^u and E^s determine codimension 1 smooth foliations $\mathcal{F}^u$ and $\mathcal{F}^s$ of M. In this case, we call the projectively Anosov flow regular.

The Anosov flows with smooth (at least C^3) stable and unstable foliations are classified by Ghys ([6], [7]). There are only the suspension flow of Anosov diffeomorphisms of the torus and quasi-Fuchsian flows on the Seifert fibred space over hyperbolic orbifolds. They are the most important examples of regular projectively Anosov flows.

For a regular projectively Anosov flow, the smooth foliations $\mathcal{F}^u$ and $\mathcal{F}^s$ may have compact leaves which are tori. In [14], the first author investigated the regular projectively Anosov flows with compact leaves, and he also gave the classification of those flows with compact leaves on torus bundles over the circle.

The torus bundles over the circle with hyperbolic monodromy and the Seifert fibred space over hyperbolic orbifolds are the 3-dimensional manifolds where the smooth foliations without compact leaves are classified up to smooth isotopy. By the results of Ghys-Sergiescu [8] and Ghys [7], the possible foliations without compact leaves of these manifolds are the unstable foliations or stable foliations of the Anosov flows.

In this paper, we look at regular projectively Anosov flows without compact leaves on these manifolds and show that they are in fact the Anosov flows. More precisely, we show the following theorems.

Theorem 1.1 *Let A be an element of $SL(2;\mathbf{Z})$ such that $|\mathrm{Tr}A| > 2$. Let M be the torus bundle over the circle with the monodromy matrix A. On M, there is the natural Anosov flow which is the suspension of the toral automorphism induced by A. Let φ_t be a regular projectively Anosov flow on M such that the unstable foliation $\mathcal{F}^u$ or the stable foliation $\mathcal{F}^s$ does not have compact leaves. Then after changing the parameter of the flow (reversing the parameter if necessary) φ_t is isotopic to the suspension Anosov flow.*

Theorem 1.2 *Let φ_t be a regular projectively Anosov flow on the unit tangent bundle M of a closed surface Σ of genus greater than 1. Assume that the unstable foliation $\mathcal{F}^u$ and the stable foliation $\mathcal{F}^s$ do not have compact* *leaves. Then after changing the parameter of the flow (reversing the parameter if necessary)* φ_t *is isotopic to a quasi-Fuchsian flow.*

Together with the result in [14], Theorem 1.1 completes the classification for the regular projectively Anosov flows on torus bundles over the circle. On the other hand, we know no examples of regular projectively Anosov flows with compact leaves on the unit tangent bundle M of a closed surface Σ of genus greater than 1.

As we mentioned, the foliations $\mathcal{F}^u$ and $\mathcal{F}^s$ are known to be those of the Anosov flow, hence our problem is whether the fact that the flow is regular projectively Anosov implies that the intersection of $\mathcal{F}^u$ and $\mathcal{F}^s$ is the same as the intersection of the Anosov flows. (In the recent paper [12], the uniqueness of the transverse intersection of $\mathcal{F}^u$ and $\mathcal{F}^s$ for the torus bundles over the circle is shown. Hence our Theorem 1.1 follows from our Lemma 3.3 and the uniqueness of the intersection. The proof is not so simple and it would be worth giving our proof valid only for our situation. In fact in [12], it is also shown that the transverse intersection of $\mathcal{F}^u$ and $\mathcal{F}^s$ for the unit tangent bundle of the hyperbolic surface is not unique. Hence the assumption that the flow is regular projectively Anosov is necessary.)

To show our results, we look at the leaf spaces of the lifted foliations $\widetilde{\mathcal{F}}^u$ and $\widetilde{\mathcal{F}}^s$ and the orbit space of the lifted flow $\widetilde{\varphi}_t$ on the universal covering space $\widetilde{M}$ with the action of $\pi_1(M)$. The key observation (which was difficult to find but is easy to prove) is that, in our case, the orbit foliation $\widetilde{\varphi}$ of $\widetilde{M}$ is Hausdorff. We give the discussion on this in Section 2.

Then we look at the action on the orbit space $\widetilde{M}/\widetilde{\varphi}$. This enable us to follow the argument of Ghys [6] to conclude that our flow is isotopic the Anosov flow after changing the parameter. In this way, we show Theorems 1.1 and 1.2 in Sections 3 and 4, respectively.

In Section 5, we give an alternative proof of Theorem 1.1, which gives the section of the flow by using the result of Schwartzman [16] in a way similar to that in a paper by Plante [15] and it may give a better understanding of the situation.

We thank Shigenori Matsumoto for pointing out the similarity of our original proof of Theorem 1.1 to the paper [15] by Plante also the members of TIT Saturday Seminar for valuable comments. The authors thank Rémi Langevin and Yakov Eliashberg for their hospitality during their stays.

2 Induced flow and foliations in the universal covering

Let φ_t be a regular projectively Anosov flow on a 3-manifold M. Let $\mathcal{F}^u$ and $\mathcal{F}^s$ be the the unstable foliation and the stable foliation for φ_t. Let

$\widetilde{\varphi}_t$, $\widetilde{\mathcal{F}}^u$ and $\widetilde{\mathcal{F}}^s$ be the induced flow and foliations on the universal covering space $\widetilde{M}$ of M.

Using the result of Tamura-Sato [17], we see that the foliations $\mathcal{F}^u$ and $\mathcal{F}^s$ do not have Reeb components and this implies that all leaves of $\widetilde{\mathcal{F}}^u$ and $\widetilde{\mathcal{F}}^s$ are diffeomorphic to planes (see [14]).

We look at the leaf spaces $Q^u = \widetilde{M}/\widetilde{\mathcal{F}}^u$ and $Q^s = \widetilde{M}/\widetilde{\mathcal{F}}^s$ ([3], [1]). The leaf spaces Q^u and Q^s may have non Hausdorff points. One can easily construct a regular projectively Anosov flow on T^3 such that Q^u is Hausdorff and Q^s is not.

For the purpose of this paper, we restrict our attention to the case where Q^u and Q^s are Hausdorff, i.e., $\widetilde{\mathcal{F}}^u$ and $\widetilde{\mathcal{F}}^s$ are diffeomorphic to the product foliation of $\mathbf{R}^3$ with leaves $\mathbf{R}^2 \times \{*\}$. Then Q^u and Q^s are diffeomorphic to the real line $\mathbf{R}$. The projections $p^u : \widetilde{M} \longrightarrow Q^u$ and $p^s : \widetilde{M} \longrightarrow Q^s$ are both $\pi_1(M)$ equivariant and determine the foliations $\mathcal{F}^u$ and $\mathcal{F}^s$.

We consider the juxtaposition map of projections;

$$p = (p^u, p^s) : \widetilde{M} \longrightarrow Q^u \times Q^s.$$

The map p to the plane is a $\pi_1(M)$ equivariant submersion and it determines the structure of the orbit foliation φ of the flow φ_t.

Now the problem is whether the orbit foliation $\widetilde{\varphi}$ of the lifted flow on $\widetilde{M}$ is Hausdorff. We have the following proposition.

Proposition 2.1 *Let φ_t be a regular projectively Anosov flow on a 3-manifold M. Assume that the lifted stable foliation $\widetilde{\mathcal{F}}^s$ of the universal covering space $\widetilde{M}$ is diffeomorphic to the product foliation of $\mathbf{R}^3$. Then the lifted orbit foliation $\widetilde{\varphi}$ restricted to each leaf $\widetilde{L}^u$ of the lifted unstable foliation $\widetilde{\mathcal{F}}^u$ is Hausdorff.*

Proof. Suppose that there are 2 orbits ℓ and ℓ' of $\widetilde{\varphi}$ on $\widetilde{L}^u$ such that a sequence $\{\ell_i\}$ of orbits of $\widetilde{\varphi}$ on $\widetilde{L}^u$ converges to them simultaneously. Let $\widetilde{L}^s_i$ denote the unstable leaf passing ℓ_i. Then by the assumption that $\widetilde{\mathcal{F}}^s$ is the product foliation, The leaf $\widetilde{L}^s_i$ converges to a leaf $\widetilde{L}^s$. Thus ℓ and ℓ' are components of the intersection of $\widetilde{L}^u$ and $\widetilde{L}^s$.

Now we take points x and x' on ℓ and ℓ', respectively. Then we take a foliated rectangles T and T' at x and x' transverse to $\widetilde{\varphi}_t$, respectively. There are curves γ^u and γ^s on $\widetilde{L}^u$ and $\widetilde{L}^s$ joining x and x'. Then we obtain holonomies $h^u_{\gamma^u}$ and $h^s_{\gamma^s}$ for the foliations $\widetilde{\mathcal{F}}^u$ and $\widetilde{\mathcal{F}}^s$, respectively. Since $\widetilde{L}^u$ and $\widetilde{L}^s$ are contractible, the holonomies do not depend on the paths on the leaves.

Now take a Riemannian metric on M adapted for the projectively Anosov flow φ. We lift it to the universal covering space $\widetilde{M}$. We look at the intersections x_i, x_i' of ℓ_i and T, T'. We may assume that $\widetilde{\varphi}_{t_i}(x_i) = x_i'$ for positive t_i. Then we see that $t_i \to \infty$ as $i \to \infty$. This implies that $\dfrac{\|T\widetilde{\varphi}_{t_i}v^u\|}{\|T\widetilde{\varphi}_{t_i}v^s\|} \geq e^{Ct_i}\dfrac{\|v^u\|}{\|v^s\|}$ for $v^u \in T\widetilde{\mathcal{F}}^u/T\widetilde{\varphi}$ and $v^s \in T\widetilde{\mathcal{F}}^s/T\widetilde{\varphi}$ at x_i. Thus this ratio tends to the infinity as $i \to \infty$.

This ratio can also be calculated as the ratio of the derivatives of the holonomies $h^u_{\gamma^u}$ and $h^s_{\gamma^s}$ at that points, and hence the ratio is bounded. This is the contradiction.

In our case, both the lifted unstable foliation $\widetilde{\mathcal{F}}^u$ and the lifted stable foliation $\widetilde{\mathcal{F}}^s$ are diffeomorphic to the product foliation of $\mathbf{R}^3$, and Proposition 2.1 clearly implies the following lemma.

Lemma 2.2 *If both the lifted unstable foliation $\widetilde{\mathcal{F}}^u$ and the lifted stable foliation $\widetilde{\mathcal{F}}^s$ are diffeomorphic to the product foliation of $\mathbf{R}^3$, then the orbit foliation $\widetilde{\varphi}$ of $\widetilde{M}$ is Hausdorff and $p : \widetilde{M} \longrightarrow Q^u \times Q^s$ is a fibration to the image with fibre being the orbit of $\widetilde{\varphi}_t$.*

Remark. We can use Lemma 2.2 to simplify the proof of several results in [14].

The first one is that T^3 does not admit regular projectively Anosov flow without compact leaves ([14], Proposition 5.6). For, if $\mathcal{F}^u$ and $\mathcal{F}^s$ are without compact leaves, they are foliations without holonomy and satisfy the assumption of Lemma 2.2. It is easy to see that $p : \widetilde{T^3} \longrightarrow Q^u \times Q^s$ is surjective and the action of $\pi_1(T^3)$ is topologically by translations. Hence there are elements $\alpha_i \in \pi_1(T^3)$ such that $\alpha_i(x, y) \to (x, y) \in Q^u \times Q^s$. On T^3, this corresponds to the Poincaré maps of transverse rectangle to the flow. The condition that the flow is projectively Anosov implies that the rectangle is distorted after some time. Since this Poincaré maps are defined on a fixed rectangle, this gives rise to nontrivial holonomy of $\mathcal{F}^u$ or $\mathcal{F}^s$.

The second one is the following. Let φ_t be a regular projectively Anosov flow on $T^2 \times [0, 1]$ such that $T^2 \times \{i\}$ $(i = 0, 1)$ are the only compact leaves of $\mathcal{F}^u$. Then we show that $\mathcal{F}^s$ is not the product foliation $\mathcal{F}^s|T^2 \times \{0\} \times [0, 1]$ ([14], Proposition 5.14). If $\mathcal{F}^s|T^2 \times \{0\}$ has a Reeb component r, we look at the holonomy along $\partial r \times \{0, 1\}$ as in [14], and the dynamics is not projectively Anosov. Otherwise, we can apply Lemma 2.2 to the foliations restricted to the interior of $T^2 \times [0, 1]$. Then $p : \widetilde{T^2 \times (0, 1)} \longrightarrow Q^u \times Q^s$ is surjective, and this gives rise to a fibration $\widetilde{T^2 \times (0, 1)}/\pi_1(T^2) \longrightarrow (Q^u \times Q^s)/\pi_1(T^2)$ with fibre being the real line. Then this admits a section. If the flow restricted to the interior has a section, the alpha limit set and the omega limit set of the flow are contained in $T^2 \times \{i\}$ $(i = 0, 1)$. At the alpha limit set this contradicts the fact that the boundaries are leaves of unstable foliation $\mathcal{F}^u$.

We mention here that, if the projectively Anosov flow φ admits a strong Anosov splitting [14], that is, TM has an invariant splitting $E^{uu} \oplus E^{ss} \oplus T\varphi$ invariant under the flow, then the orbit foliation $\widetilde{\varphi}$ of the lifted flow on $\widetilde{M}$ is Hausdorff. This is true even if Q^u or Q^s are not Hausdorff. ([3]).

3 Proof of Theorem 1.1

In [14], the first author proved the following theorem.

Theorem 3.1 ([14]) *Let φ_t be a regular projectively Anosov flow on a torus bundle over the circle. If the unstable foliation $\mathcal{F}^u$ has a compact leaf, then the stable foliation $\mathcal{F}^s$ also has a compact leaf. In fact the topological type of such flows are classified.*

Thus hereafter we assume that both $\mathcal{F}^u$ and $\mathcal{F}^s$ do not have compact leaves. In this situation, we use the following result by Ghys and Sergiescu [8].

Theorem 3.2 (Ghys-Sergiescu [8]) *Let A be an element of $SL(2;\mathbf{Z})$ such that $|\mathrm{Tr}A| > 2$. Let M be the torus bundle over the circle with the monodromy matrix A. Let ϕ_t denote the natural Anosov flow which is the suspension of the toral automorphism induced by A. Let $\mathcal{F}$ be a smooth codimension 1 foliation without compact leaves of M. Then $\mathcal{F}$ is isotopic to the Anosov unstable foliation F^u or the Anosov stable foliation F^s of M.*

For the Anosov foliations F^u and F^s of M, the pullback foliations $\widetilde{F}^u$ and $\widetilde{F}^s$ are diffeomorphic to the product foliation of $\mathbf{R}^3$ with leaves $\mathbf{R}^2 \times \{*\}$, $\widetilde{M}/\widetilde{F}^u$ and $\widetilde{M}/\widetilde{F}^s$ are the affine lines, and $\pi_1(M)$ acts on them as affine transformations determining the transversely affine structures of F^u and F^s.

Thus by Theorem 3.2, the pullback foliations $\widetilde{\mathcal{F}}^u$ and $\widetilde{\mathcal{F}}^s$ are also diffeomorphic to the product foliation of $\mathbf{R}^3$. By Lemma 2.2, we have the map $p = (p^u, p^s) : \widetilde{M} \longrightarrow Q^u \times Q^s$, which is a fibration to the image with fibre being the orbit of $\widetilde{\varphi}_t$. The map p to the plane is a $\pi_1(M)$ equivariant submersion and it determines the structure of the orbit foliation of the flow φ_t.

The first step of the proof of our theorem is the following lemma.

Lemma 3.3 *The foliations $\mathcal{F}^u$ and $\mathcal{F}^s$ are not isotopic to F^s simultaneously.*

Proof. We look at the equivariant map $p : \widetilde{M} \longrightarrow Q^u \times Q^s$. Suppose that $\mathcal{F}^u$ and $\mathcal{F}^s$ are both isotopic to F^s. We consider the family of linear foliations $\mathcal{G}_\theta$ of the affine space $Q^u \times Q^s \cong \mathbf{R} \times \mathbf{R}$ with leaves parallel to a vector $(\cos\theta, \sin\theta)$. Since $\pi_1(M)$ acts on $Q^u \times Q^s$ as on $\widetilde{M}/\widetilde{F}^s \times \widetilde{M}/\widetilde{F}^s$, it acts as similarity transformations. Hence it leaves the linear foliation $\mathcal{G}_\theta$ invariant. Since p is a submersion, $\widetilde{\mathcal{F}}_\theta = p^*\mathcal{G}_\theta$ is a foliation of $\widetilde{M}$ and it is invariant under the action of $\pi_1(M)$. Thus we obtain a family of foliations $\mathcal{F}_\theta$ of M. Since any orbit of φ is contained in a leaf of $\mathcal{F}_\theta$, $T\mathcal{F}_\theta/T\varphi$ is invariant under the action of φ_t. This contradicts the definition of the projectively Anosov flow for which there are only 2 invariant line bundles $\hat{E}^u$ and $\hat{E}^s$ in $TM/T\varphi$.

Remark. The flow with transverse similarity structure is studied by Ghys [5]. By his result, the flow with transverse similarity structure on the torus bundle over the circle is a Riemannian flow, and from which we can also deduce above lemma.

Proof of Theorem 1.1. Now we know that the foliations $\mathcal{F}^u$ and $\mathcal{F}^s$ are isotopic to F^u and F^s, respectively. Hence $\pi_1(M)$ acts on $Q^u \times Q^s$ in the same way as on $\widetilde{M}/\widetilde{F}^u \times \widetilde{M}/\widetilde{F}^s$. The image $p(\widetilde{M})$ of the map $p = (p^u, p^s) : \widetilde{M} \longrightarrow Q^u \times Q^s$ is a simply connected non empty open subset of $Q^u \times Q^s$. Any open subset of $\widetilde{M}/\widetilde{F}^u \times \widetilde{M}/\widetilde{F}^s$ contains a fixed point of the action of a non trivial element α of $\pi_1(M)$. Since $p(\widetilde{M})$ is invariant under the action of $\pi_1(M)$, $p(\widetilde{M})$ contains the stable line and the unstable line of the action of α. Note that $p(\widetilde{M})$ is also invariant under the action of the fundamental group of the fibre torus which is the translation by a lattice subgroup of $\widetilde{M}/\widetilde{F}^u \times \widetilde{M}/\widetilde{F}^s$. Since $p(\widetilde{M})$ is simply connected, $p(\widetilde{M})$ coincides with $Q^u \times Q^s$. Thus the map $p = (p^u, p^s) : \widetilde{M} \longrightarrow Q^u \times Q^s$ is surjective and it is a fibration with the fibre being the orbit of $\widetilde{\varphi}_t$.

Now we can follow the argument by Ghys [6]. Since both the Anosov flow ϕ_t and the projectively Anosov flow φ_t on M give rise to the $\pi_1(M)$ equivariant fibrations $\widetilde{M} \longrightarrow Q^u \times Q^s$ and we know that the actions on $Q^u \times Q^s$ coincide. This implies the holonomy groupoids for the orbit foliations ϕ and φ of the flows ϕ_t and φ_t are equivalent.

For a closed orbit c of φ or ϕ, we look at the lift in $\widetilde{M}$ and its image in $Q^u \times Q^s$. The curve c represents an element $\alpha \in \pi_1(M)$ and the image in $Q^u \times Q^s$ is the fixed point by the action of $\alpha \in \pi_1(M)$. Hence the holonomy covering of c is contractible.

Thus both (M, ϕ) and (M, φ) are the classifying space for the groupoid ([9]) and we have a smooth map $M \longrightarrow M$ which sends an orbit of ϕ_t to an orbit of φ_t which is transversely a diffeomorphism conjugating the transverse structures of ϕ and of φ. This smooth map lifts to a map $\widetilde{M} \longrightarrow \widetilde{M}$ which covers the identity of $Q^u \times Q^s$. As in [4], one can deform this smooth map to a diffeomorphism which sends an orbit of ϕ_t to an orbit of

φ_t. Since the resulted diffeomorphism sends the fibre of the torus bundle to an isotopic torus, it is isotopic to the identity.

4 Proof of Theorem 1.2

Let $\mathcal{F}^u$ and $\mathcal{F}^s$ denote the unstable foliation and the stable foliation of a regular projectively Anosov flow without compact leaves on the unit tangent bundle M of a closed surface Σ of genus greater than 1.

To prove Theorem 1.2, we use the following strong theorem by Ghys [7].

Theorem 4.1 (Ghys [7]) *Let $\mathcal{F}$ be a foliation of class C^3 of the unit tangent bundle M of a closed surface Σ of genus greater than 1. Suppose that $\mathcal{F}$ has no compact leaves, then there is a hyperbolic metric g on Σ such that $\mathcal{F}$ is isotopic to the stable foliation F_g^s for the geodesic flow for the metric g.*

Note that the stable foliation F_g^s and the unstable foliation F_g^u for the geodesic flow for the same metric g are isotopic.

As Anosov flows with smooth stable and unstable foliations, Ghys defined the quasi-Fuchsian flow ϕ_{g_1,g_2} in [6] and proved the following theorem.

Theorem 4.2 (Ghys [6, 7]) *Let φ_t be an Anosov flow on the unit tangent bundle M of a closed surface Σ of genus greater than 1. Suppose that the stable foliation $\mathcal{F}^s$ and the unstable foliation $\mathcal{F}^u$ are of class C^3, then there are two hyperbolic metrics g_1 and g_2 on Σ such that φ_t is isotopic to the quasi-Fuchsian flow ϕ_{g_1,g_2} whose stable and unstable foliations are isotopic to $F_{g_1}^s$ and $F_{g_2}^u$, respectively.*

For the quasi-Fuchsian flow ϕ_{g_1,g_2}, the map $p : \widetilde{M} \longrightarrow Q^u \times Q^s$ is not a surjective map.

Now in our situation, by Theorem 4.1, $\mathcal{F}^s$ and $\mathcal{F}^u$ are isotopic to $F_{g_1}^s$ and $F_{g_2}^u$ for some hyperbolic metrics g_1 and g_2 on Σ. The question is whether the intersection of $\mathcal{F}^s$ and $\mathcal{F}^u$ is isotopic to ϕ_{g_1,g_2}.

We know that $\widetilde{\mathcal{F}}^u$ and $\widetilde{\mathcal{F}}^s$ are the product foliations of $\mathbf{R}^3$. (This follows from the earlier work by Thurston [18] (see also Levitt [10]) which says that foliations without compact leaves are isotopic to the foliations transverse to the fibres.) Hence by Lemma 2.2, the map $p : \widetilde{M} \longrightarrow Q^u \times Q^s$ is a fibration to its image. The intersection of the image $p(\widetilde{M})$ and a horizontal line $Q^u \times \{*\}$ or a vertical line $\{*\} \times Q^s$ is diffeomorphic to a real line.

Theorem 4.1 also gives the information of the action of $\pi_1(M)$ on Q^u and Q^s. The generator of the centre of $\pi_1(M)$ represented by the fibre of the circle bundle acts as translation by 1. Hence any element α of $\pi_1(M)$ acts $\mathbf{Z}$ periodically on Q^u and Q^s. If α has a fixed point on Q^u, its fixed point set is the union of the attracting fixed points $\{x^s + m\}$, $(m \in \mathbf{Z})$ and the repelling fixed points $\{x^u + n\}$, $(n \in \mathbf{Z})$. Moreover, α also has the fixed point set which is the union of the attracting fixed points $\{y^s + m\}$, $(m \in \mathbf{Z})$ and the repelling fixed points $\{y^u + n\}$, $(n \in \mathbf{Z})$.

It is also important to note the following facts. The action of $\pi_1(M)$ on Q^u or Q^s is minimal, i.e., every $\pi_1(M)$ orbit is dense. The union of fixed point sets of the action of elements of $\pi_1(M)$ is dense in Q^u or Q^s. On Q^u and Q^s the notion of the length of the interval being an integer has meaning which is invariant under the action. Hence the notion of the length of the interval being greater than 1 also has meaning.

Lemma 4.3 *The length of the intersection of the image $p(\widetilde{M})$ and a horizontal line $Q^u \times \{*\}$ or a vertical line $\{*\} \times Q^s$ is at most 1.*

Proof. Assume that the length of intersection of $p(\widetilde{M})$ and $Q^u \times \{y\}$ is greater than 1. Since the image $p(\widetilde{M})$ is an open set and the union of fixed point sets is dense in Q^s, by replacing y by a nearby point, we may assume that $y \in Q^s$ is a fixed point under the action of an element $\alpha \in \pi_1(M)$. We may assume that this is an attracting fixed point y^s, for otherwise we take α^{-1}. Let y^u denote the repelling fixed point for α such that $y^s - 1 < y^u < y^s$.

Since the length of intersection of $p(\widetilde{M})$ and $Q^u \times \{y^s\}$ is greater than 1, there are at least 2 fixed points, one attracting and one repelling, on the intersection. Assume that the fixed points (x^u, y^s), (x^s, y^s), $(x^u < x^s < x^u + 1)$ are on the intersection. (The argument is similar for the case $x^s < x^u < x^s + 1$).

We first show that the image $p(\widetilde{M})$ contains neither $[x^u, x^s] \times \{y^s\} \cup \{x^s\} \times [y^s, y^u + 1]$ nor $[x^u, x^s] \times \{y^s\} \cup \{x^s\} \times [y^u, y^s]$.

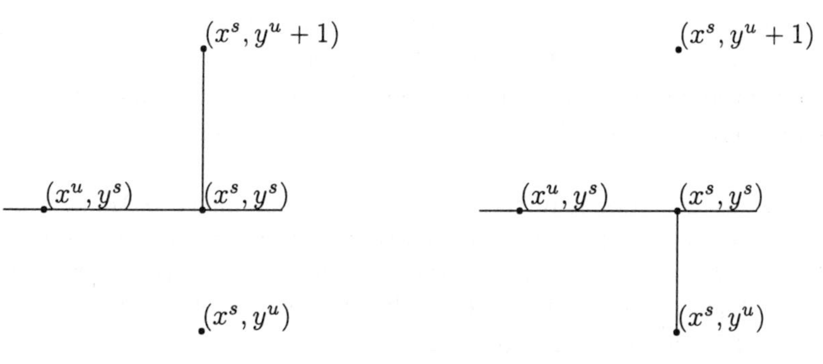

Figure 1. $[x^u, x^s] \times \{y^s\} \cup \{x^s\} \times [y^s, y^u + 1]$ and $[x^u, x^s] \times \{y^s\} \cup \{x^s\} \times [y^u, y^s]$ in $Q^u \times Q^s$.

Assume that $p(\widetilde{M})$ contains $[x^u, x^s] \times \{y^s\} \cup \{x^s\} \times [y^s, y^u+1]$. (The other case is similar.) Let L^s be the leaf of $\mathcal{F}^s$ which has a lift $(p^s)^{-1}(y^s)$ in $\widetilde{M}$. On L^s, we have two closed orbits corresponding to (x^u, y^s) and (x^s, y^s) and these two closed orbits bound an annulus with the flow which is the suspension of the action of α on the interval $[x^u, x^s] \times \{y^s\}$.

In the same way, Let L^u be the leaf of $\mathcal{F}^u$ which has a lift $(p^u)^{-1}(x^s)$ in $\widetilde{M}$. On L^u, we have two closed orbits corresponding to (x^s, y^s) and (x^s, y^u+1) and these two closed orbits bound an annulus with the flow which is the suspension of the action of α on the interval $\{x^s\} \times [y^s, y^u+1]$.

Note that since there are annuli in L^s and in L^u bounded by the closed orbits, the directions of the flow on these three closed orbits are parallel. In other words, the action of α on $p^{-1}(x^u, y^s)$, on $p^{-1}(x^s, y^s)$ and on $p^{-1}(x^s, y^s+1)$ is either simultaneously in the same direction as the flow or simultaneously in the opposite direction to the flow. (Compare the situation with that of usual geodesic flows.) Then either the action of the flow on the normal bundle $TM/T\varphi$ along the closed orbit corresponding to (x^u, y^s) or that along the closed orbit corresponding to (x^s, y^u+1) contradicts that the flow is projectively Anosov.

Now we show that the image $p(\widetilde{M})$ contains either $[x^u, x^s] \times \{y^s\} \cup \{x^s\} \times [y^s, y^u+1]$ or $[x^u, x^s] \times \{y^s\} \cup \{x^s\} \times [y^u, y^s]$.

Since the image $p(\widetilde{M})$ is invariant under the action of α and the an open interval containing $[x^u, x^s] \times \{y^s\}$ is in the image, the image $p(\widetilde{M})$ contains $(x^s-1, x^u+1) \times \{y^s\}$. By the same reason, since the image contains a neighbourhood of (x^s, y^s), the image $p(\widetilde{M})$ contains its attracting basin $(x^u, x^u+1) \times (y^u, y^u+1)$. The image also contains a neighbourhood of $\{x^u\} \times (y^u, y^u+1)$.

Then by the invariance under the action of the centre of $\pi_1(M)$, the image $p(\widetilde{M})$ contains $(x^s-2, x^u) \times \{y^s-1\}$ as well as $(x^u-1, x^u) \times (y^u-1, y^u)$ and a neighbourhood of $\{x^u-1\} \times (y^u-1, y^u)$.

Since the intersection of the image $p(\widetilde{M})$ and a vertical line $\{x\} \times Q^s$ is diffeomorphic to the real line, the image $p(\widetilde{M})$ contains $(x^s-1, x^u) \times [y^s-1, y^s]$.

Thus we see that only the intersection of the image $p(\widetilde{M})$ and $Q^u \times \{y^u+n\}$ might be of length not greater than 1. However, since the action of the $\pi_1(M)$ on Q^s is minimal, there is an element $\beta \in \pi_1(M)$ which sends y^u out of $\{y^u+n\}$ and this means that the length of the intersection of $p(\widetilde{M})$ and $Q^u \times \{y^u\}$ is greater than 1.

Since the intersection of $p(\widetilde{M})$ and $Q^u \times \{y^u\}$ contains $(x^s-1, x^u) \times \{y^u\}$, it contains either (x^s-1, y^u) or (x^s, y^u).

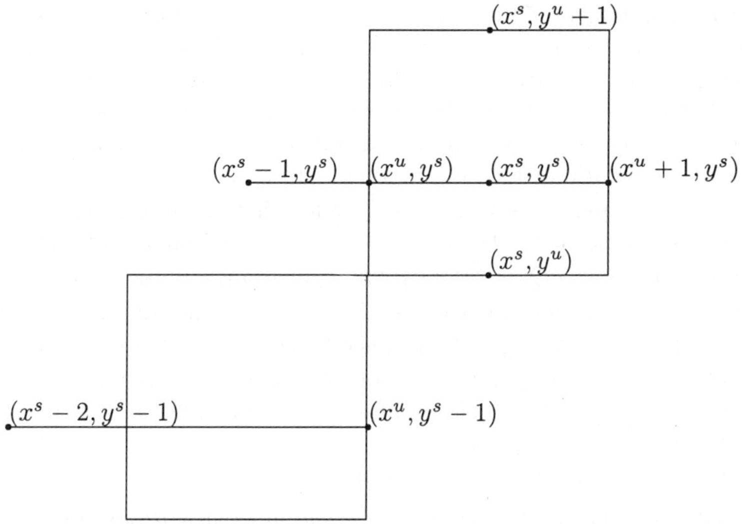

Figure 2. The image $p(\widetilde{M})$ contains $(x^s-1, x^u+1)\times\{y^s\}\cup(x^u, x^u+1)\times(y^u, y^u+1)$ and a neighbourhood of $\{x^u\}\times(y^u, y^u+1)$, and then it contains $(x^s-2, x^u)\times\{y^s-1\}\cup(x^u-1, x^u)\times(y^u-1, y^u)$ and a neighbourhood of $\{x^u-1\}\times(y^u-1, y^u)$ as well.

If the image $p(\widetilde{M})$ contains (x^s-1, y^u), then it contains (x^s, y^u+1), hence it contains $[x^u, x^s]\times\{y^s\}\cup\{x^s\}\times[y^s, y^u+1]$. If the image $p(\widetilde{M})$ contains (x^s, y^u), then it contains $[x^u, x^s]\times\{y^s\}\cup\{x^s\}\times[y^u, y^s]$.

Thus, by the contradiction, we have shown the lemma.

Let $\widehat{M}$ denote the covering corresponding to the fundamental group of the fibre. $\widehat{M}$ fibres over the Poincaré disk $\mathbb{D}$. Since the foliations $\mathcal{F}^s$ and $\mathcal{F}^u$ are isotopic to the foliations transverse to the fibres, $\widehat{\mathcal{F}}^u$ and $\widehat{\mathcal{F}}^s$ are diffeomorphic to the product foliation of $\mathbb{D}\times S^1$. We obtain the map $\widehat{p}:\widehat{M}\longrightarrow\widehat{Q}^u\times\widehat{Q}^s$, where $\widehat{Q}^u=\widehat{M}/\widehat{\mathcal{F}}^u$ and $\widehat{Q}^s=\widehat{M}/\widehat{\mathcal{F}}^s$. By Theorem 4.1, the circles $\widehat{Q}^u$ and $\widehat{Q}^s$ are with projective structures and $\pi_1(M)/\mathbf{Z}=\pi_1(\Sigma)$ acts on them as projective transformations.

In the case of the quasi-Fuchsian flow on M, as is explained in detail in [6], $\widehat{p}$ is a fibration onto $S^1\times S^1-\{(x, h(x)); x\in S^1\}$, where h is a homeomorphism of the circle.

In our case, we do not assume the existence of the invariant splitting of TM and we do not know the existence of the affine structure on of the orbit space of the $\widehat{\varphi}$ restricted to a leaf of $\widehat{\mathcal{F}}^u$ or $\widehat{\mathcal{F}}^s$. However, the transverse projective structures of $\widehat{\mathcal{F}}^u$ and $\widehat{\mathcal{F}}^s$ forces that the situation is similar.

Since the orbit foliation $\widehat{\varphi}$ restricted to a leaf of $\widehat{\mathcal{F}}^u$ or $\widehat{\mathcal{F}}^s$ is diffeomorphic to the product foliation of $\mathbf{R}^2$, by the above lemma, $\widehat{p}$ restricted to a leaf of $\widehat{\mathcal{F}}^u$ or $\widehat{\mathcal{F}}^s$ is a fibration to its image. This implies that $\widehat{p}$ is a fibration to its image. The reason is that we can lift a curve in the image of $\widehat{p}$ by using its approximation which is a union of curves in $\widehat{Q}^u \times \{*\}$ or $\{*\} \times \widehat{Q}^s$. The image is an open subset of $\widehat{Q}^u \times \widehat{Q}^s$ such that the intersection of the image and the circles $\widehat{Q}^u \times \{*\}$ or $\{*\} \times \widehat{Q}^s$ is a non empty open interval.

The map $\widehat{p} : \widehat{M} \longrightarrow \widehat{Q}^u \times \widehat{Q}^s$ is $\pi_1(M)$ equivariant and the action passes through $\pi_1(\Sigma)$. Any element α in $\pi_1(\Sigma)$ acts on $\widehat{Q}^u$ and $\widehat{Q}^s$ as projective transformations h^u_α and h^s_α with respect to the projective structures on them. The orbit of any point of $\widehat{Q}^u$ or $\widehat{Q}^s$ under the action of $\pi_1(\Sigma)$ is dense. Moreover for any two closed intervals I and J, there is an element $\alpha \in \pi_1(\Sigma)$ such that $h^u_\alpha(I) \subset \mathrm{int}J$.

We can now show the following lemma.

Lemma 4.4 $\mathcal{C} = \widehat{Q}^u \times \widehat{Q}^s - \widehat{p}(\widehat{M})$ *is a graph of an orientation preserving homeomorphism* $\mathcal{H} : \widehat{Q}^u \longrightarrow \widehat{Q}^s$.

Proof. First we show that the intersection of $\mathcal{C}$ with the circles $\widehat{Q}^u \times \{*\}$ or $\{*\} \times \widehat{Q}^s$ consists of a point.

Since the intersection of the image of $\widehat{p}$ and the circles $\widehat{Q}^u \times \{*\}$ is an open interval, the complement is a closed interval. Assume that there is a segment $J \times \{\widehat{q}^s\} = [\widehat{q}^u_1, \widehat{q}^u_2] \times \{\widehat{q}^s\}$ in $\mathcal{C}$. We take a point $(\widehat{q}^u_0, \widehat{q}^s)$ in the image of $\widehat{p}$.

Take the interval $I = [\widehat{q}^u_0, \widehat{q}^u_2] \subset \widehat{Q}^u$ which contains $J = [\widehat{q}^u_1, \widehat{q}^u_2]$. Take an element α of $\pi_1(\Sigma)$ such that $h^u_\alpha(I) \subset \mathrm{int}J$. Then we have an attractive fixed point $\widehat{q}^u_s$ of h^u_α in $\mathrm{int}J$ and a repulsive fixed point $\widehat{q}^u_u$ of h^u_α in $\widehat{Q}^u - I$. Since $(\widehat{q}^u_0, \widehat{q}^s)$ is in the image, $\widehat{q}^s$ is not a fixed point of h^s_α. Let $\widehat{q}^s_s$ and $\widehat{q}^s_u$ denote the attractive and the repulsive fixed points of h^s_α, respectively.

We look at the limit of $(h^u_\alpha)^k(J) \times \{(h^s_\alpha)^k(\widehat{q}^s)\}$ as $k \to -\infty$. Then this accumulates to $\widehat{Q}^u \times \{\widehat{q}^s_u\}$. Since $\mathcal{C}$ is a closed set, $\mathcal{C}$ contains $\widehat{Q}^u \times \{\widehat{q}^s_u\}$. This contradicts that the intersection of the image of $\widehat{p}$ and $\widehat{Q}^u \times \{*\}$ is not empty.

Thus the intersection of $\mathcal{C}$ with the circles $\widehat{Q}^u \times \{*\}$ or $\{*\} \times \widehat{Q}^s$ consists of a point.

Since $\mathcal{C}$ is closed, the above implies that $\mathcal{C}$ is a graph of homeomorphism $\mathcal{H} : \widehat{Q}^u \longrightarrow \widehat{Q}^s$. This homeomorphism is orientation preserving. For otherwise, the image of the fibre intersects $\mathcal{C}$.

Now we understand that the actions of $\pi_1(\Sigma)$ on $\widehat{Q}^u$ and $\widehat{Q}^s$ are conjugate by $\mathcal{H}$, that is, $\mathcal{H}(h^u_\alpha(\mathcal{H}^{-1}(\widehat{q}^s))) = h^s_\alpha(\widehat{q}^s)$ for $\alpha \in \pi_1(\Sigma)$.

Proof of Theorem 1.2. We may assume that $\widehat{Q}^u$ and $\widehat{Q}^s$ are identified with the standard projective line $\mathbf{R}P^1$ and the actions are both in $PSL(2;\mathbf{R})$. They determine two hyperbolic structures on Σ. As in [6], there exists a homeomorphism $H : \mathbf{R}P^1 \longrightarrow \mathbf{R}P^1$ such that

$$H(h^u_\alpha(H^{-1}(\widehat{q}^s))) = h^s_\alpha(\widehat{q}^s) \quad \text{for} \quad \alpha \in \pi_1(\Sigma).$$

We compare this homeomorphism H with our $\mathcal{H}$ given by Lemma 4.4. Since $H^{-1}\mathcal{H}$ commute with h^u_α for any $\alpha \in \pi_1(\Sigma)$, $\mathcal{H}$ coincides with H.

Let ϕ_t be the quasi-Fuchsian flow on M and F^u and F^s be the unstable foliation and stable foliation of ϕ_t. Then $\widehat{F}^u$ and $\widehat{F}^s$ determine the map $\widehat{p}_0 : \widehat{M} \longrightarrow \mathbf{R}P^1 \times \mathbf{R}P^1$ which is a fibration to the image with fibre being the orbit of $\widehat{\phi}_t$. The image of $\widehat{p}_0$ is the complement of the graph of H ([6]).

Now we can follow the argument by Ghys [6]. We compare the map $\widehat{p}_0$ with $\widehat{p}$ defined by $\widehat{\mathcal{F}}^u$ and $\widehat{\mathcal{F}}^s$. As we discussed, the images of $\widehat{p}_0$ and $\widehat{p}$ coincide. The actions of $\pi_1(M)$ on the images also coincide. They define the transverse structure of the orbit foliations φ and ϕ.

For a closed orbit c of φ or ϕ, we look at the lift in $\widehat{M}$ and its image in $\mathbf{R}P^1 \times \mathbf{R}P^1$. The curve c projected to Σ represents $\alpha \in \pi_1(\Sigma)$ and the image in $\mathbf{R}P^1 \times \mathbf{R}P^1$ is a fixed point by (h^u_α, h^s_α). Hence the holonomy covering of c is contractible.

Now as in the proof of Theorem 1.1, the situation fits to the argument by Ghys [6]. Both (M, ϕ) and (M, φ) are the classifying space for the groupoid and we see that φ_t is isotopic to ϕ_t after changing the parametrization.

5 Asymptotic cycles

In this section we give an alternative proof for Theorem 1.1.

By Lemma 3.3, we already know that $\mathcal{F}^u$ is isotopic to F^u and $\mathcal{F}^s$ is isotopic to F^s. We show that $\mathcal{F}^u$ and $\mathcal{F}^s$ are simultaneously isotopic to F^u and F^s.

To show this, we use the fact that the multiplier of the affine actions of $\pi_1(M)$ on the affine line Q^u and Q^s factors through $\pi_1(S^1)$. We can formulate this in a different way.

Let $\widehat{M}$ be the cyclic covering space of M which corresponds to the fundamental group of the torus fibre. Then we have a projection $\widehat{p} : \widehat{M} \longrightarrow \mathbf{R}$. We fix a smooth Riemannian metric on M. Then since φ_t is a projectively Anosov flow, considering the metric induced on $TM/T\varphi$, there are positive real numbers C' and K such that the following inequality holds for $t \geq 0$,

$v^u \in \hat{E}^u$ and $v^s \in \hat{E}^s$:

$$\frac{\|T\varphi_t(v^u)\|}{\|T\varphi_t(v^s)\|} \geq Ke^{C't}\frac{\|v^u\|}{\|v^s\|}.$$

We take the pullback metric on $\widehat{M}$ and the lifted flow $\widehat{\varphi}_t$, then for the induced splitting of $T\widehat{M}/T\widehat{\varphi}$, we have the same in equality as above.

Now we take account of the foliations $\mathcal{F}^u$ and $\mathcal{F}^s$.

For a curve γ on a leaf of $\mathcal{F}^u$, we look at the linear holonomy h^u_γ of the foliation $\mathcal{F}^u$ along γ. Since $\mathcal{F}^u$ is isotopic to F^u, for the induced metric on $TM/T\mathcal{F}^u$, it is contracting. It is easier to describe this fact for the curve γ on a leaf of the pullback foliation $\widehat{\mathcal{F}}^u = \widehat{p}^*\mathcal{F}^u$ of the cyclic covering space $\widehat{M}$.

Lemma 5.1 *Let $e^{\pm\lambda}$ $(\lambda > 0)$ denote the absolute value of the eigenvalues of the monodromy matrix A. Let $\widehat{p} : \widehat{M} \longrightarrow M$ be the cyclic covering. We consider a Riemannian metric on M, and on $\widehat{M}$ the pullback Riemannian metric. Let $\widehat{\mathcal{F}}^\sigma = \widehat{p}^*\mathcal{F}^\sigma$ $(\sigma = s,u)$ denote the pullback foliation. For a curve $\gamma : [0,1] \longrightarrow \widehat{\mathcal{F}}^\sigma$ on a leaf of $\widehat{\mathcal{F}}^\sigma$, let h^σ_γ denote the holonomy along γ from the germ of transverse arc at $\gamma(0)$ to the germ of transverse arc at $\gamma(1)$. Then there is a positive real number κ such that the following inequalities hold:*

$$\kappa^{-1}e^{-\lambda\{\widehat{p}(\gamma(1))-\widehat{p}(\gamma(0))\}}\|v^s\| \leq \|h^u_{\gamma *}v^s\| \leq \kappa e^{-\lambda\{\widehat{p}(\gamma(1))-\widehat{p}(\gamma(0))\}}\|v^s\|,$$

for $v^s \in T\widehat{M}/T\widehat{\mathcal{F}}^u$, and

$$\kappa^{-1}e^{\lambda\{\widehat{p}(\gamma(1))-\widehat{p}(\gamma(0))\}}\|v^s\| \leq \|h^s_{\gamma *}v^u\| \leq \kappa e^{\lambda(\widehat{p}(\gamma(1))-\widehat{p}(\gamma(0)))}\|v^u\|$$

for $v^u \in T\widehat{M}/T\widehat{\mathcal{F}}^s$. Here we are considering the induced metric on $T\widehat{M}/T\widehat{\mathcal{F}}^\sigma$.

Proof. For the Anosov foliations F^u and F^s, if we take an appropriate metric with respect to them, we have the following equalities:

$$\|h^u_{\gamma *}v^s\| = e^{-\lambda\{\widehat{p}(\gamma(1))-\widehat{p}(\gamma(0))\}}\|v^s\|$$

for $v^s \in T\widehat{M}/T\widehat{\mathcal{F}}^u$, and

$$\|h^s_{\gamma *}v^u\| = e^{\lambda(\widehat{p}(\gamma(1))-\widehat{p}(\gamma(0)))}\|v^u\|$$

for $v^u \in T\widehat{M}/T\widehat{\mathcal{F}}^s$. When we change the foliations by isotopies and change the Riemannian metric, we still have the desired inequality.

We are going to show that our projectively Anosov flow φ_t has a section. To show this the following theorem of Schwartzman plays an essential role.

Let ξ be the vector field associated with φ_t. For a signed measure on M invariant under φ_t, put $A_\varphi(\mu)(\omega) = \int_M \langle \xi, \omega \rangle d\mu$ for any closed form ω. The asymptotic cycle is the image in $H_1(M; \mathbf{R})$ of A_φ of the invariant probability measures.

A quasi-regular point x is a point such that the limit $\lim \frac{1}{T} \int_0^T f(\varphi_t(x))dt$ exists for any continuous function f. For a quasi-regular point x, $A_x = \lim \frac{\overline{C_t}}{t}$ exists, where $\overline{C_t}$ is the homology class of the orbit from x to $\varphi_t(x)$ joined with a bounded path from $\varphi_t(x)$ to x. Note that, for a quasi-regular point x, A_x is in the image of A_ϕ.

Theorem 5.2 (Schwartzman [16]) *1. For any invariant probability measure μ, the set of quasi-regular points is measure 1, and $A_\varphi(\mu) = \int_M A_x d\mu(x)$.*
2. If there exists $C \in H^1(M; \mathbf{R})$ such that $\langle A_\varphi(\mu), C \rangle > 0$ for any invariant probability measure, then φ_t has a section.

Hence, for $C \in H^1(M; \mathbf{R})$, if we show that there exist a positive real number ε such that $\langle A_x, C \rangle > \varepsilon$ for any quasi-regular point x, then φ_t has a section.

Now we prove our main theorem.

Proof of Theorem 1.1. For an orbit of $\widehat{\varphi}_t$ on $\widehat{M}$, by Lemma 5.1,

$$\kappa^{-2} e^{2\lambda \{\widehat{p}(\widehat{\varphi}_t(\widehat{x})) - \widehat{p}(\widehat{x})\}} \frac{\|v^u\|}{\|v^s\|} \leq \frac{\|T\widehat{\varphi}_t(v^u)\|}{\|T\widehat{\varphi}_t(v^s)\|} \leq \kappa^2 e^{2\lambda \{\widehat{p}(\widehat{\varphi}_t(\widehat{x})) - \widehat{p}(\widehat{x})\}} \frac{\|v^u\|}{\|v^s\|}$$

for $v^u \in \hat{E}^u$ and $v^s \in \hat{E}^s$. Since φ_t is projectively Anosov, we have

$$\frac{\|T\widehat{\varphi}_t(v^u)\|}{\|T\widehat{\varphi}_t(v^s)\|} \geq K e^{C't} \frac{\|v^u\|}{\|v^s\|}$$

for $t \geq 0$, $v^u \in \hat{E}^u$ and $v^s \in \hat{E}^s$. Hence we have

$$K e^{C't} \leq \kappa^2 e^{2\lambda \{\widehat{p}(\widehat{\varphi}_t(\widehat{x})) - \widehat{p}(\widehat{x})\}}$$

for $t \geq 0$. Hence if t is sufficiently large ($t > T_0 = (2\lambda + 2\log\kappa - \log K)/C'$), then $\widehat{p}(\widehat{\varphi}_t(\widehat{x})) - \widehat{p}(\widehat{x}) > 1$.

Let C be the cohomology class of the bundle projection. Then for any $x \in M$, we have $\langle \overline{C_T}/T, C \rangle \geq 1/T_0$. Hence for any quasi-regular point x we have $\langle A_x, C \rangle \geq 1/T_0$. Thus φ_t has a section by Theorem 5.2

The only possible section in our case is that isotopic to the fibre. Once we have a fibre transverse to φ_t, the theorem follows without difficulty.

Acknowledgments

The first author is supported in part by Research Fellowship of the Japan Society for the Promotion of Science for Young Scientists and by Grant-in-Aid for Scientific research 11-9548, Japan Society for Promotion of Science, Japan.

The second author is supported by Monbusho Zaigai Kenkyuin 1999, Ministry of Education, Science, Sports and Culture, Japan and Grant-in-Aid for Scientific research 12304003, Japan Society for Promotion of Science, Japan.

References

1. T. Barbot, *Flot d'Anosov sur les variétés graphées au sens de Waldhausen,* Ann. Inst. Fourier, **46** (1996), 1451–1517.
2. Y. Eliashberg and W. Thurston, Confoliations, University Lecture Series 13, Amer. Math. Soc., 1998.
3. S. Fenley, *Anosov flows in* 3*-manifolds,* Ann. of Math., (2) **139** (1994), 79–115.
4. É. Ghys, *Flots d'Anosov sur les* 3*-variétés fibrées en cercles,* Erg. Th. & Dynam. Sys., **4** (1984), no. 1, 67–80.
5. É. Ghys, *Flots transversalement affines et tissus feuilletés,* Analyse globale et physique mathématique (Lyon, 1989). Mém. Soc. Math. France (N.S.), **46** (1991), 123–150.
6. É. Ghys, *Déformations de flots d'Anosov et de groupes fuchsiens,* Ann. Inst. Fourier, **42** (1992), 209–247 .
7. É. Ghys, *Rigidité différentiable des groupes fuchsiens,* Inst. Hautes Études Sci. Publ. Math., **78** (1993), 163–185.
8. É. Ghys et V. Sergiescu, *Stabilité et conjugaison différentiable pour certains feuilletages,* Topology, **19** (1980), no. 2, 179–197.
9. A. Haefliger, *Groupoïde d'holonomie et classifiants,* in Structure Transverse de Feuilletages, Astérisque, **116** (1984), 70–97.
10. G. Levitt, *Feuilletages des variétés de dimension* 3 *qui sont des fibrés en cercles.* Comment. Math. Helv., **53** (1978), 572–594.
11. S. Matsumoto, *Some remarks on foliated* S^1 *bundles,* Invent. Math., **90** (1987), no. 2, 343–358.
12. S. Matsumoto and T. Tsuboi, *Transverse intersections of foliations in three-manifolds*, preprint.
13. Y. Mitsumatsu, *Anosov flows and non-Stein symplectic manifolds,* Ann. Inst. Fourier, **45** (1995), 1407–1421.
14. T. Noda, *Projectively Anosov flows with differentiable (un)stable foli-*

ations, to appear in Ann. Inst. Fourier.
15. J. Plante, *Anosov flows, transversely affine foliations, and a conjecture of Verjovsky,* J. London Math. Soc. (2), **23** (1981), 359–362.
16. S. Schwartzman, *Asymptotic cycles,* Ann. of Math. (2), **66** (1957), 270–284.
17. I. Tamura and A. Sato, *On transverse foliations,* Inst. Hautes Études Sci. Publ. Math. **54** (1981), 205–235.
18. W. Thurston, *Foliations of 3-manifolds that are circle bundles,* University of California at Berkeley, Thesis, 1972.

Received August 8, 2000.

Proceedings of
FOLIATIONS: GEOMETRY AND DYNAMICS
held in Warsaw, May 29–June 9, 2000
ed. by Paweł WALCZAK *et al.*
World Scientific, Singapore, 2002
pp. 421–440

ON THE PERFECTNESS OF GROUPS OF DIFFEOMORPHISMS OF THE INTERVAL TANGENT TO THE IDENTITY AT THE ENDPOINTS

TAKASHI TSUBOI
Graduate School of Mathematical Sciences, University of Tokyo,
Komaba Meguro, Tokyo 153, Japan,
e-mail: tsuboi@ms.u-tokyo.ac.jp

This paper is dedicated to the memory of Kikuko Arai Hudson.

This paper shows that several groups of orientation preserving homeomorphisms of the closed interval are perfect, that is, every element of the group is written as a product of commutators. Results might not be new, however the method of proof would be of interest. The method is to write an element as a product of elements which are the identity on subintervals accumulating to the end points. Then there are many cases where the latters are products of commutators. In this way, we show the perfectness of the group of homeomorphisms, of the group of Lipschitz homeomorphisms, of the group of the C^1 diffeomorphisms which are tangent to the identity at the endpoints, and of the group of the C^∞ diffeomorphisms which are infinitely tangent to the identity at the endpoints.

1 Introduction

The homology of groups of homeomorphisms has been studied in relation with the study of the topology of the classifying spaces of foliations. Even for the first homology group of orientation preserving homeomorphisms of the interval $[0, 1]$, there are several important results. The first homology group of a group is the quotient group of the group by its commutator subgroup. The group is said to be perfect if it coincides with its commutator subgroup or equivalently the first homology is zero.

First the group $\mathrm{Homeo}([0, 1])$ of orientation preserving homeomorphisms of the closed interval $[0, 1]$ is perfect. This has been shown by using the fact that Homeo([0, 1]) is isomorphic to the group Homeo((0, 1)) of orientation preserving homeomorphisms of the open interval (0, 1). Then the latter is easily shown to be perfect by using the fact that all positive topological translation of (0, 1) are conjugate.

Secondly, the first homology of the group $\mathrm{Diff}^\infty([0,1])$ of orientation preserving diffeomorphisms of the interval [0, 1] is isomorphic to $\mathbf{R} \times \mathbf{R}$ (Fukui[9]). The abelianization homomorphism is given by the evaluation of the logarithms of the slopes at the endpoints. On the other hand, the group $\mathrm{Diff}^\infty_\infty([0,1])$ of the diffeomorphisms of the interval [0, 1] which are infinitely tangent to the identity at the endpoints is perfect (Sergeraert [24]).

For groups of homeomorphisms with compact support of the open interval (0, 1), most groups are known to be perfect. For example the following groups are perfect; the group $\mathrm{Homeo}_c((0,1))$ of homeomorphisms with compact support (Anderson [3]), the group $\mathrm{Homeo}^L_c((0,1))$ of Lipschitz homeomorphisms with compact support (Anderson [3]), the group $\mathrm{Diff}^r_c((0,1))$ of C^r diffeomorphisms ($1 \leq r \leq \infty$, $r \neq 2$) with compact support (Mather [16], [17]).

In this paper, we are interested in the perfectness of several groups of homeomorphisms of the closed interval and of several groups of diffeomorphisms of the closed interval which are (infinitely) tangent to the identity the endpoints $\{0, 1\}$. We would like to show the perfectness of them under a rather simple idea.

This idea to show the perfectness is as follows. A diffeomorphism of a closed interval (infinitely) tangent to the identity at the end points would be decomposed into a product of diffeomorphisms which are the identity on subintervals accumulating to the end points. If a diffeomorphism of the open interval (0, 1) with compact support is written as a product of commutators in a controlled way, then the resulted diffeomorphisms are written as products of commutators (infinitely) tangent to the identity at the end points.

We carry out this program to show the perfectness of the group of homeomorphisms, of the group of Lipschitz homeomorphisms, of the group of C^1 diffeomorphisms which are tangent to the identity at the endpoints, and of the group of C^∞ diffeomorphisms which are infinitely tangent to the identity at the endpoints. We actually show that these groups are uniformly perfect, that is, any element is written as a product of a bounded number of commutators.

In fact, we give the proof of the perfectness of the group Homeo([0, 1]) of orientation preserving homeomorphisms of the closed interval in Section 2. This does not need any estimate and gives a good understanding for our method. Then we see in Section 3 that, by using a little careful construction with estimates on the Lipschitz constants of homeomorphisms, the proof of Section 2 also works for the group $\mathrm{Homeo}^L([0,1])$ of orientation preserving Lipschitz homeomorphisms of the closed interval.

To treat diffeomorphisms we need estimates on the norms. In Section 4, we show that the group $\mathrm{Diff}^1_1([0,1])$ of the C^1 diffeomorphisms of the closed interval which are tangent to the identity at the endpoints is perfect. For an element of $\mathrm{Diff}^1_1([0,1])$ which is the identity near the endpoint 1, we use a fragmentation with respect to a partition of unity of $(0,1]$ by 2 functions which oscillate between 0 and 1. We also use a result on small commutators of $\mathrm{Diff}^1_c((0,1))$, which we prove in Section 6. There we see that we have a good control on the C^1 norm.

In Section 5, we see that the same fragmentation as in Section 4 works for the C^∞ diffeomorphisms of a closed interval infinitely tangent to the identity at endpoints. Then we use a result in ([26]) on small commutators of $\mathrm{Diff}^\infty_c((0,1))$ to show the group $\mathrm{Diff}^\infty_\infty([0,1])$ of orientation preserving C^∞ diffeomorphisms of the interval infinitely tangent to the identity at the endpoints is perfect. This was first proved by Sergeraert [24], and reproved by Natsume-Kawabe [21]. The present proof was suggested by Francis Sergeraert in a discussion we had in 80's.

The present proof would be generalized to the cases of the groups of diffeomorphisms of a compact manifold (infinitely) tangent to the identity along the boundary isotopic to the identity. That is, the results on the small commutators would be true and we would be able to use a similar partition of unity to decompose a diffeomorphisms into a product of diffeomorphisms whose support is contained in the closure of a union of disks accumulating to the boundary. For the C^∞ case, this would give an alternative proof of the result of Masson [14].

In Section 6, as an appendix to Section 4, we prove a result on small commutators of $\mathrm{Diff}^1_c((0,1))$. We use Mather's trick ([17]) and the Denjoy-Pixton action ([22], [28]) of $\mathbf{Z}^2$ to write a C^1-diffeomorphism with compact support as a product of commutators with controlled C^1 norm.

2 Perfectness of the group of homeomorphisms of the closed interval

We review the proof of perfectness of the group of homeomorphisms of the real line with compact support (Mather [15]).

Proposition 2.1 ([15]) *The group of homeomorphisms of the real line with compact support is perfect.*

Proof. Let g be a homeomorphism of the real line with support in an interval U. Then we can find a homeomorphism h with compact support such that $g(U) \cap U = \emptyset$. We form the infinite product $G = \prod_{i=0}^{\infty} h^i g h^{-i}$ of the homeomorphisms $h^i g h^{-i}$ which have support in disjoint intervals $h^i(U)$. It is easy to see that $G = ghGh^{-1}$ and g is written as one commutator; $g = GhG^{-1}h^{-1} = [G, h]$. □

We prove the perfectness of the group of homeomorphisms of the closed interval by decomposing a homeomorphism into a composition of homeomorphisms which are the identity on subintervals accumulating to the end points.

Theorem 2.2 ([23], [11]) *The group* Homeo$([-1,1])$ *of homeomorphisms of the closed interval* $[-1,1]$ *is perfect.*

Proof. For a homeomorphism f of the interval $[-1,1]$, we choose a sequence $\{a_j\}_{j=-\infty}^{\infty}$ of points in $(-1,1)$ such that

$$\max\{a_j, f(a_j), f^{-1}(a_j)\} < a_{j+1} \text{ and } \lim_{j\to\pm\infty} a_j = \pm 1.$$

Then $[a_{4k-2}, a_{4k-1}]$ ($k \in \mathbf{Z}$) and $[f(a_{4k}), f(a_{4k+1})]$ ($k \in \mathbf{Z}$) are disjoint and let f_2 be a homeomorphism of $[-1,1]$ such that

$$f_2|[a_{4k-2}, a_{4k-1}] = \mathrm{id}_{[a_{4k-2}, a_{4k-1}]} \text{ and } f_2|[a_{4k}, a_{4k+1}] = f|[a_{4k}, a_{4k+1}].$$

The support of f_2 is contained in the union of disjoint intervals $[a_{4k-1}, a_{4k+2}]$. Put $f_1 = f{f_2}^{-1}$. Then

$$f_1|[a_{4k-2}, a_{4k-1}] = f|[a_{4k-2}, a_{4k-1}] \text{ and}$$
$$f_1|[f(a_{4k}), f(a_{4k+1})] = \mathrm{id}_{[f(a_{4k}), f(a_{4k+1})]}.$$

The support of f_1 is contained in the union of disjoint intervals $[f(a_{4k-3}), f(a_{4k})]$.

We apply the construction of the proof of Proposition 2.1 to each piece $f_1|[f(a_{4k-3}), f(a_{4k})]$ of f_1 or $f_2|[a_{4k-1}, a_{4k+2}]$ of f_2. Then the piece $f_1|[f(a_{4k-3}), f(a_{4k})]$ is written as one commutator

$$[G_{1,k}, h_{1,k}] = G_{1,k} h_{1,k} {G_{1,k}}^{-1} {h_{1,k}}^{-1}$$

with support in

$$[(f(a_{4k-2}) + f(a_{4k-3}))/2, (f(a_{4k}) + f(a_{4k+1}))/2]$$

and the piece $f_2|[a_{4k-1}, a_{4k+2}]$ is written as one commutator

$$[G_{2,k}, h_{2,k}] = G_{2,k} h_{2,k} {G_{2,k}}^{-1} {h_{2,k}}^{-1}$$

with support in

$$[(a_{4k-2} + a_{4k-1})/2, (a_{4k+2} + a_{4k+3})/2].$$

Put

$$G_1 = \prod_k G_{1,k}, \quad h_1 = \prod_k h_{1,k}, \quad G_2 = \prod_k G_{2,k}, \quad h_2 = \prod_k h_{2,k}.$$

Then these are well defined homeomorphisms of the interval $[-1,1]$, and we have $f_1 = [G_1, h_1] = G_1 h_1 G_1{}^{-1} h_1{}^{-1}$ and $f_2 = [G_2, h_2] = G_2 h_2 G_2{}^{-1} h_2{}^{-1}$ Thus $f = f_1 f_2$ is a product of two commutators.

This shows that Homeo($[-1,1]$) is uniformly perfect. □

3 Perfectness of the group of Lipschitz homeomorphisms of the closed interval

A Lipschitz homeomorphism is a homeomorphism f such that f and f^{-1} are Lipschitz maps. The logarithm of the maximum of the Lipschitz constants for f and f^{-1} measures the distance between f and the identity. This gives rise to a left invariant metric on the group $\mathrm{Homeo}^L([0,1])$ of orientation preserving Lipschitz homeomorphisms of the closed interval $[0,1]$

First we note that the proof of Proposition 2.1 also gives the perfectness of the group of Lipschitz homeomorphisms of the real line with compact support.

In the proof of Proposition 2.1, for a Lipschitz homeomorphism g supported on an interval $U_1 = [a_1, b_1]$, we can use the piecewise linear homeomorphism h with support in $U_0 = [a_0, b_0]$ $(a_0 < a_1 < b_1 < b_0)$ which is affine on $[a_0, b_1]$ and on $[b_1, b_0]$ and $h(b_1) = a_1$. Then the resulted homeomorphism $G = \prod_{i=0}^{\infty} h^i g h^{-i}$ as well as $G^{-1} = \prod_{i=0}^{\infty} h^i g^{-1} h^{-i}$ is Lipschitz with the same Lipschitz constant as that of g or g^{-1}. The Lipschitz constants of h and h^{-1} are of course $(b_0 - a_1)/(b_0 - b_1)$ and $(b_1 - a_0)/(a_1 - a_0)$, respectively. So we have the following lemma.

Lemma 3.1 *For intervals $U_0 = [a_0, b_0]$ and $U_1 = [a_1, b_1]$, $(a_0 < a_1 < b_1 < b_0)$, put $D = \max\{(b_0 - a_1)/(b_0 - b_1), (b_1 - a_0)/(a_1 - a_0)\}$. Let g be a Lipschitz homeomorphism g with support in U_1 such that the Lipschitz constants of g and g^{-1} are bounded by L. Then g is written as a commutator of Lipschitz homeomorphisms with support in $U_0 = [a_0, b_0]$; $g = [G, h] = GhG^{-1}h^{-1}$ such that the Lipschitz constants of G, G^{-1}, h, h^{-1} are bounded by* $\max\{L, D\}$.

Now we follow the proof of Theorem 2.2 taking care of the lengths of intervals and we show that perfectness of the group of Lipschitz homeomorphisms of the closed interval.

Theorem 3.2 *The group* $\mathrm{Homeo}^L([0,1])$ *of orientation preserving Lipschitz homeomorphisms of the closed interval* $[0,1]$ *is perfect.*

Proof. Any Lipschitz homeomorphism of the interval $[0,1]$ is written as a product of two Lipschitz homeomorphisms which are either the identity near 0 or near 1. It is enough to show that a Lipschitz homeomorphism of the interval $[0,1]$ which is the identity near 1 is a product of commutators.

Let f be a Lipschitz homeomorphism of the interval $[0,1]$ with support in $[0,a_0]$ ($a_0 < 1$). Since f and f^{-1} are Lipschitz, for the maximum L of Lipschitz constants for f and f^{-1}, we have $x/L \le f(x) \le Lx$. For negative integers i, put $a_i = a_0 L^{2i}$.

Then $[a_{4k-2}, a_{4k-1}]$ ($k \in \mathbf{Z}$, $k \le 0$) and $[f(a_{4k}), f(a_{4k+1})]$ ($k \in \mathbf{Z}$, $k < 0$) are disjoint and let f_2 be the homeomorphism of $[0,1]$ such that

$$f_2|[a_{4k-2}, a_{4k-1}] = \mathrm{id}_{[a_{4k-2},a_{4k-1}]}, \quad f_2|[a_{4k}, a_{4k+1}] = f|[a_{4k}, a_{4k+1}],$$

and $f_2|[a_{4k-1}, a_{4k}]$ and $f_2|[a_{4k+1}, a_{4k+2}]$ are affine. The support of f_2 is contained in the union of disjoint intervals $[a_{4k-1}, a_{4k+2}]$ as before. Since

$$a_{4k}/L \le f(a_{4k}) \le La_{4k},$$

the slope of $f_2|[a_{4k-1}, a_{4k}]$ and its inverse are not greater than

$$\frac{a_0L^{8k+1} - a_0L^{8k-2}}{a_0L^{8k} - a_0L^{8k-2}} = \frac{L^3-1}{L^2-1} = \frac{L^2+L+1}{L+1} \le L+1 \quad \text{or}$$
$$\frac{a_0L^{8k} - a_0L^{8k-2}}{a_0L^{8k-1} - a_0L^{8k-2}} = \frac{L^2-1}{L-1} = L+1.$$

In the same way, since

$$a_{4k+1}/L \le f(a_{4k+1}) \le La_{4k+1},$$

the slope of $f_2|[a_{4k+1}, a_{4k+2}]$ and its inverse are not greater than

$$\frac{a_0L^{8k+4} - a_0L^{8k+1}}{a_0L^{8k+4} - a_0L^{8k+2}} = \frac{L^3-1}{L^3-L} = \frac{L^2+L+1}{L^2+L} \le 2 \quad \text{or}$$
$$\frac{a_0L^{8k+4} - a_0L^{8k+2}}{a_0L^{8k+4} - a_0L^{8k+3}} = \frac{L^2-1}{L^2-L} = \frac{L+1}{L} \le 2.$$

Thus the Lipschitz constants of f_2 and ${f_2}^{-1}$ are estimated by $L+1$.

Put $f_1 = f{f_2}^{-1}$. Then

$$f_1|[a_{4k-2}, a_{4k-1}] = f|[a_{4k-2}, a_{4k-1}] \text{ and}$$
$$f_1|[f(a_{4k}), f(a_{4k+1})] = \mathrm{id}_{[f(a_{4k}),f(a_{4k+1})]}.$$

The support of f_1 is contained in the union of disjoint intervals $[f(a_{4k-3}), f(a_{4k})]$. The Lipschitz constants of f_1 and ${f_1}^{-1}$ are estimated by $L^2 + L$.

Now we use Lemma 3.1 and write the piece $f_1|[f(a_{4k-3}), f(a_{4k})]$ of f_1 as a commutator $[G_{1,k}, h_{1,k}] = G_{1,k}h_{1,k}G_{1,k}{}^{-1}h_{1,k}{}^{-1}$ with support in $[a_0L^{8k-7}, a_0L^{8k+1}]$. Then the Lipschitz constants of the elements in the commutator are bounded by $\max\{L^2+L, D\} = D$, where

$$D = \max\left\{\frac{a_0L^{8k+1} - a_0L^{8k-6}}{a_0L^{8k+1} - a_0L^{8k}}, \frac{a_0L^{8k} - a_0L^{8k-7}}{a_0L^{8k-6} - a_0L^{8k-7}}\right\} = \frac{L^7-1}{L-1}.$$

The piece $f_2|[a_{4k-1}, a_{4k+2}]$ of f_2 is also written as one commutator $[G_{2,k}, h_{2,k}] = G_{2,k}h_{2,k}G_{2,k}{}^{-1}h_{2,k}{}^{-1}$ with support in $[a_0L^{8k-3}, a_0L^{8k+5}]$ and the Lipschitz constants of the elements in the commutator are bounded by $\max\{L+1, D\} = D$.

Put

$$G_1 = \prod_k G_{1,k}, \quad h_1 = \prod_k h_{1,k}, \quad G_2 = \prod_k G_{2,k}, \quad h_2 = \prod_k h_{2,k}$$

as before and these are Lipschitz homeomorphisms of the interval $[0,1]$ with Lipschitz constants bounded by D. Thus

$$f = f_1f_2 = [G_1, h_1][G_2, h_2],$$

and we proved that $\mathrm{Homeo}^L([0,1])$ is uniformly perfect. □

4 Perfectness of the group of the C^1 diffeomorphisms of the closed interval which are tangent to the identity at the endpoints

Let $\mathrm{Diff}^1([0,1])$ denote the group of C^1 diffeomorphisms of the interval. For an element f of $\mathrm{Diff}^1([0,1])$, $\sup|\log f'|$ measures the distance between f and the identity. This is equivalent to $\max\{\sup|(f-\mathrm{id})'|, \sup|(f^{-1}-\mathrm{id})'|\}$. These would be referred to as C^1 norm of f. These give rise to a left invariant metric on $\mathrm{Diff}^1([0,1])$.

There is a natural homomorphism to $\mathbf{R}\times\mathbf{R}$ which is the evaluation of the logarithms of the slopes at the endpoints. That is, the homomorphism maps $f \in \mathrm{Diff}^1([0,1])$ to $(\log f'(0), \log f'(1)) \in \mathbf{R}\times\mathbf{R}$. Let $\mathrm{Diff}^1_1([0,1])$ denote the kernel of this homomorphism.

In this section, we show that $\mathrm{Diff}^1_1([0,1])$ is a perfect group.

Theorem 4.1 *Let* $\mathrm{Diff}^1_1([0,1])$ *denote the group of the* C^1 *diffeomorphisms of the interval which are tangent to the identity at the end points. The group* $\mathrm{Diff}^1_1([0,1])$ *is a perfect group.*

We will use a version of fragmentation lemma ([5], [18], [27], [7]).

Let η be a smooth monotone increasing function on $[0,1]$ such that

$\eta(x) = 0$ for $x \in [0, 1/2]$,
$\eta(x) = 1$ for $x \in [3/4, 1]$ and
$\eta'(x) \leq 8$ for $x \in [0, 1]$.

Let ν denote the smooth function on the half open interval $(0, 1]$ such that

$\nu(x) = \eta(2^{2k}x)$ for $x \in [2^{-2k-1}, 2^{-2k}]$ ($k \geq 0$, $k \in \mathbf{Z}$) and
$\nu(x) = 1 - \eta(2^{2k+1}x)$ for $x \in [2^{-2k-2}, 2^{-2k-1}]$ ($k \geq 0$, $k \in \mathbf{Z}$).

The support of ν is contained in $\bigcup_{k=0}^{\infty}[2^{-2k-1}, 2^{-2k-1}3]$ and that of $1 - \nu$ is contained in $\bigcup_{k=0}^{\infty}[2^{-2k-2}, 2^{-2k-2}3]$. Since the absolute value of $d\nu/dx$ on $[2^{-k-1}, 2^{-k}]$ is estimated by 2^{k+3}, we have the estimate $|d\nu/dx| \leq 2^3/x$. In a similar way, since the absolute value of $d^\ell\nu/dx^\ell$ ($\ell = 1, 2, \ldots$) on $[2^{-k-1}, 2^{-k}]$ is estimated by $c_\ell 2^{k\ell}$ for some constant c_ℓ, we have the estimate $|d^\ell\nu/dx^\ell| \leq c_\ell/x^\ell$ ($\ell = 1, 2, \ldots$).

Let $H : [0, 2] \times (0, 1] \longrightarrow [0, 1] \times (0, 1]$ be a continuous map defined by $H(t, x) = (t\nu(x), x)$ for $t \in [0, 1]$ and $H(t, x) = ((t-1)(1-\nu(x)) + \nu(x), x)$ for $t \in [1, 2]$. This is written as $H(t, x) = ((t - [t])\nu_{t-[t]}(x) + [t]\nu_1(x), x)$ for $t \in [1, 2)$, where $\nu_1 = \nu$ and $\nu_2 = 1 - \nu$.

For an element $f \in \mathrm{Diff}_1^1([0, 1])$, consider a foliation F of $[0, 1] \times (0, 1]$ whose leaves are line segments $\{(t, f(x) + t(x - f(x)))\ ;\ t \in [0, 1]\}$ joining $(0, f(x))$ and $(1, x)$. This is the isotopy $f_t(x) = f(x) + t(x - f(x))$ between f and the identity. The slope of the leaves are estimated by $|f(x) - x|$. Let $Z(t, x)$ be the slope of the leaf of F at (t, x). Then

$$\begin{aligned}
Z(t, x) &= \frac{\partial f_t}{\partial t}({f_t}^{-1}(x)) = {f_t}^{-1}(x) - f({f_t}^{-1}(x)) \quad \text{and} \\
\frac{\partial Z}{\partial x}(t, x) &= \frac{\partial^2 f_t}{\partial t \partial x}({f_t}^{-1}(x)) \frac{\partial {f_t}^{-1}}{\partial x}(x) \\
&= \frac{\partial^2 f_t}{\partial t \partial x}({f_t}^{-1}(x)) \Big/ \frac{\partial f_t}{\partial x}({f_t}^{-1}(x)) \\
&= (1 - \frac{df}{dx})({f_t}^{-1}(x)) / \{t + (1-t)\frac{df}{dx}({f_t}^{-1}(x))\}.
\end{aligned}$$

$Z(t, x)$ is C^∞ with respect to t and C^1 with respect to x and $Z(t, 0) = 0$. Since f is tangent to the identity at 0, $(\partial Z/\partial x)(t, 0) = 0$.

Since $Z(t, x)$ is C^1 with respect to x, we fix the modulus of continuity $\varepsilon(\delta)$ for $(\partial Z/\partial x)(t, x)$ which satisfies $\varepsilon(\delta_1 + \delta_2) \leq \varepsilon(\delta_1) + \varepsilon(\delta_2)$ and $\varepsilon(\delta) \to 0$ as $\delta \to 0$. In particular, $\big|(\partial Z/\partial x)(t, x)\big| \leq \varepsilon(x)$.

We look at the pull back H^*F of F by H.

Lemma 4.2 *If Z is small and flat along $x = 0$, then the pull back foliation H^*F is a C^1 foliation on $[i-1, i] \times [0, 1]$ ($i = 1, 2$).*

Proof. Put $H(t,x) = (u,x)$. Then the Jacobian matrix is

$$\begin{pmatrix} \frac{\partial u}{\partial t} & \frac{\partial u}{\partial x} \\ \frac{\partial x}{\partial t} & \frac{\partial x}{\partial x} \end{pmatrix} = \begin{pmatrix} \nu_{[t]+1} & (t-[t])\frac{\partial \nu_{[t]+1}}{\partial x} + [t]\frac{\partial \nu_1}{\partial x} \\ 0 & 1 \end{pmatrix}.$$

Let $X(t,x)$ denote the slope of H^*F. Then

$$\begin{pmatrix} \frac{\partial u}{\partial t} & \frac{\partial u}{\partial x} \\ \frac{\partial x}{\partial t} & \frac{\partial x}{\partial x} \end{pmatrix} \begin{pmatrix} 1 \\ X \end{pmatrix} = \begin{pmatrix} \nu_{[t]+1} + ((t-[t])\frac{\partial \nu_{[t]+1}}{\partial x} + [t]\frac{\partial \nu_1}{\partial x})X \\ X \end{pmatrix}.$$

Hence

$$Z\{\nu_{[t]+1} + ((t-[t])\frac{\partial \nu_{[t]+1}}{\partial x} + [t]\frac{\partial \nu_1}{\partial x})X\} = X.$$

Note that $\frac{\partial \nu_1}{\partial x} = -\frac{\partial \nu_2}{\partial x} = \frac{\partial \nu}{\partial x}$. If $Z\frac{\partial \nu}{\partial x}$ is small, X is given by

$$X = \frac{Z\nu_{[t]+1}}{1 - Z\left((t-[t])\frac{\partial \nu_{[t]+1}}{\partial x} + [t]\frac{\partial \nu_1}{\partial x}\right)}.$$

Thus if Z is small, X is C^1 out of a neighborhood of $x=0$. We are interested in the behavior of X along $x=0$. Since f is tangent to the identity at 0, Z near $x=0$ is estimated by $\varepsilon(x)x$ and $\partial\nu/\partial x$ is estimated by $2^3/x$. Thus $Z(\partial\nu/\partial x)$ tends to 0 as x tends to 0. Moreover since $Z(\partial\nu/\partial x)$ is estimated by $\varepsilon(x)$ and $\partial^k Z/\partial t^k$ is estimated by $\varepsilon(x)x$, X is C^∞ with respect to t. The derivative with respect to x is estimated as follows.

$$\frac{\partial X}{\partial x} = \frac{\frac{\partial Z}{\partial x}\nu_{[t]+1} + Z\frac{\partial \nu_{[t]+1}}{\partial x}}{1 - Z\left((t-[t])\frac{\partial \nu_{[t]+1}}{\partial x} + [t]\frac{\partial \nu_1}{\partial x}\right)}$$
$$-Z\nu_{[t]+1}\frac{-\frac{\partial Z}{\partial x}\left((t-[t])\frac{\partial \nu_{[t]+1}}{\partial x} + [t]\frac{\partial \nu_1}{\partial x}\right) - Z\left((t-[t])\frac{\partial^2 \nu_{[t]+1}}{\partial^2 x} + [t]\frac{\partial^2 \nu_1}{\partial^2 x}\right)}{\left(1 - Z\left((t-[t])\frac{\partial \nu_{[t]+1}}{\partial x} + [t]\frac{\partial \nu_1}{\partial x}\right)\right)^2}.$$

If x is small, then the denominators are near to 1. The numerator of the first term is estimated by $\varepsilon(x) + \varepsilon(x)x(2^3/x)$ and that of the second term is estimated by $\varepsilon(x)x\big(\varepsilon(x)(2^3/x) + \varepsilon(x)x(c_2/x^2)\big)$. Hence X is C^1 with respect to x. □

Proof of Theorem 4.1. By Lemma 4.2 H^*F is a C^1 foliation of $[i-1,i]\times[0,1]$ $(i = 1,\ 2)$. Let f_i $(i = 1,\ 2)$ denote the holonomy of H^*F from $\{i\}\times[0,1]$ to $\{i-1\}\times[0,1]$. Then we see that $f = f_1 f_2$, and f_1 and f_2 are C^1 diffeomorphisms with support contained in the closures of the unions of disjoint intervals $\bigcup_{k=0}^{\infty}[2^{-2k-1}, 2^{-2k-1}3]$ and $\bigcup_{k=0}^{\infty}[2^{-2k-2}, 2^{-2k-2}3]$, respectively.

Since f_1 is C^1 and of course tangent to the identity at the end point 0, the C^1 norm of the piece $f_1'|[2^{-2k-1}, 2^{-2k-1}3]$ tends to zero as k tends to the infinity.

In Section 6, we show that each piece $f_1|[2^{-2k-1}, 2^{-2k-1}3]$ of f_1 is written as a product $[g_{1,1,k}, g_{2,1,k}][g_{3,1,k}, g_{4,1,k}][g_{5,1,k}, g_{6,1,k}]$ of 3 commutators with support in $[2^{-2k-4}7, 2^{-2k-2}7]$ and the C^1 norms of the elements in the commutators are estimated by $(\sup \log f_1'|[2^{-2k-1}, 2^{-2k-1}3])^{1/5}$. In fact, by taking the affine map A_k which sends $[2^{-2k-1}, 2^{-2k-1}3]$ to $[2^{-2}, 2^{-2}3]$. Theorem 6.1 is applied to $A_k(f_1|[2^{-2k-1}, 2^{-2k-1}3])A_k{}^{-1}$. The C^1 norm of it is the same as that of $f_1|[2^{-2k-1}, 2^{-2k-1}3]$ and it is written as a product of 3 commutators with support in $[2^{-5}7, 2^{-3}7]$ and the C^1 norms of the elements in the commutators are estimated by $(\sup \log f_1'|[2^{-2k-1}, 2^{-2k-1}3])^{1/5}$. Then the elements conjugated by $A_k{}^{-1}$ has support in $[2^{-2k-4}7, 2^{-2k-2}7]$ and their C^1 norms are estimated by $(\sup \log f_1'|[2^{-2k-1}, 2^{-2k-1}3])^{1/5}$.

Since the C^1 norm of the piece $f_1'|[2^{-2k-1}, 2^{-2k-1}3]$ tends to zero as k tends to the infinity, the composition $G_{i,1} = \prod_k g_{i,1,k}$ $(i = 1, \ldots, 6)$ is a C^1 diffeomorphism of the interval and

$$f_1 = [G_{1,1}, G_{2,1}][G_{3,1}, G_{4,1}][G_{5,1}, G_{6,1}].$$

In the same way, f_2 is written as $[G_{1,2}, G_{2,2}][G_{3,2}, G_{4,2}][G_{5,2}, G_{6,2}]$. This completes the proof of Theorem 4.1

Note that an element of $\mathrm{Diff}_1^1([0,1])$ is written as a composition of a diffeomorphism with support in $(0,1)$ and 2 diffeomorphisms with small C^1 norms which are identity on neighborhoods of 0 or 1. Since $\mathrm{Diff}_c^1((0,1))$ is uniformly perfect ([17], [28]), we showed $\mathrm{Diff}_1^1([0,1])$ is uniformly perfect.

□

5 Perfectness of the group of the C^∞ diffeomorphisms of the closed interval which are infinitely tangent to the identity at the endpoints

Let $\mathrm{Diff}^\infty([0,1])$ denote the group of C^∞ diffeomorphisms of the interval. In the group $\mathrm{Diff}^\infty([0,1])$, neighborhoods of the identity are described by the semi norms $\sup|(f - \mathrm{id})^{(n)}|$ $(n = 0, 1, 2, \ldots)$.

There is a natural homomorphism to $J^\infty \times J^\infty$, where J^∞ is the group of infinite jets at the endpoint 0 or 1 of the diffeomorphisms fixing 0 or 1. Let $\mathrm{Diff}_\infty^\infty([0,1])$ denote the kernel of this homomorphism.

In this section, we show that $\mathrm{Diff}_\infty^\infty([0,1])$ is a perfect group. This was first shown by Sergeraert [24] and reproved by Natsume-Kawabe [21]. The present proof in the same strategy as in the previous sections was suggested by Francis Sergeraert in a discussion we had in 80's.

Theorem 5.1 *Let* $\mathrm{Diff}^\infty_\infty([0,1])$ *denote the group of the* C^∞ *diffeomorphisms of the interval which are infinitely tangent to the identity at the end points. The group* $\mathrm{Diff}^\infty_\infty([0,1])$ *is a perfect group.*

We assume that $f \in \mathrm{Diff}^\infty_\infty([0,1])$ is the identity near the end point 1 and we look at the same foliation on F of $[0,1]\times[0,1]$ as in Section 4 whose leaves are line segments

$$\{(t, f(x) + t(x - f(x)))\ ;\ t \in [0,1]\}$$

joining $(0, f(x))$ and $(1, x)$. Then $Z(t,x)$ is C^∞ in t and x, and $Z(t,x)$ is infinitely flat at $x = 0$.

We use the same partition of unity $\nu_1 = \nu$, $\nu_2 = 1 - \nu$ on $(0,1]$ as in Section 4, and define the map $H : [0,2]\times(0,1] \longrightarrow [0,1]\times(0,1]$.

We look at the pull back H^*F of F by H. We show that this is a C^∞ foliation on each $[i-1,i]\times[0,1]$.

Lemma 5.2 *If* Z *is small and infinitely flat along* $x = 0$, *then the pull back foliation* H^*F *is a* C^∞ *foliation on* $[i-1,i]\times[0,1]$ $(i = 1,2)$.

Then we apply the following theorem on the small commutators for $\mathrm{Diff}^\infty_c((0,1))$.

Theorem 5.3 ([26], Theorem(8.1)) *Let* f *be a* C^∞ *diffeomorphism of* $[0,1]$ *with support in* $(1/8, 7/8)$. *For any positive integer* m, *there exist a positive integer* n *and a positive real number* c *such that if* $\sup|(f - \mathrm{id})^{(n)}| < c$ *then* f *is written as a composition of 4 commutators of elements of* $\mathrm{Diff}^\infty_c((0,1))$; $f = [h_1,h_2][h_3,h_4][h_5,h_6][h_7,h_8]$, *where for a constant depending only on* m *and* n,

$$\sup|(h_i - \mathrm{id})^{(m)}| \le K_{m.n}\big(\sup|(f-\mathrm{id})^{(n)}|\big)^{1/(m+2)}.$$

Proof of Lemma 5.2. We use the same notation as in the proof of Lemma 4.2. Put

$$Y_0 = Z\nu_{[t]+1} \text{ and } Y_1 = Z\big((t-[t])\frac{\partial \nu_{[t]+1}}{\partial x} + [t]\frac{\partial \nu_1}{\partial x}\big).$$

Then we have

$$Y_0 + Y_1 X = X.$$

Since Z is flat along $x = 0$, by using $\left|\partial^m Z/\partial x^m\right| \le K_m$ $(m \ge 2)$, we have the estimate $\left|\partial^k Z/\partial x^k\right| \le (K_m/(m-k)!)x^{m-k}$. By using $\left|\partial^k \nu_{[t]+1}/\partial x^k\right| \le$

c_k/x^k, for $0 \le \ell < m$, Y_0 and Y_1, we have the following estimates.

$$\begin{aligned}\left|\frac{\partial^\ell Y_0}{\partial x^\ell}\right| &\le \sum_{k=0}^{\ell}\binom{\ell}{k}\left|\frac{\partial^{\ell-k}Z}{\partial x^{\ell-k}}\frac{\partial^k \nu_{[t]+1}}{\partial x^k}\right| \\ &\le \sum_{k=0}^{\ell}\binom{\ell}{k}\frac{K_m}{(m-k)!}x^{m-k}\frac{c_k}{x^k} \le 2^\ell c_\ell K_m x^m, \\ \left|\frac{\partial^\ell Y_1}{\partial x^\ell}\right| &\le \sum_{k=0}^{\ell}\binom{\ell}{k}\left|\frac{\partial^{\ell-k}Z}{\partial x^{\ell-k}}\frac{\partial^k}{\partial x^k}\left((t-[t])\frac{\partial \nu_{[t]+1}}{\partial x}+[t]\frac{\partial \nu_1}{\partial x}\right)\right| \\ &\le \sum_{k=0}^{\ell}\binom{\ell}{k}\frac{K_m}{(m-k)!}x^{m-k}2\frac{c_{k+1}}{x^{k+1}} \le 2^{\ell+1}c_{\ell+1}K_m x^{m-1}.\end{aligned}$$

In particular, since $Y_0 + Y_1 X = X$ and $|Y_1| \le 2c_1 K_m x^{m-1} \le 2^{-1}$ for $2^5 K_m x^{m-1} \le 1$, we have $|X| \le 2K_m x^m \le c_1 K_m x^m$, where $c_1 = 2^3$. This shows that if $2^5 K_m x^{m-1} \le 1$, X is C^∞ on $[i-1, i] \times (0, 1]$ $(i = 1, 2)$.

We look at the behavior of X along $x = 0$. We show that

$$\left|\frac{\partial^k X}{\partial x^k}\right| \le 2^{k^2}\left(\prod_{j=1}^{k+1} c_j\right) K_m x^m$$

for $2^5 K_m x^{m-1} \le 1$ by an induction of k. Note that by Leibniz' formula, we have

$$\frac{\partial^\ell Y_0}{\partial x^\ell} + \sum_{k=0}^{\ell-1}\binom{\ell}{k}\frac{\partial^{\ell-k}Y_1}{\partial x^{\ell-k}}\frac{\partial^k X}{\partial x^k} + Y_1\frac{\partial^\ell X}{\partial x^\ell} = \frac{\partial^\ell X}{\partial x^\ell}.$$

If the inequality is true for $k < \ell$, then

$$\begin{aligned}\left|\frac{\partial^\ell X}{\partial x^\ell}\right| &\le 2\left\{\left|\frac{\partial^\ell Y_0}{\partial x^\ell}\right| + \sum_{k=0}^{\ell-1}\binom{\ell}{k}\left|\frac{\partial^{\ell-k}Y_1}{\partial x^{\ell-k}}\frac{\partial^k X}{\partial x^k}\right|\right\} \\ &\le 2^{\ell+1}K_m c_\ell x^m + 2\sum_{k=0}^{\ell-1}\binom{\ell}{k}2^{\ell-k+1}x^{m-1}K_m c_{\ell-k+1}2^{k^2}\left(\prod_{j=1}^{k+1}c_j\right)K_m x^m \\ &\le 2^{\ell+1}K_m c_\ell x^m + \sum_{k=0}^{\ell-1}\binom{\ell}{k}2^{k^2+\ell-k-3}c_{\ell-k+1}\left(\prod_{j=1}^{k+1}c_j\right)K_m x^m\end{aligned}$$

$$\leq 2^{\ell+1}K_m c_\ell x^m + \sum_{k=0}^{\ell-1}\binom{\ell}{k}2^{(\ell-1)^2-2}\left(\prod_{j=1}^{\ell+1}c_j\right)K_m x^m$$
$$\leq 2^\ell 2^{(\ell-1)^2-2}\left(\prod_{j=1}^{\ell+1}c_j\right)K_m x^m \leq 2^{\ell^2}\left(\prod_{j=1}^{\ell+1}c_j\right)K_m x^m$$

This shows the inequality.

By this inequality, for any ℓ, $\left|\frac{\partial^\ell X}{\partial x^\ell}\right|$ tends to 0 as x tends to 0. Hence X is C^∞ along $x = 0$. □

Now we use Theorem 5.3 to prove Theorem 5.1.

Proof of Theorem 5.1. The proof goes as in that of Theorem 4.1. By Lemma 5.2, H^*F is a C^∞ foliation of $[i-1, i] \times [0,1]$ $(i = 1, 2)$. Let f_i $(i = 1, 2)$ denote the holonomy of H^*F from $\{i\} \times [0,1]$ to $\{i-1\} \times [0,1]$. Then we see that $f = f_1 f_2$, and f_1 and f_2 are C^∞ diffeomorphisms with support contained in the closures of the unions of disjoint intervals $\bigcup_{k=0}^\infty [2^{-2k-1}, 2^{-2k-1}3]$ and $\bigcup_{k=0}^\infty [2^{-2k-2}, 2^{-2k-2}3]$, respectively.

Since f_1 is of course infinitely tangent to the identity at the endpoint 0, for $n \leq p$ we have

$$\sup|(f_1 - \mathrm{id})^{(n)}| \leq \frac{K'_p}{(p-n)!}x^{p-n}.$$

To apply Theorem 5.3 to each piece $f_1|[2^{-2k-1}, 2^{-2k-1}3]$, we conjugate it by the affine map A_k which sends $[2^{-2k-1}, 2^{-2k-1}3]$ to $[2^{-2}, 2^{-2}3]$. Then

$$\begin{aligned}
&\sup|(A_k(f_1|[2^{-2k-1}, 2^{-2k-1}3])A_k{}^{-1} - \mathrm{id})^{(n)}| \\
\leq\;& (2^{2k-1})^{1-n}\sup|(f_1 - \mathrm{id})^{(n)}| \\
\leq\;& (2^{2k-1})^{1-n}\frac{K'_p}{(p-n)!}(2^{-2k-1}3)^{p-n} \\
\leq\;& \frac{K'_p}{(p-n)!}2^{(2k-1)(1-p)+2n-2p}3^{p-n}.
\end{aligned}$$

If k is big, then we use Theorem 5.3 to obtain diffeomorphisms $h_{j,1,k}$ $(j = 1, \ldots, 8)$ with support in $[2^{-5}7, 2^{-3}7]$ such that

$$A_k(f_1|[2^{-2k-1}, 2^{-2k-1}3])A_k{}^{-1} = \prod_{j=1}^{4}[h_{2j-1}, h_{2j}]$$

and

$$\sup|(h_{j,1,k}-\mathrm{id})^{(m)}| \le K_{m,n}\left(\frac{K'_p}{(p-n)!}2^{(2k-1)(1-p)+2n-2p}3^{p-n}\right)^{1/(m+2)}$$
$$\le K'_{m,n,p}2^{(2k-1)(1-p)/(m+2)},$$

where $K'_{m,n,p}$ is a constant depending on m, n and p. Then the conjugated piece $A_k{}^{-1}h_{j,1,k}A_k$ has support in $[2^{-2k-4}7, 2^{-2k-2}7]$ and satisfies the following estimate.

$$\sup|(A_k{}^{-1}h_{j,1,k}A_k-\mathrm{id})^{(m)}| \le (2^{2k-1})^{m-1}K_{m,n,p}2^{(2k-1)(1-p)/(m+2)}$$
$$\le K_{m,n,p}2^{(2k-1)((m-1)-(p-1)/(m+2))}.$$

Hence if $p-1>(m-1)(m+2)$, then this tends to 0 as k tends to ∞. Put

$$G_{j,1}=\prod_k A_k{}^{-1}h_{j,1,k}A_k \quad (j=1,\ldots,8).$$

Then this is a C^∞ diffeomorphism infinitely tangent to the identity at the endpoint 0 and

$$f_1=\prod_{j=1}^{4}[G_{2j-1,1},G_{2j,1}].$$

In the same way, f_2 is written as $\prod_{j=1}^4[G_{2j-1,2},G_{2j,2}]$. Thus $f=f_1f_2$ is written as a product of 8 commutators in $\mathrm{Diff}^\infty_\infty([0,1])$. □

*Remark.*Note that an element of $\mathrm{Diff}^\infty_\infty([0,1])$ is written as a composition of a diffeomorphism with support in $(0,1)$ and 2 diffeomorphisms with small semi norms which are identity on neighborhoods of 0 or 1. The fact that $\mathrm{Diff}^\infty_c((0,1))$ is uniformly perfect is shown by using the theorem of Herman [13] on the uniform perfectness of $\mathrm{Diff}^\infty(S^1)$ and the proof of $H_1(\mathrm{Diff}^\infty_c((0,1)))\cong H_1(\mathrm{Diff}^\infty(S^1))$ (see for example [5], [7]). Thus we showed $\mathrm{Diff}^1_1([0,1])$ is uniformly perfect.

6 Appendix. Small commutators of C^1 diffeomorphisms

In this section, as an appendix to Section 4, we show the following theorem which is used in Section 4 to show the perfectness of $\mathrm{Diff}^1_1([0,1])$.

Theorem 6.1 *Let f be a C^1 diffeomorphism of $[0,1]$ with support in $[a,b]$, where $0<a<b<1$. f is written as product of 2 commutators $f=[g_1,g_2][g_3,g_4][g_5,g_6]$ of C^1 diffeomorphisms with support in $[0,1]$ such that $\sup|\log g_i'|$ is estimated by $(\sup|\log f'|)^{1/5}$.*

We use Mather's trick [17] to obtain a diffeomorphism with small support in the same first homology class. Then this diffeomorphism with small support can be written as a product of 2 commutators by a result of [28].

For the group $\mathrm{Diff}_c^1(\mathbf{R})$ of C^1 diffeomorphisms of the real line with compact support, we know that it is perfect [17] and is in fact acyclic [28]. We need the construction in [28] of the diffeomorphisms to write an element of $\mathrm{Diff}_c^1(\mathbf{R})$ to be a product of commutators. This construction uses an action of $\mathbf{Z}^2$ given by Pixton [22].

Proposition 6.2 ([28], Proposition(1.1) for $n = 1$, $N = 2$) *There are a homomorphism $\overline{\Phi}_2 : \mathbf{Z}^2 \longrightarrow \mathrm{Diff}_c^1(\mathbf{R})$ and open intervals U_1, U_2 in $\mathbf{R}$ with the following properties.*

(i) $U_1 \supset U_2$.

(ii) $\overline{\Phi}_2(\lambda)(U_k)$ *for* $\lambda \in \mathbf{Z}^k \times \{0\}^{2-k}$ *are disjoint* ($k = 1, 2$).

(iii) *For* $\lambda \in \{0\}^k \times \mathbf{Z}^{2-k}$, *the support of* $\overline{\Phi}_2(\lambda)$ *is contained in the closure of* $\bigcup \overline{\Phi}_2(\lambda')(U_k)$, *where the union is taken over* $\lambda' \in \mathbf{Z}^k \times \{0\}^{2-k}$.

(iv) *For* $\lambda \in \mathbf{Z}^2$, *the restriction* $\overline{\Phi}_2(\lambda)|U_2 : U_2 \longrightarrow \overline{\Phi}_2(\lambda)(U_2)$ *is of class* C^∞.

(v) *There exists a positive real number C_2 such that, for any $\lambda \in \mathbf{Z}^2$ and any vector field ξ of class C^1 with support in U_2, we have $|\overline{\Phi}_2(\lambda)_*\xi|_1 \leq C_2|\xi|_1$, where $|\ |_1$ denotes the C^1-norm.*

If the support of f is contained in U_2, then by a result of [28], f is written as a product of 2 commutators. At the end of Section 3 of [28], for the isotopy Q from f to the identity, we constructed isotopies $\mathbf{I}(\Lambda, \{1\})Q$, $s_\emptyset\mathbf{I}(\Lambda, \{1\})Q$, $s_{\{1\}}\mathbf{I}(\Lambda, \{1\})Q$, $\mathbf{I}'(\Lambda, \{1\})Q$, $s'_\emptyset\mathbf{I}'(\Lambda, \{1\})Q$, $s'_{\{1\}}\mathbf{I}'(\Lambda, \{1\})Q$. If they are the isotopies from F, $F_\emptyset$, $F_{\{1\}}$, F', $F'_\emptyset$, $F'_{\{1\}}$ to the identity, respectively, we showed that

$$F = F_\emptyset F_{\{1\}}, \qquad F' = F'_\emptyset F'_{\{1\}}, \quad F = fF',$$
$$g_\emptyset F_\emptyset {g_\emptyset}^{-1} = F'_{\{1\}}, \text{ and } g_{\{1\}} F_{\{1\}} {g_{\{1\}}}^{-1} = F'_\emptyset,$$

where $g_\emptyset = \overline{\Phi}_2(\gamma_\emptyset) = \overline{\Phi}_2(1,0)$ and $g_{\{1\}} = \overline{\Phi}_2(\gamma_{\{1\}}) = \overline{\Phi}_2(1,1)$. Then we have

$$\begin{aligned} f = FF'^{-1} &= F_\emptyset F_{\{1\}} {F'_{\{1\}}}^{-1} {F'_\emptyset}^{-1} = F_\emptyset F_{\{1\}} g_\emptyset {F_\emptyset}^{-1} {g_\emptyset}^{-1} g_{\{1\}} {F_{\{1\}}}^{-1} {g_{\{1\}}}^{-1} \\ &= [F_\emptyset, F_{\{1\}} g_\emptyset][F_{\{1\}}, g_{\{1\}}]. \end{aligned}$$

The C^1 norms of F, $F_\emptyset$, $F_{\{1\}}$, F', $F'_\emptyset$, $F'_{\{1\}}$ are estimated by that of f. The C^1 norms of $g_\emptyset$ and $g_{\{1\}}$ are estimated by the C^1 norm of $\overline{\Phi}_1$ and the constants a_λ ($\lambda = (i_1, i_2) \in \mathbf{Z}^2$) in the construction of $\overline{\Phi}_2$. For a big real number M if we assume C^1 norm of $\overline{\Phi}_1$ is estimated by M^{-1} and put the constant $a_\lambda = (|i_1| + |i_2| + M)^{-1}$, then the C^1 norms of $g_\emptyset$ and $g_{\{1\}}$ are estimated by M^{-1}. Then the length of the interval U_2 is as big as M^{-2}. Thus if f is a C^1 diffeomorphism with support in U_2 such that $\sup|\log f'| \leq M^{-1}$, f is written as a product of 2 commutators of C^1 diffeomorphisms with norms less than M^{-1}.

Thus we showed the following lemma.

Lemma 6.3 *For a large positive real number M, a C^1 diffeomorphism f of $[0,1]$ with support in an interval of length M^{-2} near the point $1/2$is written as a product of 2 commutators in* $\mathrm{Diff}_c((0,1))$*; $f = [g_1, g_2][g_3, g_4]$, where the C^1 norms of g_j ($j = 1, 2, 3, 4$) are estimated by the C^1 norm of f and M^{-1}.*

We use Mather's trick [17] to obtain a diffeomorphism with small support in the same first homology class. We follow Section 8 of ([26]), where we proved a theorem on the small commutators for $\mathrm{Diff}_c((0,1))$ which is used in Section 5.

Lemma 6.4 *Let f be a C^1 diffeomorphism with support in $[1/8, 7/8]$ near to the identity. For a small real number β which is sufficiently large with respect to $\alpha = \sup|\log f'|$, there exist C^1 diffeomorphisms g, g_5, g_6 of $[0,1]$ such that $f = g[g_5, g_6]$, the support of g is contained in an interval of length 3β near the point $1/2$, and the C^1 norms of g, g_5, g_6 are estimated by α/β^2.*

Proof. We assume $\alpha = \sup|\log f'|$ is small and we have $\sup|f(x) - x| \leq \alpha$.

Let μ be a vector field on $[0,1]$, such that the support of μ is contained in $(0,1)$ and $\mu(x) = \frac{\partial}{\partial x}$ for $x \in [1/16, 15/16]$. Let A be the time 1 map of $\beta\mu$. If β is small, but sufficiently large with respect to α, Af is a small perturbation of A which coincides with the translation T_β by β on $[1/16, 7/8]$. Put $a = (7/8) + 2\beta$. If α is small, Af has no fixed points in $[1/8, 7/8]$. Therefore, for any point $x \in [1/16, (1/16) + \beta)$, there exist a unique integer N and a point $y \in [a, a + \beta)$ such that $y = (Af)^N(x)$. Let $\sigma_a(f)$ denote this map; $y = \sigma_a(f)(x)$.

Since for $x \in [1/16, (1/16) + \beta)$,

$$\sigma_a(f)(x) - \sigma_a(\mathrm{id})(x) = (Tf)^N(x) - T^N(x),$$

$$\begin{aligned}\sup\{|\sigma_a(f)(x) - \sigma_a(\mathrm{id})(x)|, x \in [1/16, (1/16) + \beta)\} &\leq N \sup|f(x) - x| \\ &\leq N\alpha \leq (7/8)(\alpha/\beta).\end{aligned}$$

Since

$$\log(\sigma_a(f))'(x) = \log((Tf)^N)'(x) = \sum_{i=0}^{N-1} \log f'((Tf)^i(x)),$$

$$\sup\{|\log(\sigma_a(f))'(x)|, x \in [1/16,(1/16)+\beta)\} \leq N\alpha \leq (7/8)(\alpha/\beta).$$

By using $B : [1/16,(1/16)+\beta) \longrightarrow S^1 = \mathbf{R}/\mathbf{Z}$ defined by $B(x) = (x-(1/16))/\beta$ and $B_a : [a, a+\beta) \longrightarrow S^1$ defined by $B_a(x) = (x-a)/\beta$, we put $\Gamma_a(f) = B_a\sigma_a(f)B^{-1}$. Then $\Gamma_a(f)$ is a well defined C^1 diffeomorphism of S^1 with norm estimated by α/β^2.

Let $\overline{\nu}_1$, $\overline{\nu}_2$ be a partition of unity on S^1 such that $\overline{\nu}_1(x) = 0$ for $x \in [-1/8,1/8]$ and $\overline{\nu}_1(x) = 1$ for $x \in [3/8,5/8]$. We consider the fragmentation of $\Gamma_a(f)$ with respect to $\overline{\nu}_1$ and $\overline{\nu}_2$. That is, as in Section 4, we consider the foliation F of $[0,1] \times S^1$ whose leaves are line segments

$$\{(t, \Gamma_a(f)(x) + t(x - \Gamma_a(f)(x))) \ ; \ t \in [0,1]\}$$

joining $(0,\Gamma_a(f)(x))$ and $(1,x)$. We define the map $H : [0,2] \times S^1 \longrightarrow [0,1] \times S^1$ as in Section 4 using $\overline{\nu}_1$ and $\overline{\nu}_2$, and the holonomy of the pull back foliation H^*F gives the decomposition $\Gamma_a(f) = k_1k_2$ such that k_1 is the identity on $[-1/8,1/8]$ and k_2 is the identity on $[3/8,5/8]$. The norms of k_1 and k_2 are estimated by α/β^2.

Let $\widetilde{k}_1$ and $\widetilde{k}_2$ be the diffeomorphisms of $[0,1]$ defined by

$$\widetilde{k}_1(x) = \begin{cases} A^{-1-[N/2]}B_ak_1{B_a}^{-1}A^{1+[N/2]}(x) & \text{for } x \in [a-\beta, a) \\ x & \text{for } x \in [0,1]-[a-\beta,a), \end{cases}$$

$$\widetilde{k}_2(x) = \begin{cases} A^{-5/2-[N/2]}B_aR_{1/2}k_2R_{1/2}{B_a}^{-1}A^{5/2+[N/2]}(x) & \\ & \text{for } x \in [a-(5/2)\beta, a-(3/2)\beta) \\ x & \text{for } x \in [0,1]-[a-(5/2)\beta, a-(3/2)\beta), \end{cases}$$

where $[N/2]$ is the largest integer not greater than $N/2$ and $R_{1/2}$ is the half rotation of S^1. Then C^1 norms of $\widetilde{k}_1$ and $\widetilde{k}_2$ are estimated by α/β^2. Put $g = \widetilde{k}_1\widetilde{k}_2$.

Since $\Gamma_a(f) = \Gamma_a(g)$, according to Mather [17], Ag and Af are conjugate by a diffeomorphism H defined by $H(x) = (Ag)^N(Af)^{-N}(x)$ for $x \in [1/16,15/16]$; $HAf = AgH$. For the estimate of the C^1 norm of H, we have the following.

$$\begin{aligned} \sup|(Ag)^N(Af)^{-N}(x) - x| &\leq N\sup|f(x)-x| + 3\sup|g(x)-x| \\ &\leq N\alpha + 3(\alpha/\beta^2) \\ &\leq (7/8)(\alpha/\beta) + 3(\alpha/\beta^2) \leq 4\alpha/\beta^2. \end{aligned}$$

Since

$$\log((Ag)^N(Af)^{-N})'(x) = \sum_{i=0}^{N-1} \log f'((Tf)^{-i}(x)) + \sum_{i=0}^{2} \log g'((Ag)^i(A)^{[N/2]}((Af)^{-N}(x)),$$

$$\sup|\log((Ag)^N(Af)^{-N})'(x)| \leq N\alpha + 3(\alpha/\beta^2) \leq (7/8)(\alpha/\beta) + 3(\alpha/\beta^2) \leq 4\alpha/\beta^2.$$

Thus

$$f = A^{-1}H^{-1}AgH = g[(Ag)^{-1}, H^{-1}],$$

and the C^1 norms of g, $(Ag)^{-1}$ and H^{-1} are estimated by α/β^2 and the support of g is contained in a interval of length 3β near the point $1/2$. □

Proof of Theorem 6.1. Put $\alpha = \log|f'(x)|$. Put $\beta = \alpha^{2/5}$. Then by Lemma 6.4 we can write $f = g[g_5, g_6]$, where C^1 norms of g, g_5 and g_6 are estimated by $\alpha/\beta^2 = \alpha^{1/5}$ and the support of g is contained in an interval of length $3\beta = 3\alpha^{2/5}$.

We can choose the number M in Lemma 6.3 as large as $\alpha^{-1/5}$ and we can define $\overline{\Phi}_2$ such that the C^1 norms of generators are estimated by $\alpha^{1/5}$ and U_2 contains the support of g. Then $g = [g_1, g_2][g_3, g_4]$, where C^1 norms of g_1, g_2, g_3 and g_4 are estimated by $M^{-1} = \alpha^{1/5}$. □

Acknowledgments

The author would like to thank Kojun Abe and Kazuhiko Fukui for their interest taken for this work. He also thanks Yakov Eliashberg for his warm hospitality during his stay at Stanford University in 1999.

References

1. K. Abe and K. Fukui, *On commutators of equivariant diffeomorphisms,* Proc. Japan Acad., **54** (1978), 52–54.
2. K. Abe and K. Fukui, *On the structure of the group of Lipschitz homeomorphisms and its subgroup,* preprint.
3. R. Anderson *The algebraic simplicity of certain groups of homeomorphisms,* Amer. J. Math., **80** (1958), 955–963.
4. A. Banyaga, *On the structure of the group of equivariant diffeomorphisms,* Topology, **16** (1977), 279–283.

5. A. Banyaga, *Sur la structure du groupe des difféomorphismes*, preprint, Genève, 1977.
6. A. Banyaga, *Sur la structure du groupe des difféomorphismes qui préservent une forme symplectique,* Comm. Math. Helv., **53** (1978), 174–227.
7. A. Banyaga, The structure of classical diffeomorphism groups, Mathematics and its Applications, 400. Kluwer Academic Publishers Group, Dordrecht, 1997.
8. D.B.A. Epstein, *The simplicity of certain groups of homeomorphisms*, Compositio Math., **22** (1970) 165–173.
9. K. Fukui, *Homology of the group* $\mathrm{Diff}^\infty(\mathbf{R}^n, 0)$ *and its subgroups*, J. Math. Kyoto Univ., **20** (1980), 475–487.
10. K. Fukui, *Commutators of foliation preserving homeomorphisms for certain compact foliations*, Publ. RIMS, Kyoto Univ., **34** (1998), 65–73.
11. K. Fukui and H. Imanishi, *On commutators of foliation preserving homeomorphisms*, J. Math. Soc. Japan, **51** (1999), 227–236.
12. M. Herman, *Simplicité du groupe des difféomorphismes de classe* C^∞*, isotopes à l'identité, du tore de dimension* n*,* C. R. Acad. Sci. Paris, **273** (1971), 232–234.
13. M. Herman, *Sur la conjugaison différentiable des difféomorphismes du cercle à des rotations,* Inst. Hautes Études Sci. Publ. Math., **49** (1979), 5–233.
14. A. Masson, *Sur la perfection du groupe de difféomorphismes d'une variété à bord, infiniment tangents à l'identité sur le bord,* C. R. Acad. Sci. Paris, **285** (1977), 837–839.
15. J. Mather, *The vanishing of the homology of certain groups of homeomorphisms,* Topology, **10** (1971), 297–298.
16. J. Mather, *Integrability in codimension* 1, Comment. Math. Helv., **48** (1973), 195–233.
17. J. Mather, *Commutators of diffeomorphisms I, II and III*, Comment. Math. Helvetici, **49** (1974), 512–528, **50** (1975), 33–40 and **60** (1985), 122–124.
18. J. Mather, *On the homology of Haefliger's classifying space*, C.I.M.E., Differential Topology, (1976), 71–116.
19. D. McDuff, *The homology of some groups of diffeomorphisms*, Comm. Math. Helv., **55** (1980), 97–129.
20. D. McDuff, *Local homology of groups of volume preserving diffeomorphisms I, II and III,* Ann. Sci. Ec. Norm. Sup. 4^e ser., **15** (1982), 609–648, Comm. Math. Helv., **58** (1983), 135–165 and Ann. Sci. Ec.

Norm. Sup. 4^e ser., **16** (1983), 529–540.

21. H. Natsume-Kawabe, *On the conjugation of local diffeomorphisms infinitely tangent to the identity*, Advanced Studies in Pure Math. **5**, Foliations, (1985), 461–481.
22. D. Pixton, *Nonsmoothable, unstable group actions*, Trans. Amer. Math. Soc., **229** (1977), 259–268.
23. G. Segal, *Classifying spaces related to foliations*, Topology, **17** (1978), 367-382.
24. F. Sergeraert, *Feuilletages et difféomorphismes infiniment tangents à l'identité*, Invent. Math., **39** (1977), 253–275.
25. W. Thurston, *Foliations and groups of diffeomorphism*, Bull. Amer. Math. Soc., **80** (1974), 304–307.
26. T. Tsuboi, *On 2-cycles of* $B\mathrm{Diff}(S^1)$ *which are represented by foliated* S^1*-bundles over* T^2, Ann. Inst. Fourier, **31** (2) (1981) 1–59.
27. T. Tsuboi, *On the homology of classifying spaces for foliated products*, Advanced Studies in Pure Math., **5**, Foliations, (1985), 37–120.
28. T. Tsuboi, *On the foliated products of class* C^1, Ann. of Math., **130** (1989), 227–271.
29. T. Tsuboi, *Homological and dynamical study on certain groups of Lipschitz homeomorphisms of the circle*, J. Math. Soc. Japan, **47** (1995), 1-30.
30. T. Tsuboi, *Small commutators in piecewise linear homeomorphisms of the real line*, Topology, **34** (1995), 815-857.

Received June 7, 2000.

Proceedings of
FOLIATIONS: GEOMETRY AND DYNAMICS
held in Warsaw, May 29–June 9, 2000
ed. by Paweł WALCZAK *et al.*
World Scientific, Singapore, 2002
pp. 441–448

BASIC DISTRIBUTION FOR SINGULAR RIEMANNIAN FOLIATIONS

ROBERT A. WOLAK
Instytut Matematyki, Uniwersytet Jagielloński,
Wł. Reymonta 4, 30-059 Kraków, Poland,
e-mail: wolak@im.uj.edu.pl

Pierre Molino introduced the notion of a basic foliation in his study of transversely parallelisable (TP) foliations, cf. [3]. The key result he proved is that the leaves of the basic foliation are the closures of leaves of the initial foliation. In our paper we propose to study the basic foliation of a Riemannian foliation, regular or singular. Our aim is to prove that for these foliations the leaves of the basic foliation are the closures of leaves of the initial one. The proof of this property would ensure that the conjecture proposed by P. Molino in [5] is true. However, we cannot demonstrate our theorem in all its generality. It is not surprising, as some years after announcing the conjecture Pierre Molino himself began to doubt about its truth and in his lecture at the Tokyo meeting, cf. [6], he introduced the orbit–like foliation for which he claims that the closures of leaves form a singular Riemannian foliation (SRF).

Our approach is based on Molino's proof of the fact that the closures of leaves of a regular Riemannian foliation form a SRF. We refine the notion of the commuting sheaf of a RF using global basic functions, and in some cases we are able to show that the orbits of this sheaf are the closures of leaves.

Let $C_b^\infty(M,\mathcal{F})$ be the algebra of smooth global basic functions on $(M,\mathcal{F})$. The sheaf $\mathcal{B}(M,\mathcal{F})$, called the basic sheaf, we define as follows:

$$\mathcal{B}(M,\mathcal{F})(U) = \{X \in \mathcal{X}(U) \colon df(X) = 0 \text{ for any } f \in C_b^\infty(M,\mathcal{F})\}.$$

It is not difficult to check that $\mathcal{B}$ is a sheaf of Lie algebras. In fact, for any vector fields X, Y of the sheaf $\mathcal{B}(M,\mathcal{F})$ and any global basic function f

$$0 = d^2 f(X,Y) = 1/2\{X df(Y) - Y df(X) - df([X,Y])\} = -1/2 df([X,Y])$$

thus indeed $[X, Y]$ is a vector field of the basic sheaf.

Any sheaf $\mathcal{S}$ of germs of vector fields on M defines a distribution $\mathcal{D}_{\mathcal{S}}$:

$$\mathcal{D}_{\mathcal{S}}(x) = \{v \in T_x M \colon \exists X \in \mathcal{S} \colon X(x) = v\}.$$

The distribution $\mathcal{D}_{\mathcal{B}(M,\mathcal{F})}$ we call the basic distribution of the foliation $\mathcal{F}$. The basic distribution is involutive but a priori it is neither completely integrable nor even differentiable, cf. [8].

We cannot consider just the following subbundle F_B of TM:

$$F_{Bx} = \{v \in TM_x \colon df(v) = 0 \text{ for any } f \in C_b^\infty(M, \mathcal{F})\}.$$

In the case of TP foliations the tangent bundle to the leaves of the basic foliation is precisely the bundle F_B. For RF-s these bundles are different as the following example indicates.

Example Let $A \in SO(3)$ and

$$A = \begin{pmatrix} 1 & 0 & 0 \\ 0 & cos\theta & -sin\theta \\ 0 & sin\theta & cos\theta \end{pmatrix}$$

with $\theta/\pi \notin Q$. The transformation A induces a diffeomorphism A of S^2 with two fixed points. The closures of orbits are circles. Suspending this diffeomorphism A we obtain a RF $\mathcal{F}$ of $S^1 \times S^2$ with two compact leaves L_1 and L_2 corresponding to the fixed points of A. It is not difficult to verify that at any point y of L_1 or L_2 $\mathcal{D}_{\mathcal{B}(M,\mathcal{F})}(y) = T\mathcal{F}_y$ and $F_{By} = T(S^1 \times S^2)_y$.

1 The case of regular Riemannian foliations

Regular Riemannian foliations are relatively well understood. Using P. Molino's results we are able to demonstrate that the basic distribution is totally integrable and its leaves are the closures of leaves of our foliation. First we need to recall some facts about regular Riemannian foliations.

The total space $E_T^1(M, \mathcal{F}, g_T)$ of the bundle of transverse orthonormal frames , i.e. the bundle of orthonormal frames of the normal bundle $N(M, \mathcal{F}) = TM/T\mathcal{F}$ of the foliation $\mathcal{F}$ with the induced Riemannian metric g_T, admits a foliation $\mathcal{F}_1$ whose leaves are holonomy coverings of the corresponding leaves of $\mathcal{F}$ and which is transversely parallelisable (TP). For example, let the foliation $\mathcal{F}$ is given by a cocycle $\mathcal{U} = \{U_i, f_i, g_{ij}\}$ modelled on a manifold N_0, i.e.

i) $\{U_i\}$ is an open covering of M,

ii) $f_i \colon U_i \longrightarrow N_0$ are submersions with connected fibres defining $\mathcal{F}$,

iii) g_{ij} are local diffeomorphisms of N_0 and $g_{ij} \circ f_j = f_i$ on $U_i \cap U_j$.

The image of the submersion f_i is an open subset of the manifold N_0.The disjoint union $N = \coprod f_i(U_i)$ (also a q–manifold) we call the transverse manifold of $\mathcal{F}$ associated to the cocycle $\mathcal{U}$ and the pseudogroup $\mathcal{H}$ generated by g_{ij} the holonomy pseudogroup (representative) on the transverse manifold N. As our foliation is Riemannian the transverse manifold admits a Riemannian metric g_N such that the submersions f_i are Riemannian and the holonomy pseudogroup is a pseudogroup of local isometries of the Riemannian manifold (N, g_N). In fact we have some freedom of choice for the Riemannian metric on M. Only the Riemannian metric g_T induced in the normal bundle $N(M, \mathcal{F})$ of $\mathcal{F}$ is determined by the Riemannian metric g_N, and vice versa. Then the submersions f_i define submersions $\tilde{f}_i\colon E^1_T(U_i, \mathcal{F}, g_T) \to E^1_T(N, g_N)$ and local isometries g_{ij} induce local diffeomorphisms $\tilde{g_{ij}}$ of $E^1_T(N)$ - the bundle of orthonormal frames of the the Riemannian manifold (N, g_N). If we denote $E^1_T(U_i, \mathcal{F}, g_T)$ by V_i, the cocycle $\mathcal{V} = \{V_i, \tilde{f}_i, \tilde{g_{ij}}\}$ defines the foliation $\mathcal{F}_1$.

Let fix our attention on the TP foliation $\mathcal{F}_1$ of $E^1_T(M, \mathcal{F})$. Any local foliated vector field $\tilde{X}$ which commutes with the transverse parallelism of $\mathcal{F}_1$ (i.e. the bracket of $\tilde{X}$ with any vector field of the parallelism is a vector field tangent to the foliation $\mathcal{F}_1$) is the lift of a local foliated vector field X on $(M, \mathcal{F})$ which additionally is a Killing vector field the Riemannian metric g_T in the normal bundle $N(M, \mathcal{F})$. These local foliated Killing vector fields form a locally constant sheaf $\mathcal{C}(M, \mathcal{F})$ of germs of vector fields called the commuting sheaf of the Riemannian foliation $\mathcal{F}$.

We begin with the comparison of the basic sheaf and the commuting sheaf.

Lemma 1 *Let $\mathcal{C}$ be the commuting sheaf of $(M, \mathcal{F})$. Then $\mathcal{C} \subset \mathcal{B}(M, \mathcal{F})/\mathcal{X}_\mathcal{F}$ where $\mathcal{X}_\mathcal{F}$ is the sheaf of germs of vector fields tangent to $\mathcal{F}$.*

Proof. We have to show that for a given open set U any foliated vector field from the commuting sheaf can be represented by a vector field annihilated by all global basic functions. Let $X \in \mathcal{C}(U)$. The foliated vector field X has a representative $\bar{X}$ which is orthogonal to $\mathcal{F}$ at any point. The vector field $\bar{X}$ is an infinitesimal automorphism of $\mathcal{F}$. Let $\tilde{X}$ be the lift of $\bar{X}$ to the total space of the bundle of transverse orthonormal frames $E^1_T(M, \mathcal{F})$. From the very definition $[\tilde{X}, Y] \in T\mathcal{F}_1$ for any global infinitesimal automorphism Y of $\mathcal{F}_1$. Let $f \in C^\infty_b(M, \mathcal{F})$, then $f_1 = f\pi$ is a basic function for $\mathcal{F}_1$ and

$$[\tilde{X}, f_1 Y] = \tilde{X}(f_1)Y + f_1[\tilde{X}, Y] \in T\mathcal{F}_1.$$

Thus $\tilde{X}(f_1) \equiv 0$ as there exist global infinitesimal automorphisms of $\mathcal{F}_1$

which are nonvanishing sections of $T\mathcal{F}_1^{\perp}$. Therefore

$$0 = \tilde{X}(f_1) = \tilde{X}(f\pi) = (d(f\pi))(\tilde{X}) = df(\bar{X})$$

So the vector field $\bar{X}$ is an element of $\mathcal{B}(M,\mathcal{F})$. □

The above lemma permits us to demonstrate the main result of this section

Theorem 1 *Let $\mathcal{F}$ be a regular Riemannian foliation of a compact manifold M. Then the basic distribution $\mathcal{D}_{\mathcal{B}(M,\mathcal{F})}$ is completely integrable and its leaves are the closures of leaves of $\mathcal{F}$.*

Proof of Theorem. First we will settle the relation between the basic and commuting sheaves. We know that the foliated orbits of the commuting sheaf are the closures of leaves, cf. [4, 5], so the orbits of $\mathcal{B}(M,\mathcal{F})$ contain these closures. On the other hand vector fields from $\mathcal{B}(M,\mathcal{F})$ must be tangent to the closures of leaves, cf. [7], or Lemma 4 of this note for the more general singular case. So the basic distribution is the distribution tangent to the closures of leaves.

Lemma 1 assures that for any point we can find an open neighbourhood on which the basic distribution is spanned by a finite number of vector fields tangent to the foliation and a finite number of infinitesimal automorphisms of the foliation which transversely define Killing vector fields of the commuting sheaf. So our distribution is smooth and integrable. As the flows of all vector fields mentioned above are tangent to the closures of leaves our distribution satisfies the assumption of a generalised Frobenius theorem, cf. [8], so the distribution is completely integrable as well. □

2 Basic functions

Now we turn our attention to singular Riemannian foliations. At the beginning of this section we will recall the natural stratification of M induced by our singular foliation, cf. [5].

The manifold M is stratified by the dimension of leaves, i.e. let for any $x \in M$ denote by L_x the leaf of $\mathcal{F}$ passing through x. Then for i = 0,... n=dimM let $M_r = \{x \in M : dimL_x = i\}$. Obviously there exist r_{min} and r_{max} such that $M_r = \emptyset$ for $r < i_{min}$ or $r > r_{max}$. r_{min} is the smallest dimension of leaves of the foliation $\mathcal{F}$ and r_{max} is the greatest dimension of leaves of this foliation. The set $M_0 = M_{r_{max}}$ is open and dense in M. The set $\Sigma_0 = M_{r_{min}}$ is a closed submanifold of M called the minimal stratum of $(M,\mathcal{F})$. Moreover, for any r $\bar{M}_r \subset \bigcup_{j\leq r} M_j$. In each stratum the foliation $\mathcal{F}$ induces a regular Riemannian foliation. By blowing up the minimal stratum Σ_0 we obtain a compact manifold M^1 foliated by a SRF without leaves of dimension r_{min}. After a finite number of such blow ups we obtain a compact manifold foliated by a regular Riemannian foliation, cf. [5].

Let S be a compact submanifold of M foliated by $\mathcal{F}$, S can be with the nonempty boundary provided that the boundary is a foliated submanifold as well. Let $N(S)$ be the normal bundle of S in M. The exponential mapping exp_S defined by the Riemannian metric defines a tubular neighbourhood N_S of S. The foliation $\mathcal{F}$ on N_S is invariant by homotheties defined by orthogonal geodesics, cf. [2]. Let denote by $B_\epsilon(S)$ the subbundle $\{v \in N(S)\colon \|v\| \leq \epsilon\}$ of $N(S)$ and by $Sp_\epsilon(S)$ the subbundle $\{v \in N(S)\colon \|v\| = \epsilon\}$. The exponential mapping defined by orthogonal geodesics is a diffeomorphism onto the image when restricted to $B_\epsilon(S)$ for some $\epsilon > 0$. Its image we denote $B(S,\epsilon)$. The image of the corresponding sphere bundle we denote by $Sp(S,\epsilon)$. Leaves of $\mathcal{F}$ are contained in $Sp(S,\delta)$ $(\delta < \epsilon)$ and the natural projection $B(S,\epsilon) \longrightarrow S$ maps leaves of $\mathcal{F}$ onto leaves of $\mathcal{F}$ and therefore the closures of leaves onto the closures . Using the exponential mapping it is quite easy to prove a following lemma.

Lemma 2 *For any $0 < \delta_1 \leq \delta_2 \leq \epsilon$ there exists a basic smooth function*

$$\lambda(\delta_1,\delta_2)\colon B(S,\epsilon) \to [0,1]$$

such that $supp\lambda(\delta_1,\delta_2) \subset B(S,\delta_2)$ and $\lambda(\delta_1,\delta_2)|B(S,\delta_1) \equiv 1$.

Proposition 1 *Let S be foliated compact submanifold of $(M,\mathcal{F})$. Any basic function on $(S,\mathcal{F})$ can be extended to a basic function on $(M,\mathcal{F})$.*

Proof. The natural projection $p\colon N_S \to S$ maps leaves of $\mathcal{F}$ into leaves of $\mathcal{F}$. Therefore for any basic function f on $(S,\mathcal{F})$ fp is a basic function for $\mathcal{F}$ on the saturated neighbourhood N_S of S. We have to extend the function fp to the whole manifold M. Multiplying the function fp by a suitable function from Lemma 2 we obtain a function we have been looking for. □

Remark. Proposition 1 can be used to extend any basic function from the minimal stratum to the whole manifold. Unfortunately other strata are not compact submanifolds.

Our next task is to extend basic functions from other strata to the whole manifold. The next lemma provide us with a sufficient tool.

Lemma 3 *Let f be a basic function on an open foliated subset U of a stratum Σ. Then for any point x of U there exist an open foliated neighbourhood $V \subset U$ and a global basic function $\bar{f}$ such that $f|V = \bar{f}|V$.*

Proof. Using a function from Lemma 2 we obtain a basic function f' defined on the whole stratum and equal to f on a tubular neighbourhood of the leaf $L(x)$ passing through the point x. Its support is compact, contained in $B(\bar{L(x)}, \epsilon)$ for some $\epsilon > 0$. Let $p: N(\Sigma) \to \Sigma$ be a tubular neighbourhood of the stratum Σ. Then $f'p$ is a foliated function on $N(\Sigma)$. Multiplying $f'p$ by a suitable function from Lemma 2 we get a global basic function we have been looking for. □

To complete our study of basic functions we prove a lemma which says that there are sufficiently many of them.

Lemma 4 *For any vector $X \in T_x\Sigma$ orthogonal to the closure S of a leaf L in Σ, there exists a global basic function f such that $df(X) \neq 0$.*

Proof. Lemma 2 permits us to reduce the problem to a local one. There exists $\epsilon > 0$ such that the mapping $exp_S: B_\epsilon(S) \to M$ is an embedding. The geodesic with the initial condition X is tangent to the stratum Σ, cf. [5]. Then in the stratum Σ there is a leaf L', with the closure S', on the geodesic with the initial condition X and at the distance less than ϵ from L such that the mapping $exp_{S'}: B_\epsilon(S') \to M$ is an embedding. Then the function $f_{S'}(y) = d(y, S')^2$ is a smooth basic function on $exp_{S'}(B_\epsilon(S'))$ for which $df_{S'}(X) \neq 0$. $f_{S'}$ can be easily extended to a global basic function. □

Corollary 1 *For any singular Riemannian foliation the basic distribution is tangent to the closures of leaves.*

3 The case of singular Riemannian foliations

At the very beginning let us recall the facts about the basic sheaf $\mathcal{B}(M, \mathcal{F})$.

The basic sheaf is a sheaf of Lie algebras, the basic distribution is tangent to the closures of leaves. For any regular Riemannian foliation the basic distribution is completely integrable and its leaves are the closures of leaves of the foliation $\mathcal{F}$. In particular, for any stratum Σ the basic distribution $\mathcal{B}(\Sigma, \mathcal{F})$ of the foliated manifold $(\Sigma, \mathcal{F})$ is completely integrable and its leaves are the closures of leaves of $\mathcal{F}$. We have to elucidate the relations between the sheaves $\mathcal{B}(M, \mathcal{F})$ and $\mathcal{B}(\Sigma, \mathcal{F})$.

The restriction of any vector field X of the sheaf $\mathcal{B}(M, \mathcal{F})$ to the stratum Σ is a vector field of the basic sheaf $\mathcal{B}(\Sigma, \mathcal{F})$. It is a simple consequence of Lemma 3. We would like to know whether this restriction is surjective.

Let S be the closure of a leaf L of the foliation $\mathcal{F}$. S is foliated regularly by $\mathcal{F}$, therefore, according to [2] the foliation has the same transverse model along S. For any point x of S there exist open neighbourhoods U and V of x in M and in Σ, respectively, such that U is diffeomorphic to $V \times R^p$ ($p = codim_M S$) and the restriction of $\mathcal{F}$ is the product foliation of $\mathcal{F}|V$ and a singular foliation of R^p with just one 0-dimensional leaf. If X is a basic vector field on V then its trivial extension to U is a basic vector field on U. So indeed the restriction mapping is surjective. In this way we have proved that the basic distribution is the distribution tangent to the closures of leaves but we could not assure that it is differentiable and completely integrable; that would ensure that the Molino conjecture is true.

Theorem 2 *Let $\mathcal{F}$ be a SRF on a compact manifold M. Then its basic distribution is the distribution tangent to the closures of leaves of $\mathcal{F}$.*

Now let us look at some examples and see whether in these cases the basic distribution is completely integrable.

Example 1 Let H be a Lie group of isometries of a Riemannian manifold (M, g). Then the orbits of H define a SRF $\mathcal{F}_H$. If the group $G = \bar{H}_e \subset Isom(M, g)$, ($H_e$ the component of e in H) is compact, then the foliation $\mathcal{F}_G$, defined by the group G, is the basic foliation of the foliation $\mathcal{F}_H$. In fact, the leaves of $\mathcal{F}_H$ are the connected components of the orbits of H, so the leaves are the orbits of H_e. Any basic function is an H_e-invariant function, so it is also an $\bar{H}_e$-invariant function, so the fundamental vector fields of the G-action are basic and they span the distribution tangent to the closures of leaves, so both distributions are equal and completely integrable.

Example 2 Let H be a finitely generated subgroup of $Isom(M, g)$. Then there exists a surface Σ and a locally trivial fibre bundle $B(\Sigma, M)$ of base Σ and fibre M with a foliation $\mathcal{F}$ transverse to the fibres such that the trace of any of its leaves on any fibre is diffeomorphic to the corresponding orbit of the group H. Any fibre, diffeomorphic to M, is a complete transverse submanifold, so any global basic function is determined by its restriction to the fibre. Therefore global basic functions are in one-to-one correspondence with H-invariant functions on the fibre M, or with $\bar{H}$-invariant ones, $\bar{H} \subset Isom(M, g)$. Therefore the closures of leaves are in one-to-one correspondence with the orbits of $\bar{H}$, and thus the distribution tangent to the closures of leaves when restricted to the fibre is spanned by the fundamental vector fields of the H-action on M. So the basic distribution is, indeed, completely integrable and its leaves are the closures of leaves of the foliation $\mathcal{F}_H$.

Example 2 is particularly instructive. One can easily check that two facts were essential to obtain the conclusion.

1. the existence of a complete section,
2. any local holonomy transformation starting and ending at points of the section can be extended to a global isometry of the section.

Small saturated (foliated) tubular neighbourhoods of the closure of any leaf of an OLF, cf. [6], and in particular of a transversely integrable SRF, cf. [1], have these two properties. Therefore we can formulate the following theorem.

Theorem 3 *Let $\mathcal{F}$ be an OLF of a compact manifold M, then the basic distribution of $(M, \mathcal{F})$ is completely integrable and its leaves are the closures of leaves of the foliation $\mathcal{F}$.*

References

1. H. Boualem, *Feuilletages riemanniens singuliers transversalement intégrables*, Comp. Math., **95** (1995), 101–125.
2. H. Boualem and P. Molino, *Modèle locaux saturés de feuilletages riemanniens singuliers*, C. R. Acad. Sci. Paris, **316** (1993), 913–916.
3. P. Molino, *Etude des feuilletages transversalement complets et applications* Ann. Sci. Ecole Norm. Sup., **10** (1977), 289–307.
4. P. Molino, *Géométrie globale des feuilletages riemanniens*, Proc. Kon. Neder. Akad., **85** (1982), 45–76.
5. P. Molino, Riemannian Foliations, Progress in Math 73, Birkhäuser, Boston – Basel – Berlin 1988.
6. P. Molino, *Orbit-like foliations*, in Geometric Study of Foliations, T. Mizutani et al. (eds.) Tokyo 1993, World Sci., Singapore 1994, 97–119.
7. M. Pierrot, *Orbites des champs feuilletés pour un feuilletages riemanniens sur une variété compacte*, C. R. Acad. Sc. Paris, **301** (1985), 443–445.
8. I. Vaisman, Lectures on the Geometry of Poisson Manifolds, Progress in Math 118, Birkhäuser, Boston-Basel-Berlin 1994.
9. R. Wolak, *Pierrot theorem for singular Riemannian foliations*, Publ. Math. UAB, **38** (1994), 433–439.

Received October 31, 2000.

LIST OF PARTICIPANTS

Álvarez López, Jesus (*U. de Santiago*)
Asuke, Taro (*École Norm. Sup. Lyon*)
Baditoiu, Gabriel (*Inst. Math. Rom. Acad.*)
Badura, Marek (*U. Łódzki*)
Bartoszek, Adam (*U. Łódzki*)
Biś, Andrzej (*U. Łódzki*)
Blachowska, Dorota (*U. Łódzki*)
Blachowski, Konrad (*U. Łódzki*)
Borisenko, Alexander (*Kharkov U.*)
Brito, Fabiano (*U. de São Paulo*)
Brittenham, Mark (*U. of Nebrasca*)
Bufetov, Alexander (*U. of Moscow*)
Calegari, Danny (*U. of California*)
Colman, Hellen (*U. of Illinois at Chicago*)
Conlon, Lawrence (*Washington U.*)
Czarnecki, Maciej (*U. Łódzki*)
Dobrowolski, Tadeusz (*Pittsburg State U.*)
Fenley, Sergio (*Washington U.*)
Florek, Wojciech (*U. of Illinois at Chicago*)
Francaviglia, Stefano (*Sc. Norm. Sup. Pisa*)
Frydrych, Mariusz (*U. Łódzki*)
Ghiggini, Palolo (*Scuola Norm. Sup. Pisa*)
Glazunov, Nikolaj (*Glushkov Inst. NAS, Kiev*)
Goetz, Arek (*San Francisco State U.*)
Góra, Paweł (*Concordia U., Montreal*)
Grines, Viacheslav (*Nizhny Novgorod U.*)
Haefliger, André (*U. de Genéve*)
Heitsch, James (*U. of Illinois at Chicago*)
Hilsum, Michel (*U. Pierre Marie Curie, Paris*)
Hoffoss, Diane (*Rice U.*)
Honda, Ko (*U. of Georgia*)
Hurder, Steven (*U. of Illinois at Chicago*)
Inaba, Takashi (*Chiba U.*)
Kaimanovich, Vadim (*U. Rennes 1*)
Kalina, Jerzy (*Politechnika Łódzka*)
King, Simon (*Inst. Rech. Math., Strassbourg*)
Kodama, Hiroki (*U. of Tokyo*)
Kubarski, Jan (*Politechnika Łódzka*)
Küssner, Thilo (*U. Tübingen*)
Langevin, Rémi (*U. de Bourgogne, Dijon*)
Leichtman, Eric (*École Norm. Sup. Paris*)
Matsumoto, Shigenori (*Nihon U., Tokyo*)
Mikami, Kentaro (*Akita U.*)
Mitsumatsu, Yoshihiko (*Chuo U., Tokyo*)
Mizutani, Tadayoshi (*Saitama U.*)
Moriyoshi, Hitoshi (*Keio U., Jokohama*)
Movasati, Hossein (*IMPA*)
Nakae, Yasuharu (*U. of Tokyo*)
Nakayama, Hiromichi (*Hiroshima U.*)
Noda, Takeo (*U. of Tokyo*)
Pierzchalski, Antoni (*U. Łódzki*)
Plachta, Leonid (*Lvov U.*)
Przytycki, Feliks (*Inst. Mat. Pol. Akad. Nauk*)
Rebelo, Julio (*PUC, Rio de Janeiro*)
Roberts, Rachel (*Washungton U.*)
Rogowski, Jacek (*Politechnika Łódzka*)
Rybicki, Tomasz (*AGH, Kraków*)
Salem, Eliane (*U. Paris VI*)
Scardua, Bruno (*IMPA*)
Schweitzer, Paul (*PUC, Rio de Janeiro*)
Shive, Joseph (*U. of Illinois at Chicago*)
Tsuboi, Takashi (*U. of Tokyo*)
Turakulov, Zafar (*Tashkent Inst. Nucl. Phys.*)
Vogt, Elmar (*Freie U. Berlin*)
Walczak, Paweł (*U. Łódzki*)
Walczak, Zofia (*U. Łódzki*)
Wardetzky, Max (*U. of Maryland*)
Wolak, Robert (*U. Jagielloński, Kraków*)
Zeghib, Abdelghani (*École Norm. Sup. Lyon*)
Zhuzhoma, Evgeny (*N. Novgorod Tech. U.*)

PROGRAM

May 29 (Monday)
Opening meeting.
T. Tsuboi: *Transverse intersection of foliations in 3-manifolds*
S. Hurder: *Dynamics of C^1-foliations*
S. Fenley: *Topology and geometry of foliations in 3-manifolds (Part 1)*

May 30 (Tuesday)
L. Conlon: *Foliation cones and the Thurston norm*
T. Inaba and H. Nakayama: *Invariant fibre measures of angular flows and the Ruelle invariant*
R. Roberts: *Essential laminations in 3-manifolds (Part 1)*
B. Scardua: *Complex foliations having polynomial or non-exponential growth*
M. Badura: *Prescribing growth types*
E. Zhuzhoma: *On Anosov-Weil problem for surface foliations*

May 31 (Wednesday)
A. Haefliger: *Foliations and compactly generated pseudogroups*
M. Brittenham: *Sutured handlebodies and depth of knots*
S. Fenley: *Topology and geometry of foliations in 3-manifolds (Part 2)*
M. Czarnecki: *Hadamard foliations*
N. Glazunov: *Application of algebraic geometry and ergodic theory to dynamics of foliations*
F. Brito: *Volume of vector fields and plane fields on unit spheres*

June 1 (Thursday)
D. Calegari: *Foliations, circles and hyperbolic geometry*
P. Schweitzer: *Codimension-one foliations, Reeb components and an extension of Novikov's Theorem*
S. Fenley: *Topology and geometry of foliations in 3-manifolds (Part 3)*
V. Kaimanovich: *Conformal measures on laminations associated with rational maps*
J. Shive: *Regularity of Hirsch foliations*
T. Rybicki: *The leaf preserving diffeomorphisms group as a Lie group*
M. Hilsum: *Riemannian foliations with positive longitudinal scalar curvature*
A. Bufetov: *Topological entropy for free semigroup and group actions*

June 2 (Friday)
A. Zeghib: *Global linearizations of group actions*
E. Vogt: *Tangential Lusternik-Scnirelmann category of foliations*
R. Roberts: *Essential laminations in 3-manifolds (Part 2)*
H. Colman: *LS-category of compact Hausdorff foliations*
A. Aranson and E. Zhuzhoma: *Geometry and topology of foliations and 2-webs on closed surfaces*
T. Dobrowolski: *Failure of Sard's Theorem and existence of strange bump functions in infinite dimensions*

June 3 (Saturday)
E. Leichtnam: *A local formula for the index of a contact transformation*
R. Roberts: *Essential laminations in 3-manifolds (Part 3)*

June 5 (Monday)
J. Heitsch: *Traces and invariants for non-compact manifolds*
H. Moriyoshi: *Operator algebras and the index theorem on foliated manifolds (Part 1)*
Y. Mitsumatsu: *Foliations and contact structures in dimension 3 (Part 1)*
A. Goetz: *New results in dynamics of piecewise isometries*
A. Pierzchalski: *U(n)-invariant differential operators*
D. Bolotov and A. Borisenko: *Submanifolds and foliations*
R. Wolak: *A few remarks on singular foliations*
T. Noda: *Regular projectively Anosov flows on 3-manifolds*
Z. Turakulov: *Simplest foliations of Minkowski space-time and their applications*

June 6 (Tuesday)
P. Walczak: *Prescribing mean curvature for foliations of codimension* ≥ 1
S. Matsumoto: *Leafwise cohomology and rigidity of certain Lie group actions*
H. Moriyoshi: *Operator algebras and the index theorem on foliated manifolds (Part 2)*
J. Álvarez López: *Distributional Betti numbers of Riemannian foliations*
G. Baditoiu: *Semi-Riemannian submersions from real and complex pseudo-hyperbolic spaces*
Problem session (organized by P. Schweitzer)

June 7 (Wednesday)
J. Rebelo: *On global behaviour of polynomial ODE's*
R. Langevin: *Some integral geometric results on foliations*
H. Moriyoshi: *Operator algebras and the index theorem on foliated manifolds (Part 3)*
K. Honda: *Tight contact structures and taut foliations*
H. Kodama: *Holomorphic contact structures and Legendrian flows*
A. Bartoszek: *Conformal compactification and non-linear wave equations*
V. Grines: *Codimension-one laminations and classification of A-diffeomorphisms*
A. Biś: *Entropies of a semigroup of mappings*

June 8 (Thursday)
T. Asuke: *Secondary characteristic classes of transversely holomorphic foliations*
T. Mizutani: *Foliations associated with Nambu structures*
Y. Mitsumatsu: *Foliations and contact structures in dimension 3 (Part 2)*
A. Borisenko: *Foliations with extrinsic negative curvatures*
M. Frydrych: *Partially holomorphic foliations*
Y. Nakae: *Foliation cones associated to some pretzel links*
H. Movasati: *On the space of holomorphic foliations with a centre singularity*
V. Medvedev, E. Zhuzhoma: *On codimension-one Plykin's attractors on 3-manifolds*

June 9 (Friday)
L. Plachta: *The study of incompressible surfaces in link components via the natural foliations on them*
T. Küssner: *Efficient fundamental cycles and foliations of cusped hyperbolic manifoldss*
Y. Mitsumatsu: *Foliations and contact structures in dimension 3 (Part 3)*
Closing meeting